NORTHERN HORIZON

DRACO
URSA MAJOR "BIG DIPPER"
URSA MINOR "LITTLE DIPPER"
POLARIS, NORTH STAR
CEPHEUS
DENEB
CYGNUS
CASSIOPEIA
PEGASUS
ANDROMEDA
TRIANGULUM
PERSEUS
CAPELLA
AURIGA
CASTOR
POLLUX
GEMINI
CANCER
LEO
REGULUS
HYDRA
ARIES
PISCES
PLEIADES
TAURUS
ALDEBARAN
CANIS MINOR
PROCYON
BETELGEUSE
ORION
RIGEL
CETUS
SIRIUS
LEPUS
CANIS MAJOR
COLUMBA
CANOPUS

EASTERN HORIZON

WESTERN HORIZON

SOUTHERN HORIZON

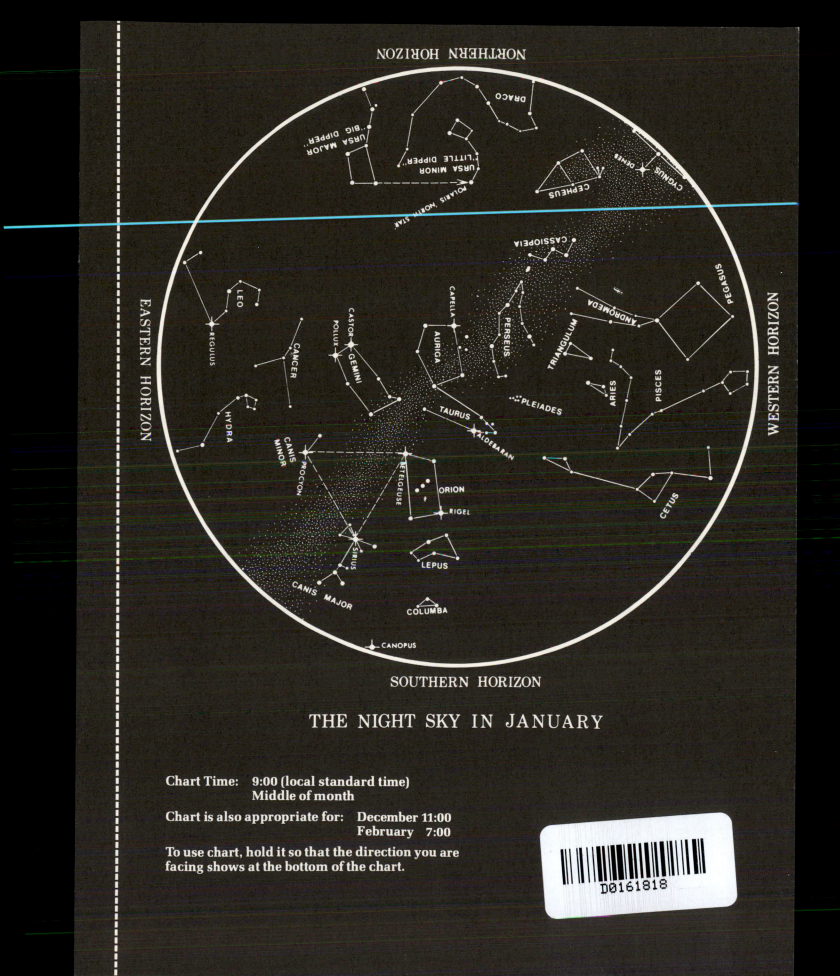

THE NIGHT SKY IN JANUARY

Chart Time: 9:00 (local standard time)
Middle of month

Chart is also appropriate for: December 11:00
February 7:00

To use chart, hold it so that the direction you are facing shows at the bottom of the chart.

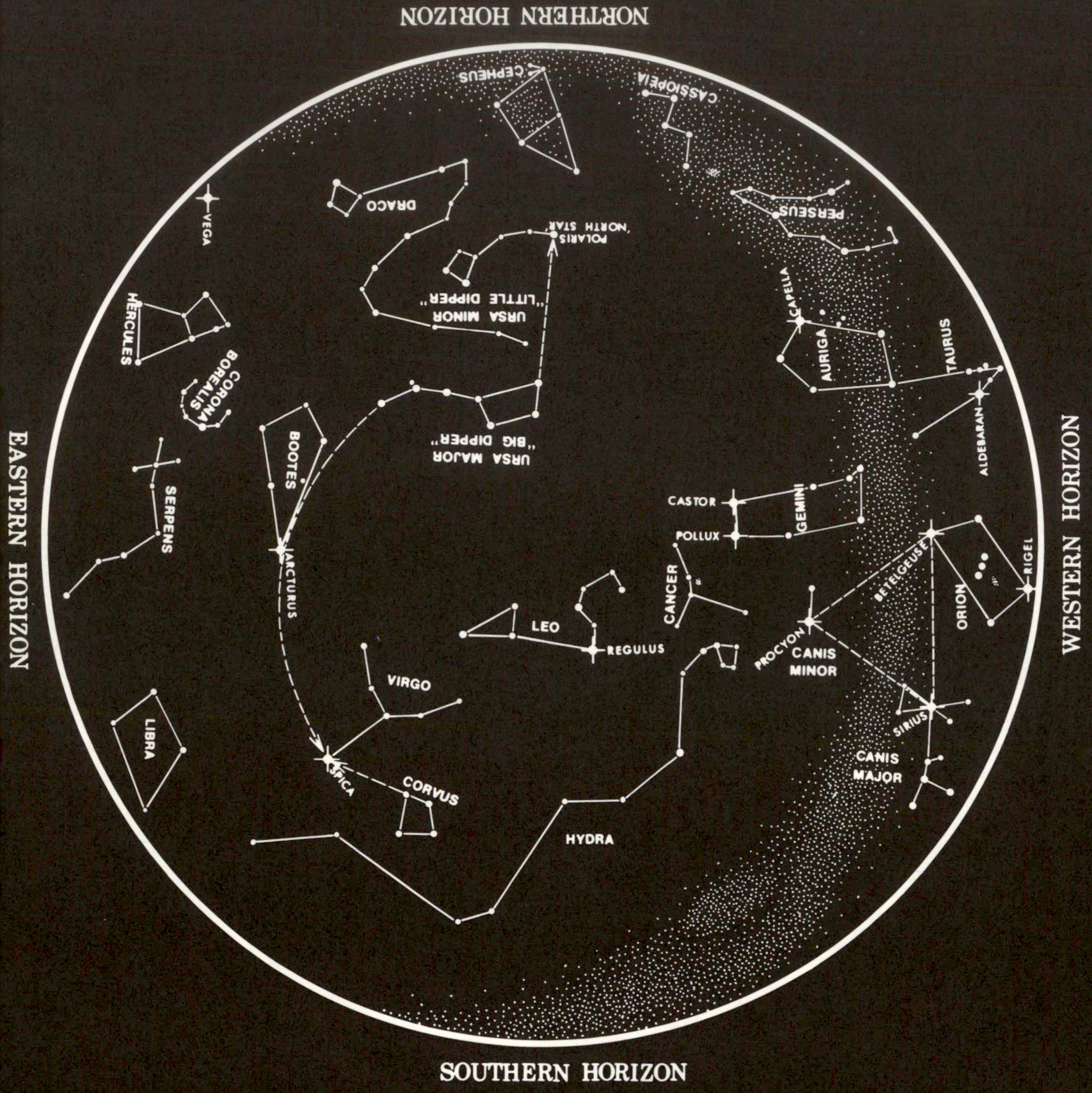

NORTHERN HORIZON

EASTERN HORIZON

WESTERN HORIZON

SOUTHERN HORIZON

THE NIGHT SKY IN APRIL

Chart Time: **9:00 (local standard time)**
Middle of month

Chart is also appropriate for: **March 11:00**
May 7:00

To use chart, hold it so that the direction you are facing shows at the bottom of the chart.

Concepts of the Cosmos
An Introduction to Astronomy

Concepts of the Cosmos
An Introduction to Astronomy

Barry R. Parker

Idaho State University

HARCOURT BRACE JOVANOVICH, PUBLISHERS
San Diego New York Chicago Washington, D.C. Atlanta
London Sydney Toronto

ISBN: 0-15-512850-7

Library of Congress Catalog
Card Number: 83-81329

Printed in the
United States of America

preface

Many students have been attracted to astronomy in the last few years because of the publicity it has received in the media. They want to learn more about the new planetary discoveries, about black holes, pulsars, and quasars. My purpose in writing this book is to present a clear, simple, and interesting introduction to these discoveries, and at the same time to give a descriptive overview of all of astronomy.

The book is divided into eight parts, arranged in what I believe is the most natural sequence. In arriving at this sequence I have used the following three guidelines:

a. The smallest and most familiar objects are introduced first, with a gradual transition to increasingly larger objects and distances.
b. Evolution—of the solar system, the stars, and the universe—receives special treatment.
c. The history of astronomy is emphasized.

The book begins with an overview that should be helpful in giving a feeling for our place in the universe. A chapter is devoted to the early history of astronomy, beginning with the early Babylonians and continuing through the time of Newton. We see how the early astronomers struggled to understand what they saw, and we gain an appreciation of their efforts and ingenuity.

The last chapter of Part 1 introduces the phenomena we see as we look into space from Earth, including phases of the moon, eclipses, and movement of the stars; also included is a detailed discussion of coordinate systems. Many instructors now use planetariums in the presentation of this material, and the chapter has been written with that in mind.

The same chapter also contains a section on the constellations. For many students, one of the most exciting aspects of an astronomy class is a group session in which the constellations are identified and various objects are viewed through a telescope. This section gives instructions for finding the constellations and shows the positions of some of the more interesting telescopic objects.

Part 2 introduces the basic tools needed by the astronomer and also gives some of the background needed to understand concepts that will be introduced later. An understanding of light is important, but even more important is the realization that light is just one small section of the electromagnetic spectrum and that many different branches of astronomy are now associated with this spectrum.

Part 3 begins with an overview of the origin of the solar system and its subsequent evolution. Each of the objects in the solar system is studied utilizing the latest results from the Voyager, Viking, and Mariner explorations.

In Part 4 the focus turns from the solar system to the stars and their evolution. Chapter 16 deals with the birth and early evolution of stars; Chapter 17 and later chapters continue the story up to the death of a star. Most of the "exotic" objects of astronomy are associated with the death of a star, and these objects are given considerable emphasis. In Part 5 an entire chapter is devoted to black holes and to quasars, and a considerable portion of a chapter to pulsars.

Part 6 discusses the largest units in the universe—galaxies. Our own galaxy, the Milky Way, is discussed in detail, and the numerous galaxies around us are examined.

The three chapters of Part 7 constitute a detailed treatment of cosmology. The first chapter is primarily historical; chapter 25, the heart of the section, goes into considerable detail regarding the age, beginning, and end of the universe. Although the Big Bang theory is accepted by most astronomers, there are a number of interesting rival theories, and an entire chapter is devoted to them.

Part 8 deals with the possibilities of extraterrestrial life. Chapter 27 discusses discoveries related to the basic molecules of life and the origin of life on Earth. The Epilogue deals with the difficulties of contacting, and the possibility of visiting, any advanced civilizations that may exist.

The book is designed for a one-semester course, but with supplementary material it can be used in longer courses. A particularly helpful feature, I believe, will be the numerous teaching aids. Each chapter begins with a statement of its aim (on a separate page) and concludes with a summary and questions. Summary items are numbered for easy reference, and the question sections include two types: review questions, and thought and discussion questions designed to challenge students. Easy-to-perform projects are also suggested in many of the chapters. Finally, a number of references are given so that further study can be pursued.

A study guide, written by Ronald J. Bieniek of the University of Missouri, accompanies the text and is a particularly useful supplement.

I am particularly grateful to the following people for their assistance in reviewing the various stages of this project: Ronald J. Bieniek (University of Missouri-Rolla), Robert C. Bless (University of Wisconsin), George S. Greenstein (Amherst College), Raymond T. Grenchik (Louisiana State University), Frank J. Harmon (Idaho State University), Paul D. Lane (College of St. Thomas), Robert Marksteiner (Triton College), Dave McIntyre (Idaho State University), Gerald H. Newsom (Ohio State University), Alexander G. Smith (University of Florida), Harding E. Smith (University of California, San Diego), Sumner G. Starrfield (Arizona State University), David Theison (University of Maryland), Raymond E. White (Steward Observatory, University of Arizona), and Jan C. Yutzy (Boston College High School).

Finally, I would like to thank the staff of Harcourt Brace Jovanovich for their patience and hard work. Marilyn Davis merits special thanks; without her help the book would not have come into being. Others to whom I am grateful are: Judy Burke, Geri Davis, Don Fujimoto, Tricia Griffith, John Holland, Mike McKinley, Phillip Ressner, and Richard Wallis.

Last, but not least, I must thank my wife for putting up with an absentee husband.

contents

part 6
GALAXIES AND QUASARS 371

part 7
COSMOLOGY AND THE UNIVERSE 423

This book contains both black-and-white and full-color illustrations. All the color illustrations are grouped together between pages 52 and 53, regardless of where they are discussed in the text. A textual reference such as *Figure 1.9, color* indicates that the illustration will be found in the color section.

Concepts of the Cosmos
An Introduction to Astronomy

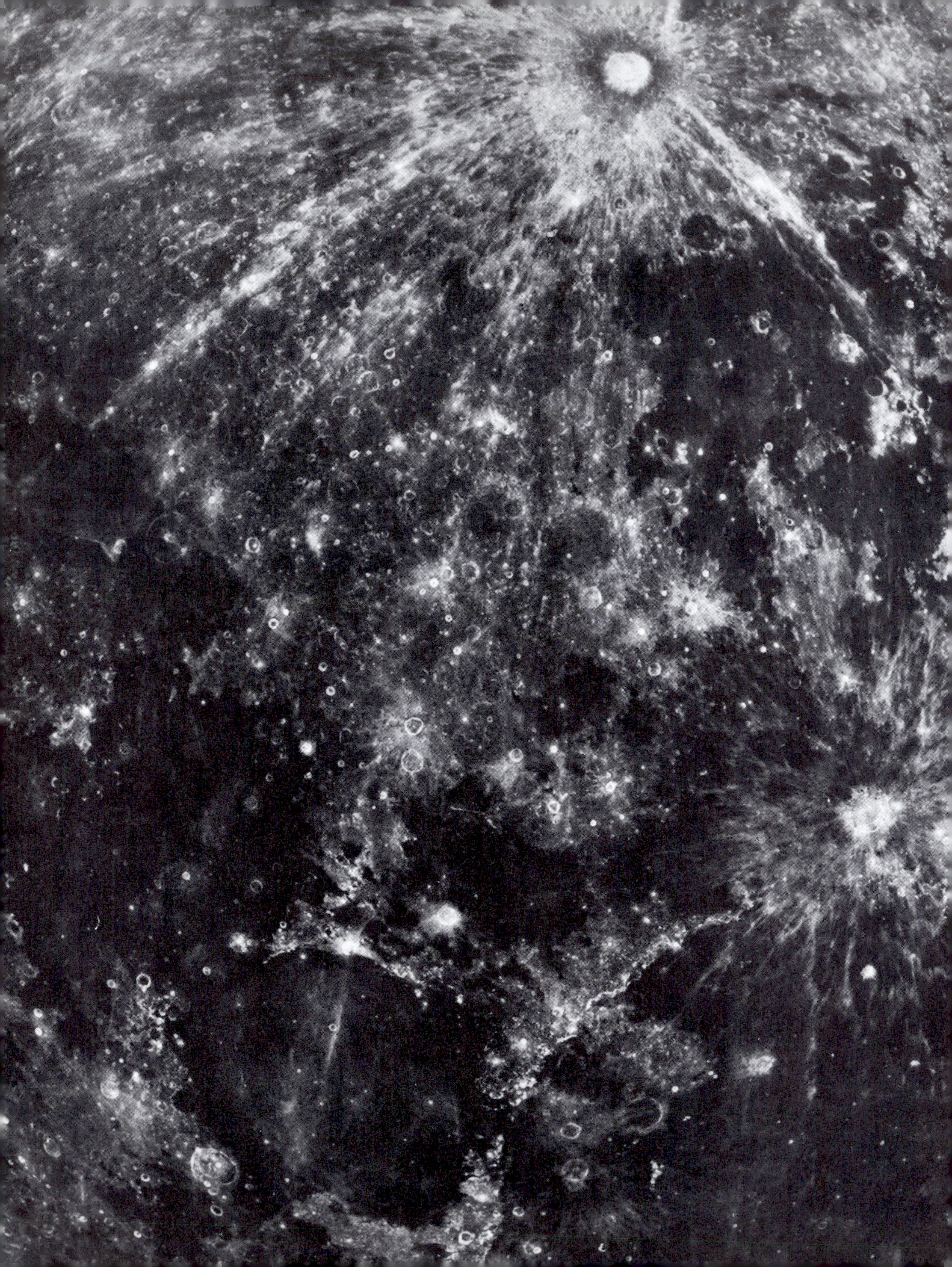

part 1

Introduction: Man's Place in the Universe

The first people on earth, looking up at the stars, saw about the same configuration of stars as we see today. There have been few changes, even over thousands of years. Relative to the ancients, who could do little more than guess about it, we may feel sophisticated and smug about our knowledge of the universe, but even today few people can really visualize the tremendous distances involved in astronomy. Our nearest interstellar neighbor, α Centauri, is over 24 trillion miles away; a round trip to it at, say, 25,000 miles per hour, would take about 225,000 years.

We will begin in Chapter 1 with an overview of the universe. The aim here is to give a sense of the objects we will encounter later and of the scale involved, and we will answer questions such as: What does the universe consist of? How many planets are there? How much farther away are the stars than the planets? What is a galaxy?

In Chapter 2 we will trace the course of astronomy from the Babylonians up through the time of Newton. Misconceptions were, of course, common in early civilizations. Nevertheless, when we consider the means available to them, we must acknowledge that they were as ingenious as we are today. We study their ideas of the cosmos—many strange and incorrect—so that, recognizing the pitfalls encountered, we can better chart our search. By starting at the beginning, we gain a much better perspective on modern astronomy. (And in addition to all this, the history of astronomy has some of the most fascinating personalities in all of science.)

In Chapter 3 we will look at the universe from a modern point of view. Beginning with the solar system we will discuss the rotation and revolution of the earth, the seasons, the motions of the sun and moon, and eclipses. Then, turning to the stars, we will explain how astronomers locate objects in the sky. The chapter concludes with a brief discussion of the constellations.

Aim of the Chapter

To introduce the planets, galaxies, and other objects encountered in astronomy and to give a sense of the scale involved.

chapter 1

Overview of the Universe

On a clear, dark night the sky appears at first to be a chaotic array of stars. Some are dim, some bright; some sparkle, a very few seem fuzzy, and indeed are not individual stars. After some examination it will be evident that all is not chaos and that patterns can be seen. However, the patterns observable with the naked eye do not reveal much of what is known about the composition and scale of the universe and of our place in it. Figure 1.1 shows some of the elements involved.

Some features visible with the naked eye are probably already familiar; for example, most people can recognize the Big Dipper, which is part of a constellation (stars arbitrarily designated a group) known as Ursa Major. The constellation Orion (Fig. 1.2) is also well-known, and is easily recognizable in the winter sky.

Some of the stars twinkle and even change color. (This is a result of turbulence in the air, which causes the light from the star to be bent back and forth slightly as it passes through the air.) The few brighter objects that do not twinkle are the planets. From earth we can see as many as five of the other eight planets at one time or another without the aid of a telescope, but usually no more than two or three are visible on a given evening.

Another well-known feature, the Milky Way, is visible as a faint ribbon of light stretching across the sky (Fig. 1.3). In the summer it starts at the southern horizon, passes overhead, and ends at the northern horizon. In the south it is bright and dense, then is interrupted by a dark region (called the Great Rift), continues, still quite bright, and gradually dims as it approaches the northern horizon.

Through binoculars or a small telescope the Milky Way will seem to be composed of thousands of faint stars; actually it is made up of *billions* of stars (and other objects), which are part of an island universe of stars called a galaxy. The Milky Way galaxy is the one of which the earth is a part, and it is sometimes called simply "the Galaxy" with a capital G to distinguish it from the millions of other galaxies in space. One of these other galaxies—the only one that can

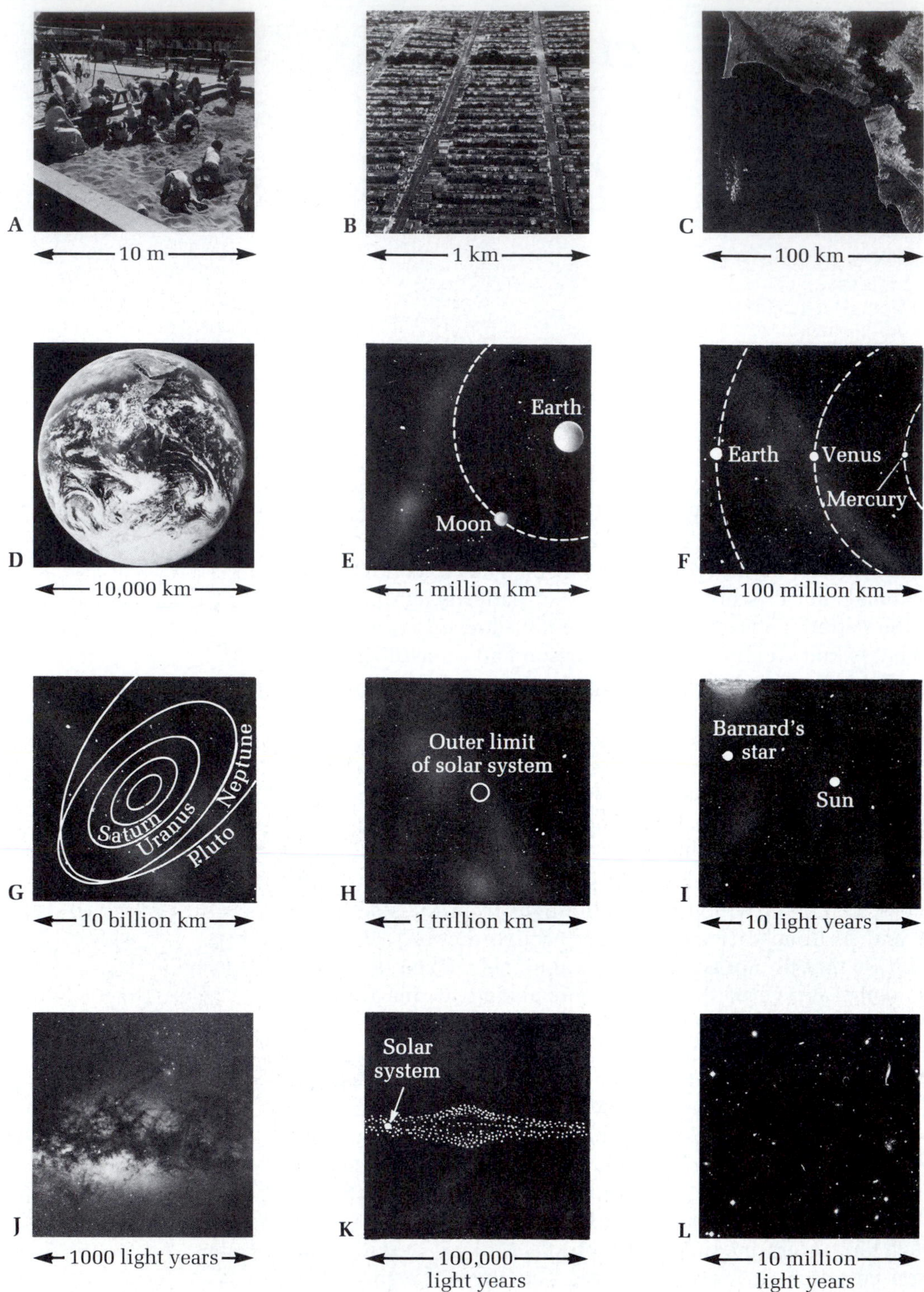
A
10 m
B
1 km
C
100 km
D
10,000 km
Earth
Moon
E
1 million km
Earth
Venus
Mercury
F
100 million km
Neptune
Saturn
Uranus
Pluto
G
10 billion km
Outer limit
of solar system
H
1 trillion km
Barnard's
star
Sun
I
10 light years
J
1000 light years
Solar
system
K
100,000
light years
L
10 million
light years

Figure 1.1 Views of the universe, beginning with a playground (*A*), and increasing by a factor of about 100 (on a side):
B is about a quarter square-mile urban area;
C, a satellite view of the San Francisco area;
D, a satellite view of the earth;
E, Earth and the moon;
F, Earth, Venus, and Mercury;
G, the entire solar system;
H, the outer limit of the solar system, with surrounding space;
I, the sun in relation to the nearest other star;
J, a section of our galaxy, the Milky Way;
K, an edge-on view of the Milky Way;
L, a cluster of galaxies.

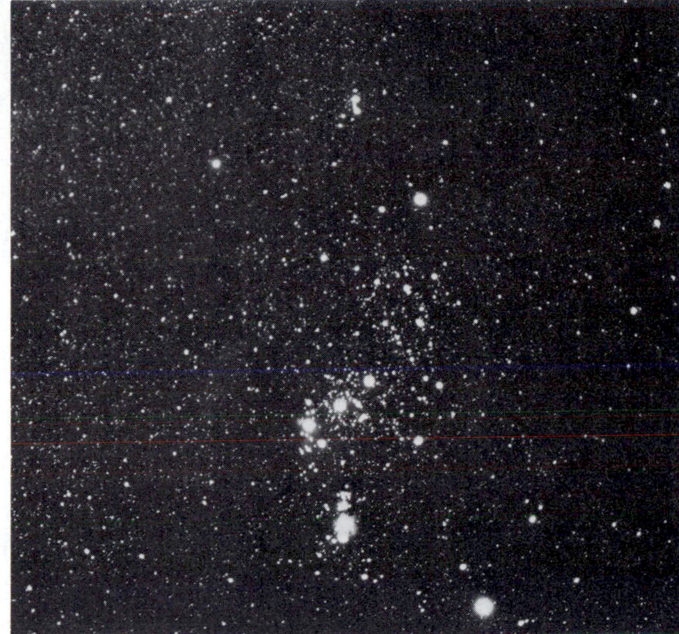

Figure 1.2 A section of the sky in the constellation Orion.

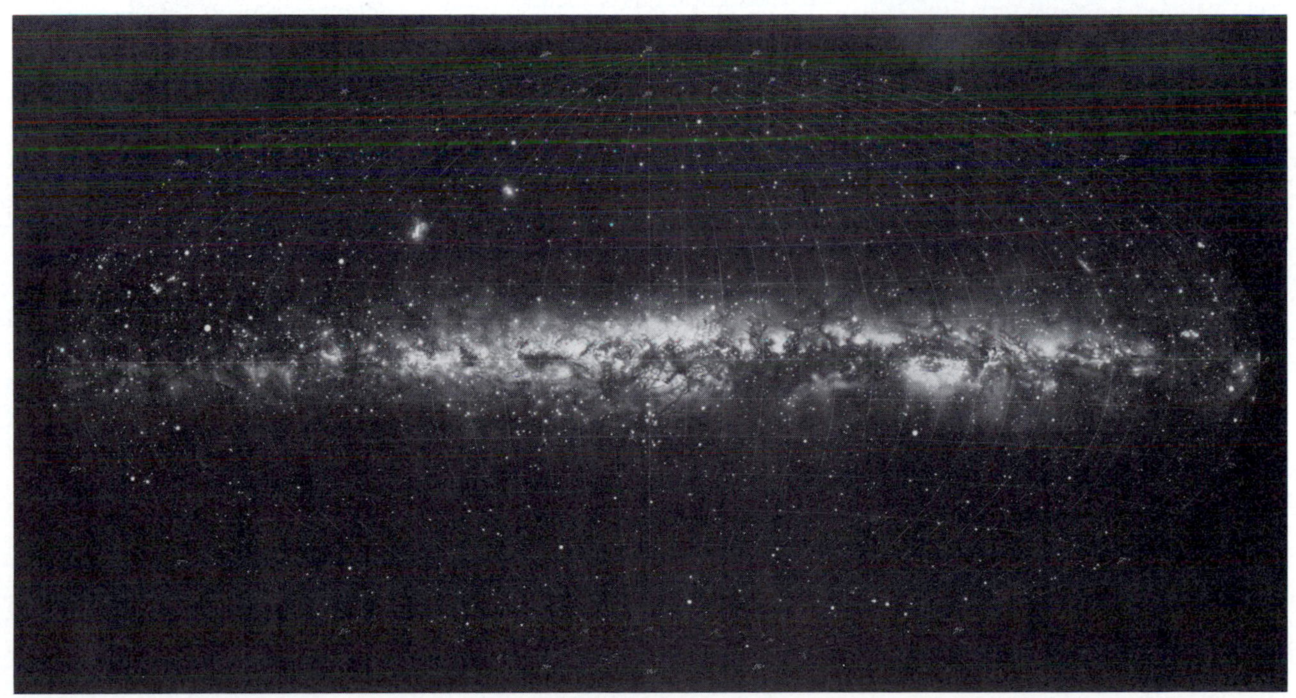

Figure 1.3 A painting of the Milky Way. In this view the center (in the direction of the constellation Sagittarius) is the actual center of our galaxy. The Great Rift can be seen a bit to the left of the center, and the two small cloudlike features to the lower right are the Magellanic Clouds. (The painting is by Martin and Tatjana Keskula, under the direction of Knut Lundmark, Lund Observatory, Sweden.)

be seen readily in the Northern Hemisphere without a telescope—is visible on clear autumn evenings as a soft, fuzzy glow in the constellation Andromeda.

Our galaxy has the form of a giant disk in space. It appears dense on the southern horizon because this is the direction of its center and we are relatively close to its outer edge. Thus, we see more of the Galaxy when we look in this direction. The Great Rift is caused by small intervening particles in space—dust—that block our view. The Milky Way is dimmer in the north because, when looking to the north, we are looking radially outward from our position near the edge, so that there are fewer stars to be seen in this direction. This also explains why we see only a few stars in any direction away from the Milky Way.

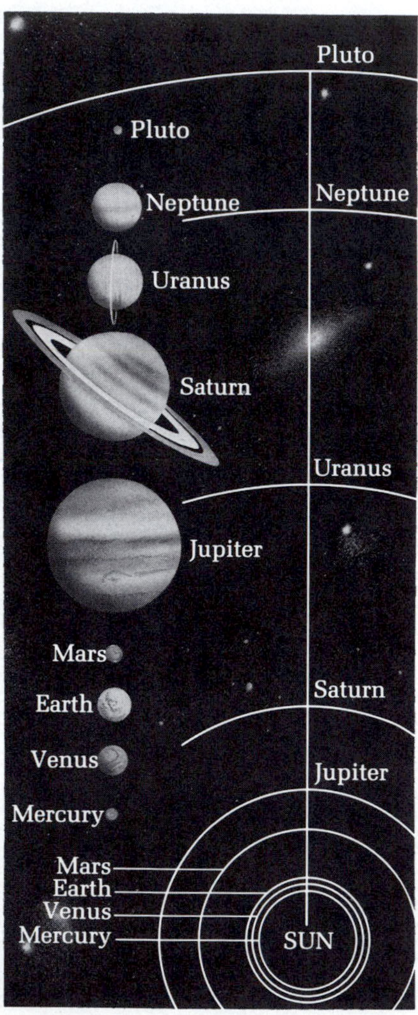

Figure 1.4 The solar system, with the planets shown in relative size at left, and their orbits, to scale, at right.

THE SOLAR SYSTEM

The solar system (Fig. 1.4) consists of nine planets (including Earth) revolving around our star, the sun, in the Galaxy. Between these planets is an amazing array of objects ranging from boulders over a hundred miles across to tiny specks of dust. Chunks of ice and dust, like giant, dirty snowballs a mile or so in diameter, also move through this region. When one approaches the sun, it develops a long, spectacular tail of gas; we call them comets.

The sun is a seething inferno of gas, so hot that we would be immediately consumed in flame were we to approach its surface. Yet it supplies us, directly or indirectly, with all we need to sustain life. The planet closest to the sun, and the second smallest in the solar system, is Mercury. Like our moon, Mercury is a dead world dotted with craters.

Next outward from the sun is Venus. For many years it was believed that a civilization might lie beneath its dense layer of clouds. We now know, however, that its high temperatures make this virtually impossible. Venus is about the same size as Earth, but as we shall see later, in most other respects it is quite different.

The third planet from the sun is Earth. Lying approximately 150 million km from the sun, it appears mostly blue and white from space, and its vast oceans and lush green vegetation contrast sharply with what we believe is on the other planets.

Beyond Earth is Mars. Because it can easily be seen with a small telescope, Mars has been carefully studied over the years. It appeared to be largely covered with deserts, yet there were other regions. Were they vegetation? Recent space probes have shown that they are merely outcroppings of dark rock. There are also canyons, volcanoes, old river beds, and craters.

Between Mars and Jupiter (the next planet) is a region of space containing scattered boulders—some many miles across, others small. This is the asteroid belt. Jupiter, largest of the planets, has a greater mass and volume than all the other planets *combined*. Its surface consists of colored bands, and a large red spot is also usually visible. The colored bands are clouds carried around the planet by strong winds caused by its rapid rotation.

Beyond Jupiter is Saturn, one of the most striking planets in the solar system. With a ring system that extends 79,000 km out from its surface, it is indeed a beautiful sight in a telescope. Although it is a giant, it is incredibly light relative to its size, for, like Jupiter, it is composed mostly of gas (hydrogen and helium). Put in water, it would float.

Two more giants, Uranus and Neptune, lie beyond Saturn. Neither can be seen with the naked eye (though some claim Uranus can be seen occasionally). Like Saturn, they are also composed mostly of gas. Finally, at the outer edge of the solar system, almost 6 billion km from the sun, lies Pluto, the smallest of the planets. With its solid surface, it seems somehow out of place. Some astronomers believe that it was at one time a moon of Neptune.

Because the distances and sizes involved in the solar system (and even more so in the universe) are so great that they are hard to grasp, it can be useful to scale them down—that is, to envision them with all parts reduced in proportion. With Earth the size, say, of an orange (about 8 cm in diameter), the moon in proportion would be the size of a cherry (about 2 cm across) and the sun a gigantic ball (about 8.7 m

The Metric System

There are two major systems of measurement in use throughout the world: the English and the metric. Almost every country except the United States now uses the metric system, and its gradual adoption is now under way here. Science has used the metric system for many years, the main reason being the system's simplicity, which derives from its being based on the unit 10.

In the metric system, units of length and mass that correspond to the feet and ounces of the English system are meters and grams. Units expressing time are the same in both systems.

Subdivisions and multiples of the basic units of meters and grams are as follows:

1 meter (m) = 100 centimeters (cm)
= 1000 millimeters (mm)
1000 meters (m) = 1 kilometer (km)
1000 grams (gm) = 1 kilogram (kg)

Equivalences of these metric units in the English system are:

1 mm = .039 inches
1 cm = .3937 inches
1 m = 39.37 inches
1 km = .6214 miles

1 gram (gm) = .0353 ounces
1 kilogram (kg) = 2.2046 pounds

Conversely,

1 inch = 2.53 cm
1 ounce = 28.3495 gm
1 pound = 0.4536 kg = 453.6 gm

Most distances in this book will be given in kilometers (km). Those who more readily visualize distances in miles convert kilometers to miles (approximately) by multiplying kilometers by .62. For example, 1000 km is approximately 620 miles. Miles can be converted to kilometers by multiplying by 1.6093.

across) lying about 930 m (.93 km) away. Mars would be 4.25 cm in diameter and approximately 350 m from Earth at its closest approach. Jupiter would be 89 cm across and approximately $3\frac{1}{2}$ km away, and Pluto, a tiny sphere about 2 cm in diameter, almost 35 km from Earth.

STARS AND NEBULAE IN OUR GALAXY

Outside the solar system the great vastness of space soon becomes even more evident. The nearest star, though still in our galaxy, is 40 trillion km, or about 4.3 light years, away.[1] Even on the scaled-down model used above for the solar system, its distance would be 250,000 km. That star is α Centauri[2] (actually a three-star system). One of its three stars is quite similar in size and color to our sun; the other two are smaller and orange. It is the smallest, Proxima Centauri, that has the distinction of passing closer to us than any known star other than the sun.

Our second nearest neighbor is a recently discovered dwarf star at a distance of 4.4 light years. Just beyond it is the dim red star known as Barnard's star. Despite its small size, Barnard's star has recently attracted considerable attention, since there is some evidence that it has two dark companions—possibly planets—orbiting it.

Beyond Barnard's star is Wolf 359 at 7.6 light years, and a small double-star system at 8.1 light years. Finally, at 8.6 light years is the brilliant blue star Sirius, with its faint white-dwarf companion. Sirius is seen as the brightest star in our sky.

Out to 10 light years from Earth there are 8 star systems; to 13 light years, 25 systems; and to 17 light years, 43 systems. One of our most startling discoveries concerning these stars was that many are indeed systems—that is, rather than being single, like our sun, they are double and triple—hence, systems.

Many other types of stars occur in our galaxy. Some appear to vary in brightness over days, even over months and years. Others brighten erratically, then fade quickly; some explode (Fig. 1.5). There are also pulsars and, perhaps, black holes (p. 353). (Though it is not certain that there is a black hole in our galaxy, there is at least one excellent candidate.) Both black holes and pulsars are associated with the final death throes of a star.

The Galaxy also contains large patches of diffuse light, called nebulae (Fig. 1.5 and 1.6). These are enormous clouds of gas, but the density of matter in them is so small that most of these regions constitute a better vacuum than any we can create on Earth.

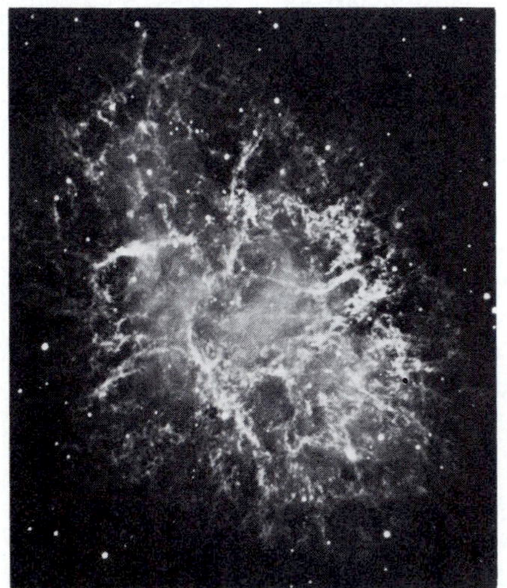

Figure 1.5 The Crab Nebula, an exploding star.

[1]Such large distances are given by astronomers in light years and in parsecs (defined later). A light year is the distance that light travels in one year. It is the speed of light (3×10^8 m/sec) multiplied by the number of seconds in a year—roughly 9.3 trillion km.

[2]In a given constellation the brightest star is normally designated by the Greek letter α (alpha), the second brightest by β (beta), and so on through the Greek alphabet.

Figure 1.6 The gaseous cloud known as the Great Nebula of Orion.

THE GALAXY ITSELF

The giant disk of stars that is our galaxy, when viewed edge-on, has a bulging center. It contains about 200 billion stars and is 100,000 light years across. Thus, even if we could travel at the speed of light, it would take 100,000 years to cross it.

With so many stars it might seem that collisions would be frequent. However, it is likely that there have been few collisions of individual stars since the birth of our galaxy. This is explainable by the facts that the stars are separated by enormous distances and that they move, not at random, but in orbits (well-defined paths largely parallel to the main plane of the disk of the Galaxy), in much the same way that the planets orbit the sun.

OUTSIDE OUR GALAXY

Figure 1.7 A globular cluster, composed of more than 1 million stars.

As we approach the edge of our galaxy, stars become fewer and fewer. Beyond is the void of intergalactic space. Between that void and the Galaxy are clusters of stars, called globular clusters (Fig. 1.7), which surround the Galaxy and also orbit it, but not generally in the plane of the disk. These clusters, like a swarm of bees around a beehive, are small compared to our galaxy, yet some contain over a million stars.

Vast distances away lie other island universes of stars, a trip to any of which would be exceedingly long. Some are spirals, like our Milky Way, but other forms are also evident: barred spirals (spirals with barlike formations through their center), ellipticals, and a few that appear irregular in form.

One of the largest galaxies in our region is the Andromeda galaxy, approximately 2 million light years away. Like the Milky Way, it is a spiral; in fact, in many ways it is quite similar to our galaxy, so that looking at it (Fig. 1.8) gives a good idea of what our galaxy would look like from a point outside. (Note, though, that its disk is tilted from our line of sight.)

Another large spiral, which can be seen in the same region of the sky, is the Triangulum galaxy (Fig. 1.9, color). Although smaller than the Andromeda galaxy, it can be seen almost face-on, and so we can easily observe its long, trailing arms.

The two galaxies closest to us are irregulars called the Magellanic Clouds (Fig. 1.3). From Earth they are seen only in the Southern Hemisphere.

Figure 1.8 The Andromeda galaxy.

Figure 1.10 The Coma cluster, about 300 million light years away. Because most of the objects seen here are galaxies, they are not—as stars would be—round.

THE UNIVERSE

Our galaxy, the Andromeda galaxy, the galaxy in Triangulum, and the Magellanic Clouds are all part of a cluster of galaxies called the Local Group, which consists of about 30 galaxies, most of which are relatively dim. Many similar groups exist, and some are much larger than the Local Group. For example, the giant Coma cluster of galaxies (Fig. 1.10), with about 10,000 members, dwarfs our group. The Virgo cluster (in the direction of the constellation Virgo) is also much larger; it contains over 2500 galaxies.

Clusters of galaxies may not be the largest things in the universe, for there appear to be clusters of clusters, called by most astronomers superclusters. Our Local Group is believed to be part of one of these superclusters.

While our Local Group of galaxies travels as a unit through space, all other galaxies and other clusters of galaxies appear to be moving away from us. In other words, the universe seems to be expanding. However, the galaxies themselves are not changing; their size remains the same; it is the distance *between* them that is increasing. In fact, the more distant the galaxy, the more quickly it is receding from us.

But where does the expansion end? Does it continue until we reach the end of the universe? Indeed, what do we mean by "the end of the universe"? It is questions like these that make cosmology (the study of the universe) fascinating.

SUMMARY

1. Our solar system is part of a galaxy called the Milky Way. The parts of the Galaxy that surround us can be seen on clear, dark nights as a faint band of light stretching across the sky. It consists of approximately 200 billion stars as well as other objects. Our sun is one star—a not unusual one—in this galaxy.
2. Earth is the third planet out from the sun in a system of nine planets. The sun is approximately 150 million km from Earth.
3. Distances outside the solar system are usually expressed in light years—the distance that light travels in one year.
4. Stars are grouped in galaxies. Our galaxy has the form of a spiral disk with long, trailing arms and is about 100,000 light years across. The nearest star (in our galaxy) is 4.3 light years away.
5. Galaxies, in turn, are found in clusters of galaxies. These clusters are moving away from each other as the universe expands.

REVIEW QUESTIONS

1. Why do stars twinkle?
2. When we look at our galaxy, the Milky Way, from Earth, we see different densities of stars in the various directions. Why?
3. Name the planets in order outward from the sun.
4. What is the smallest planet? The largest?
5. What is the difference between a nebula and a galaxy?
6. Approximately how many stars are there in a globular cluster?
7. Describe the different types of galaxies mentioned in this chapter.
8. Why are stellar collisions infrequent?
9. If the universe is expanding, must the size of the galaxies also be increasing?
10. How many kilometers are there in one light year? How many miles?

THOUGHT AND DISCUSSION QUESTIONS

1. If α Centauri is 4.3 light years away, how far away is it in kilometers? In miles?
2. Using the table in Chapter 7 and the box on exponential numbers in Chapter 4, scale the sun and planets in our solar system—that is, give their proportional sizes—were the earth the size of a basketball (about 32 cm across).
3. List all the kinds of astronomical objects mentioned in this chapter in order of size (smallest to largest). Describe any types of astronomical object you can imagine that might be larger.

Aim of the Chapter

To show how our view of the solar system and universe has evolved, to illustrate the pitfalls and problems encountered, and finally to show how some of the dimensions were first determined.

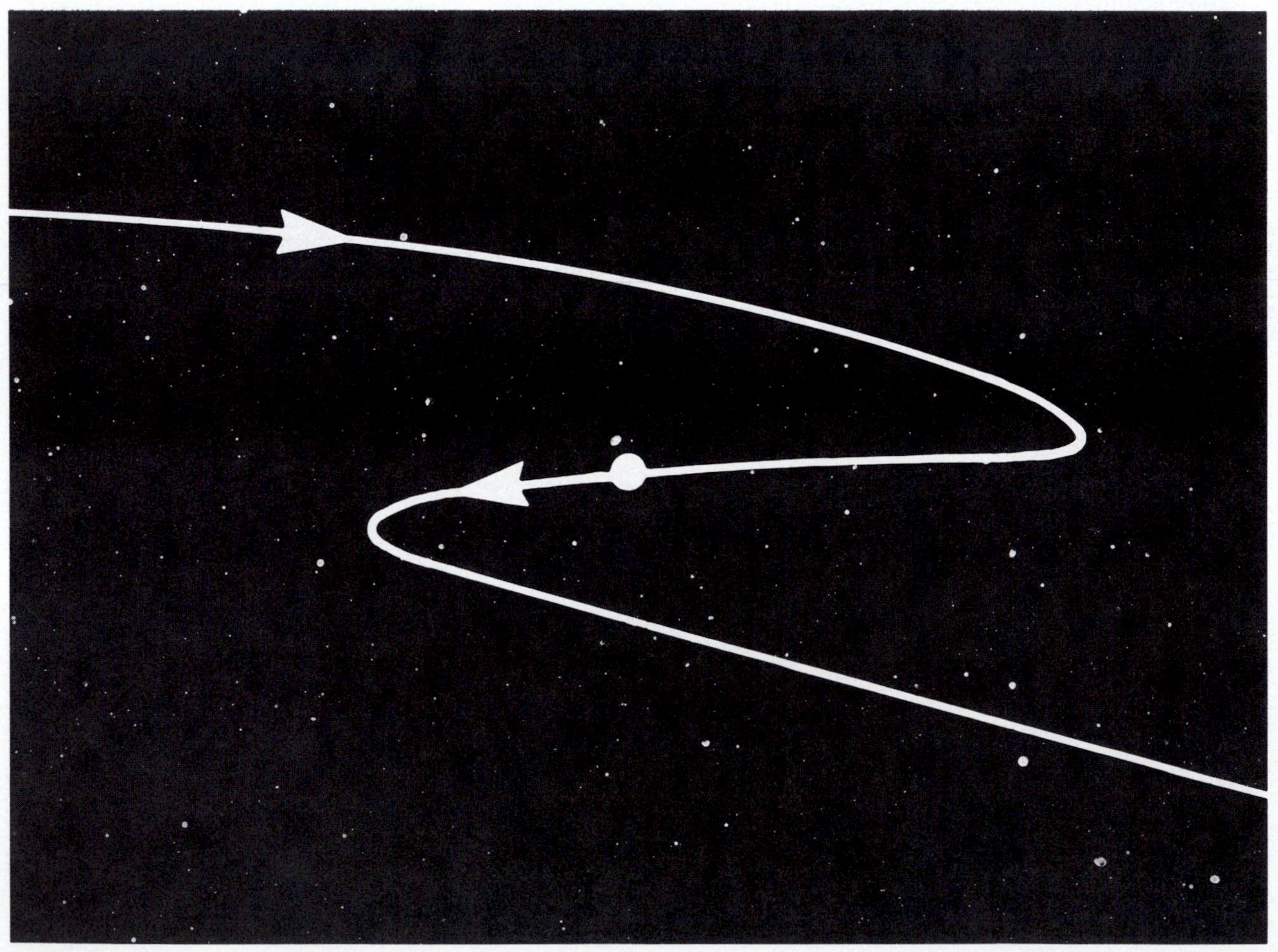

chapter 2

Chaotic Beginnings

Over thousands of years, astronomers have struggled to understand the solar system. Each generation made its contribution to our understanding; succeeding generations built on the work of predecessors. Gradually the pieces of the puzzle began to fit; little by little, year by year, the model was perfected until finally a comprehensive and accurate picture evolved. Today, thanks to the efforts of many, we feel confident that we understand most of the complexities of our solar system.

It is difficult to say where and when astronomy began. There is fragmentary evidence that some prehistoric peoples made astronomical observations; in the case of Stonehenge, in England, it is certain that the circles of huge stones erected beginning as early as 1900 B.C. did constitute an astronomical observatory and, indeed, what could even be called an astronomical computer, by which many significant events, such as eclipses, could be predicted. There is also evidence of an early interest in both China and Central America, but it is generally believed that the first serious work was done by Babylonians, who began recording planetary positions as early as 2000 B.C.

BABYLON AND EGYPT

Though the first measurements made by the Babylonians were crude, by 750 B.C. their observations were sophisticated enough to detect retrograde motion, the occasional seeming backward motion of the planets (Fig. 2.1). Although they were able to predict when this motion would occur, they were satisfied that it was the work of the gods and sought no other explanation. This prescientific view, of course, characterized all their astronomic activity. Thus, though they

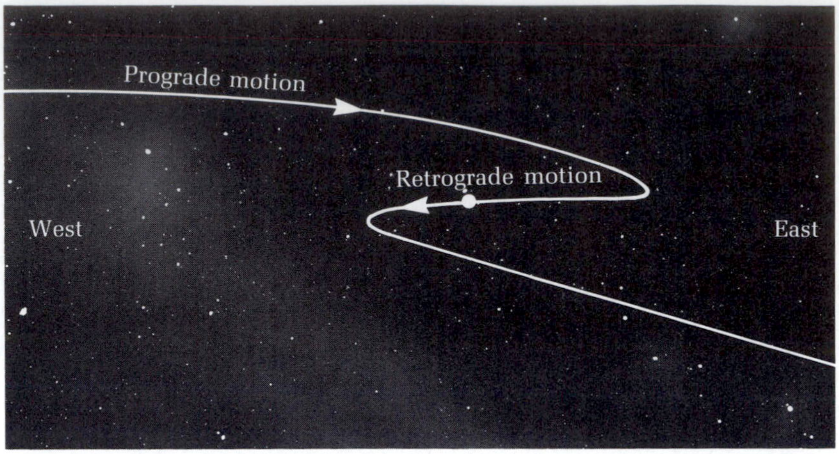

Figure 2.1 **Retrograde motion of Mars. The westward (retrograde) motion occurs for several days.**

had extensive records and sought cycles, sequences, and repetitions, they produced no geometrical models to explain what they had observed. Their main interests lay in devising a basis for a calendar (quite similar to the one we use today) and for horoscopes.

The Babylonians also recorded the motions of the stars, grouping them into constellations. Though of course different from ours, these constellations constitute the beginnings of our present-day Zodiac.

The early Egyptians were also interested in astronomy but, unlike the Babylonians, left no records of planetary positions, eclipses, or other astronomical phenomena. Their astronomy was centered on the development of a calendar. Their first, a lunar calendar that was out of step with the seasons, was replaced with one resembling ours—twelve months of 30 days each, with five additional days set aside at the end of the year for festivities.

The Egyptians had little idea of the extent and structure of the universe. Their cosmology, as with the Babylonians, reflected their religious beliefs (Fig. 2.2).

EARLY GREEK ASTRONOMY

Thales and Pythagoras

Where the Babylonians' strengths and interests lay in observational astronomy, the Greeks, in the first efforts that could be called scientific, pondered causation and structure: What was the nature of the planets? Why did they move? How big were the sun, the earth, and moon? Though they relied heavily on the observations of the Babylonians, they devised a model[1] of the universe such as the

[1]A scientific "model" is a mental picture based on known (and presumably well-understood) concepts that allow you to explain what you see in nature.

Babylonians had never dreamed of—and their model came amazingly close to predicting all that could be observed.

The first Greek natural philosopher of importance, Thales of Miletus (born 640 B.C.), is often called the father of Greek astronomy. His contributions were numerous; besides his development of astronomical navigation he is frequently credited with predicting an eclipse that occurred in 585 B.C.—if true, an amazing feat in view of the meagerness of the astronomical records of the time.

Pythagoras (born about 590 B.C.), an important contemporary of Thales, is generally considered one of the greatest of the early Greek teachers. Though only eight celestial bodies were known at the time, Pythagoras believed there should be ten—the five known planets, the sun, the moon, Earth, and a so-called counter-Earth. Amazingly for the time, not the earth, but a "central fire" was at the center of his system. Around it, each attached to its own crystal sphere, moved the known planets, Earth, the counter-Earth, the moon, and the sun.

Aristotle: Ultimate Authority

Artistotle (384–322 B.C.), most famous (to us) of the early Greek philosophers, assumed, with Pythagoras, that the earth, sun, moon and planets must be spheres. Aristotle differed from Pythagoras, however, in basing his assumption of a spherical earth on something tangible. He noticed, for example, that during lunar eclipses the shadow cast on the moon by the earth was round; he also pointed out that, as a person travels north, the polar stars move higher in the sky and other stars become visible along the horizon. This could happen only if the earth were round.

Aristotle had produced a model of the universe that was a

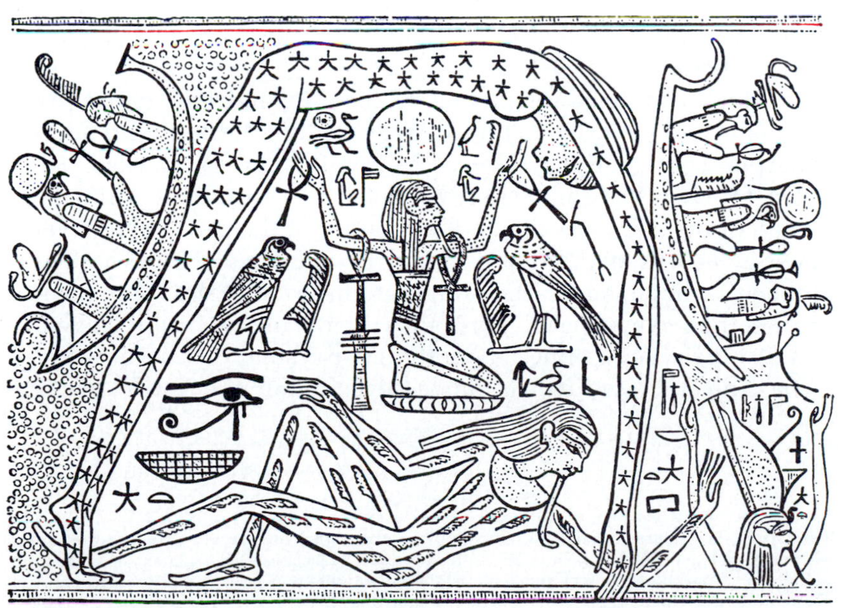

Figure 2.2 An early Egyptian concept of the universe. Here Nut, goddess of the heavens, arches her body over the earth. A river, which flows over her, is navigated daily by Ra, the sun god, who appears each dawn (left) and disappears at dusk (right).

modification of one, devised by Eudoxus (c. 370 B.C.), in which the sun, moon, known planets, and stars were attached to one or more of a series of invisible concentric spheres, with the earth at the center and the stars on the outermost sphere. The celestial sphere, which contained the stars, drove all the others through an invisible connection among them.

Although Aristotle's model gave reasonably accurate positions and could predict retrograde motion, he was dissatisfied with certain aspects of it. Why, for example, did the planets seem to vary in brightness? (Like Eudoxus, Aristotle had placed the earth at the center, and, if this was the case, the planets—always more or less at constant distances from Earth—should not have varied in brightness.) He considered a heliocentric system (one in which the planets move around the sun) but found it unacceptable, for there was no parallax effect observable (Fig. 2.3).[2]

But did the earth move? A contemporary of Aristotle, Hericlides, pointed out that if the earth rotated on its axis, the celestial sphere—containing all the surrounding stars—need not move in order for us to see the motions we do. But Aristotle could not accept this. His reason: if the earth rotated on its axis, we would all be thrown off. At best there would be a force acting continuously to lift us from the earth. Furthermore, an object thrown directly upward would be at a different spot when it fell back to the earth. For, while the object was aloft, the earth would rotate under it. This did not appear to happen. And finally, Aristotle argued, the earth would break up as a result of the rotational forces that would be generated by its spin.

The Alexandrian School

About 350 B.C., Alexander the Great established the city of Alexandria in what is now modern Egypt. With its gigantic museum and library, Alexandria eventually became the learning capital of the world, and most of the great scholars of the next few centuries worked there.

The first famous astronomer of the Alexandrian school, Aristarchus of Samos (310–230 B.C.), is generally credited with several important contributions: He was the first to express a serious belief in the heliocentric (sun-centered) model; he assumed that the daily motion of the stars was due to the rotation of the earth; and he used rather ingenious methods to estimate the relative distances and sizes of the sun, moon, and earth. Although these estimates were far less precise than we expect today, they were an important advance, establishing several things about the solar system that we now know to be true: that the sun is much bigger than the earth, that the

[2]Parallax is the apparent change in location of an object that occurs when the location of the viewer is changed (Fig. 2.3). Thus, Aristotle—particularly because he believed the stars to be equidistant from the earth (and much closer than they are)—expected that, if the earth moved, the stars should appear to shift position, that is, show the effect of parallax.

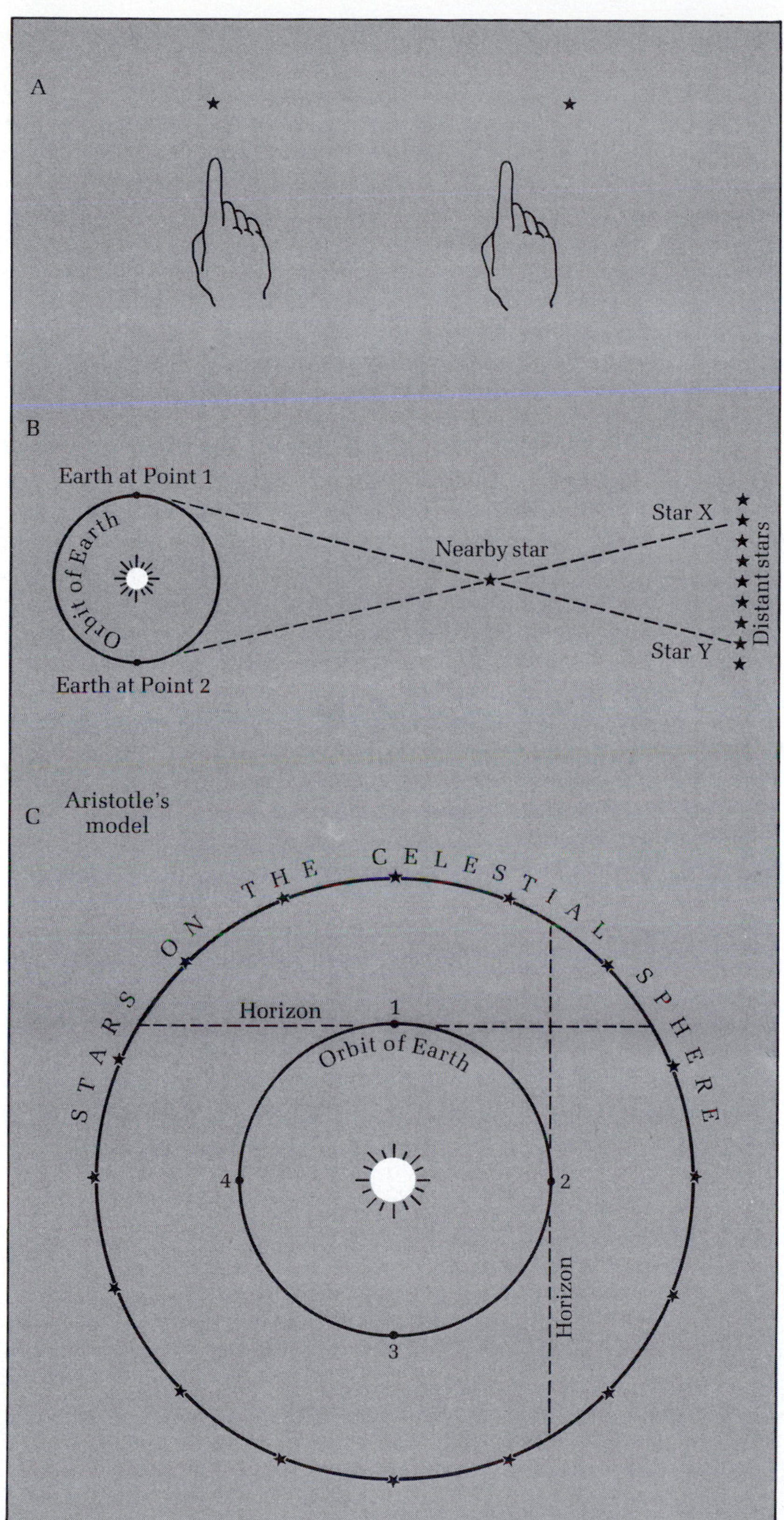

Figure 2.3 Three examples of parallax: (*A*) Using the left eye (right eye closed), align one finger with a mark on a wall. Without moving the finger, close the left eye and open the right; the mark on the wall seems to have shifted and is no longer aligned with the finger. (*B*) An observer at Point 1 in Earth's orbit will see a nearby star aligned with Star Y, while the observer at Point 2 sees it aligned with Star X. (*C*) In Aristotle's heliocentric model all stars were equidistant from Earth, which would make the parallax effect impossible, though all the stars would shift as Earth's position in orbit changed.

moon is much smaller, and that the sun is much farther away than the moon.

Still a central problem was the earth's size. Eratosthenes (276–194 B.C.) of Alexandria, reading a discussion of the differences in the lengths of shadows on different days, noted that at noon on June 21, a vertical stick in Syene, an outpost in the south, would cast no shadow. (This meant that the sun was directly overhead—i.e., at the zenith—at that time.) However, he found that, according to the shadows in Alexandria on the same date and time, the sun was a little more than 7° from the zenith.

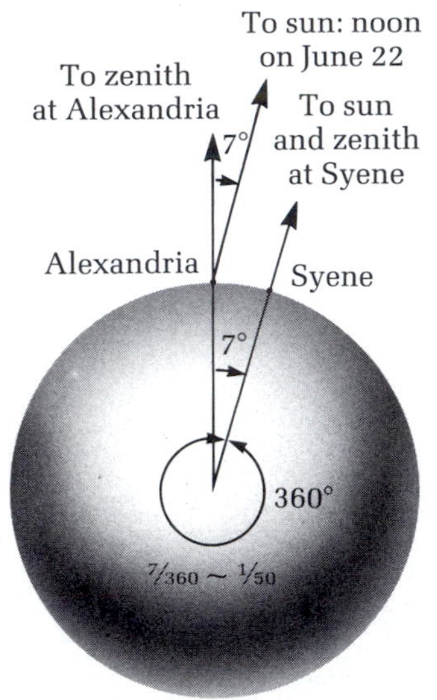

Figure 2.4 Eratosthenes' method of measuring the circumference of the earth.

Eratosthenes assumed from this that the earth was a sphere, for, if the earth were flat, the sun would have been directly overhead at both Syene and Alexandria. Then, using simple geometry (Fig. 2.4), he calculated the earth's circumference, as follows: Since 7° (the sun's angle from the zenith in Alexandria) is approximately $\frac{1}{50}$ of a circle, the circumference of the earth should be about 50 times the distance from Alexandria to Syene. Eratosthenes gave this distance as 5000 stadia, making the circumference of the earth 250,000 stadia (50×5000).

Unfortunately, we cannot verify Eratosthenes' accuracy, for we are uncertain of the length of the stadium. But there are indications that it was the length of the conventional Greek stadium or arena. If so, Eratosthenes' figure was incorrect by about 20%. On the other hand, were it $\frac{1}{10}$ of a mile, as suggested by other evidence, then his calculation was within 1% of the value accepted today.

With a reasonably good estimate of the size of the earth available, and knowing the relative distances of the moon and sun (determined by Aristarchus), the Greeks could determine the sizes of the sun and moon. These estimates were, of course, still crude, the moon's diameter being given as .33 of the earth's diameter (present value, .27), and the sun's diameter too small by a factor of 15.

A basic tool of modern astronomy—trigonometry—was still missing, but it was soon to come. Its inventor, Hipparchus (born *c.* 175 B.C. in Nicacea), generally considered one of the greatest of the early Greek astronomers, was the first of the Greeks to carry out systematic and extensive observations of the planets and stars.

The culmination of his efforts, carried out in an observatory he set up on the Island of Rhodes, was a star catalog containing about 850 entries; each star was labeled according to its celestial coordinates (like latitude and longitude on earth) and its magnitude, or apparent brightness. Hipparchus also devised a scale in which the brightest stars were designated as having magnitude 1, and the dimmest, magnitude 6.

Hipparchus determined the moon's distance to be $29\frac{1}{2}$ times the earth's diameter—amazingly close to the currently accepted value of 30—and also arrived at a better figure for the moon's size. Finally, starting with Aristotle's model of the universe, he added a new feature, the epicycle, which had been suggested as early as the end of the third century B.C. by Apollonius of Perga. Using the epicycle, a small, circular orbit superimposed on a larger one, called the def-

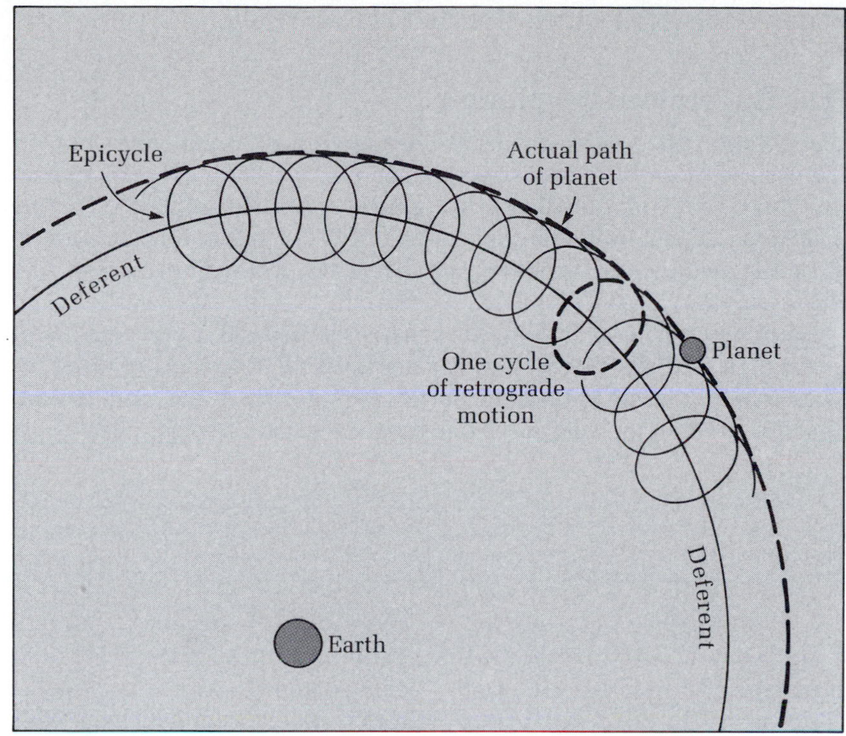

Figure 2.5 Aristotle's epicycle explanation of retrograde motion. The deferent is the line that would result from connecting the centers of all epicycles.

erent (Fig. 2.5). Hipparchus was able to explain both retrograde motion (p. 17) and the varying brightness of the planets.

Ptolemy: The Pinnacle

Most recent of the ancient Greek astronomers was Ptolemy, who lived about 125 A.D. Included in his prodigious scholarly output was a series of 13 volumes on astronomy known as the *Almagest*. Much of what we know about Greek astronomy has come to us through these volumes. Some were based on Ptolemy's own work, but most were based on the work of Hipparchus and others.

Although some of Ptolemy's techniques are now sometimes ridiculed, the system he devised is ingenious at the least. With it he predicted to within a few degrees the future and past positions of all the known planets—an amazing feat, considering the age in which he lived. This Ptolemaic system remained unchallenged for over a thousand years—far longer than our current system has been in use.

It should be noted that Ptolemy did not necessarily believe the solar system was physically of the form postulated in his system. To him it was simply a geometrical scheme that worked, one by means of which reasonably correct predictions could be made.

Although Ptolemy is generally remembered as a model builder (cosmologist), he also published a catalog of 1022 stars grouped into 48 constellations and made careful measurements of the positions of the sun, moon, and planets.

RENAISSANCE AND ENLIGHTENMENT

The Copernican Revolution

For about 1,000 years—from the destruction of Rome until the fifteenth century—little was added to the world's knowledge of astronomy. The end of the fifteenth century saw the birth of Nicolaus Copernicus, who is sometimes given credit for being the first to suggest the heliocentric system. However, the heliocentric view had been seriously considered by Aristarchus, as we saw earlier, and Aristotle refers to such a system in his writings. But circumstances were not then favorable to such an idea: Observation techniques were still crude, concepts of size were extremely limited, and explanations based on the actions of the gods were firmly entrenched.

Copernicus

Nicholaus Copernicus, born Mikolaj Kopernik on February 19, 1473, in Torun, Poland, was barely ten years old when his father died and he was sent to live with an uncle. After graduation from the local schools in Torun, he attended the University of Cracow and, later, the Universities of Bologna and Padua in Italy. Canon (church) law and medicine were his major areas of study, but he also found time for mathematics and astronomy.

After receiving a doctorate in canon law from the University of Ferrara, he became canon of the diocesan cathedral of Frauenberg, Denmark. By this time Copernicus had developed a considerable interest in astronomy. Some of the early texts he studied still survive, and his extensive notes in their margins clearly show his enthusiasm for the subject. He was particularly interested in the calculations of planetary positions, but the surprising discrepancies with observation annoyed him. Why did they occur? What caused them? And the Ptolemaic system seemed, in general, far too complex. Copernicus saw that the explanation of retrograde motion could be simplified by placing the sun rather than the earth at the center of the solar system, and thus began his great contribution to astronomy. Other features of such a system were particularly satisfying to him. He hoped, by using it, to reduce the number of circles

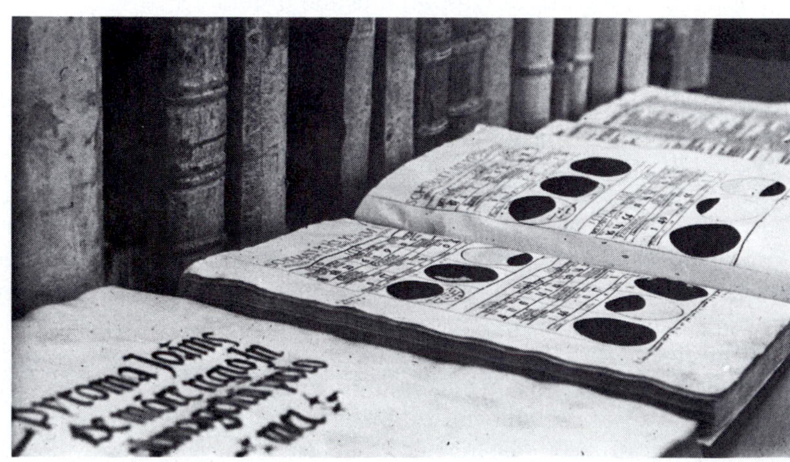

Copernicus spent almost 20 years working out the details of his model, systematically considering such things as the shape and size of the earth and of the orbits, the positions of the planets, and the size of the universe.

Aristotle had argued that a rotating earth would break apart under the stress of rotational forces (p. 20). Copernicus countered by questioning the rotation of Aristotle's celestial sphere. Since it, too, was assumed to rotate, wouldn't it also break apart for the same reasons? In fact, since the celestial sphere had a much larger radius than the earth, wouldn't it have had to move that much faster than the earth and, therefore, be under an even greater rotational strain?

An important feature of the Copernican system was the simple way in which retrograde motion could be explained. Figure 2.6 shows the explanation for the case of Mars. Because Earth is travel-

that Ptolemy had used, but strangely, he ended with more. He also found that his system required retention of small epicycles—though not to explain retrograde motion (p. 17).

This greater complexity of Copernicus's system—despite the fact that his system is much closer to reality than Ptolemy's—is largely attributable to Copernicus's attempt to account for a phenomenon that did not exist. Errors and inconsistencies in measurements had, over the years, led to the idea that there was a small "oscillation" of the equinoxes. Later, when it was realized that this was not the case, the Copernican system was considerably simplified.

Though Copernicus's contributions to astronomy are certainly his most important, they are only some among many, for he was also a diplomat, a medical practitioner, and an artist: A self-portrait still survives. His extensive training in medicine was of great value to him in later years, and he is said to have spent considerable time treating the poor.

The publication of Copernicus's account of his astronomical system was an event that took over 20 years to happen. Certainly this would not be the case today. But scientific progress was much slower in Copernicus's time, and fewer people worked on a given problem; thus there was little danger of being "scooped" on a big discovery.

Though encouraged by the prominent mathematician Joachim Rheticus, who had at first publicly opposed the new model, Copernicus was reluctant to publish, for he knew that any system in which the earth was not the center of the universe would be regarded by the Church as blasphemous.

When the entire manuscript was finally published, it contained two significant deviations from the original: The title had been changed from *De Revolutionibus (On the Revolutions)* to *De Revolutionibus Orbium Coelestium (On the Revolutions of the Heavenly Spheres)*. This was doubtless done to take the emphasis away from the earth. Second, a preface had been added stating that, although the system set forth was a heliocentric one, it was a model, and not necessarily to be taken as a true representation of reality. The preface, it was discovered a few years later, had been written not by Copernicus but by Andrew Osiander, a friend of Rheticus.

Though at first there was little concern or interest in the work—it had little effect during the latter half of the sixteenth century—its influence began to spread little by little and eventually its effect was immense.

The book was published just before Copernicus's death at the age of 70. A copy of it was brought to him as he lay on his deathbed.

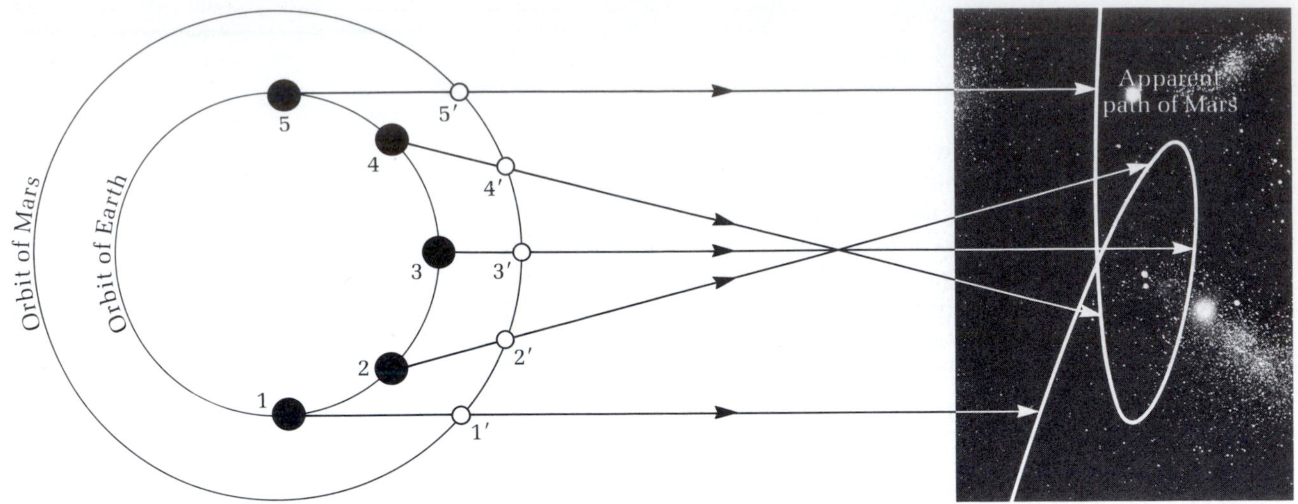

Figure 2.6 Copernicus's explanation of retrograde motion. Arrows indicate lines of sight of an observer on Earth.

ing faster in its orbit than is Mars, it, in effect, repeatedly overtakes (Position 3) and passes Mars. Here arrows drawn through pairs of positions (1–1′, 2–2′, etc.), which are, respectively, Earth and Mars at various points in their orbits, indicate how Mars would appear to shift its direction (in relation to the background of fixed stars), depending on Earth's position in relation to it. Thus, at positions 1–1′ and 2–2′ Mars appears to us to be moving to the left (*up*ward in the diagram at the right). Shortly thereafter (by Position 3–3′) it seems to have changed direction and begun to move to the right (*down*ward in the diagram). We can easily see why: Earth has caught up with and passed Mars. (The phenomenon is similar to what occurs when you look out the window of a train that overtakes and passes a slower-moving train on an adjacent track—the slower train appears to move backwards.) Finally, as Position 4–4′ is passed, another reversal occurs and Mars appears to move to the left again.

Tycho

In 1572 Danish astronomer Tycho Brahe observed a new "star" in the sky. He studied it until it disappeared from view 16 months later. When he attempted to measure its parallax, he found that there was none; the object was indeed on the celestial sphere. This was of particular significance to him, for this appearance of a body where there had been none meant that the celestial sphere was not the unchangeable entity it was thought to be—here was change.

About five years later, Tycho observed a comet and established that it was definitely neither on the celestial sphere nor within the earth's atmosphere. Before Tycho's time many had believed that comets were atmospheric phenomena. Finally, he showed that the comet was beyond the moon and—most important of all—that it moved between the planets. At last, here was proof that the spheres that drove the planets could not possibly exist.

Despite his extensive experience and training in astronomy, Tycho was not immediately attracted to the Copernican system. He

Tycho Brahe

Tycho Brahe, born at Knudstrup Manor (Denmark) in 1546, was of noble lineage. Although his parents were living, he was raised by a childless uncle who apparently wanted an heir. Tycho's interest in astronomy began early. A partial eclipse of the sun, witnessed when he was 14, left a lasting impression. That astronomers were able to predict such things was amazing to him, and he soon began to observe and measure the planets and stars.

Sent off to the University of Copenhagen to study law, he spent almost every evening observing the heavens. Much of his allowance was spent on astronomy books.

Tycho was arrogant, scrappy, and always ready for a duel—in one of which he lost his nose. Its replacement, made at his direction, was of silver and gold and made him as colorful in appearance as in character. Tycho's taking a peasant woman as a common-law wife was, for his time, consistent with his reputation.

Tycho was an established astronomer by the early 1570s, but an event occurred on November 11, 1572 that would make him famous throughout Europe. Strolling homeward that evening, he noticed a bright new star in the constellation Cassiopeia. He asked his servant and some peasants to look also, and all agreed it was there. That evening Tycho watched the new star far into the night.

The book he published on his observations made him famous, and Fredrick II of Denmark built Tycho an observatory on the Island of Hveen, just off the coast of Denmark. The most elaborate and best equipped observatory of the day, it consisted of a main observatory, a library, a laboratory, living quarters, servant quarters, and a printing room; it even had a jail. Tycho called it Uraniborg (hills of Uranus). The telescope had not yet been invented, but Tycho saw to it that the observatory was equipped with the best and largest instruments of the day. He also designed a clock that was the most accurate in the world. As his fame spread, students flocked to study under him, and with their help he compiled extensive records. These records would, in fact, become the tool that would finally overthrow the Ptolemaic system. Tycho was established as the feudal lord of the island, and taxes collected from its inhabitants went to the support of his work.

The peasants of Hveen were in awe of this strange and impressive figure with bright red hair, long flowing robes, and a gold and silver nose. Indeed, Tycho ruled the island with an iron—at times ruthless—hand.

Christian IV, successor of Frederick II, became increasingly annoyed with Tycho's arrogance, and eventually cut off his support. Forced to leave Hveen, Tycho took his records and some of his instruments to Prague, where he remained until his death.

In 1600, a year before his death, Tycho took on Johannes Kepler as an assistant. Shortly before his death, Tycho turned over all his data to Kepler with the intention that Kepler would use his observations to prove the Tychonic model; indeed, Kepler did use the data—but not as Tycho had hoped.

had measured no parallax in the case of his "new star"; the earth must therefore be stationary. Besides, how could an object as heavy and sluggish as the earth move? Tycho could not accept such an idea. Instead, he developed his own model, one in which the other planets moved around the sun, the sun and moon moved around the earth, and the earth itself stood still.

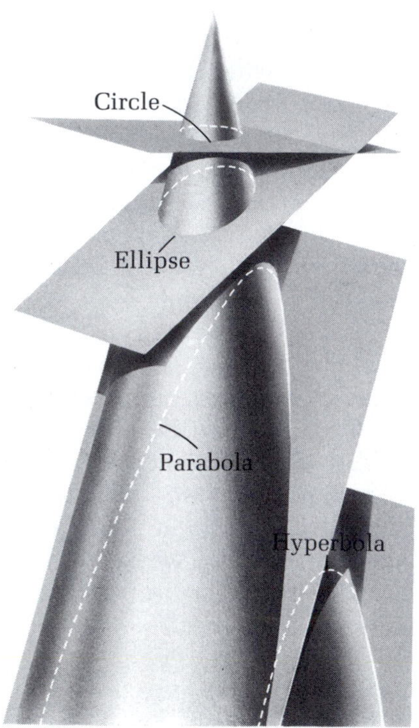

Figure 2.7 Conic sections. An ellipse is any section (a plane cutting through) of a cone that does not cut through the base of the cone. Dotted lines in each case correspond to the unseen section of the cut, on the far side of the cone.

Kepler

Johannes Kepler, in many ways Tycho's successor (see box), is known particularly for three "laws" that bear his name. An understanding of Kepler's laws requires some familiarity with the properties of the ellipse. As Fig. 2.7 shows, the ellipse is one of four basic conic curves—that is, curves formed by the intersection of a cone with planes at various angles. Because ellipses can vary considerably in shape, some means of describing each particular ellipse is needed. To arrive at this "parameter," one can begin by constructing an ellipse, as shown in Fig. 2.8. Here F_1 and F_2 are points called foci; a line from either of these foci to any point on the ellipse is called a radius vector. *V* is a vertex, or intersection of the major axis with the curve. In the simplest method of drawing an ellipse, attach the ends of a piece of string (longer than the distance from F_1 to F_2) at F_1 and F_2. With a pencil, draw out the string to its limit and trace the curve that results as the pencil is moved while keeping the string taut. The shorter the distance between F_1 and F_2, the more nearly circular the ellipse will be, and, correspondingly, the greater the separation of the foci, the greater its elongation. Thus any parameter set up to describe the shape of the ellipse would have to involve this distance. The parameter we are seeking is called the eccentricity *(e)* of the ellipse. It is expressed:

$$e = \frac{F_1O}{VO}$$

where F_1O is the distance from a focus to the center *(O)* and *VO* is

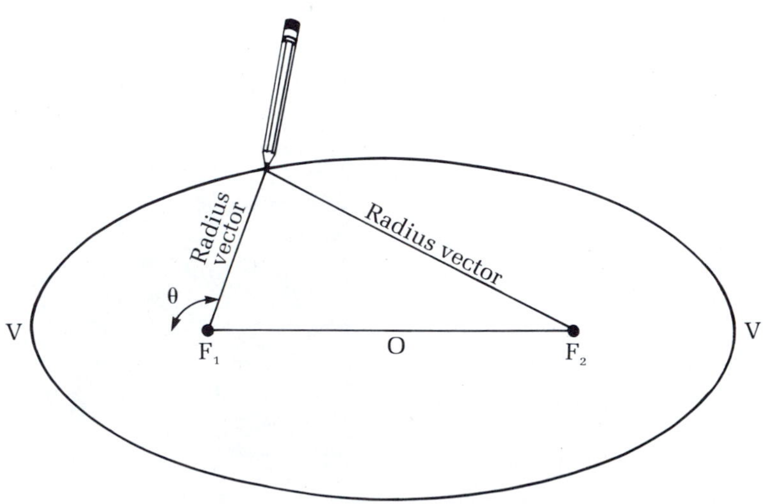

Figure 2.8 Mechanical construction of an ellipse.

Kepler

The Copernican system had been born. Many were interested in it, but few were convinced of its validity. With its circular orbits and epicycles it was, indeed, still quite crude and cumbersome—but the stage was set. Tycho had accumulated extensive data. All that was needed was someone with keen insight, a little ingenuity, and considerable patience to sort things out; this someone was Johannes Kepler.

Born in Weil, Germany, in 1571, Kepler is a biographer's delight. He led a tragic but highly eventful life. Sickly and frail from birth, he was, without a doubt, slightly odd. Although he dabbled in mysticism and the occult, he was a mathematician of the first rank and is now considered to be one of the greatest theoretical astronomers.

Kepler described his parents as unusual—his father a vicious, quarrelsome mercenary soldier who was unemployed much of the time, his mother, a small woman who shared her husband's bad temper. (In later life Kepler rescued her from a burning at the stake when she was accused of witchcraft.)

At 18, Kepler entered the University of Tübingen, where he studied theology, mathematics, and astronomy and got his first glimpse of the Copernican system. He had planned to become a Lutheran minister, but his teachers, who saw him as unsuited for the role, discreetly encouraged him to pursue his outstanding mathematical abilities. A position as a teacher of mathematics was therefore obtained for him at an academy in Graz, Austria. Though brilliant, Kepler was a disaster as a teacher, and students avoided his classes. But Kepler enjoyed his position, for it left time for his new interest: astronomy. Soon he was preparing astronomical tables, and later he became heavily involved in projecting horoscopes.

In 1596 Kepler published his first book, *Mysterium Cosmographicum (Cosmic Mystery)*, which was a lucid and simple account of the Copernican theory. Other than this, however, the book offered little of scientific value, being concerned mainly with numerical relationships, mysticism, and planetary positions. It was bold and original, however, and eventually attracted the attention of Tycho (who was now at Benatky castle near Prague). Invited by Tycho to join him, Kepler accepted. But the Tycho-Kepler alliance did not flourish: Tycho's arrogance had not lessened, and, with Kepler an avowed Copernican, his life under Tycho was anything but joyful. And to Kepler's dismay, Tycho would reveal little of his observational data.

But Tycho's death, in 1601, made Kepler heir to both his position and his records. The next ten years were one of the most fruitful periods of Kepler's life. He began his famous "War with Mars," in which, for years, he tried to fit its orbit to a circle. But there were always discrepancies, until he tried an ellipse (Fig. 2.7), which provided a model that seemed to agree with observation. Kepler published his results in 1609 in the book *Astronomia Nova (New Astronomy)*.

More years of concentrated effort brought new discoveries, among them three that are now referred to as Kepler's laws (pp. 28, 30). Kepler perceived his laws largely through trial and error; for each correct step he usually made at least three incorrect ones. He was proud of his blunders, however, and described them at length in his books.

Kepler's history, in many ways tragic, continued so after death. After a long and trying journey to Bavaria to collect overdue wages, he fell ill and died. He was buried at Regensburg. Three years later, during the Thirty Years War, the cemetary was laid waste. As a result, the exact site of his grave is unknown.

the distance from a vertex to the center. Figure 2.9 shows the eccentricities of several curves. Note that a circle is an ellipse of eccentricity zero, for, in the case of a circle, $F_1O = 0$, and so, regardless of the value of VO, $e = 0$.

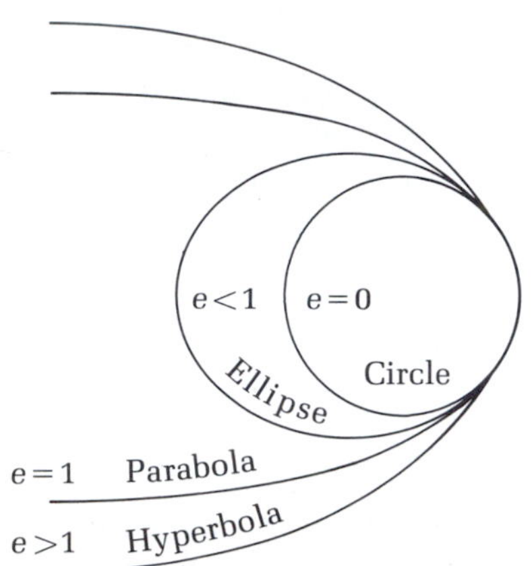

Figure 2.9 Eccentricities for various curves.

Kepler's first law states: *The orbit of each planet is an ellipse with the sun at one of the foci.* (The other focus is in space and is not centered on any body.) Nevertheless, the orbits of most of the planets are actually very nearly circular—that is, their eccentricities are very close to zero. In fact, the orbits of most planets, if drawn with a compass on a sheet of paper about this size, would differ in shape from the true orbit only by about the thickness of the pencil line.

Kepler's second law states: *The radius vector of each planet sweeps out equal areas in equal times.* If the radius vector (a line from the sun to a planet) traces out equal areas in equal times then, as a consequence, the speed of the planet must vary from point to point around the orbit. As Fig. 2.10 shows, Earth's speed in orbit must be greatest near perihelion. This is because the radius vector is shortest there, and so, were Earth's speed constant, it would sweep out a smaller area than where the radius vector is larger. Conversely, at aphelion, where the radius vector is greatest, the planet moves slowest. This is easily seen by comparing the distances A–B and G–H in Fig. 2.11.

In the case of the earth, perihelion is passed in early January, and therefore its orbital speed is greatest at that time. From January to June its speed decreases until it reaches a minimum early in July; the speed then begins to increase until it reaches its maximum again in early January.

Kepler's third law states: *The squares of the orbital periods of the planets are proportional to the cubes of their mean (average) distances from the sun.* The period of a planet is the time it takes to make one orbit around the sun. Kepler's third law tells us that the ratio of the square of that period (T^2) to the cube of the mean distance (a^3) is a constant for all orbits.

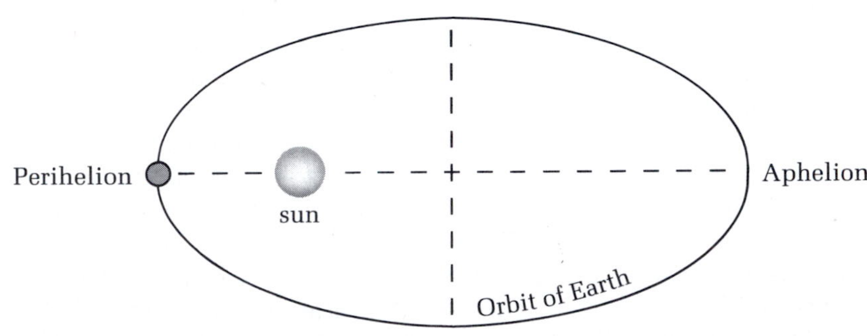

Figure 2.10 Orbit of Earth, exaggerated to emphasize aphelion (farthest point from the sun) and perihelion (nearest point to the sun).

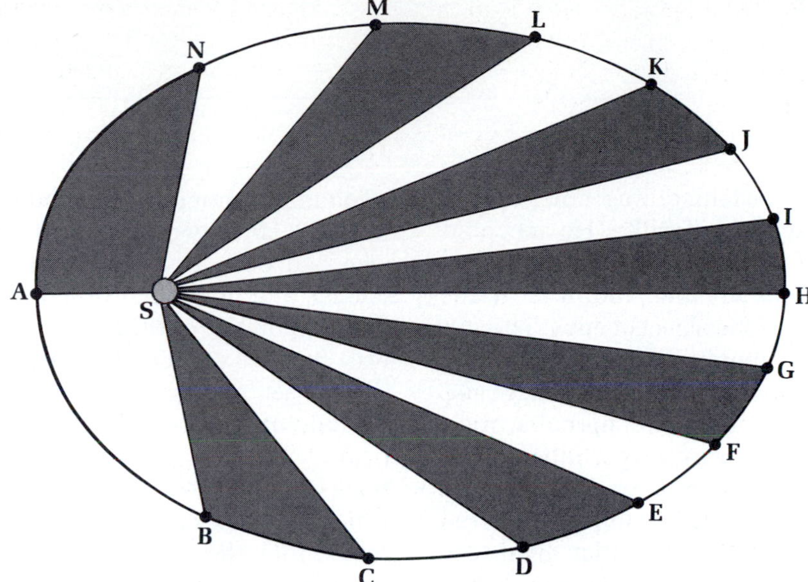

Figure 2.11 Graphic representation of Kepler's second law (equal areas are traced out in equal times). A planet travels most rapidly at Position *A* and least rapidly at Position *H*. The shaded areas are equal.

Galileo

Galileo played an important role in the development of both astronomy and physics. His major contribution was the conclusive establishment of the scientific method. He also discovered many of the basic laws of mechanics, studied falling bodies, and formulated the law of inertia. (Inertia is the persistence of a body in its motion or state of rest.) According to this law, any object orbiting the earth tended to remain in motion (at a constant height) unless acted on by some external force. The establishment of this law was a relief to many, for astronomers had wondered for years why the planets remained in motion as they did. What caused them to move? Would they ever stop? According to the law of inertia they were in a state of "natural motion" and would remain that way until forced to do otherwise.

Through his telescope Galileo discovered the craters on the moon (Fig. 2.12), the phases of Venus, and clouds of stars along the Milky Way. But what impressed him most was his first view of Jupiter, for orbiting it were four tiny moons—a solar system in miniature (Fig. 2.12). His discoveries overcame many of the apparent shortcomings of the Copernican system. For example, astronomers had wondered at the seeming anomaly of the moon revolving around Earth, whereas everthing else revolved around the sun. Galileo's discovery of the moons of Jupiter established that other such systems could exist.

Another problem was the variation in brightness of Venus. Why was there such a large change? Galileo's discovery that it presents phases, just as our moon does, resolved this. Furthermore, phases of this nature could occur only in the Copernican system. As for the

Galileo

Galileo Galilei was born in the city of Pisa, Italy, in 1564. He was the eldest of seven children, whom his once wealthy parents struggled to support. At 17 he entered the University of Pisa to pursue a career in medicine, but he soon turned to mathematics and science. When he was 21 years old, financial problems forced him to leave the university to help support his family. To do so, he began designing and building mechanical devices. His abilities eventually found him a wealthy patron, who, in 1589, secured him a position in mathematics at his old Alma Mater, the University of Pisa. In 1592, he accepted a much better position at the larger and more prestigious University of Padua.

Galileo was not popular with the faculty. He accepted little without proof, was forever arguing, and, it is likely, was the object of envy. The students loved him, and his lectures drew large audiences, unlike the case for his colleagues. He thoroughly enjoyed demolishing a well-established belief—for example, the Aristotelian view that light objects fell slower than heavy ones—with a public demonstration giving experimental proof to the contrary.

Galileo was about 45 years old when he constructed his first telescope. Within a short time he had made several important astronomical discoveries, some of which, he was sure, provided the proof he needed to vindicate the Copernican system.

But the sudden flood of discovery caused concern. Galileo felt he now had proof that the Copernican system was correct, and the Church was alarmed. As early as 1611 it began to keep tabs on his activities. At first, told politely to keep his views on the Copernican system to himself, he did. But in 1623, when his friend Cardinal Barberini was elected Pope, Galileo felt that his chance to establish the authenticity of the Copernican system was at hand. (In fact, it was Barberini who eventually summoned Galileo to trial.) However, Barberini, who had been an admirer of science and of Galileo in particular, pointed out that there was not yet sufficient proof. Galileo set out to get it.

His attempted proof appeared in the now-famous book *Dialogue on the Two Chief Systems of the World,* which took the form of a discussion between three philosophers: Salviato, the most intelligent and mouthpiece of Galileo's view; Sagredo, quickly convinced by Salviato's arguments; and Simplicio, stubborn and a bit of a simpleton. Simplicio argues for the Ptolemaic system, and indeed, his arguments are those of Barberini.

The Pope, shocked and angered, ordered publication stopped immediately. Galileo was summoned to Rome and, in 1633, was brought before the Inquisition. There was fear that he might be tortured, and indeed, he was at least twice shown torture devices. In the end Galileo avoided torture by pleading guilty and signing a document denying his views on the Copernican system.

The rest of his days were spent under house arrest,where his health rapidly declined. In 1638 he lost his sight; in 1642 he died.

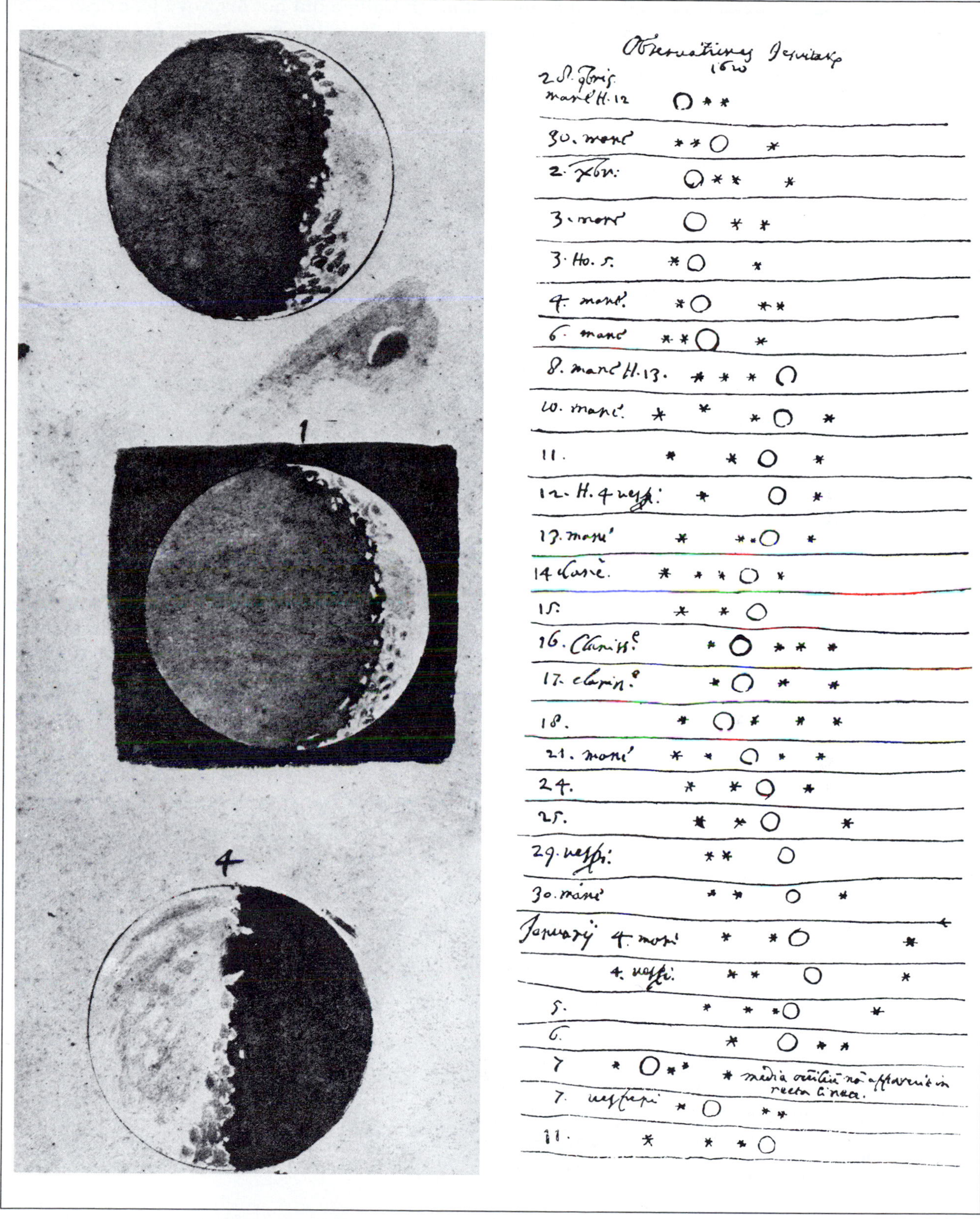

Figure 2.12 Galileo's drawings of (at left) the moon, and (at right) Jupiter and its satellites.

celestial sphere, although Galileo did not disprove its existence, he hastened its disappearance from the model with his discovery of numerous faint stars that did not appear to increase in size with increasing telescopic power (implying that they were extremely far away). This, of course, also explained the absence of observed parallax in the case of the stars.

Galileo also studied sunspots and determined the rotational period of the sun. His contributions were, indeed, numerous, but most of all he is remembered for the development of the "experimental method." A new door was now open, and science would begin to advance rapidly.

Newton

One of Isaac Newton's most important contributions was his formulation of the law of gravity. Basic to an understanding of this law are four concepts: velocity, acceleration, force, and mass. *Velocity* is speed in a particular direction. *Acceleration* is change in velocity per unit time. For example, a car that changes velocity from 20 km/hr to 60 km/hr in 6 seconds (the unit time in this case) has accelerated. *Force* is what causes acceleration, and *mass* is what gives an object inertia (p. 31). For example, assume that you are in space (free of the earth's gravitational field). Objects placed in front of you just sit there—that is, they do not "fall." If the objects are of the same material, but of different sizes, then pushing each with the same effort (the same *force*) would move the larger ones at a slower rate than the smaller; or, conversely, to move the larger ones at the same rate would require a greater force than that necessary to move the smaller. This is because the larger objects have more mass.

Newton used the concepts just discussed in explaining planetary motion. Assume that a planet moves in a circle. (We know, of course, that it actually moves in an ellipse.) Its velocity (speed *and* direction) is continuously changing as it moves around the sun, as the changing

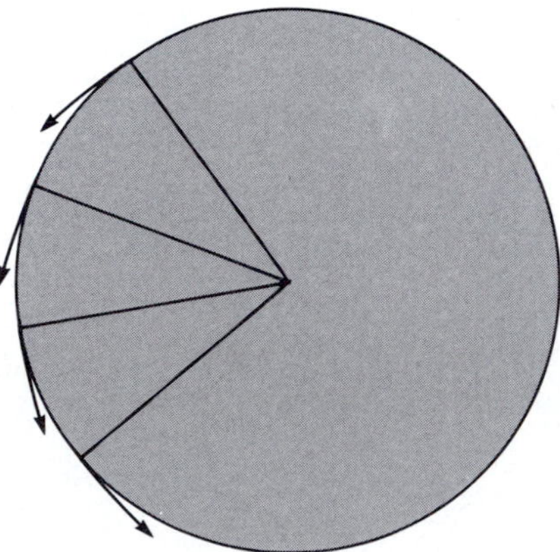

Figure 2.13 A body moving in a circle. Arrows indicate changing direction, hence acceleration. The change in direction of the arrows is toward the center.

direction of the arrow (Fig 2.13) indicates. This means that it must be accelerating and therefore that there must be a force acting on it. It is easy to show that the force is directed toward the sun; but what causes it, and what is its magnitude? Newton, the story goes, was pondering this problem one day as he sat in his garden. An apple dropped before him. Why did it fall, he wondered. Was there a force between it and the earth? Did this same force perhaps also hold the moon in its orbit? From such ruminations came Newton's formulation of a new and fundamental law of motion, the law of gravity, which can be stated as follows: *Every particle of matter in the universe attracts every other particle of matter with a force that is proportional to the product of their masses and inversely proportional to the square of the distance between them.* This is summed up in the formula

$$\text{Force} = G\frac{mm'}{r^2}$$

where G is the universal constant of gravity (6.66×10^{-8} dyne · cm^2/gm^2), m and m' are the masses of any two bodies, and r^2 is the square of the distance between them.

After Newton formulated this law, he set out to test it. He knew the approximate value of the acceleration of gravity (g) at the surface of the earth and used it to calculate the value of gravity at the moon. This was a particularly bold leap, for it assumed that celestial and terrestrial motions—heretofore considered basically different—were of the same nature.

The moon is 60 times farther from the center of the earth than is the earth's surface. If, as Newton's law postulates, gravitational force is inversely proportional to the square of the distance, then at the moon it must be ($1 \div 60^2$), or $\frac{1}{3600}$, of the force at the earth's surface. We know that $g = 980$ cm/sec^2 at the surface of the earth; therefore, at the moon it is $\frac{980}{3600} = .272$ cm/sec^2. This, in turn, should be numerically equal to its acceleration, and this acceleration can easily be calculated from the period of the moon, which we know to be 27.3 days.

Newton made the calculation. Although he did not find exact agreement, he was pleased with the results, and we know today that if he had had other than approximate values for the constants, these two numbers, indeed, would have been equal.

Newton's theory of gravity was a milestone in the history of science, and progress was rapid after its discovery. Two planets were eventually found as a result of predictions based on the theory. The mystery of the reappearing comets (Chapter 1) was solved; it was now possible to calculate their orbits and predict when they would reappear.

Though Newton's law is still the basis of most celestial calculations, it has been superceded by another monument of science, the Theory of Relativity, which will be discussed in a later chapter.

Isaac Newton was born in Woolsthorpe, England, on Christmas Day, 1642, the year of Galileo's death. Sent to live with a grandmother after his father's death, the young Newton was quiet and lonely and kept to himself much of the time. The genius that was later to reveal itself so strongly seems not to have been evident in his youth. In later life he admitted being very inattentive in school, and from most indications he was not outstanding in any way. Like Galileo, however, he loved to work with his hands and built windmills, waterclocks, and numerous other mechanical toys. It eventually became his passion.

In 1656 Newton returned to his home in Woolsthorpe, where his mother hoped to make a farmer of him. It soon became obvious that he was not suited for the role. Whenever possible he would slip off to read or to work on one of his mechanical toys. An academic position somehow seemed more in line with his abilities, so, after considerable persuasion from his teachers, his mother allowed him to register at Trinity College, Cambridge.

Little is known of Newton at Cambridge except that he took the usual course of instruction. The seeds of his genius went unrecognized by all but one—a teacher, Isaac Barrow. Barrow did much to encourage Newton and is sometimes referred to as Newton's intellectual father.

Newton took his bachelor's degree in 1665 and was planning to continue toward a master's degree when Cambridge was closed because of the plague. The next two years (1665–67) spent at Woolsthorpe, were the most productive of his life, and perhaps the most scientifically productive two years in the history of mankind. During this time Newton discovered the laws of motion, the law of gravity, many of the basic properties of light, and—to top it off—he invented calculus.

Newton did not publish reports of his work until many years later. Indeed, without considerable persuasion, he might never have published some of them. In 1669 he gave a manuscript to Isaac Barrow outlining his ideas on calculus or "fluxions," as it was then called. Barrow was impressed and, two years later, when he resigned his chair, saw to it that Newton succeeded him. The manuscript was not published until 1711, and by then Leibniz (in Germany) had also invented calculus. There has been much controversy as to who was really the first.

Newton's first paper to the Royal Society was on optics. In it he explained his discoveries in detail: White light was actually composed of light of all colors and could be separated into them by the use of a prism. He discussed the logic and reasoning behind his theory, but some were not convinced, and lengthy argument and criticism ensued.

Newton began patiently and carefully to answer the criticisms, but eventually he became exasperated and resolved never to publish again; he kept his resolve for almost ten years.

In 1684 Halley, in asking Newton's help with a problem concerning orbits, was astonished to discover that Newton had solved the problem many years before but had never bothered to publish his solution.

Subsequently, Halley also learned—to his dismay—that

Newton had a virtual goldmine of discoveries, all of which had remained unpublished. Among them was a manuscript expounding the law of gravity. At Halley's urging, Newton finally agreed to publish.

After many vicissitudes, Newton's major work was published in 1687. Now popularly referred to as the *Principia*, its complete title is *Philosophiae Naturalis Principia Mathematica*. The *Principia* was highly mathematical (for its time) and quite difficult to read. This was intentional, for Newton, weary of criticisms from those barely able to understand his arguments, had aimed to make it accessible only to those who could. Despite its complexity, the *Principia* was soon recognized as a masterpiece; it is now considered one of the most important scientific books.

With the publication of the *Principia*, Newton's stature increased rapidly. In 1681 he was elected to Parliament, in 1696 he was appointed Warden of the Mint, and, within a few years, he was elected to the presidency of the Royal Society; in 1705 he was knighted.

The importance of Newton's contributions cannot be exaggerated. His reputation was, in fact, so well established by the time of his death, in 1727, that he was buried in Westminster Abbey, traditional place of burial for England's great and celebrated.

SUMMARY

1. The Babylonians, excellent observers, kept extensive records of the positions of the planets. The Egyptians, though poor observers of the skies, who kept few records, did develop a reasonably accurate calendar.
2. The first significant advances in astronomy were made by the ancient Greeks. They developed the first "models" of the universe, introducing the idea of crystal spheres, the epicycle, and the deferent. They also determined, to a fair degree of accuracy, the sizes of the earth and the moon and the distance to the moon. The culmination of their efforts was the (earth-centered) Ptolemaic system.
3. Copernicus replaced the Ptolemaic system with the heliocentric (sun-centered) system. However, there was little interest in the new system at first.
4. Tycho set up an elaborate observatory on Hveen and accumulated a tremendous amount of accurate planetary (and other) data. These data eventually helped to discredit the Ptolemaic system. Tycho himself did not accept the Copernician system and proposed instead a model in which all the planets but Earth revolved around the sun.
5. Kepler used Tycho's records to discover three laws of planetary motion, now known as Kepler's laws. He helped establish the Copernican system.
6. Galileo was the first to turn the telescope to the heavens. What he learned would eventually confirm the Copernican system. His greatest contribution, however, was his introduction of the scientific method.
7. Newton discovered the laws of motion and the law of gravity, which had an enormous effect on astronomy.

REVIEW QUESTIONS

1. Describe the cosmologies of the Babylonians and Egyptians. What contributions did each of these civilizations make to astronomy?
2. Why did Pythagoras think there was a planet hidden from our view?
3. List the major contributions of Hipparchus.
4. Compare and contrast the models of Pythagorus, Aristotle, Ptolemy, and Copernicus. Outline the strong points and weaknesses of each.
5. Compare the explanations of retrograde motion in the Ptolemaic and Copernican systems.
6. Outline Aristotle's arguments against a rotating earth.
7. How did Aristotle account for the movement of the stars? What basic flaw pointed out by Copernicus did this point of view contain?
8. Discuss Aristotle's arguments for a spherical earth. Were they valid?
9. Explain how Eratosthenes measured the circumference of the earth.
10. What was Copernicus striving for in developing his sun-centered system? Did he achieve it?
11. What was Tycho's major contribution to astronomy?
12. Why did Tycho not believe in the Copernican system?
13. Write Kepler's three laws in your own words.
14. Where does the earth move the fastest in its orbit? Why?
15. Show how an ellipse can be drawn with two tacks and a piece of string. Explain how ellipses of different eccentricities can be obtained.

16. List some of the discoveries made by Galileo with his telescope.
17. Write Newton's law of gravity in your own words.
18. Explain the terms: velocity, acceleration, force, and mass.
19. Explain how Newton confirmed his law of gravity.

THOUGHT AND DISCUSSION QUESTIONS

1. Had Eratosthenes found the angle between the sun and the zenith at Alexandria on the first day of summer to be 10° rather than 7°, what would he have determined the circumference of the earth to be?
2. The retrograde loops made by the same planet vary in shape and size. How is this explained (a) using the Ptolemaic system? (b) using the Copernican system?
3. List all the observations that can be explained by Kepler's three laws.
4. Galileo argued that the phases of Venus supported the Copernican system. Show the cycle of phases you would expect to see (a) if reasoning from the Ptolemaic system; (b) if reasoning from the Copernican system.
5. Does the fact that an object dropped in space does not fall imply a defect in Newton's law of gravity? Explain.

FURTHER READING

Abetti, G. *The History of Astronomy.* Schuman (1952).
Berry. A. A. *A Short History of Astronomy.* Dover, New York (1961).
Friedemann, C. et al. *Astronomy: A Popular History.* Van Nostrand Reinhold (1975).
Gingerich, O. "Copernicus and Tycho," *Scientific American* (Dec. 1973).
Koestler, A. *The Watershed: A Biography of Kepler.* Doubleday, New York (1960).
Pannekoek, A. A. *A History of Astronomy.* Rowman and Littlefield (1961).
Wilson, C. "How Did Kepler Discover his First Two Laws?" *Scientific American* (Mar. 1972).

Aim of the Chapter
To give an understanding of why things in space appear as they do.

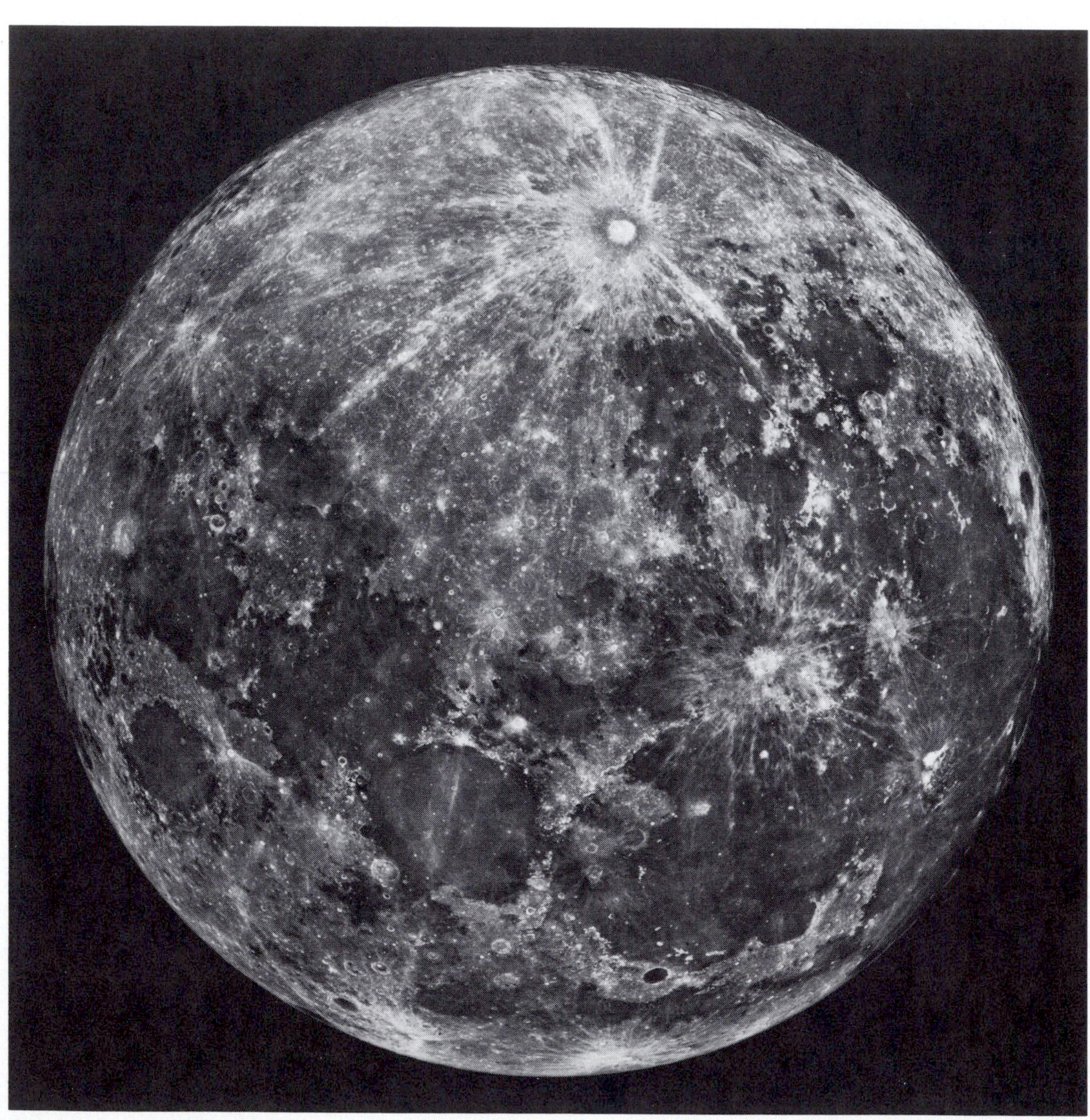

chapter 3

The View from Earth

Why do the sun, moon, and planets appear to move differently? Why do they look different? Why does the moon present phases? Why are the positions of the constellations continually changing? What causes eclipses? All of these questions can be lumped into the single comprehensive question: Why do things in space appear as they do? This is the question we will attempt to answer in this chapter.

MOTION OF THE EARTH

Many of the things we see as we look outward are actually a result of the earth's motion. It is important, then, that we thoroughly understand this complex motion. That the earth rotates on its axis in 24 hours and revolves around the sun in approximately 365 days is fairly common knowledge. It is important to keep in mind the distinction between the two terms: *rotation* (associated with spin, and consequently our day), and *revolution* (associated with orbital motion and our year).

Spin

To see one of the effects of the earth's spin, consider a nonrigid spherical ball. If spun fast enough the ball will begin to bulge in a direction perpendicular to the axis of spin; in other words, it will become oblate (Fig. 3.1). This is the result of the action of centrifugal force, by which the material of a spinning object tends to move away from the line around which it spins. Since the earth is a spinning object (and is also somewhat elastic), it is also oblate. A further complication, and one associated (in its effects) with the revolution of the earth, is the inclination of the spin axis at an angle of approximately 23.5° (Fig. 3.2) from the perpendicular to the plane of its orbit. (Such a tilt is not peculiar to the earth; all the planets have their spin axes

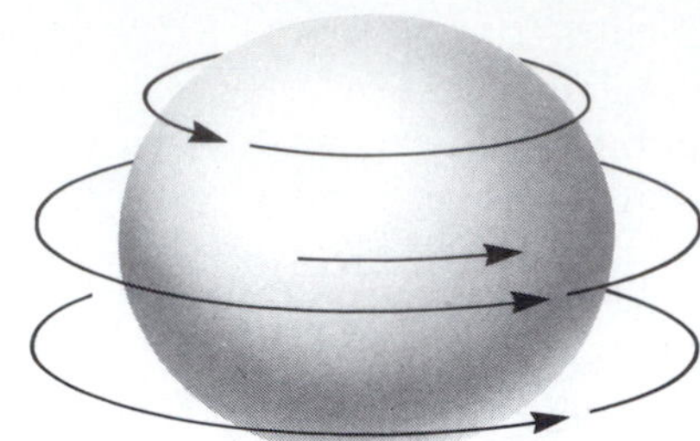

Figure 3.1 A stationary, nonrigid sphere (left) becomes oblate (right) under the influence of spin forces.

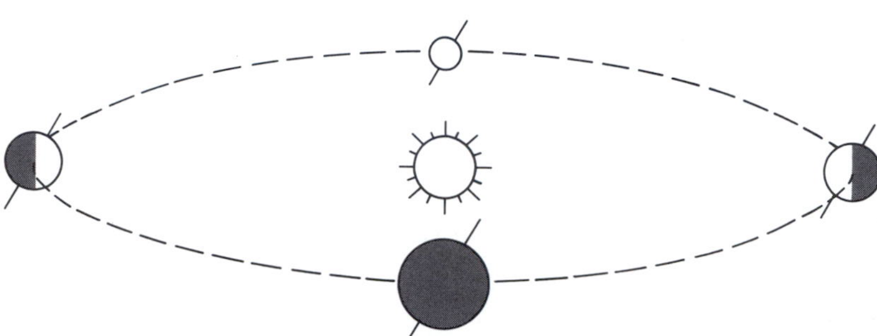

Figure 3.2 The constant 23.5° tilt of Earth's axis regardless of position in orbit. In position at left it is summer in the northern hemisphere, winter in the southern.

inclined to the plane of their orbit.) As the earth moves around the sun, its axis remains at the same angle (see Fig. 3.2). As a result, during part of the year the Northern Hemisphere is tilted toward the sun, and six months later it is tilted away. (This is, of course, true for the Southern Hemisphere as well, though at times of the year that differ by six months from those for the Northern Hemisphere.) Also, as we saw earlier, as the earth moves through its orbit, its distance from the sun varies. At perihelion this distance is 147.2 million km; at aphelion, 152.1 million km. Which of these two factors, then—tilt of the axis or variation in distance from the sun—is responsible for the seasons? It is the tilt of the axis, for the following reasons. First, as Fig. 3.3 shows, the sun's rays when perpendicular to the earth's surface are more concentrated than when they strike at an angle. The amount of light falling on 1 square foot of surface, for example, is spread out over $1\frac{1}{4}$ square feet when the angle of incidence is 35° from the vertical. This means that the heating effect of the sun's rays striking at an angle is considerably less than for those rays that strike more directly.

The second reason that tilt is the critical factor in seasonality is associated with the length of the day, which is greater the higher the sun is in the sky—as it is during the summer. This means that, on a daily basis, the sun's rays heat the air and ground for a longer period; futhermore, the shorter nights give less time for cooling.

Precession The tilt of the axis is also related to the phenomenon called precession, which can be seen in the action of a spinning top. If placed at some angle to the vertical, a stationary top, naturally, falls to one side. If set at the same angle while spinning, it does not fall but holds (at least for a while) its angle of inclination. There is, of course, a force (gravity) acting to tip it—the same force that made it fall when it was not spinning. But the spin resists this force and keeps the top from falling—until the speed of the spin falls below a certain rate. During the spin the top shows another motion: Its axis—always at the same angle—revolves around the vertical so that its axis eventually traces out a cone (Fig. 3.4). We call this motion *precession*.

Like the top, the earth is spinning, and, like the top, it is also oblate and has a force acting on it (in this case the gravitational pull of the sun and the moon). This means that it should also precess, and indeed it does—but slowly. So slowly, in fact, that it takes 26,000 years for the north celestial pole to trace out a complete circle.

It is easy to see that this precession affects how we see things from Earth. First, it slowly changes the position of the north polar axis relative to the stars. At present our North Star is Polaris, but this will not always be the case; gradually the north axis will leave this position, and for thousands of years we will not have a North Star (Fig. 3.5). Eventually (in about 12,000 years) it will approach the bright star Vega, and Vega will be our North Star.

Like the North Star, the other stars also show a small apparent motion as a result of precession. In fact, because of this motion, their published coordinates, which give their location in the sky (p. 57),

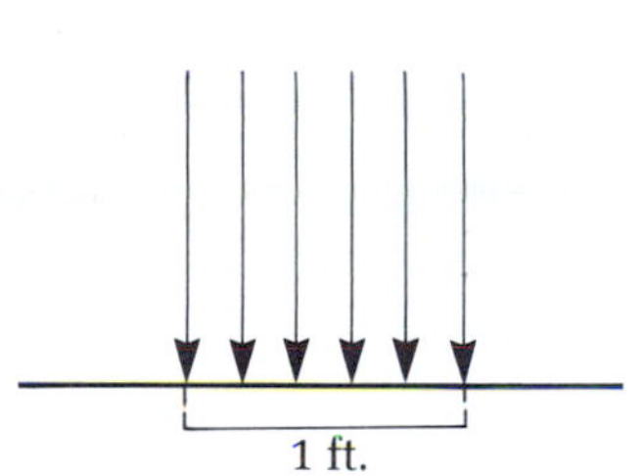

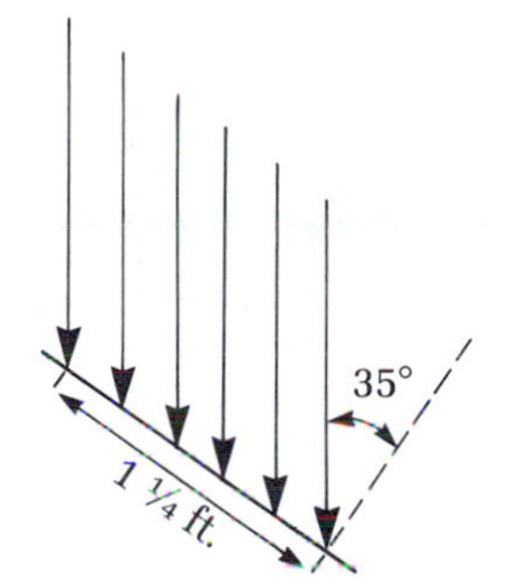

Figure 3.3 Sunlight striking a surface perpendicularly (left) and at an oblique angle. The same beam (represented by the six arrows) is spread out—in this case by 25%—when it strikes other than perpendicularly.

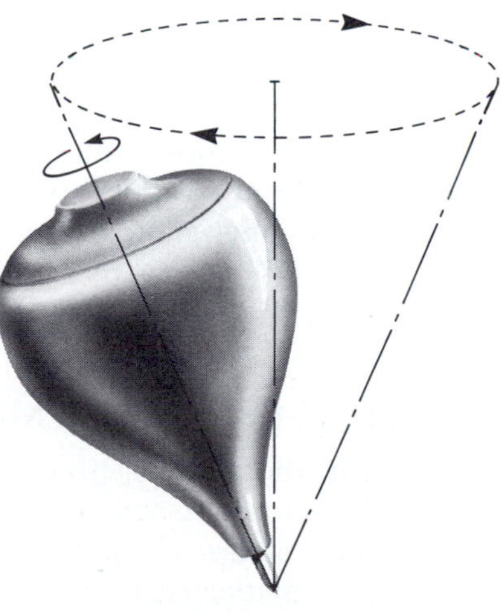

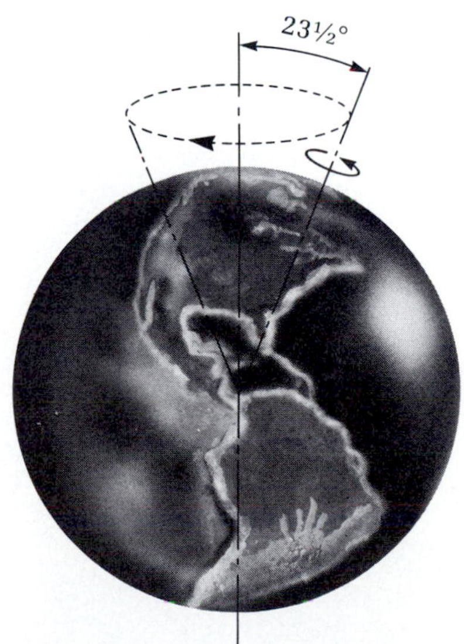

Figure 3.4 The earth's rotation compared to that of a spinning top. The dashed arrow indicates the path of the precession of both.

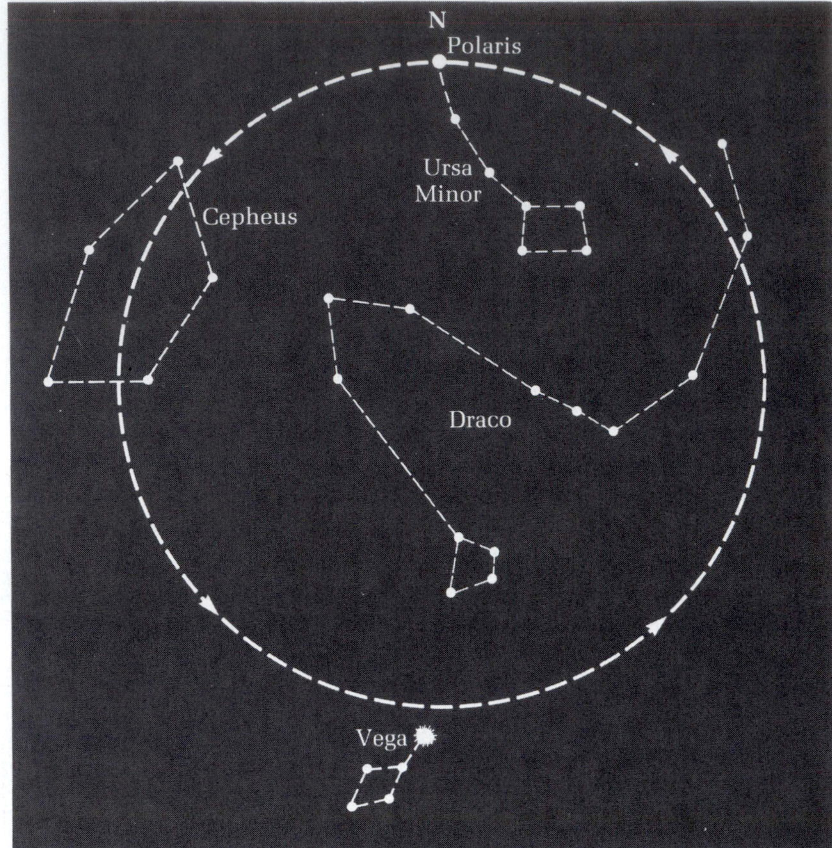

Figure 3.5 The circle traced out by Earth's precession, and the constellations to which our north polar axis will consequently change direction. Eventually Vega will be our North Star.

are readjusted every so often. Most published coordinates now in use are based on the direction of the polar axis in the year 1950. Furthermore, the position of the sun at a given time of year relative to the constellations also changes. For example, it is currently in the constellation Pisces on the first day of spring, but in a few years it will be in Aquarius.

Nutation

Another small motion, one superimposed on the precessional motion, is a regular wobble called *nutation* (Fig. 3.6). This is caused by the slowly changing orientation of the moon's orbit (the plane of the moon's orbit is not exactly aligned with that of the earth) and the consequent variation in the gravitational pull exerted as a result of it.

Sun- and Moon-related Motions

The earth, then, spins, precesses, wobbles slightly, and of course, revolves around the sun. This last means that it also must follow the sun through space. Thus, for a complete discussion of Earth's motion, we must consider that of the sun. First, there is the sun's "random motion," much like the random motion of the molecules of a gas. At

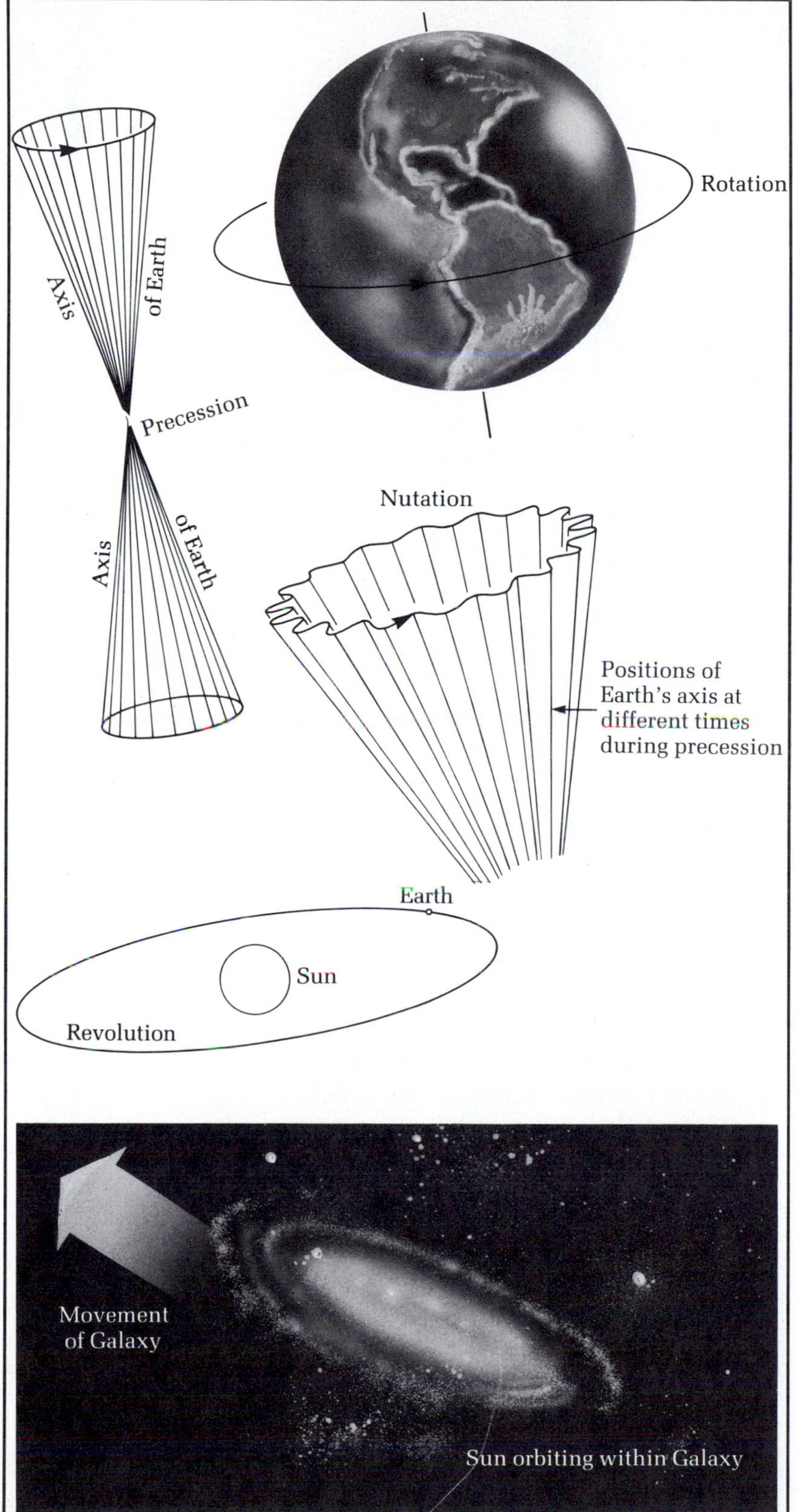

Figure 3.6 Major motions of the earth.

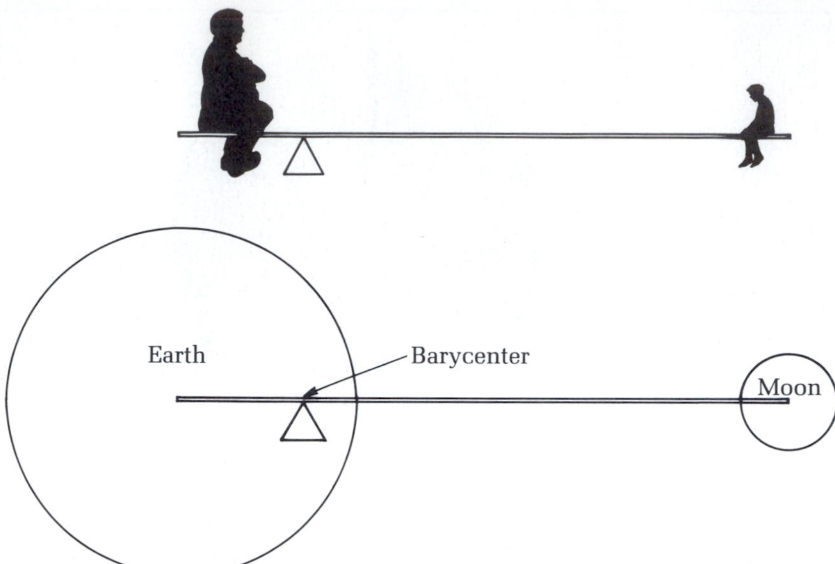

Figure 3.7 Analogy of the Earth-moon barycenter system to two people of unequal weights on a seesaw. The balance is at the point where mass times distance of one object is equal to that of the other.

present the sun's random motion is carrying it toward the star Vega at a speed of 19.3 km/sec. Second, there is its orbit around our galaxy, one of which takes approximately 200 million years. Finally, there is its movement with the Galaxy, which itself is moving through space at approximately 500 km/sec. All of these motions taken together produce an overall spiraling motion through space.

The moon also affects the earth's orbit to a small degree. Although the moon seems to revolve around the earth, actually the two bodies revolve around their common center of mass (the barycenter), which is inside the earth. This can be illustrated by analogy with a seesaw (Fig. 3.7). If one person is much heavier than the other, he must sit much closer to the turning axis or balance point, if they are to balance. This means that both distance from the axis *and* mass are important; the product of these (mass × distance) must be the same for both if they are to balance. Using this principle, we can easily calculate the center of mass of the combined mass of Earth and moon: it is 4667 km from Earth's center.

This relation of the earth and the moon results in a slight oscillation or wobble in the earth's path as it moves around the sun.

THE MOON AND ITS EFFECTS

Many of the motions so far discussed are so small that we notice no effect from them in our day-to-day living here on Earth. However, there are phenomena in the sky that are caused by the motions of Earth (and of other objects) that we do easily notice; one example is the phases of the moon, a complete cycle of which takes 29.5 days.

The moon, of course, shines only by reflected light; it gives off no light of its own and is lighted by (and, therefore, only on the side facing) the sun. As Fig. 3.8 shows, the moon in its "new-moon" phase

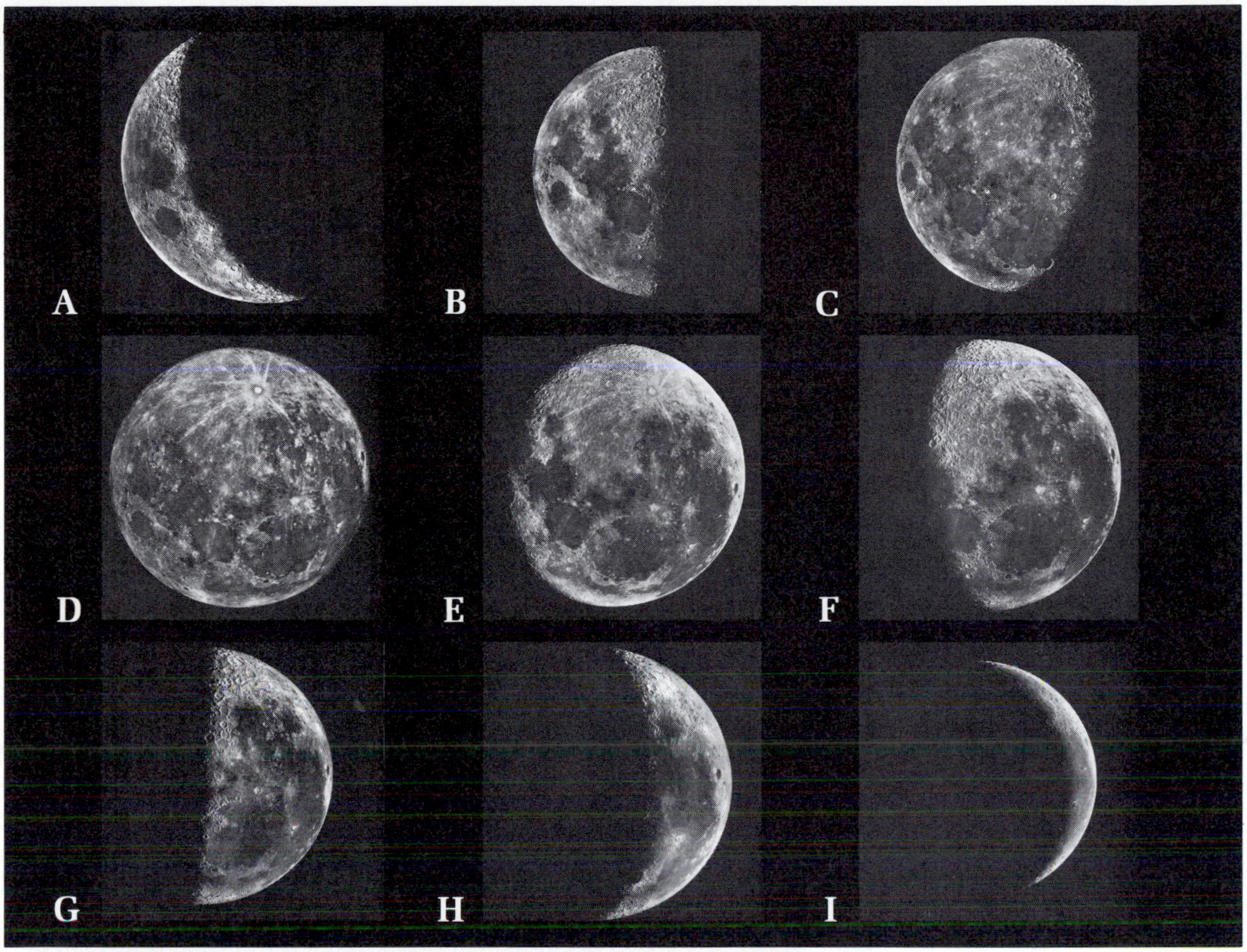

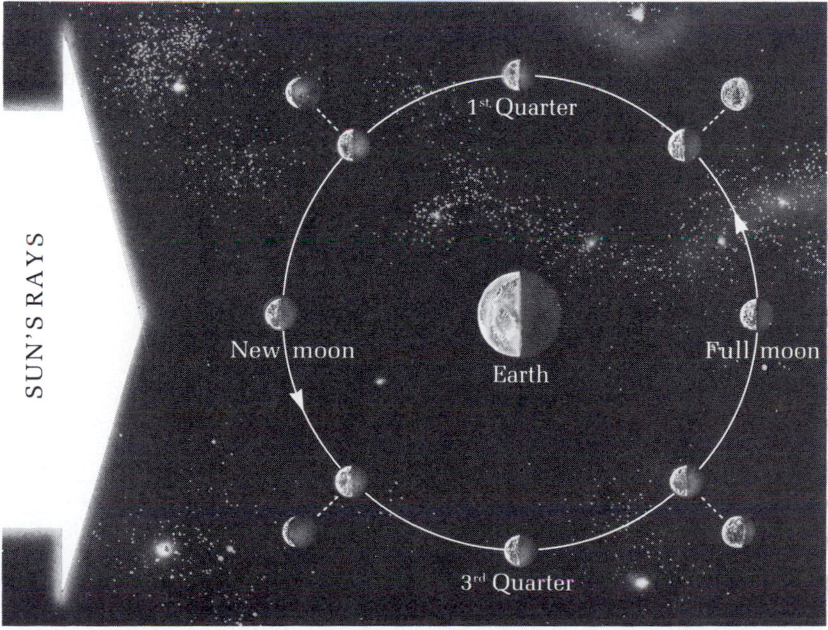

Figure 3.8 The phases of the moon. At left, the four projected moons (the outer ones of the pairs connected by dashed lines) are as the moon would appear at those points from the earth. Above is the moon seen from Earth, beginning with *A*, four days after new moon, then continuing through *B*, first quarter (seven days); *C*, an intermediate stage called the gibbous moon; *D*, full; *E*, the beginning of the waning sequence; and *F* to *I*, the opening (waxing) sequence in reverse order.

(at left) will not be visible on Earth, since the sun is then directly behind it. A day or so after the moon passes this point, however, a small crescent is visible. Then, after about a week half the moon is lighted; this is the first quarter. (Note that, at this point, lines drawn from the moon to the earth and to the sun form a right angle.) After another week the moon is directly opposite the sun in the sky and is said to be "full."

To this point, through the first half of the month, as the moon appears to grow, it is said to *wax;* through the second half, as it declines, it is said to *wane.*

Tides

In addition to the above-mentioned wobble in the earth's orbital motion, the moon has another important effect on the earth: It causes tides.

We know that the gravitational pull between two objects depends partly on the extent of their separation (p. 35). But if one (or both) of the objects is so large that it cannot be regarded as a point mass, obviously some of its part will be much closer to the second object than are other parts. In Fig. 3.9, for example, the distance *CM* (from point C to the moon) is less than the distances *AM* and *BM,* and these, in turn, are less than *DM.* This causes different gravitational pulls on different parts of the earth—strongest near C and weakest near *D.*

The net result of this differential pull is tides. Consider point *C,* for example: if there are land masses at this point, they are pulled toward the moon to a small degree—a few inches. But when there is water here (i.e., oceans), the displacement can be as large as several feet. Of course, the ocean doesn't stretch—there is an actual flow of water into the area at this point. Note also that there is a similar bulge at *D.* It is caused because the gravitational pull is much weaker here. Or, putting it another way, the gravitational pull is stronger everywhere else, and this water is being left behind as the rest of the earth is pulled toward the moon.

A detailed and accurate description of tides is quite complex, but the mechanisms involved can be described with a simple model. Assume the earth is spherical and completely covered with water. High tide will then be in the direction of the moon, and of course there will be a second high tide exactly opposite it. The sun also pulls

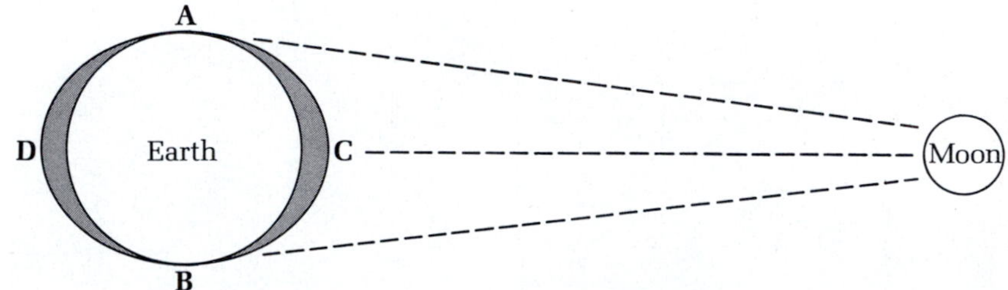

Figure 3.9 The moon in relation to Earth tides. Since the moon is closest to Point C, its gravitational pull will be greatest there. Shaded portion indicates, in greatly simplified form, the height of the tides. The higher tides at Point *D* result from the fact that the moon's pull is least here, so that water is, in effect, left behind.

Figure 3.10 Alignment of the earth, sun, and moon that produces the exceptionally high spring tides.

on the earth; indeed, its gravitational pull is 150 times as great as the moon's. In this case, though, it is not the total magnitude of the effect that is important, but the difference in gravitational forces from point to point. Thus, since the moon is much closer than the sun, those differences are much larger. However, the sun *has* an effect, but it is only half that of the moon.

The sun's influence is important when the sun, moon, and earth are lined up and both the sun and moon are influencing the tides (Fig. 3.10). The combined effect can be considerable; the resulting high tides are called *spring tides*. The opposite occurs when the sun, moon, and earth are in a right-angle configuration (Fig. 3.11), and the sun diminishes the effect of the moon. The resulting low tides are called *neap tides*.

The picture that has been presented is, of course, much simplified; when such factors as the shape and size of land masses and the nonspherical form of the earth are taken into consideration, things become much more complicated.

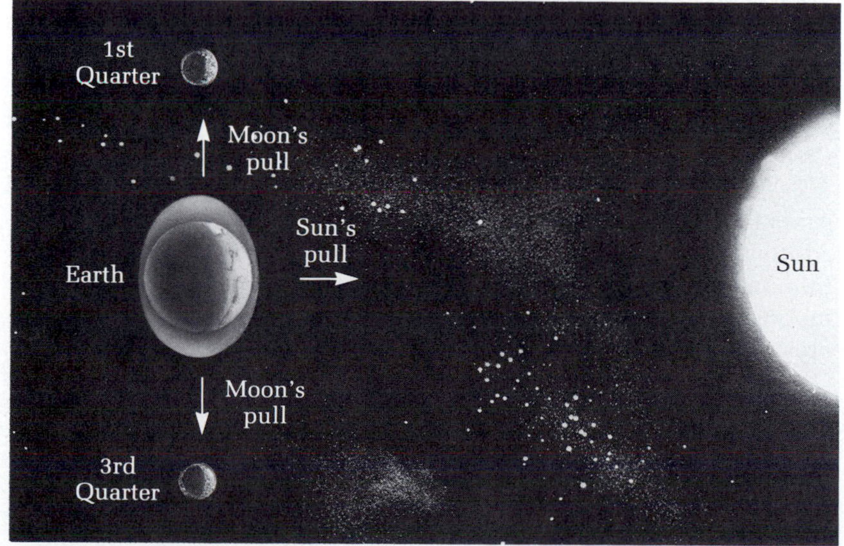

Figure 3.11 Alignment of the earth, sun, and moon (sun's pull at right angle to the moon's) that produces the exceptionally low neap tides.

THE CALENDAR

As noted in Chapter 2, the Egyptians took twelve lunar cycles to constitute a year. Problems were inevitable, since twelve cycles of the moon total only 354 days—about $11\frac{1}{4}$ days short of the actual year. The Egyptians later determined that there were 365 days in the year and adjusted their calendar accordingly. Because of their neglect of the $\frac{1}{4}$ day, by 46 B.C. the seasons had drifted out of step with the calendar by almost three months. On the advice of his astronomer Julius Caesar both readjusted the date to bring things back in line and introduced the leap year. We now refer to this system as the Julian Calendar.

The year, however, is actually 11.3 seconds short of $365\frac{1}{4}$ days, and, as time passed, the calendar began to fall out of step again. By 1582 it was ten days off. This time it was brought into line by Pope Gregory, who decreed that there would be no leap year on the turn-of-the-century years that were not divisible by 400 (e.g., 1700, 1800, 1900). This calendar, known as the Gregorian Calendar, gives a discrepancy of only one day in 3300 years. Our present calendar, a slightly revised version of the Gregorian (4000, 8000, 12,000, etc. are not leap years), gives a discrepancy of only about one day in 20,000 years.

ECLIPSES

Solar

An awe-inspiring spectacle is the solar eclipse[1] (Fig. 3.12). About an hour before totality, the moon, which cannot be seen (it is a new moon), begins to encroach on the sun, and the eclipse (partial phase) has begun. The disk of the moon gradually cuts off more and more of the sun until, about 15 minutes before totality, the sky begins to darken. In the last minutes, sparkling beads of light, called *Baily's beads,* appear near the edge of the sun's disk. They are caused by sunlight streaming through the numerous valleys on the moon. The final sparkle of light gives the impression of a glittering diamond ring; indeed, the phenomenom is called the diamond ring effect.

When totality is reached, the corona (an envelope of ionized gases that extends out from the surface of the sun) bursts into view. Although at this stage the eclipse can be viewed directly, in a few minutes the sun will reappear at the trailing edge. This time the diamond ring effect will come first, then Baily's beads, then a slim silver crescent, and finally the full disk of the sun again.

Not the least amazing feature of total solar eclipses such as this is that they occur. The probability that two objects of vastly different sizes would subtend the same angle in space is obviously very small.

[1] Viewing a solar eclipse without a protective filter *will damage the vision.* A good filter can be made of two sheets of black-and-white film that have been exposed to the light and developed.

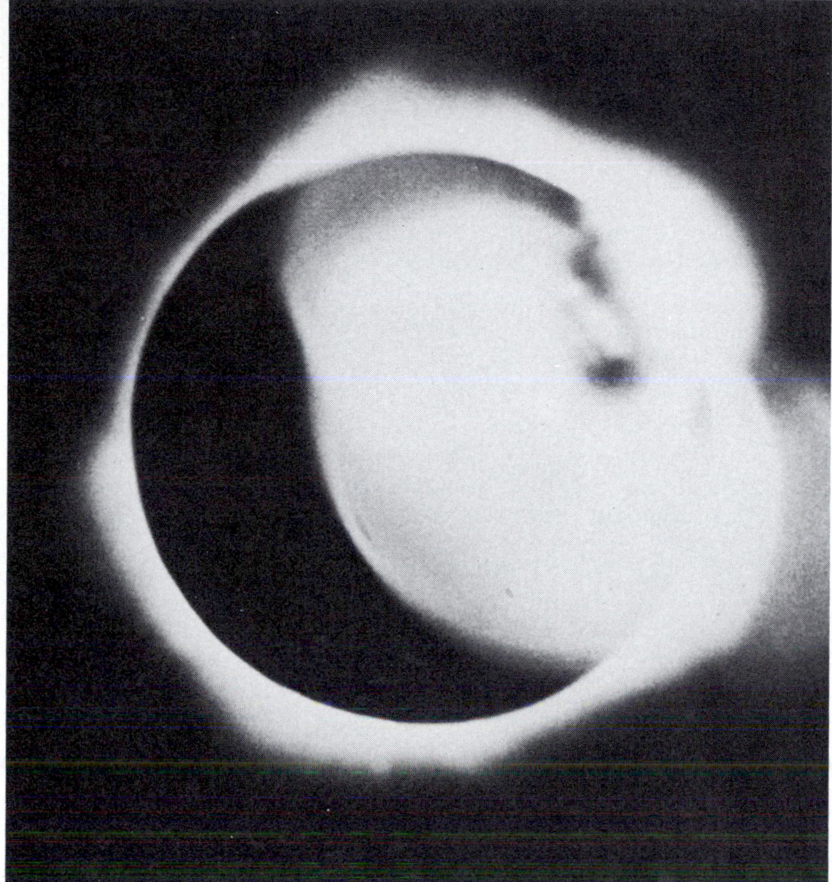

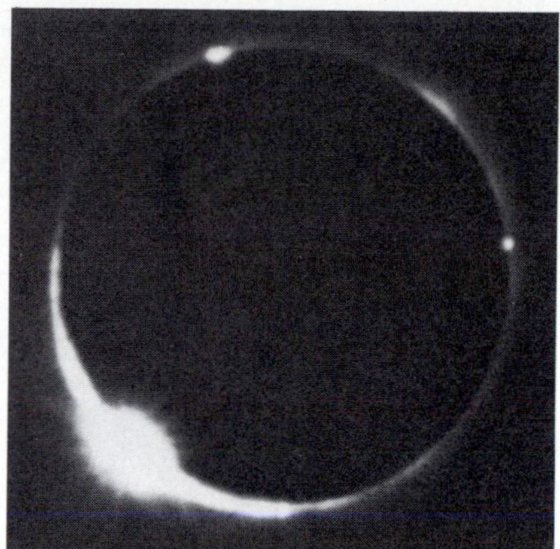

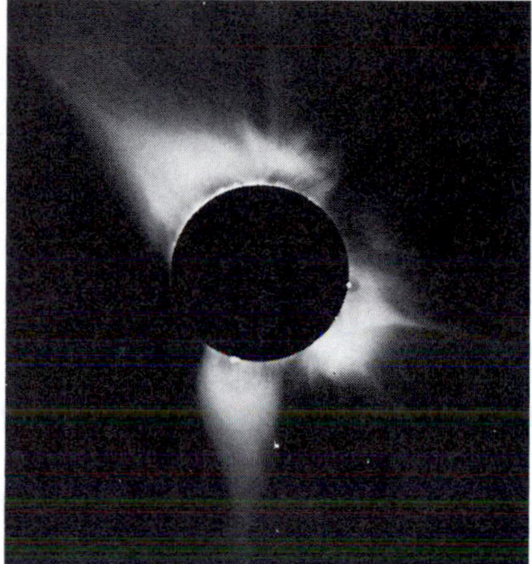

Figure 3.12 Stages in the eclipse of the sun: at left, partial eclipse; at upper right, the diamond effect at the moment before totality; and at lower right, totality, with the corona bursting into view.

Our sun is 400 times larger than our moon; this means that, for the eclipse to be possible, the moon must be almost exactly 400 times closer, which, of course, it is.

The mechanism of the solar eclipse is as follows. Although each time the moon is new it is between Earth and the sun, an eclipse does not occur at each new moon. This is because the orbits of Earth and the moon are inclined at an angle of 5.5°, so that there are only two points (called nodes) at which the moon passes through the orbit of the earth and, so, is in a position where it blocks (eclipses) our view of the sun. The line through these points (*AB* in Fig. 3.13) is called the line of nodes. If a new moon occurs at either *A* or *B*, a solar eclipse will result.

Five solar eclipses can occur in a single year. Of course, this is exceedingly rare, but at least two (not necessarily total) are visible from somewhere on Earth each calendar year.

There are two types of solar eclipse other than the total eclipse. The moon traces out an elliptical orbit and, therefore, varies in distance from the earth; at its closest point (perigee) it is 355,800 km away; at its farthest point (apogee) it is 407,400 km away. This causes a slight change in its apparent size in the sky. The total eclipse discussed above occurs only when the moon is near perigee. If it is near

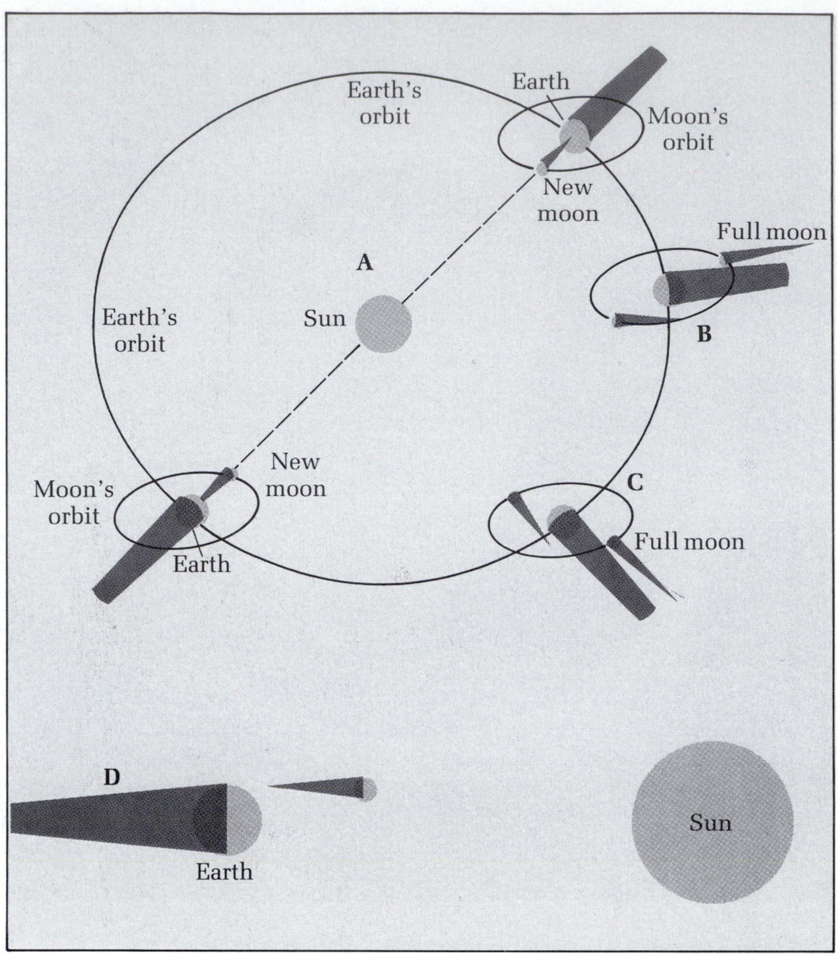

Figure 3.13 *A*, **the alignment of the earth and moon (called the line of nodes) at which a solar eclipse occurs.** ***B*** **and** ***C*** **show other positions in which, although the moon is new (and so between the sun and the earth), eclipses will not occur.** ***D*** **shows an edge-on view of the arrangement at** ***C*****.**

apogee, its disk is not large enough to cover the disk of the sun and the result is an annular eclipse, in which the sun appears to be a ring of light, or annulus.

As shown in Fig. 3.14, in the total eclipse, the umbra—the dark, cone-shaped shadow cast by the moon (and the earth)—just reaches the surface of the earth. As the earth and the moon move, the umbra traces out a path on the earth that is called the eclipse path. (The path varies considerably in width, up to a maximum of 270 km.) This is the path of totality, and you can see a total eclipse only if you are on this path. If the moon is at or near apogee, its umbra does not reach the surface of the earth, there is no eclipse path, the entire disk of the sun is therefore not cut off from our view, and we are still able to see an annulus of sunlight around the moon's edge. An observer who was in the region of the penumbra (Fig. 3.14) would see the moon pass over only part of the sun—a partial eclipse (Fig. 3.12). As compared to the path of totality, the region of the penumbra is an area of less shadow.

In ancient times the prediction of eclipses was a highly honored function. Incorrect predictions were taken equally seriously: Hsi and Ho, ministers of the board of astronomy in China in 2150 B.C., lost their heads for such a failure. Today we can predict with considerable accuracy all the eclipses for thousands of years into the future.

The moon, Saturn (near right of moon), and Jupiter over a city skyline.

Figure 1.9 The Triangulum galaxy.

Figure 5.21 The University of Wyoming's new infrared telescope. The primary is 2.3 m in diameter.

Figure 6.1 The three basic types of spectra: In the continuous spectrum, all the colors appear, running smoothly together from red to blue; in the bright-line, emission lines of specific colors appear; and in the dark-line, dark absorption lines appear against a background of continuous spectrum.

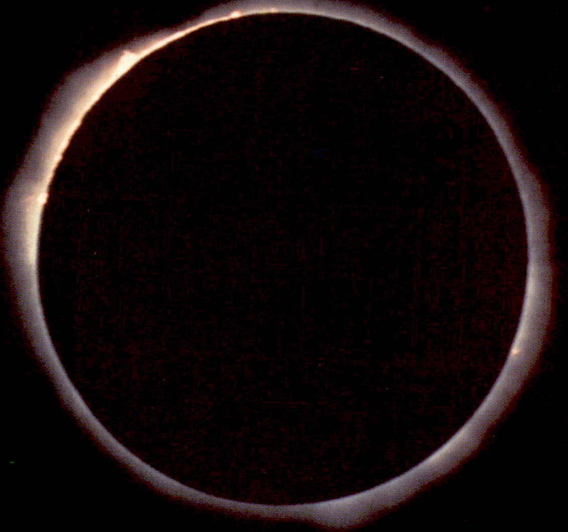

Figure 8.2 Coronal streamers in t[illegible]
solar eclipse of 1966.

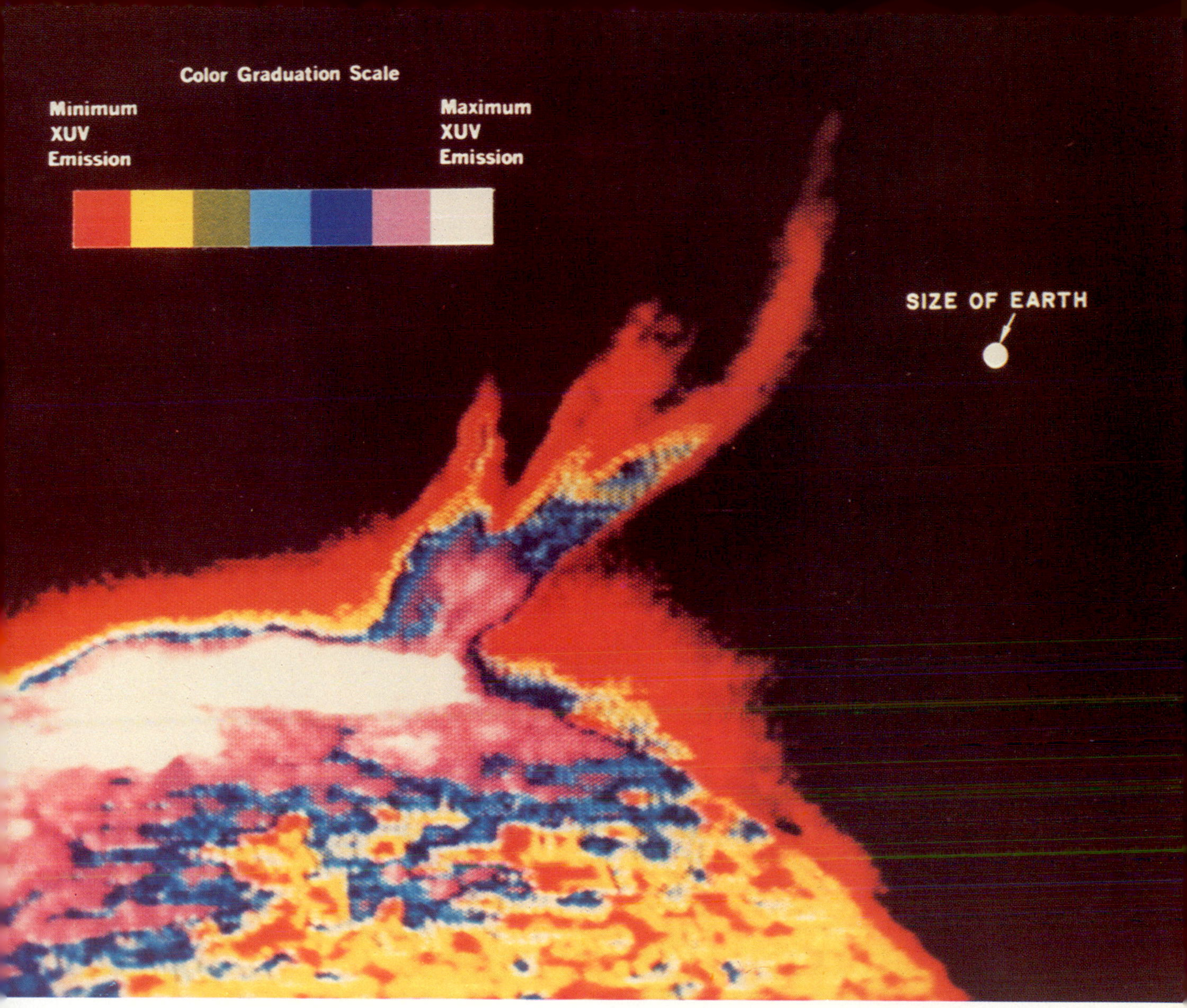

Color Graduation Scale

Minimum XUV Emission

Maximum XUV Emission

SIZE OF EARTH

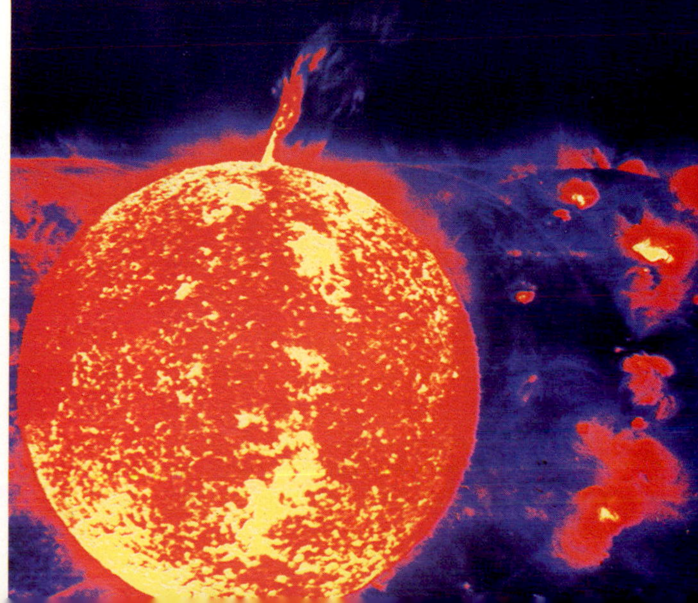

Figure 8.17 Prominences, flamelike projections from the sun. In the enhanced photograph above, the color scale shows the degree of ultraviolet emission.

Figure 10.1 The earth from space.

Figure 11.5 (opposite) Venus, photographed from a distance of 700,000 km (through ultraviolet filters to enhance cloud structure).

Figure 11.24 At lower left (opposite), the dawn side of Mars, taken by Viking Orbiter 2; at lower right, the first panorama of the Martian surface, taken by Viking 1, a portion of which appears in the foreground.

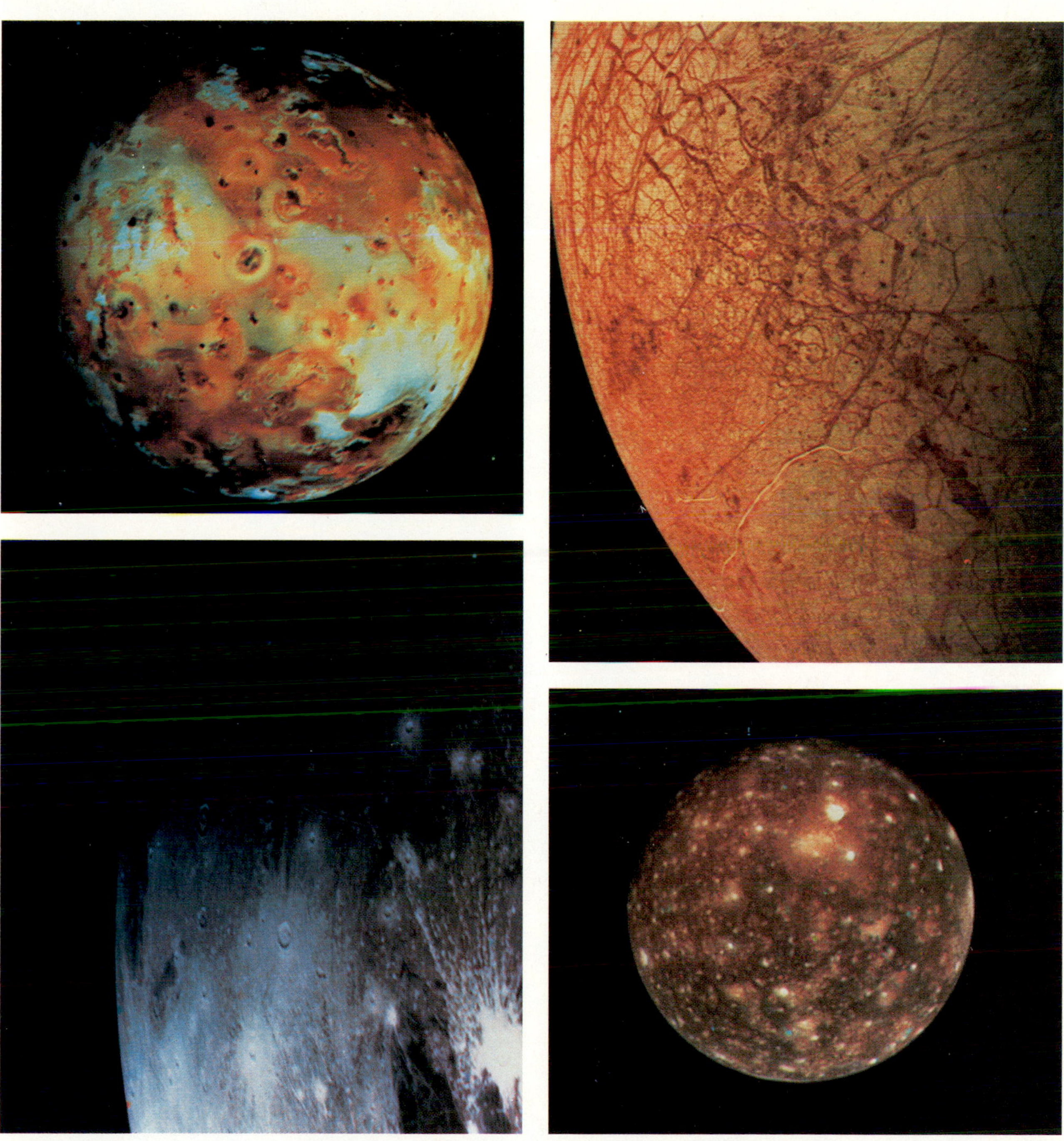

Figure 12.5 Jupiter and its four largest moons. Above, from left to right are: Io, Europa, Ganymede, Callisto. In montage at left, individual photos of the moons (not to scale) are assembled in their relative positions: Io, closest of the four to Jupiter, is at upper left; Europa is at center, Ganymede at lower left, and Callisto is in the foreground at right.

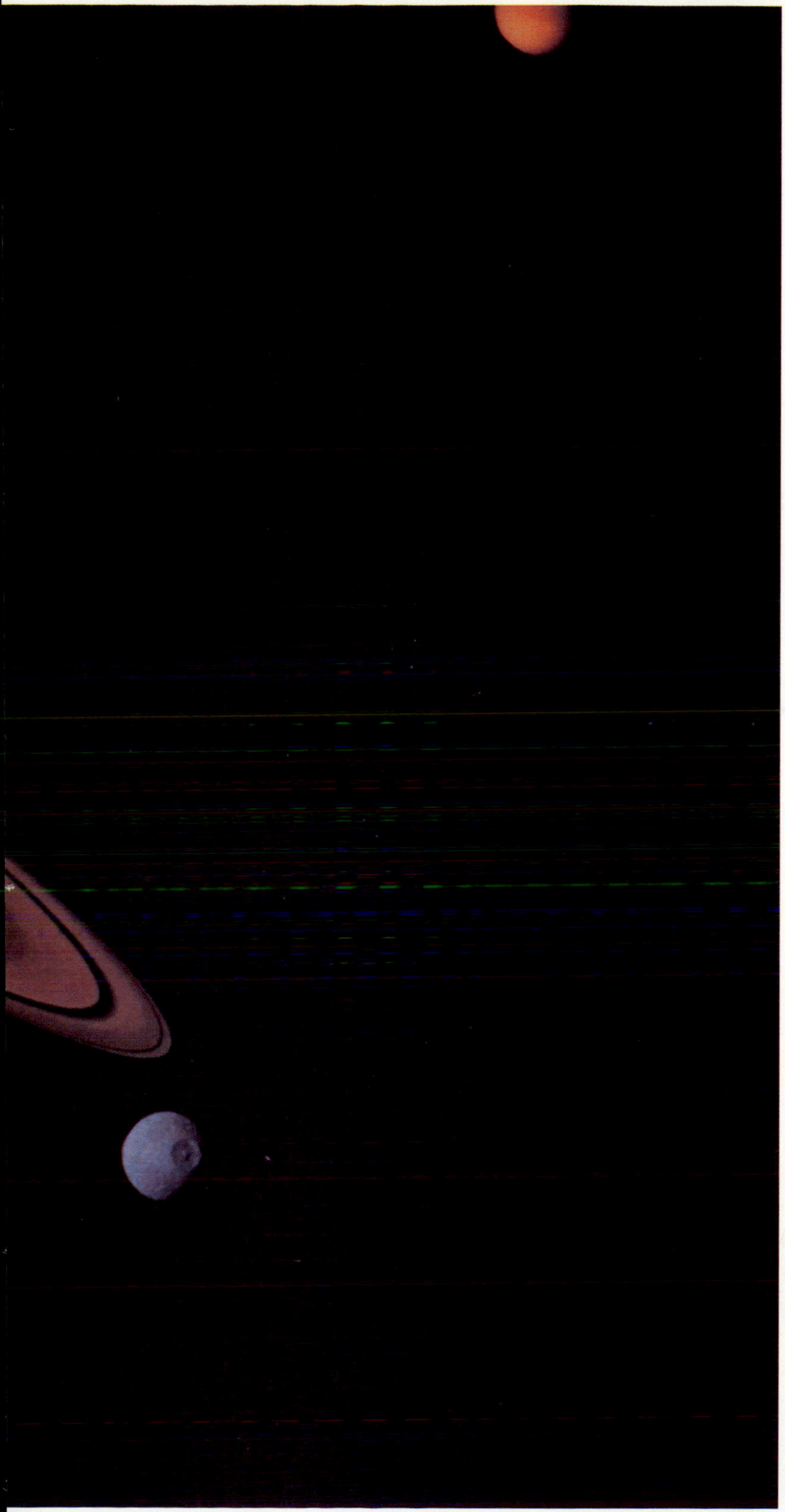

Saturn and five of its moons (see Chapter 12). This montage of photographs (not to scale) shows Dione in foreground, Tethys and Mimas in the distance at right, Enceladus and Rhea at left, and Titan in its distant orbit at upper right.

A view of Saturn's rings, taken by Voyager 1.

Figure 16.3 The Rosette Nebula in Monoceros.

Figure 17.11 The Pleiades, a young cluster formed about 60 million years ago. Note the nebulosity.

Figure 18.4 The Crab nebula, in a photograph made in the red light of hydrogen.

Figure 20.12 An artist's conception of Cygnus X-1. At the right is the accretion disk around the black hole. (From Griffith Observatory by Lois Cohen.)

Figure 21.10 The Lagoon Nebula (above) and Trifid Nebula (lower right).

Figure 21.12 The Great Nebula of Orion.

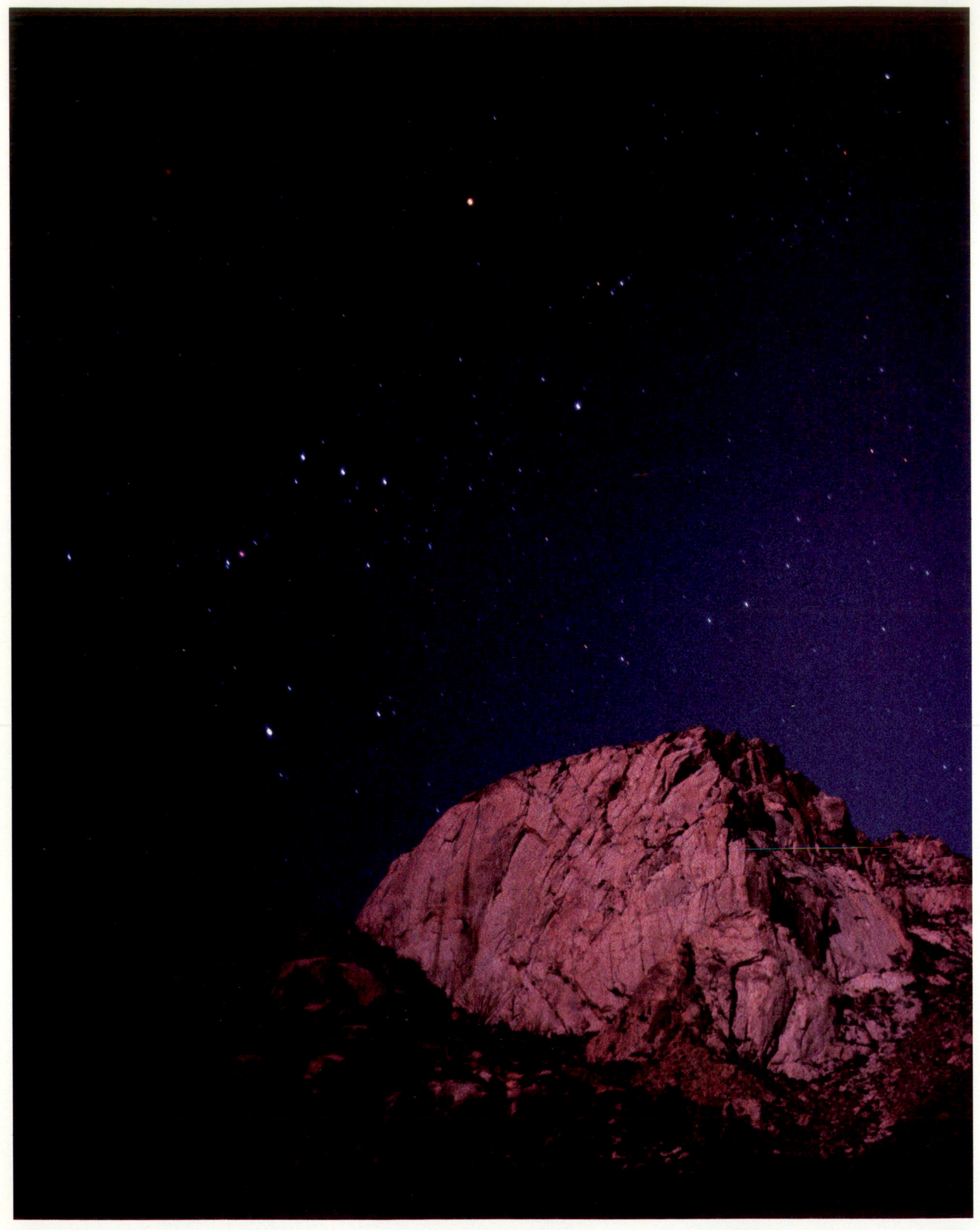

A view of Orion, over a desert horizon.

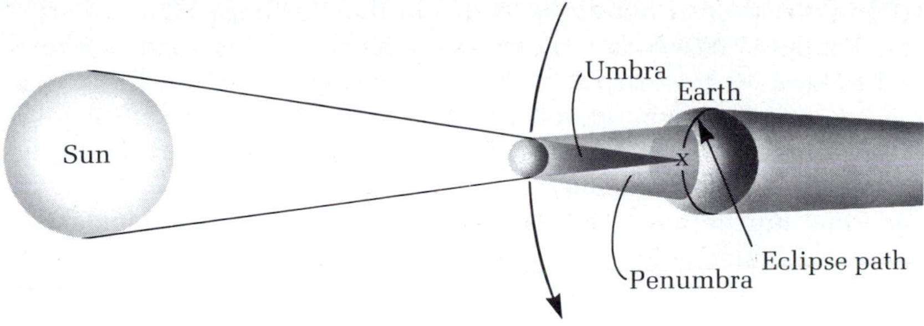

Figure 3.14 Geometry of a total solar eclipse. Diagram is not to scale.

Lunar

Another type of eclipse—that of the moon—can occur when the moon passes through the earth's shadow. This happens at nodes *A* and *B* (already illustrated in Fig. 3.13 and discussed above with respect to solar eclipses). Since the moon is full at this stage, it will darken as the earth's shadow passes over it; this is a lunar eclipse. Although lunar eclipses are actually less common than solar eclipses, they can be seen by more people—everyone on the dark side of the earth (unlike the case in a solar eclipse, which can be seen only by those along the eclipse path). However, most pass with little fanfare and, indeed, are much less impressive than solar eclipses: The disk of the full moon darkens for a few hours, then returns to normal as the shadow passes.

APPARENT MOVEMENT OF STARS

If watched long enough, most stars will seem to rise in the east and set in the west. However, stars in the direction of the pole star (Polaris) do not rise and set; they are always visible. Actually, when photographed with a long exposure they show paths that are relatively short arcs of circles. Figure 3.15 diagrams the phenomenon. On the

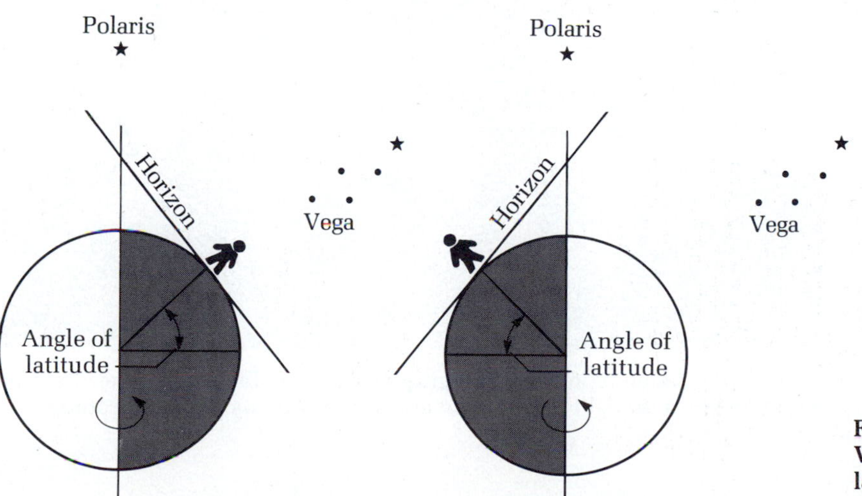

Figure 3.15 Relation to Polaris and Vega of an observer at middle latitudes at two rotational positions of the earth.

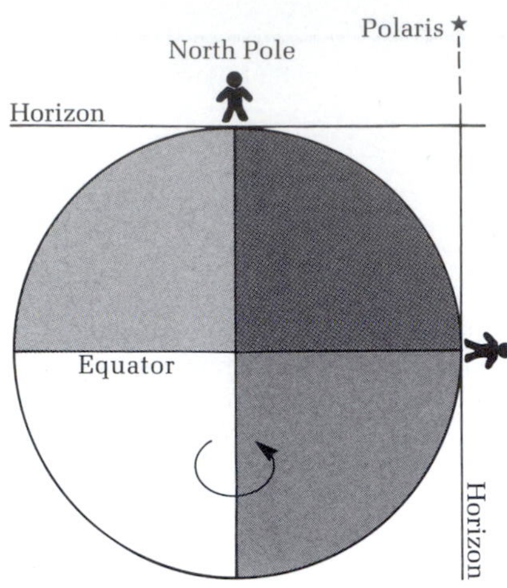

Figure 3.16 Relation to Polaris of an observer at the North Pole (Polaris overhead) and at the equator (Polaris on the horizon). It is assumed to be night at both positions. In the diagram Polaris does not appear directly overhead for the observer at the North Pole because of distortions of scale. Were Polaris at its correct distance from Earth relative to the scale at which Earth is shown here, it would appear to be virtually overhead.

left, an observer at middle latitudes in the Northern Hemisphere—say at about 10:00 P.M. in summer—sees the bright star Vega overhead and Polaris off toward the north. Everything below the horizon is hidden, and, as the earth rotates, the observer is moved until Vega disappears below the horizon. Polaris remains visible, as do all the stars around Polaris that are within the cone traced out by the horizon line. Furthermore, as the earth has rotated, these stars have traced out their short arcs. (Because they seem to revolve about Polaris, they are referred to as the circumpolar stars.) All other stars, such as Vega, have moved across the sky and are now below the horizon, having traced out *large* arcs. To an observer at the North Pole (Fig. 3.16) Polaris would seem directly overhead, and, as the earth rotated, all the other stars, down to the horizon, would appear to rotate around Polaris. None rises or sets, and the same stars are visible throughout the night. A camera pointed directly overhead and with a long exposure would produce a photograph similar to that shown in Fig. 3.17.

To an observer at the equator (Fig. 3.16), Polaris is on the horizon. Therefore, only a few of the same stars will be visible to him throughout the night. Those near the eastern horizon in the evening will move completely across the sky and be near the western horizon at dawn, their paths straight lines perpendicular to the horizon. Thus a long-exposure photograph of the sky directly overhead will be similar to the one shown in Fig. 3.18.

As described above, at middle latitudes an observer will see a continuous procession of stars rising in the east and setting in the west because of the rotation of the earth. But at intervals of a few months the set of stars he sees will be different. This is because, as the earth revolves in its orbit, its dark side points in continuously dif-

Figure 3.17 Long-exposure photograph of the sky directly overhead at the North Pole shows arcs traced by stars as Earth rotates.

Figure 3.18 Long-exposure photograph of the sky directly overhead at the equator shows straight lines traced out by the stars as the earth rotates.

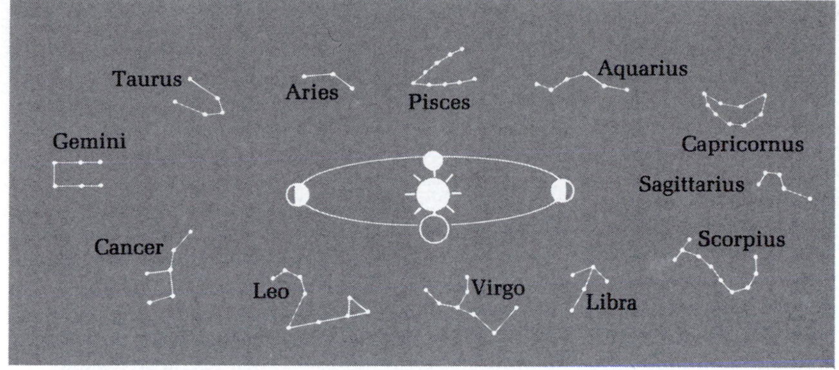

Figure 3.19 The earth in various positions in orbit and some of the constellations (those composing the Zodiac) visible from various positions.

ferent directions (Fig. 3.19), from which different stars will be visible. Over the period of a year the total span of 360° is traced out, and all the stars usually visible in a given hemisphere will be visible at one time or another. However, an observer in the northern latitudes will never be able to see the stars near the southern celestial pole, and, similarly, one in the south, will not be able to see the stars near the north celestial pole.

THE CELESTIAL SPHERE

It is important to be able to give the location of stars, as we are able to designate a point on Earth. In the case of the latter, we do so by giving its latitude and longitude (Fig. 3.20). Lines of latitude are parallel to the equator, with the equator itself taken as zero latitude. Positions north of it are positive, those south of it negative, and the range is

Figure 3.20 Lines of latitude and longitude on the earth.

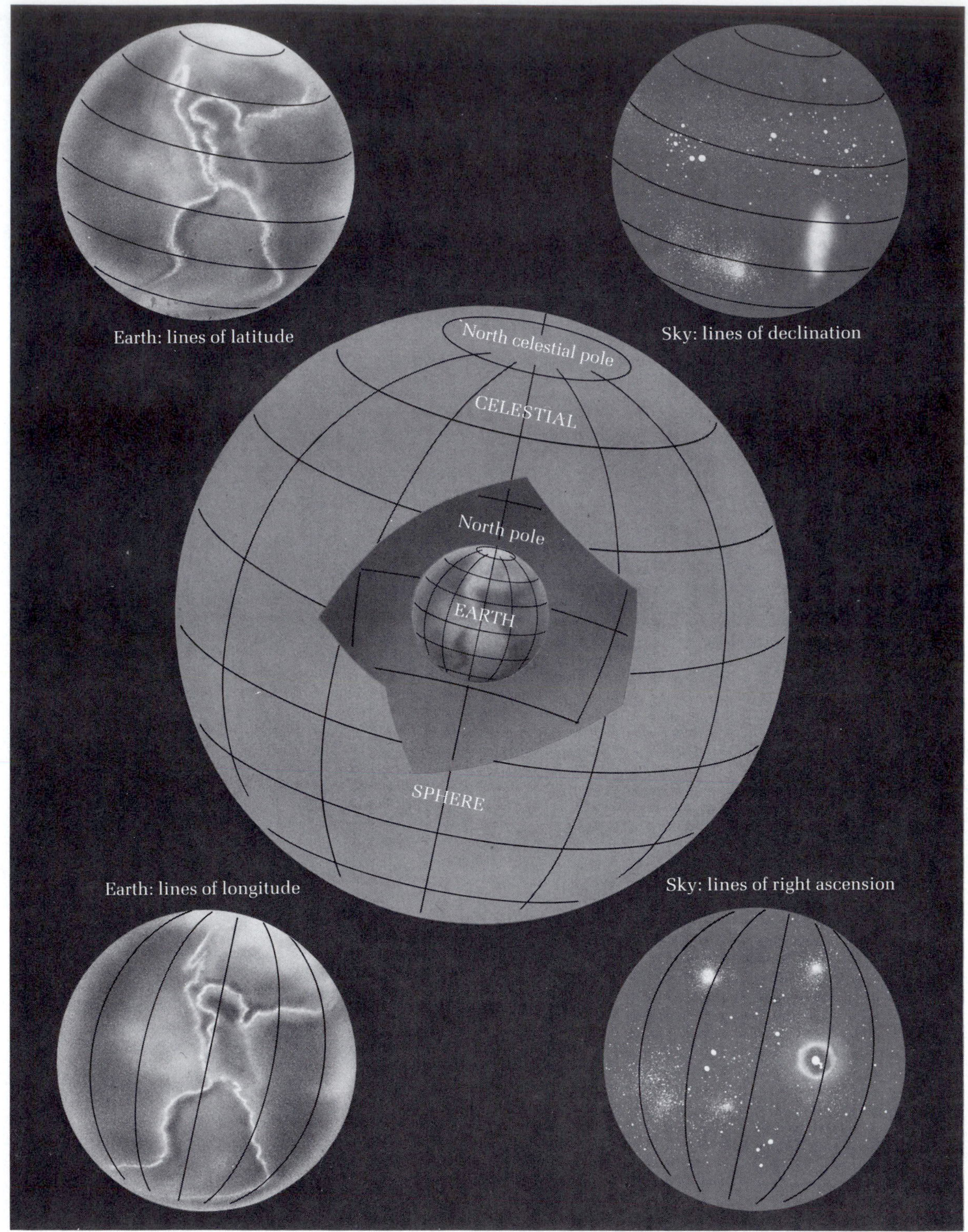

Figure 3.21 Relation of declination and right ascension to latitude and longitude on the earth.

from 0° to 90°. New York, for example, has a latitude just over +40°; Auckland, New Zealand, has a latitude of −37°. As Fig. 3.20 also shows, lines of longitude, running north and south, range from 0° to 180°, with the zero line passing through Greenwich, England. Using these two coordinates, we can locate to a fraction of a second (degrees are subdivided into minutes and seconds) any point on the surface of the earth.

A similar network has been established for designating locations in space. The terms corresponding to latitude and longitude are, respectively, declination (Dec.) and right ascension (R.A.). In this system an imaginary sphere (Fig. 3.21), called the celestial sphere, surrounds the earth. The earth's equator is projected[2] onto this sphere, where it is called the celestial equator. Another circle on the sphere is a projection of the apparent path of the sun (Fig. 3.22); this is inclined at an angle of 23.5° (the tilt of Earth's axis) to the celestial equator and is called the ecliptic. The two points where the ecliptic and the celestial equator cross are called equinoxes (equal nights and days). The two points at which the ecliptic and the celestial equator are most widely separated are called solstices (Fig. 3.22).

[2] Imagine a transparent Earth with a light at its center and an actual line drawn at the equator. The line would cast a shadow (be *projected*) onto the enclosing sphere.

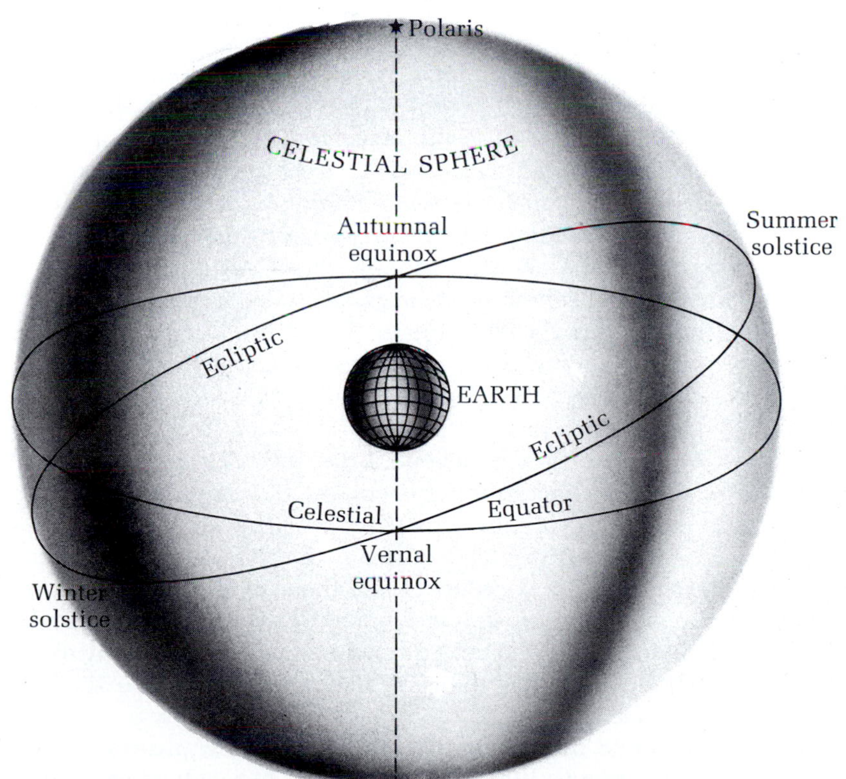

Figure 3.22 The ecliptic (apparent path of the sun on the celestial sphere) and four important points on it. The point of intersection of the ecliptic and the celestial equator (vernal equinox) is the zero point for right ascension.

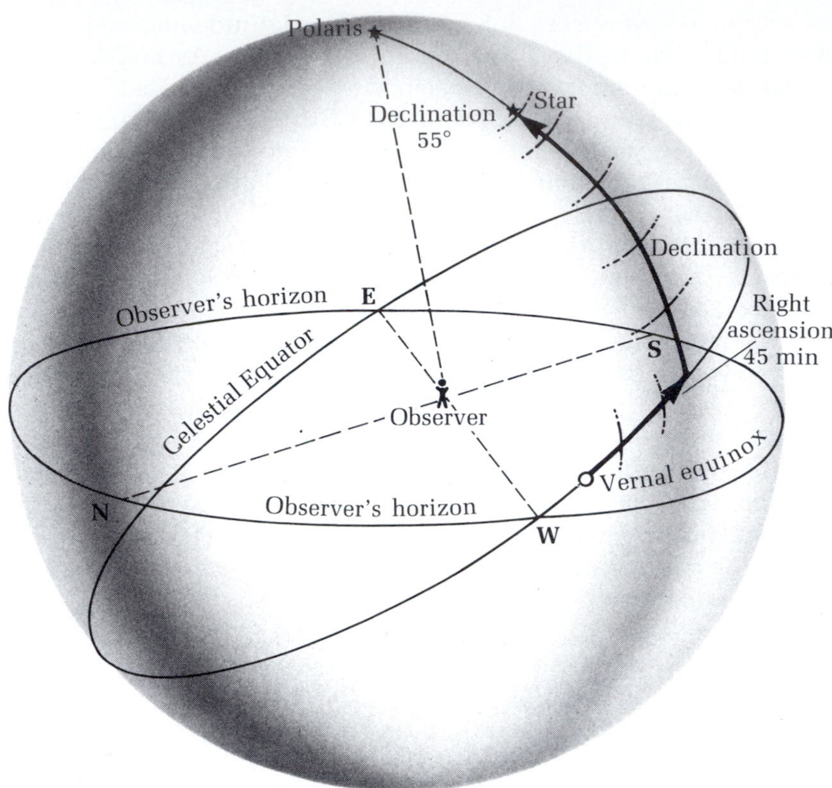

Figure 3.23 Determining right ascension and declination of a star.

The declination of a star, then, is its angular distance north or south of the celestial equator (Fig. 3.23)—analogous, as described above, to latitude on Earth. Right ascension (R.A.) is obtained by projection of the lines of longitude. Zero longitude on earth is arbitrarily designated to be at Greenwich, England. However, Earth revolves with respect to the celestial sphere (from Earth, of course, it would appear as if the celestial sphere were rotating). Therefore a point on the celestial sphere must be chosen to represent zero longitude; the point is the vernal equinox (Fig. 3.22). R.A. differs from longitude in one other way: It is expressed not in degrees but in hours, minutes, and seconds. The range is 0 to 24 hours; in 360° there are 24 hours, so each hour corresponds to 15°.

The R.A. of a star is the basis of a measure of elapsed time referred to as sidereal time. In this system, the day is the time it takes the earth to make one rotation with respect to the stars: 23 hours, 56 minutes. This differs from the familiar 24-hour solar day for these reasons: The solar day is the time it takes the earth to make one rotation with respect to the sun, that is, from one day in which the sun is, say, directly overhead to the next day when the sun is directly overhead. But, while the earth rotates, it also revolves around the sun (Fig. 3.24) and so at the end of one rotation it is at a new position in its orbit (and, hence, in relation to the sun). This new position is one in which it must rotate a little further in order to reach its preceding day's position relative to the sun.

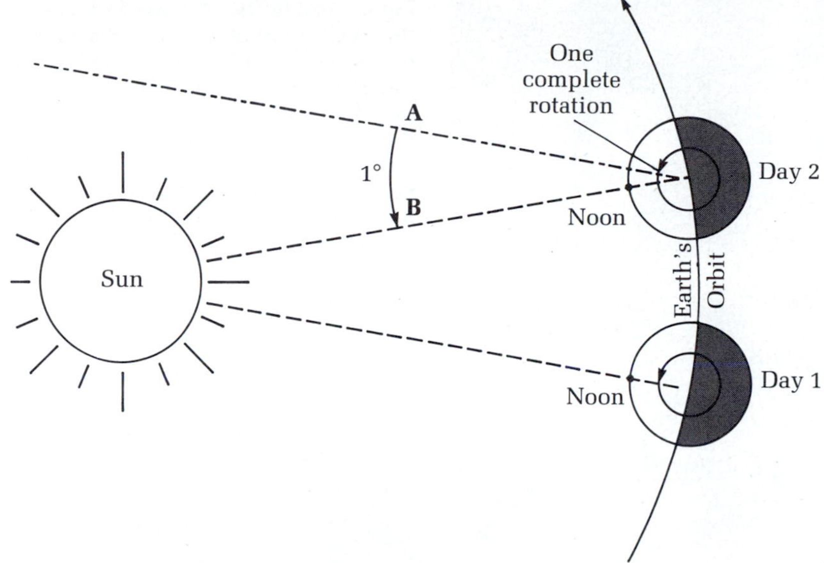

Figure 3.24 Cause of difference between solar and sidereal day. One sidereal rotation is complete where the arrow intersects line A. *AB* is the additional rotation required to reach point where Earth is again in "noon" relation to sun.

Because of this difference the stars seem to set approximately four minutes earlier each successive night. For this reason most observatories have, in addition to the usual clock, one that shows sidereal time.

THE CONSTELLATIONS

Eighty-eight constellations are now recognized by astronomers worldwide, and it is quite likely that many of the same groups were recognized by the ancients. It was not until 1928, however, that constellation boundaries were unambiguously defined. Before 1928 many stars and objects could be defined only as being near or in the vicinity of a particular constellation, but not within its boundaries.

Constellations are given designations of many different things: religious and mythological figures, animals, insects, and even scientific instruments. In many cases, however, considerable imagination is required to see the resemblance. Libra, for example, looks little like the scales it represents, and Pisces looks nothing like a pair of fish. On the other hand, Scorpius does have somewhat the shape of a scorpion.

The discussion that follows deals with some of the constellations visible each season. (A constellation chart, covering the four seasons, is included on the inside of the covers of this book.)

Spring

The major constellations of spring are Leo, Bootes, and Virgo, with Gemini still visible just above the horizon in the west. (The circumpolar constellations will not be listed since they are visible all year.)

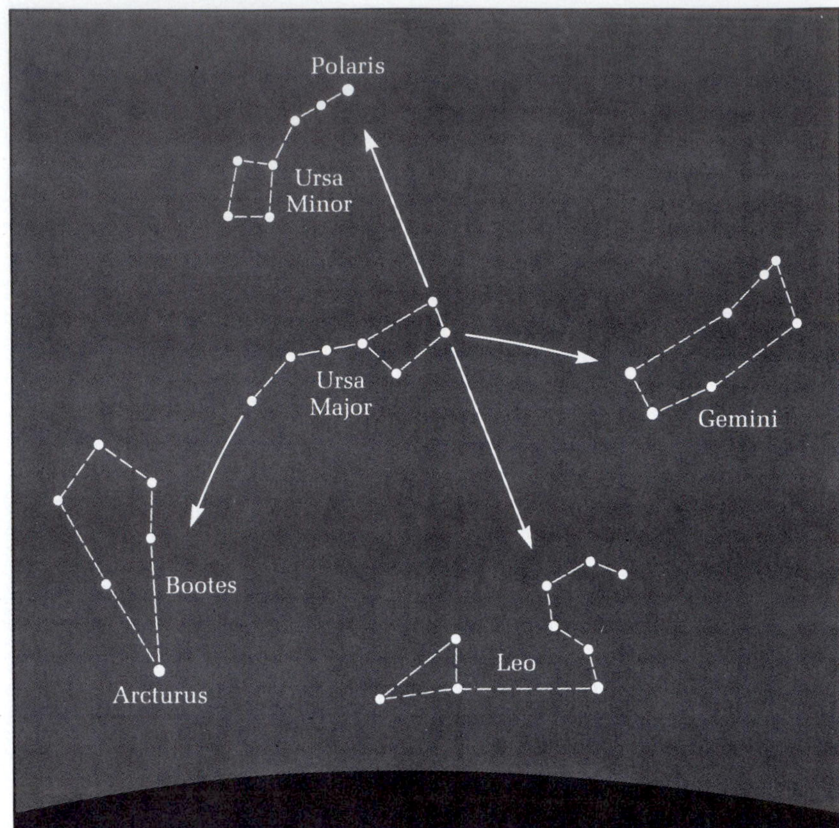

Figure 3.25 (left) Arrows indicate the position of other constellations, relative to Ursa Major. The bottom of the diagram represents the area of sky that is directly overhead during Spring. Thus, to the observer, Leo is directly overhead, and Ursa Major is higher in the sky (further above the horizon) than Polaris. North is at the top of the diagram, east is at the left.

Figure 3.26 (below) The Ring Nebula in Lyra.

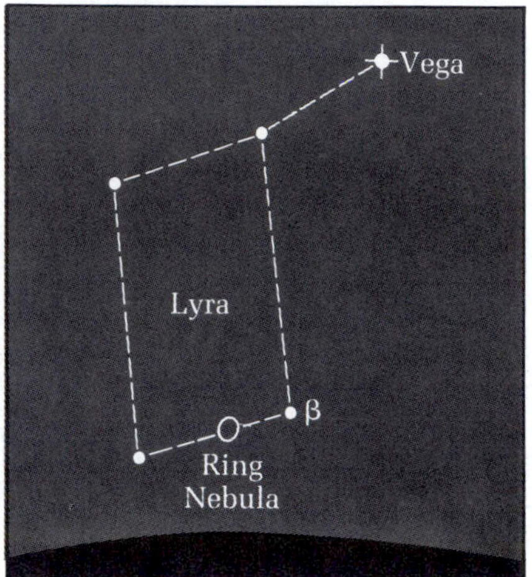

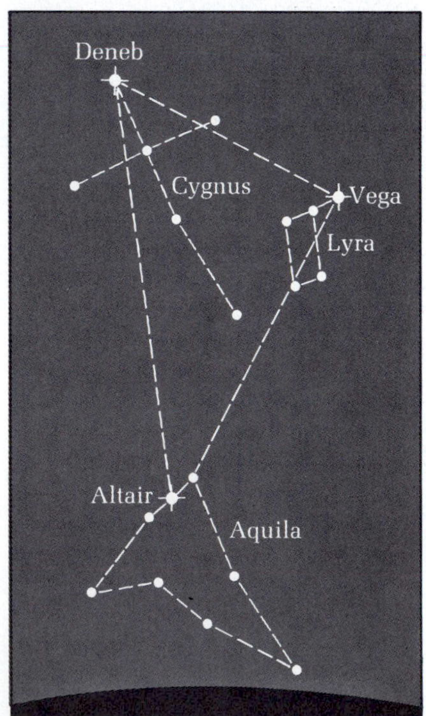

Figure 3.27 The Summer Triangle.

To find these constellations, it is best first to locate the Big Dipper in the constellation Ursa Major (Fig. 3.25). The stars composing its "handle" form a line that points to the constellation Bootes, the bright bottom star of which is called Arcturus. Leo, Gemini, and Ursa Minor can also be found readily on the basis of their association with the Big Dipper. Note that a line through the two end stars of the "bucket" points toward Polaris, our North Star.

The constellation Coma Berenices can be detected as a small, hazy region on a line that would join Arcturus and Denebola. A small telescope shows the haze to be a swarm of stars.

Slightly north of Coma Berenices is the constellation Canes Venatici, itself not very impressive, but important because it contains one of the most spectacular galaxies in the sky—the Whirlpool galaxy. In fact, this whole general area (and in particular the region of Ursa Major) is dotted with galaxies, some of which can be seen with a relatively small telescope. Numerous galaxies can be seen here because we are looking in a direction away from our galactic disk.

Summer

Although the summer sky does not have the numerous bright stars of the early spring sky, it does have many constellations, the main ones being Cygnus, Hercules, Lyra, Scorpius, Corona Borealis, and

Bootes. Most are easily found. Bootes, recognizable from the spring sky, is setting in the west; in a direction toward the zenith are Corona Borealis and Hercules. Continuing in that direction discloses Lyra, with its bright star Vega in a position almost directly overhead, and in which is found the Ring Nebula (Fig. 3.26), a gaseous cloud, located between the two bottom stars on the small, rectangular part of the constellation. The right-hand one of these two stars, β Lyra, is a double star system (eclipsing binary). In the middle of the Milky Way is Cygnus (also known as the "Northern Cross"), with its bright star Deneb. Deneb and Vega, along with Altair of the constellation Aquila, form the three corners of the Summer Triangle (Fig. 3.27), which is helpful in locating Cygnus, Lyra, and Aquila (the eagle); the latter can also be located by continuing south along the Milky Way, where Sagittarius can also be found. Almost directly south is Scorpius, with its bright central star Antares.

The constellation Hercules contains the well-known Hercules globular cluster (Fig. 3.28) in its upper right arm. The North American and Veil nebulae (both difficult to detect in a small telescope) are in the region of Cygnus along with a number of clusters (Fig. 3.29).

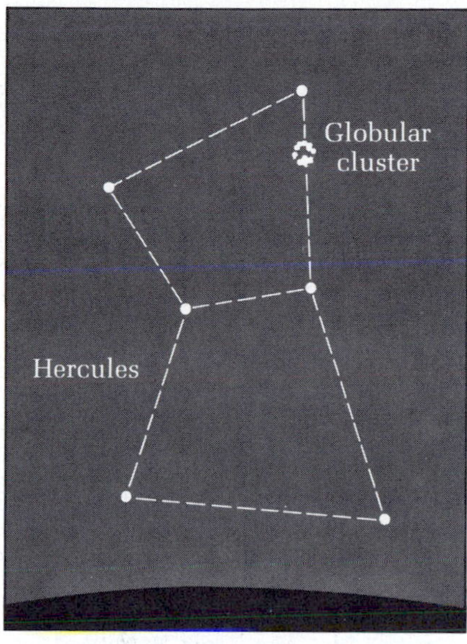

Figure 3.28 The Hercules Cluster.

Autumn

The main constellations of autumn are Pegasus, Andromeda, Pisces, Cassiopeia, and Perseus. Almost directly overhead during October is the great square of Pegasus (Fig 3.30), its center almost devoid of stars to the naked eye. On one corner of the great square is a large "handle"—the constellation Andromeda. Within Andromeda is the well-known Andromeda galaxy (Fig. 1.8), and not far below, the spiral in the constellation Triangulum (Fig. 1.9, color).

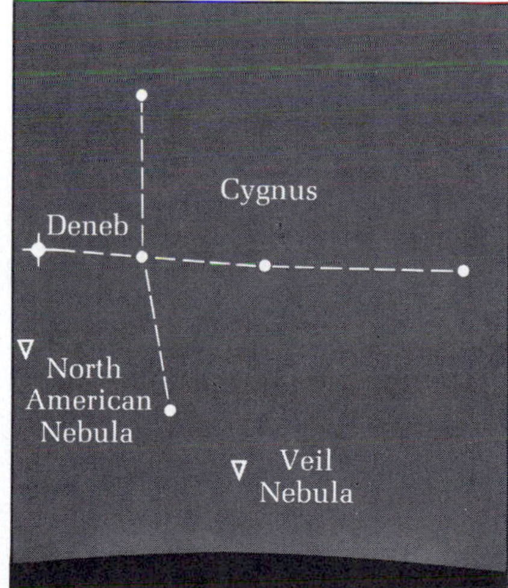

Figure 3.29 (above) Cygnus and vicinity, showing position of the North American and Veil nebulae.

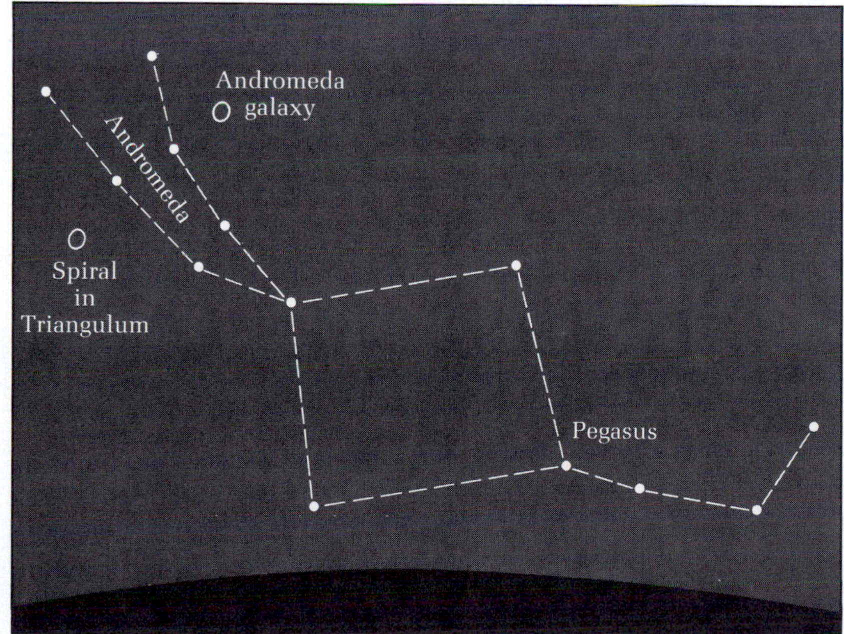

Figure 3.30 Galaxies in the vicinity of Andromeda.

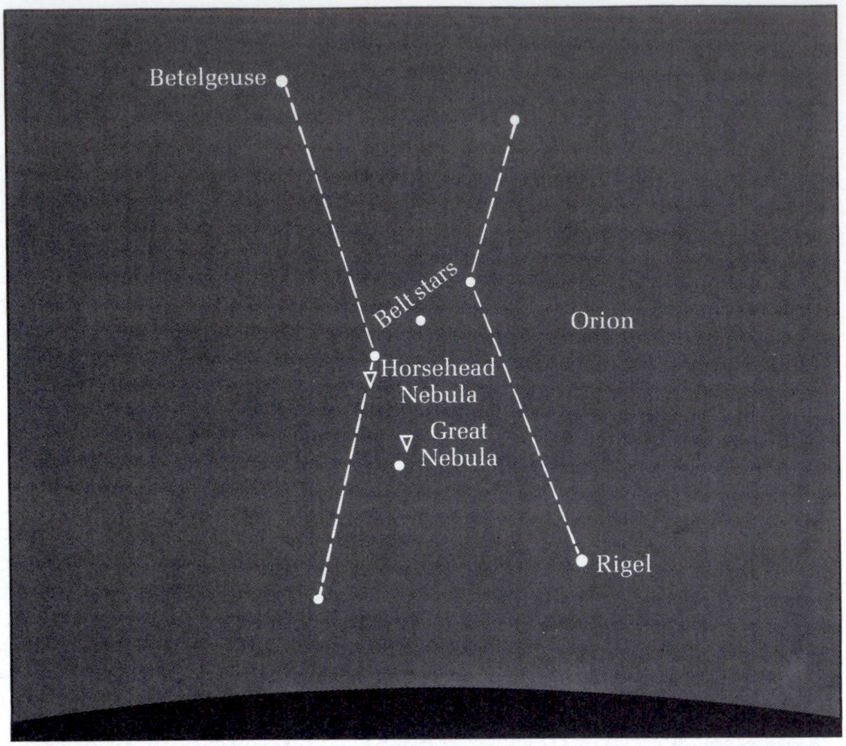

Figure 3.31 The Great Nebula of Orion and the Horsehead Nebula.

Further south is the constellation Aries, significant because it is a zodiacal constellation. Passing through Aries, then through Pisces (which contains the vernal equinox), we come to Cetus, largest of the autumn constellations. Cetus is significant because it contains Mira, a long-period, large, red, variable star that at its brightest is easily visible to the naked eye; at its dimmest, however, a telescope is required.

Winter

The most prominent constellations of the winter are Auriga, Gemini, Orion, Canis Major, Canis Minor, and Taurus. The most spectacular of these, Orion (the Hunter), looms large and impressive, high in the southern sky. The red star in the upper left-hand corner is Betelgeuse, and the blue star in the right-hand corner is Rigel.

Situated in the line of three faint stars that make up Orion's sword is the gaseous nebula called the Great Nebula (Fig. 3.31), which to the naked eye appears to be a star. Also within Orion is the Horsehead Nebula, which is slightly below the left-hand star of the three bright stars that make up Orion's belt.

Directly to the upper right of Orion is the constellation Taurus (the Bull). Taurus contains two interesting clusters: the Pleiades (the face of Taurus) and Hyades, both visible to the unaided eye, and discussed in detail later. Both are clusters of several hundred stars, only a few of which (six in the Pleiades) are visible to the unaided eye. Also within Taurus (upper left-hand corner) is the famed Crab Nebula (Fig. 3.32), and directly connected to Taurus is the constellation Auriga, with its bright yellowish star Capella (Fig. 3.33).

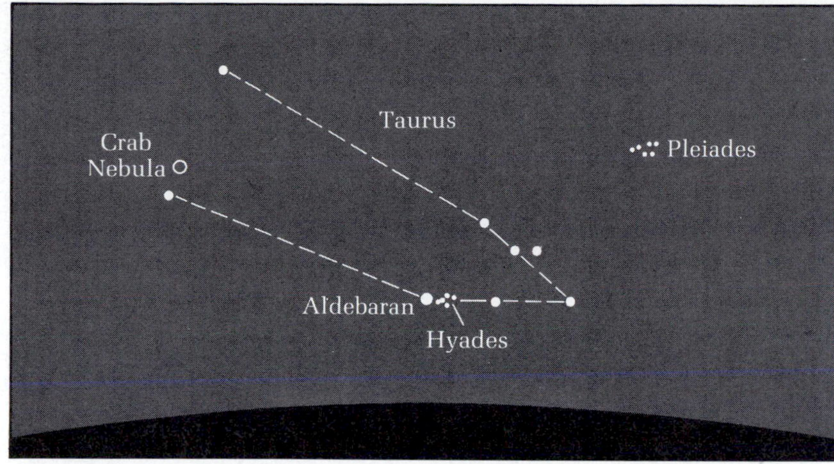

Figure 3.32 The Crab Nebula, and the Pleiades and Hyades in Taurus.

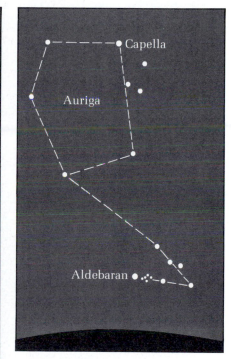

Figure 3.33 (above) Auriga with its bright star, Capella.

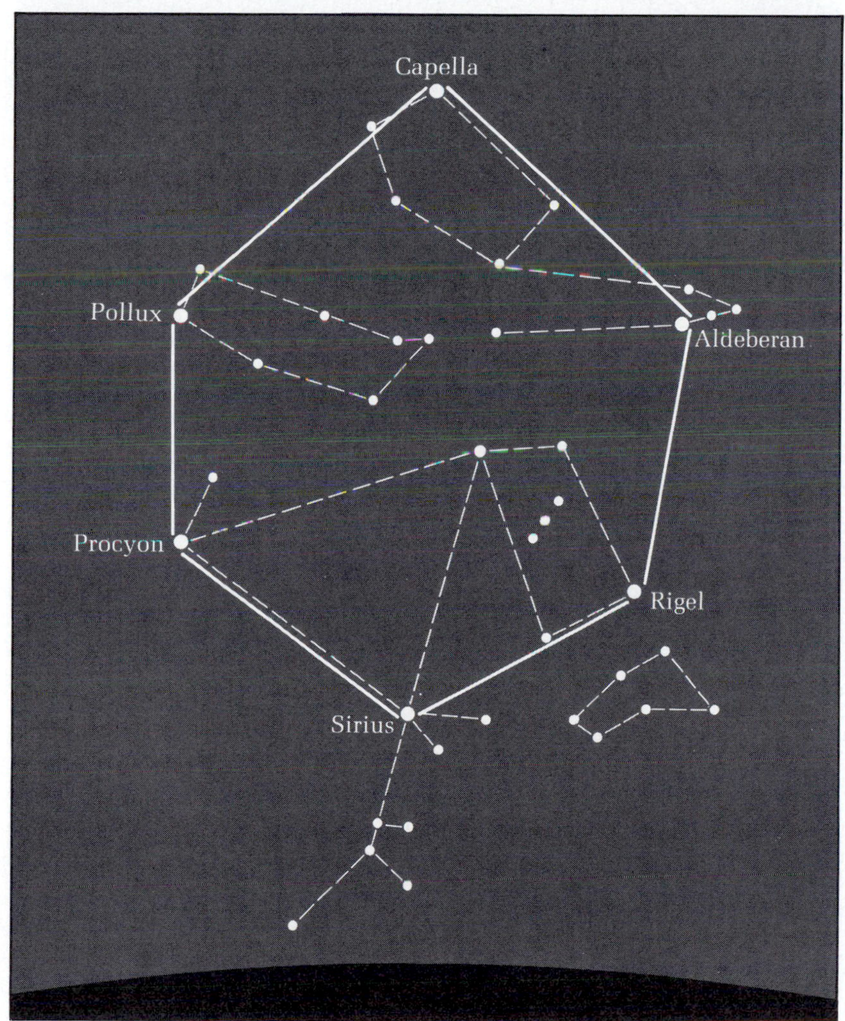

Figure 3.34 The Great Hexagon. East is on the left in the diagram, west on the right. Capella is almost overhead during the winter months.

A useful feature visible in the winter–early spring sky is the Great Hexagon (Fig. 3.34), formed from six of the brightest winter stars and helpful in locating various constellations.

SUMMARY

1. The earth rotates on its axis in 24 hours and revolves around the sun in approximately $365\frac{1}{4}$ days.
2. The spin axis of the earth is inclined at 23.5° to the plane of its orbit. This causes the seasons.
3. The earth precesses, its rotational axis tracing out a complete cone in 26,000 years. The position of our north polar axis changes as a result.
4. Superimposed on the precession of the earth is a regular wobble called nutation. Other motions of the earth are: random motion, orbital motion around our galaxy, and linear motion through intergalactic space.
5. The earth and the moon move around their common center of mass (which is just inside the earth). This causes the earth to wobble slightly as it moves around the sun.
6. The moon goes through a series of phases. Starting with new moon, it passes in about a week through first quarter; then, by about the second week, it is full; at the end of the third week it is at third quarter; at about a month it is back to new moon.
7. Tides are caused by both the moon and sun. Because the differential pull of the moon is much greater than that of the sun, the lunar tides are about twice as high as the solar tides.
8. There are three types of solar eclipses: total, annular, and partial. Solar eclipses occur only at new moon, when the moon and sun are along the line of nodes. Lunar eclipses occur when the full moon passes through the shadow of the earth.
9. At middle latitudes most stars rise in the east and set in the west. In the Northern Hemisphere some, the circumpolar stars, are always visible. At the North Pole only the circumpolar stars are visible—and these are visible throughout the night. At the equator only a few of the same stars are visible throughout the night.
10. Stars can be located by their declination and right ascension. Declination is analogous to latitude on Earth, right ascension to terrestrial longitude. They can be visualized as a projection of latitude and longitude onto a celestial sphere that surrounds the earth.
11. The main constellations are, for spring: Leo, Bootes, Virgo, and Gemini (Ursa Major is useful in locating them); for summer: Cygnus, Hercules, Lyra, Scorpius, Corona Borealis, and Bootes (the Summer Triangle is useful in locating several of them); for autumn: Pegasus, Andromeda, Perseus, and Pisces (the Andromeda galaxy is visible at this time); for winter: Auriga, Gemini, Orion, Canis Major and Minor, and Taurus (the Great Nebula of Orion is visible with the aid of binoculars.

REVIEW QUESTIONS

1. Explain the terms *rotation* and *revolution.*
2. What causes seasons? How?
3. What is precession? What do we observe as a result of precession in the earth?
4. What causes nutation?
5. Summarize the motions of the earth through space.
6. Does the moon revolve around the earth? Explain.
7. Where is the barycenter of the earth–moon system? How is its position calculated?
8. Explain why the phases of the moon occur.

9. Explain how tides are caused. Which produces the higher tides, the effects of the sun or the moon? Why?
10. Explain why there are high tides on the side of the earth away from the moon.
11. What is the difference between spring tides and neap tides?
12. How does the Gregorian calendar differ from the Julian calendar?
13. Explain the difference between a total, an annular, and a partial eclipse. Show the geometry of each.
14. What are Baily's beads? The diamond ring effect?
15. What happens to the moon during a lunar eclipse? Why are we likely to see more lunar eclipses, although solar eclipses are more common? Why doesn't a solar eclipse occur at each new moon?
16. For an observer at middle latitudes, some stars never rise or set. Why?
17. What does an observer at the North Pole see if he watches the stars throughout the night? An observer at the equator?
18. Why do we see different groups of constellations at different times of the year? List some of the major constellations of each of the seasons.
19. What is the celestial sphere? Why is it useful? Discuss the analogy between terrestrial latitude and longitude and declination and right ascension.
20. Explain: ecliptic, celestial equator, vernal equinox, autumnal equinox.

THOUGHT AND DISCUSSION QUESTIONS AND PROJECTS

1. What is the maximum angle by which Polaris can deviate from true north as a result of precession? (It differs at present by 1° from true north.)
2. How will the positions of the constellations differ because of precession (a) in 5,000 years; (b) in 10,000 years? (Use the star charts shown on the inside covers. Give approximate positions relative to those in the charts.)
3. Where is the moon in the sky at sunset when it is (a) full, (b) at first quarter, (c) new?
4. Assume you are on the dark side of the moon during a solar eclipse (taking place on Earth). What would you see if you looked toward Earth? Assume you are on the side of the moon that is in earth's shadow during a lunar eclipse. What would you see if you looked toward Earth?
5. What would the elevation of Polaris be above your horizon if you were in (a) Tokyo, (b) London, (c) Mexico City, (d) Stockholm?
6. There are places on the earth where the sun does not set for long periods. Where are they? Explain why this occurs.
7. If the angle between the celestial equator and the plane of the ecliptic is 23.5°, what is the maximum (minimum) angle from the horizon at noon that the sun will pass for your latitude?
8. Measure the angle of Polaris above the horizon. Use it to determine your latitude.
9. Using the star charts, identify as many of the constellations as you can.

FURTHER READING

Berry, R. "Celestial Software: Part 1, Right Ascension and Declination," *Astronomy*, p. 42 (May 1978); "Celestial Software: Part 2, Telling Time," *Astronomy*, p. 47 (July 1978).

McNally, D. *Positional Astronomy*. Wiley, New York (1975).

Norton, A. P. *Norton Star Atlas*. Sky Publishing, Cambridge, Mass. (1973).

Whitney, C. A. *Whitney's Star Finder*. Knopf, New York (1977).

part 2

Astrophysical Tools

In this part we will examine the various tools used by the astronomer. Chapter 4 begins with a discussion of the properties of light—its velocity, its wave nature, its membership in the same family—the electromagnetic spectrum—as X-rays and radio waves.

In Chapter 5 we turn to the many types of telescopes: reflectors, refractors, Schmidts, radio, infrared, and so on.

Many people picture astronomers as working only at night, perched high on a platform, studying the stars and galaxies through a telescope that could pass for an H. G. Wells rocket ship. Early astronomers such as William Herschel did sit patiently night after night with their eyes glued to the eyepiece of a telescope, but the modern astronomer actually spends perhaps a day or two a month at the telescope, and the rest of the month (days) analyzing the data obtained, usually with the aid of a computer.

Working with the telescope is, however, one of the most enjoyable parts of an astronomer's routine. An astronomer on an observatory staff is assigned certain nights for observing the skies, during which he or she obtains as much data as possible. First, one decides what objects to study. Then, the telescope is aimed, which is referred to as "centering" the object. It is only in this brief time that astronomers actually look through a telescope. For the most part stellar objects—or rather, their "spectra"—are photographed. (As we will see, these spectra give much more information than a photograph of the actual object.)

Once the telescope is set on the star the astronomer has to see that it remains centered. The machinery in the base of the telescope drives it to compensate for the earth's rotation, but additional guiding is also usually needed to compensate for small fluctuations due to our atmosphere. On some nights an astronomer may work on 20 or more objects; on others the entire night may be spent collecting light from a single dim star or galaxy.

Of course, not all astronomers work at optical observatories; radio observatories are now almost as common as their optical counterparts. Astronomers who do not work at an observatory—for example, professors of astronomy—usually can gain access to a telescope by submitting a proposal to a selection committee at one of the large observatories, such as Kitt Peak or Hale. If the proposal is deemed worthy the applicant is assigned a few nights with one of the telescopes.

Not all astronomical research requires an observatory; numerous observations made from the various orbiting observatories need analyzing, and many astronomers work with these data. Others work with the National Geographic-Palomar survey plates (Fig. 5.14); many useful projects have been completed merely by ordering some of these plates and taking measurements from them. Last, but certainly not least, is the theoretical astronomer, who rarely, if ever, uses a telescope. He builds mathematical models, then waits for verification or rejection by the observational astronomer.

Chapter 6 deals with the rather more difficult—but essential—subjects such as radiation laws, atomic structure, and line broadening.

Aim of the Chapter

To define the properties of light and the electromagnetic spectrum, and to show how they are related.

chapter 4

Light and the Electromagnetic Spectrum

The stars we see—whether through a telescope or with the unaided eye—are those stars as they appeared many years ago—as many *years* as it has taken their light to reach us. Indeed, this light is all we will ever be able to examine of them; unlike most other scientists, who can study the objects of their specialty in the laboratory, the astronomer must be satisfied with studying only the light that emanates from them.

It might seem therefore that the task is as hopeless as a frog in a bowl of molasses. Fortunately, things are not quite as bad as they appear; there is actually much more—indeed, very much more—to what light can tell us than there might seem. To understand this, we need to examine the properties of light.

LIGHT

Properties

Light is, of course, blocked by opaque objects, but the surface of most objects will reflect light in different degrees (depending on the properties of the material); in some cases part of the light is transmitted—that is, passes through. Note that a beam of light reflects at an angle corresponding to the angle at which it strikes the object's surface (Fig. 4.1).

In any transparent, homogeneous (uniform) medium, such as air, light travels in a straight line; upon entering another transparent medium of different density, such as water, it changes direction slightly (Fig. 4.1). (This phenomenon, called *refraction*, will be discussed in more detail later.)

Everything we see is, of course, light reflected from or emitted

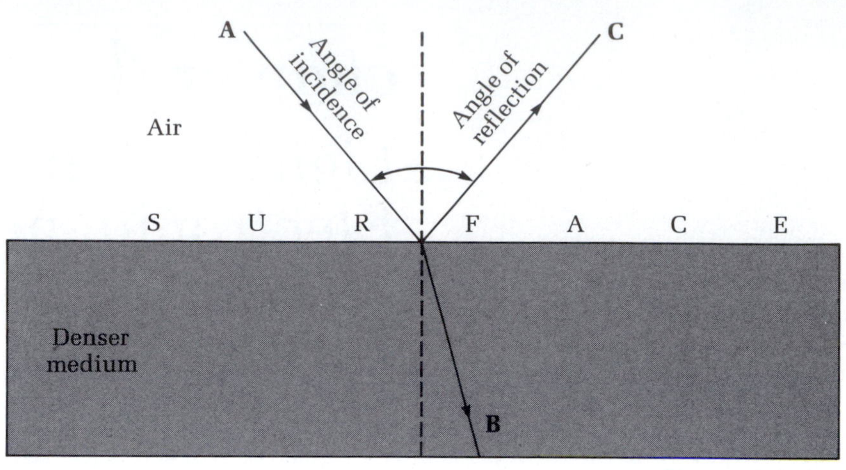

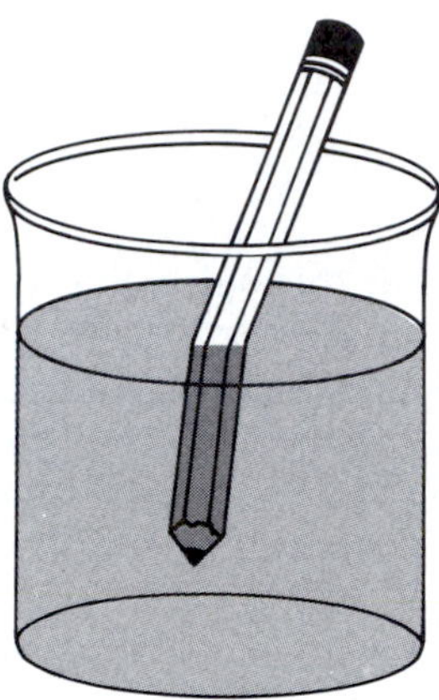

Figure 4.1 Top right shows refraction *(AB)* and reflection *(AC)* of a beam of light. Lower right shows the refractive effect of water.

by what we see. For example, the page you now see before you is seen only because of the light reflected from it into your eye, where it passes through the lens onto the retina. Cells in the retina, in turn, convert the light into electrical impulses that, transmitted to the brain, tell you the form of the page and of the words you are reading.

Light has many properties not immediately obvious. It might seem, for instance, to have infinite velocity. For example, the light that comes to your eyes seemingly instantaneously when a light is turned on actually arrives at your eye after a small—extremely small—delay. The first person to determine correctly (or nearly correctly) the velocity of light was the seventeenth-century Danish astronomer Ole Roemer. Roemer had discovered that eclipses of Jupiter's moons occurred sometimes earlier and at other times later than predicted in eclipse tables that had been published about 1675. Roemer assumed that it was the different distances of Jupiter (at different times of the year) that caused the discrepancy. This seemed to be the only answer and, with it as a basis, he arrived at a velocity of 212,000 km/sec for light in a vacuum. Indeed, this is amazingly close to the value we accept today (299,792 km/sec).

Nature

Though its velocity and other basic properties are established, the actual nature of light—what it *is*—will probably never be known. The best we can do is describe how it acts and, on the basis of this, formulate a theory based on these observed properties.

One of the first theories was put forward by Newton, who assumed that a beam of light was composed of millions of tiny particles he called corpuscles. When these corpuscles struck an object, they bounced back into our eyes, thereby giving us an image of the object. At about the same time, Dutch scientist Christian Huygens showed that light could also be interpreted as a wave. In many ways Huygens's theory was better than Newton's, but Newton's prestige and influence were so strong that most physicists ignored Huygens's theory. In addition, there seemed to be problems with it: Waves can pass around corners, so that if light was a wave, we should be able to see around corners. And of course we cannot—or can we? In 1789 English scientist Thomas Young discovered that if a beam of light was passed through two side-by-side slits (Fig. 4.2), the emerging beams seemed to interfere with one another; indeed, the combination could be either enhanced or canceled altogether as a result of the interference. The explanation of the phenomenon is as follows. As the light waves move outward from the two slits, some of the crests of the waves intersect. "*Constructive* interference" occurs at these points, and the light is enhanced, making a bright line on the screen. Where crests and troughs intersect, *destructive* interference occurs, with the result that dark lines appear on the screen. This could not be explained by a particle (corpuscular) theory. Soon afterward French scientist Augustin Fresnel showed that, even in the case of light passed through a single slit, there occurred a phenomenon, which he called diffraction, that could not be explained by particle theory. Close examination of a shadow will reveal the effect (Fig. 4.3). At the edge of a shadow there is a series of dark and bright lines, and some of the latter occur within the usual shadow boundary—which means, of course, that light *can* bend around corners. With this evidence, supporters of Newton's theory soon conceded the superiority of Huygens's wave theory.

But if light is a wave, what is it that is waving (or, more precisely, oscillating)? Consider the case of a wave on a pond. The medium transmitting the wave is water, and without the water there can be no wave. Therefore, if light occurred as waves, a medium was needed. Scientists answered the need by assuming the existence of something they called the aether—a mysterious substance that permeated the universe; but the concept eventually had to be dropped.

The solution to the problem lay in the theory of electromagnetic waves, which was developed by English physicist James Clerk Maxwell in the mid-1800s. Maxwell's theory is explainable as follows. An electron, one of the particles of which every atom is composed, has a negative electrical charge. Because of this charge the electron is surrounded by an electric field, which can be pictured as lines of force radiating outward from the electron. (Another electron,

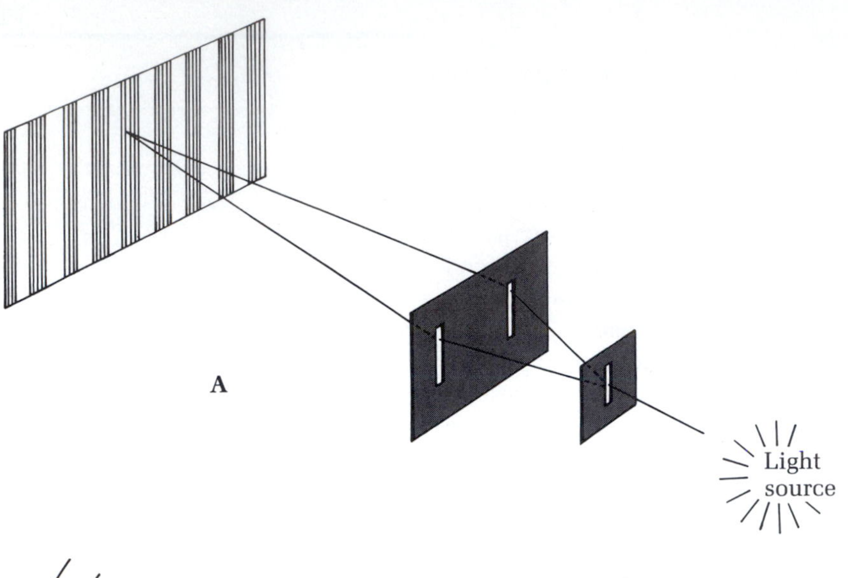

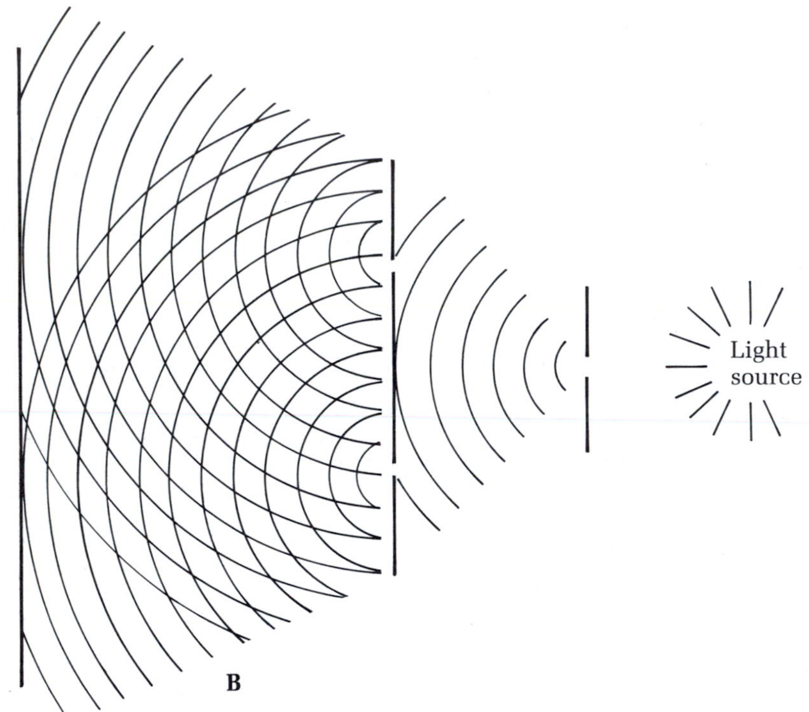

Figure 4.2 How interference patterns result when a beam of light is passed through two closely spaced openings. *A* shows a light path that produces one of the non-interfered portions of the resulting pattern (bars of light and dark at left). *B*, which is *A* as if seen from above, shows how wave fronts emerging from the slits can interfere (intersect) with one another.

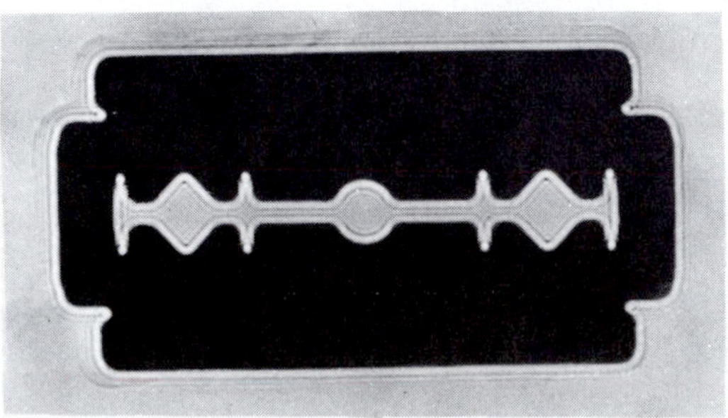

Figure 4.3 Diffraction patterns of light passing through slots in razor blade.

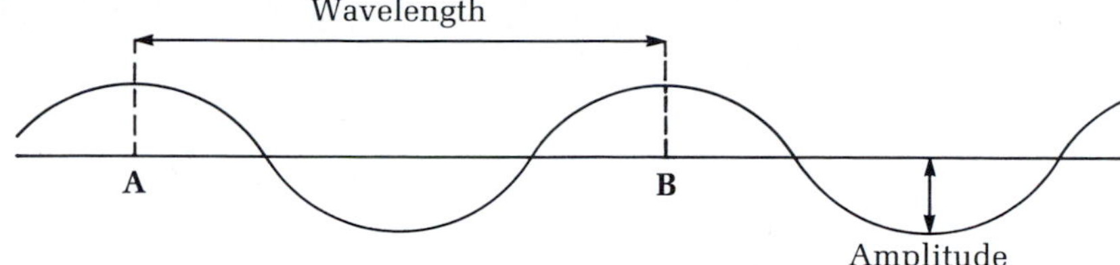

Figure 4.4 Properties of a wave. ***A*** **and** ***B*** **indicate crests.**

brought near this one, will be repelled by these lines of force. A proton—a particle with a positive charge—will be attracted to the electron.) A rapidly reciprocating electron creates a changing or oscillating electric field, which leaves the particle and moves off into space. Thus a continuously moving particle generates a continuously emitted electric wave. In the 1800s scientists had shown that a changing electric field generates a corresponding magnetic field, such as those familiarly associated with the common horseshoe and bar magnets. (Conversely, a changing magnetic field generates an electric field.) The combination of an electric wave train and oscillating magnetic field is referred to as an electromagnetic wave. When Maxwell calculated the velocity of these waves through empty space, he found, much to his surprise, that it was the same as the velocity of light—confirming that light was an electromagnetic wave.

LIGHT AS WAVES AND ENERGY

Waves, a familiar sight in whatever form they occur, consist of troughs and crests. A cross section of one would have roughly the profile shown in Fig. 4.4. Moving one end of a rope up and down will generate waves in which the humps move as they do in water waves. A particular rate of movement of the rope will produce waves that do not move along the rope; these are called standing waves. (Note that, while the part of the rope forming the humps moves up and down, the rope does not move along its length; this is an important characteristic of waves.)

As Fig. 4.4 shows, wavelength is the distance between any two consecutive points that have the same position relative to the wave (e.g., *A* and *B*). Closely related to wavelength is frequency (Fig. 4.5),

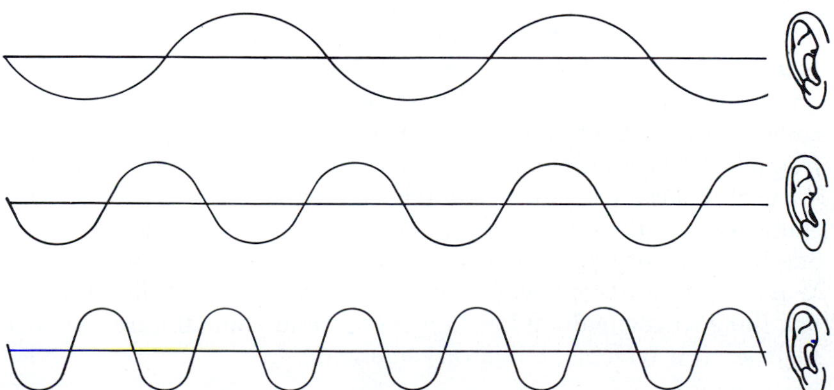

Figure 4.5 Waves of increasing (from top to bottom) frequency. The greater the frequency, the shorter the wavelength.

Powers-of-Ten Notation

A convenient, widely applied shorthand way of expressing very large and very small numbers uses the number 10 and an exponent, or index. Thus, $10^2 = 100$, with the index *2* denoting the number of zeros after the numeral *1*. Similarly, 1000 is expressed as 10^3, 10,000 as 10^4, and 93,000,000 as 93×10^6 (or 9.3×10^7). For extremely small numbers—fractions—the index is preceded by a minus sign. Thus 1/1000 (or .001) is 10^{-3}. (For $1000 = 10^3$, and $1/1000 = 1/10^3$, which we write as 10^{-3}.) Similarly, $1/10{,}000 = 10^{-4}$, $1/100{,}000 = 10^{-5}$, and so on.

Multiplication and division in this system require only adding or subtracting indices, respectively: $10^3 \times 10^2 = 10^{3+2} = 10^5$ and $10^3 \div 10^2 = 10^{3-2} = 10$.

the number of humps that pass a given point in one second. Finally, there is amplitude—half the vertical distance from a crest to a trough.

Since light is a wave, these terms apply to it also, although the way they manifest themselves is different from the way they do in the case, say, of sound. In the case of light, different wavelengths produce different colors. ("White" light passing through a prism is separated into all the colors. Newton, in 1666, explained that white light was composed of all colors and that the prism merely spread them out. We now know that each color has a different wavelength.)

Red light has a wavelength of approximately .00007 cm. Since the centimeter is a cumbersome unit for expressing wavelength, it is written in the more compact form of an exponent of 10 (see box) or in an entirely new unit. As an exponent of 10, the wavelength of red light is approximately 7.0×10^{-5} cm; in the new unit—called an angstrom (Å), which is equal to 10^{-8} cm—it is approximately 7000 Å. Yellow light has a wavelength of approximately 5800 Å, and blue light has a wavelength of approximately 4700 Å.

Light is also a form of energy. In fact, the higher its frequency (and, consequently, the smaller its wavelength) the greater its energy. If we consider some everyday aspects of life, it is not surprising that electric and magnetic waves carry energy through space. When we lie on the beach, for example, we feel energy in the form of heat from the sun. What *is* surprising is the way it appears to be carried: It is as if it is concentrated in small packets or particles. The first clue that this was the case came in 1887 with the discovery, by Heinrich Hertz, of the photoelectric effect—that electrons are ejected from the surface of a metal by light falling on it. To explain this, Einstein proposed that light was a hail of particles, or bundles of energy, now called photons. Each photon had a specific energy depending on its frequency. The relationship is given in the formula $E = h\nu$ where E is the energy, ν is the frequency, and h is a constant called Planck's constant that has the value 6.62×10^{-27} erg sec. (The erg is a unit of energy.) This formula can also be written in terms of the wavelength, λ, as $E = hc/\lambda$. It tells us that when the frequency of the photon is low (long wavelength), it has low energy, and when the frequency is high (short wavelength), it is very energetic.

Our eyes are sensitive to light of wavelengths from about 3800 Å to about 7600 Å; this corresponds to the colors from violet to red. (All other colors are, of course, between these in wavelength.) In the regions immediately above and below the range of light are photons of infrared waves (associated with heat) and ultraviolet (UV) waves (linked to skin cancer). Beyond these regions are other electromagnetic waves, such as those—with photons of particularly long wavelengths—that make radio and TV possible, and, near the short-wavelength end of the spectrum, X-rays. X-rays, with their much shorter wavelength (and consequently higher frequency) than ordinary light, are much more energetic. (As mentioned above, the higher the frequency, the greater the energy.) This is why these photons are generally so much more dangerous to us than light: The energy they carry destroys living tissue.

We cannot, of course, see any of these; it takes special equipment to detect such radiations. It is, in fact, good that our eyes are not sensitive to them; were they sensitive to X-rays, for example (instead of ordinary light), we might see a black sky.

X-rays, radio waves, and other radiation, then, are closely related to ordinary light. Like light, they are also photons, the only difference being their wavelength. And, of course all travel at the same speed—the speed of light. Taken together, they constitute what is called the electromagnetic spectrum (Fig. 4.6).

Astronomers are interested in photons of all these different wavelengths because stars and other celestial objects emit photons of all wavelengths. For some time, only the visible ones were studied; this was because most gamma rays, X-rays, and UV and infrared radiations do not pass through our atmosphere, as does visible light. Most of these radiations are dangerous to us, and our atmosphere forms a protective blanket screening them from us. As Fig. 4.6 shows, there are only two highly transparent "windows" through this layer of air—the visible window (which extends partly into the infrared and a little way into the UV) and the radio window.

Infrared is stopped mainly by water vapor; because some penetrates the atmosphere, infrared astronomers usually refer to the infrared window as a "murky" one. Ultraviolet waves are stopped by the ozone (a gas) layer, which apparently has been diminished recently by the action of gases once used as a propellant in most spray cans.

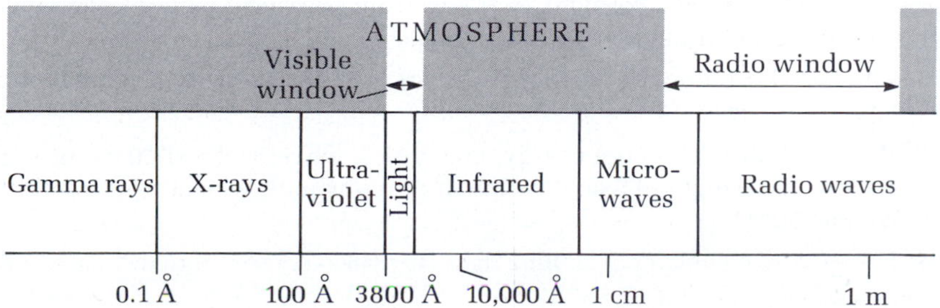

Figure 4.6 The electromagnetic spectrum. Wavelengths are shown at bottom.

SUMMARY

1. When a light ray passes from a less dense to a more dense medium (or vice versa), it is refracted (bent).
2. Ole Roemer was the first person to calculate a value for the velocity of light that was reasonably close to the one we accept today (299,792 km/sec).
3. Light exhibits phenomena—called interference and diffraction—that indicate that it has wave properties. Both of these phenomena give a pattern of alternating dark and bright lines.
4. Elecromagnetic waves consist of a combination of changing, or oscillating, electric and magnetic fields. Light is an electromagnetic wave.
5. The major properties of waves are wavelength, frequency, and amplitude. Wavelength is the distance along the wave between corresponding points; frequency is the number of humps that pass a given point per second, and amplitude is half the vertical distance from a crest to a trough.
6. Light also has properties that indicate that it has particle characteristics. The particle is called a photon.
7. Photons occur over an entire spectrum of wavelengths—the electromagnetic spectrum, which ranges from X- and gamma-ray photons (of very short wavelength) to radio waves (of very long wavelength). Light is only one small section of this electromagnetic spectrum.
8. Astronomers now study cosmic photons of all wavelengths (light, radio waves, X-rays, gamma rays, infrared, and UV waves).

REVIEW QUESTIONS

1. Explain how Roemer arrived at a value for the velocity of light.
2. What is a wave? Give some examples.
3. Explain the terms wavelength, frequency, amplitude.
4. What is the difference between refraction and diffraction?
5. Explain how light manifests itself in two ways.
6. What is an electromagnetic wave?
7. Define the angstrom (Å). Why is it used?
8. Can we see photons of all wavelengths? Why?
9. What is the main difference between light photons and X-ray photons?
10. What region of the electromagnetic spectrum contains the longest wavelengths? What region contains the shortest?
11. What regions of the electromagnetic spectrum lie immediately to either side of visible light?

THOUGHT AND DISCUSSION QUESTIONS AND PROJECTS

1. How long does it take light to travel from the sun to the earth? How long from the sun to Mars, Jupiter, Pluto?
2. Convert the following to angstroms: (a) 6×10^{-5} cm, (b) 3×10^{-4} cm, (c) 8×10^{-7} cm.
3. Which photon has more energy, one with a wavelength of 10 cm or one with a wavelength of 5 cm? Which of photons with wavelengths of 10^{-10} cm and 10^{-5}?
4. How many wavelengths of blue light (say, 4800 Å) are there in 1 cm?

5. How much energy is associated with a photon of frequency 10^{10} oscillations per second?
6. *Project:* Shine the light from a small laser through a single slit, then through a double slit. Sketch the patterns they produce on a screen. What do they tell you?

FURTHER READING

Adler, I. *The Story of Light*. Harvey House, New York (1971).
Minnaert, M. *The Nature of Light and Color in the Open Air*. Dover, New York (1954).

Aim of the Chapter

To introduce the basic tools of the astronomer, and to explain how some of them work.

chapter 5
Telescopes

Chapter 4 described how a light beam bends as it passes from one medium to another. Why does this occur? Suppose a ray of light passes through a piece of glass with parallel sides, such as a windowpane. Its path in the glass depends on the angle it makes with the surface of the glass (p. 69). A ray entering perpendicular (at an angle of 90°) to the surface passes through undeflected. To see what happens when a ray enters at an angle (other than 90°), we must examine the position of the wave fronts (Fig. 5.1) as they approach the glass. In the case of the ray entering at an angle, the bottom section of the wave front enters the glass first. Because light travels slightly slower in glass than it does in air,[1] this part of the wave front slows, lagging behind the upper section, and so the ray bends. As the ray emerges from the glass, however, it bends back and emerges parallel to the

[1]A number called the index of refraction designates the speed of light in different materials.

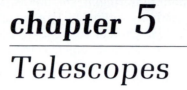

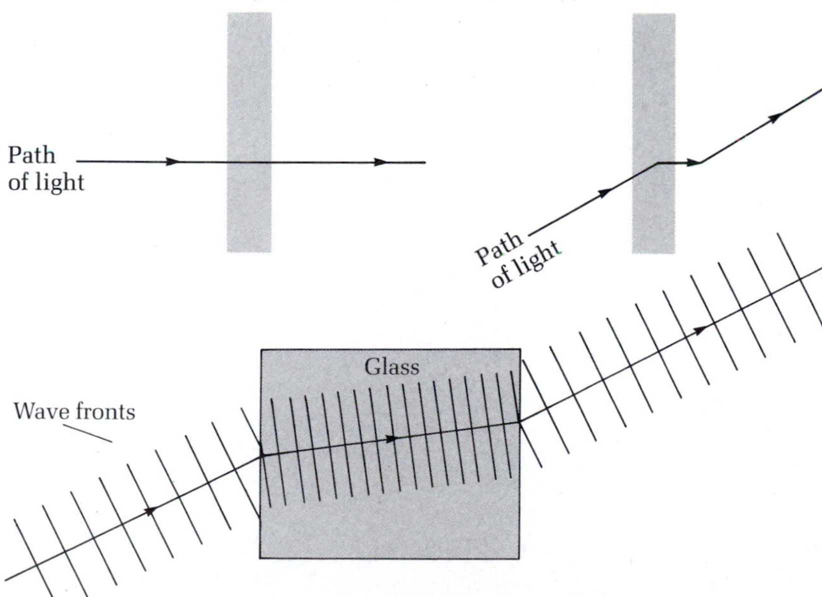

Figure 5.1 At top, typical path of light through a piece of glass. The ray entering at an angle is deflected; the one entering perpendicular to the surface is not. Below, an enlargement shows how the lower portion of the wave front is slowed when it enters the glass, allowing upper portion to overtake it. The reverse occurs as the light ray emerges from the glass.

incident (entering) beam, but shifted slightly to one side. (This shift can easily be verified by looking through a piece of plain glass held at an angle in front of your eyes. When it is moved quickly in and out of your line of vision, objects beyond it will appear to move back and forth.) A simple analogy may be made between the bending of the beam and the effect on a line of soldiers advancing side by side on solid ground who encounter a swampy area (Fig. 5.2). The soldiers at the end of the line that reach the swamp first find the going more difficult and, while the others continue their original pace, begin to fall behind. The result is a bend in the line. When the rest of the troops have reached the swamp and are also mired in it, the line they

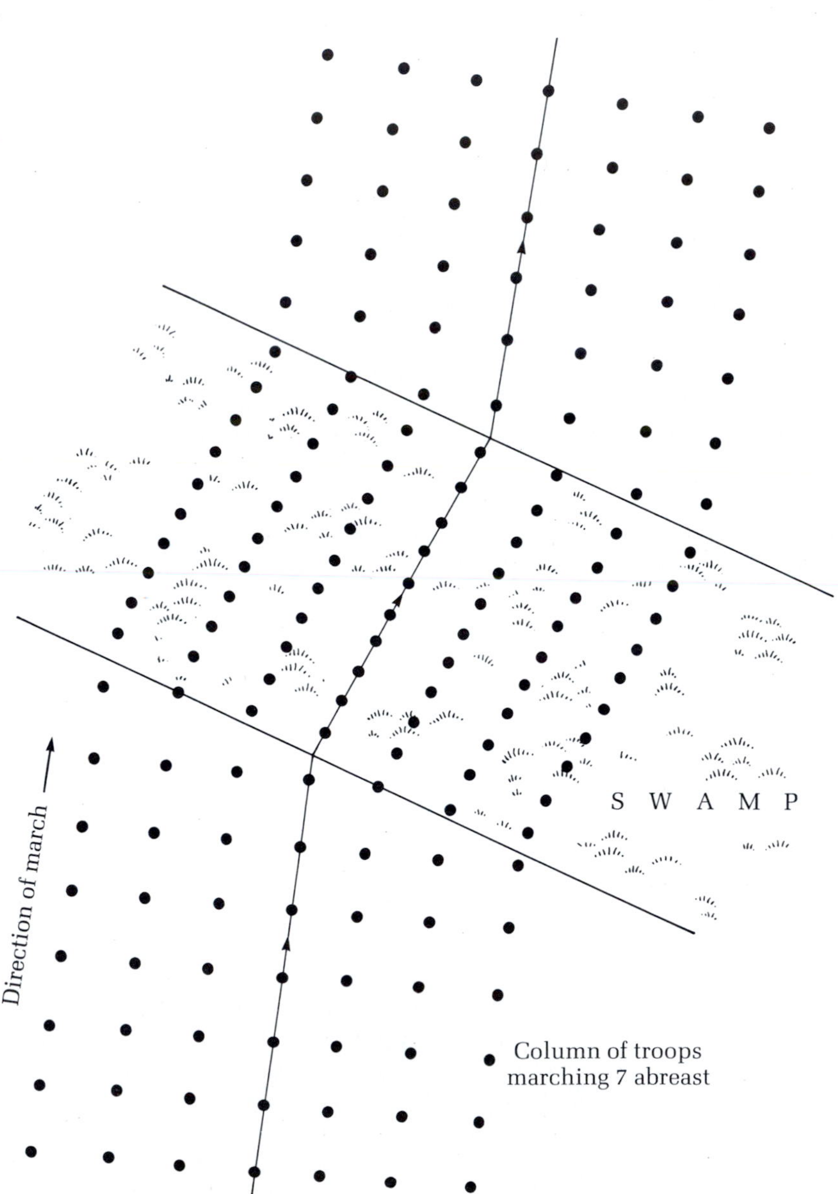

Figure 5.2 Column of troops passing through a swamp in analogy with ray of light passing from air and into and through a denser medium. Soldiers at right edge of column are first to be slowed by swamp, and first to resume speed on leaving it.

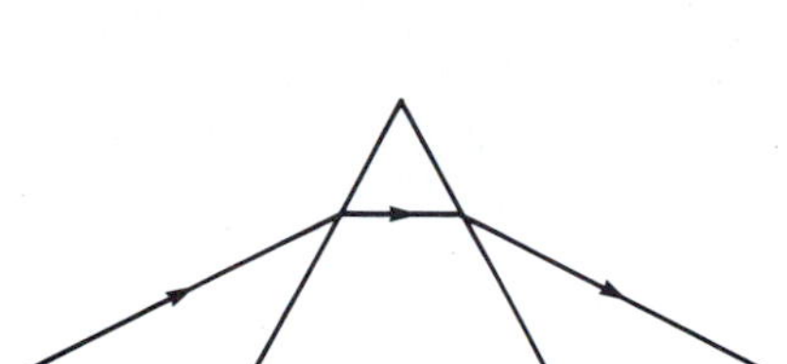

Figure 5.3 A ray of light passing through a prism. Note that it emerges at an angle to the incident beam.

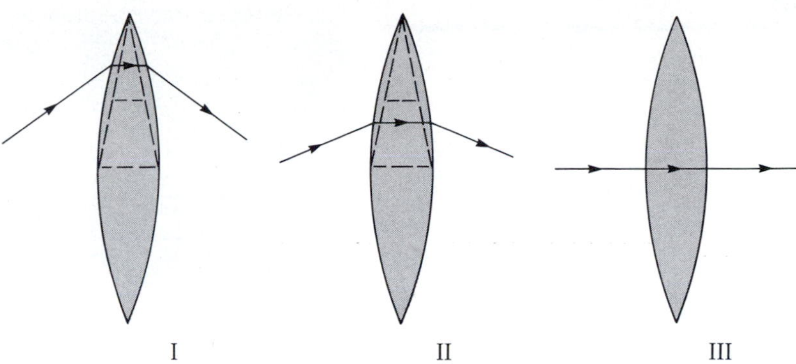

Figure 5.4 Effects on a beam of light of passing through different parts of a simple convex lens. Dotted lines in first lens (at left) indicate how lens is, in effect, two prisms base to base.

form straightens out but is now moving at a different angle. When the line of troops begins to emerge from the swamp (which has parallel sides, like the glass), those that emerge first begin to move more quickly, beginning another change of direction that will eventually bring the line around to its original direction.

If the sides of the glass through which a ray passes are not parallel, as in a prism, the result will be different. (Since a prism disperses light to form a spectrum of color, we will assume we are using light of a single color.) As Fig. 5.3 shows, a ray approaching the surface of a prism at an angle will bend, as in the previous case, toward the normal (perpendicular to surface) as it enters the glass. And when it emerges from the glass, it will bend again; but, unlike the case with the glass having parallel sides, this time the emerging ray is *not* parallel to the incident ray. As we will see, this is an important difference, one that is the basis of lenses and, hence, of telescopes, which we discuss below.

The simple lens, such as that found in hand magnifiers, has a double convex surface (Fig. 5.4). In cross section (Fig. 5.4) the lens looks like a small prism (with slightly curved sides)—or rather, two prisms base to base. This means that light that enters at an angle near the top will pass through as it does in a prism; in other words, it will be bent. A ray passing through a bit lower (closer to the middle) is bent slightly less. (This portion is, in effect, the bottom section of a prism, with its apex omitted.) Finally, the ray that enters perpendicularly along the center goes through undeflected, as before. (The sides are approximately parallel here over a small distance.)

These rays taken together (and with the corresponding ones for the bottom of the lens), make the pattern shown in Fig. 5.5. Because, in this type of lens (convex), the emerging rays converge, it is usually called a converging lens. In a concave lens (Fig. 5.6) the same rays would diverge.

Now suppose the rays entering the lens are parallel, which is the case when they come from a source at infinity; a star is a good ex-

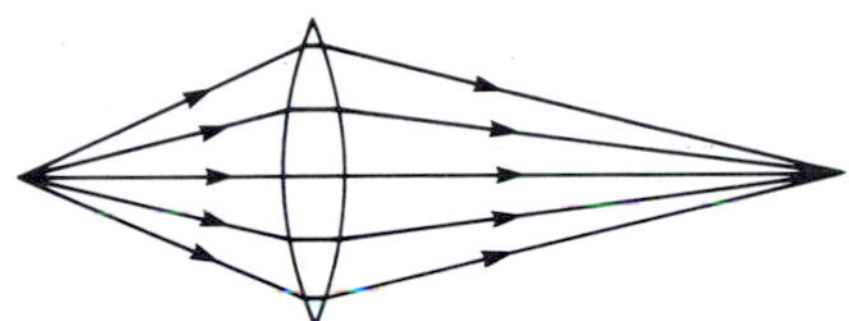

Figure 5.5 Typical convergence pattern of light rays passing through a converging (convex) lens.

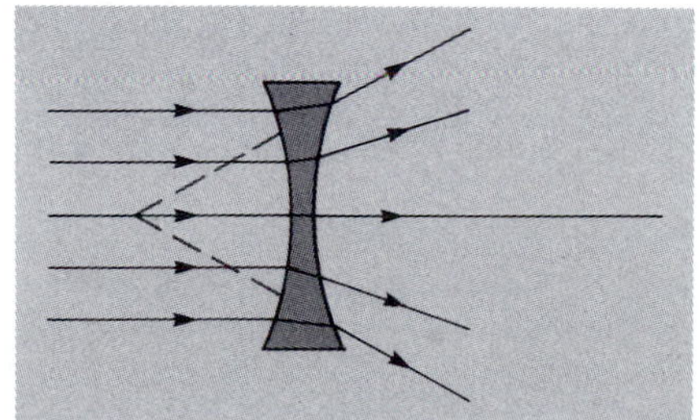

Figure 5.6 Pattern of light rays passing through a concave lens. Junction of the dashed lines is the apparent convergence point.

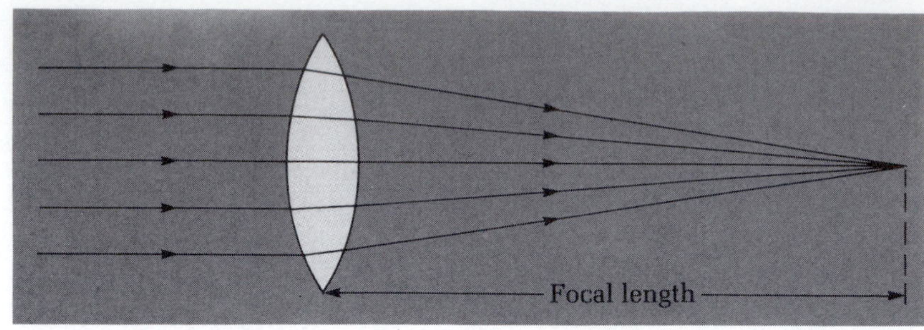

Figure 5.7 Pattern of parallel light rays passing through a convex lens.

ample of a source that is near enough to infinity for our purposes. Although in a convex lens such rays again converge, in this case the distance from the convergence point to the center of the lens is called the focal length of the lens (Fig. 5.7). An image of the star is formed here. Figure 5.6 shows the corresponding point for a concave lens.

EARLY DESIGNS

The telescope was invented by Dutch optician Hans Lippershey about 1604. It is said that his children were playing with lenses he had just ground and noticed that, when they held them apart in just the right way, they magnified distant objects. The Dutch government, fearing that the discovery would become an instrument of war, kept it secret for several years. About five years after the discovery, however, Galileo heard of it and immediately began constructing one. Within a few days he had a three-power telescope (the image was three times larger than that seen with the naked-eye). When he had improved this to ten power, he demonstrated it for the navy, showing them how it could be used to spot ships at sea. Finally, he devised a much improved instrument of about fifteen power, which he turned to the heavens; with that the telescope passed from an interesting novelty to an important research tool. Today, most astronomers dream of making a single great discovery; Galileo, with his new telescope, made such night after night.

He discovered craters on the moon, myriad stars along the Milky Way, four moons of Jupiter, the phases of Venus, the structure of sunspots, and the strange oblong shape of Saturn. He never found out that Saturn's shape was caused by its rings. In 1610 he published his results and, within months, was a celebrity. So, although Galileo did not invent the telescope, he helped perfect it and was the first to use it for astronomical purposes.

REFRACTOR TELESCOPES

The main component of a refractor telescope, the simplest type, is a convex lens, referred to as the objective (Fig. 5.11). Light passes through this lens and forms an image in front of another, smaller lens called the occular, or eyepiece, which magnifies this image. When we say a telescope has a diameter of, say, 15 cm, we mean that the objective is this size. The objective is, indeed, the most important part of the astronomical telescope; it determines the light-gathering power of the instrument—its main function.

Light Gathering

Magnification is of secondary importance in astronomical telescopes. Theoretically (but not in practice) any magnification can be obtained from any size objective lens. Magnification is determined by the ratio of the focal length of the objective to the focal length of the eyepiece; thus the smaller the focal length of an eyepiece (for a given objective), the greater the power of the telescope. A telescope, with, say, a 15-cm objective and a magnification of five or six hundred will give very poor results; this is because there is only so much light (so many photons) getting through the objective, and as this is spread more and more (by magnification) the resulting image becomes increasingly dimmer.

A telescope with a large objective, however, has greater light-gathering power, so that the image can be magnified with less weakening. Generally, maximum useful magnification is limited to about 20 times the objective size (in centimeters). The light-gathering power of a telescope is proportional to the area of its objective. Since circular area varies as the square of the diameter, a 30-cm telescope (area, 900 cm) has not twice, but four times the light-gathering power of a 15-cm telescope (area, 225 cm).

Astronomers (and others who use optical instruments) refer to the f-ratio, or f-number, of a lens, which is the ratio of the focal length to the diameter of the objective. The lower the f-number, the "faster" the lens; this is because the image of an object is brighter in a low-f-number instrument and, hence, requires a shorter ("faster") exposure.

Resolving Power and Lens Aberrations

Another important feature of a telescope is its ability to resolve, or to show as separate, objects that are close to one another—two stars, for example. This is called resolving power. Close inspection of the images of point sources of light in a short-exposure photograph will show that they appear as bright spots surrounded by several bright rings. These rings are the diffraction pattern of the source.[2] This pattern is important because it is the basis for determining whether two sources (e.g., two stars) can be considered to be resolved. They

[2]Most photographs of stars will not show diffraction patterns; when photographing bright stars, astronomers generally overexpose, and this gives an image that is a uniform disk. In the case of dim stars, the diffraction patterns are so faint that they are usually obliterated by atmospheric interference.

Figure 5.8 Typical diffraction pattern for a star. At right, figure shows degree of intersection of patterns of two stars that are at the minimum threshold of resolution.

are if they are separated at least to the extent that the innermost ring of one pattern coincides with the center of the other (Fig. 5.8). The diffraction pattern of a star seen in a large telescope will be smaller than the pattern seen in a small telescope, reflecting the better resolution of which the larger is capable.

Two common defects of telescopes are chromatic and spherical aberration. A star seen through a telescope with chromatic aberration looks like a diamond, sparkling with all the colors of the rainbow. This is because, as mentioned above, a convex lens has some of the properties of a prism and so tends to separate—to disperse—white light into its component colors (Fig. 5.9). To overcome this effect, the rays composing the various colors must be brought back to a single focus. This can be done by placing a second lens of different shape and different refractive index (p. 79[1]) at the back of the convex lens—the primary objective. Actually, no lens can be made completely achromatic, but many are nearly free of this defect.

In the case of spherical aberration (Fig. 5.9), which occurs in all simple lenses, rays passing through the lens near the center do not focus at the same point as those that pass through the lens closer to its rim. This defect can be corrected in the same way as chromatic aberration.

There are other defects, such as astigmatism and coma, all of which can be reduced by the addition of appropriate lenses. It is the use of such correcting and compensating measures that accounts for the complexity of so many modern lens systems.

Large Refractors

The world's largest refractor (Fig. 5.10), built at Yerkes Observatory in Wisconsin about 1895, has an objective 100 cm in diameter. There are several reasons why, in over 80 years, a larger one has never been built. The main reason is the difficulty of casting a slab of optical-quality glass of such a size that is free of bubbles and other defects. Furthermore, the enormous weight of a disk this large causes it to sag as it is moved from position to position, with resulting changes in its curvature and loss of precision. Until solutions are found to this problem, it is unlikely that larger refractors will be built.

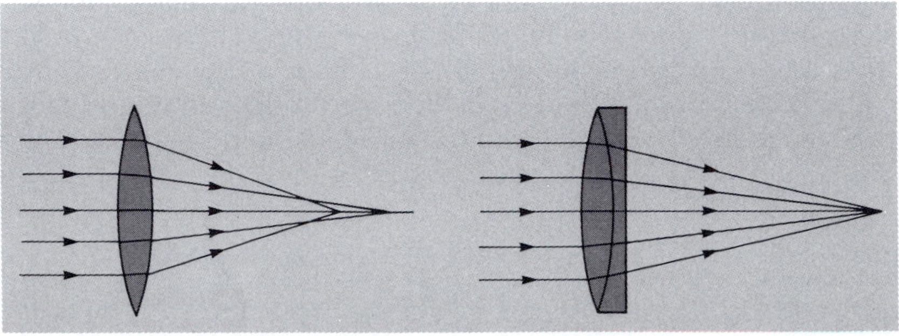

Figure 5.9 Because of spherical aberration, rays emerging from uncorrected lens at left have different focal lengths; corrected lens, at right, brings all rays to one focus.

Figure 5.10 World's largest refractor telescope, at Yerkes Observatory. The objective is 100 cm in diameter.

REFLECTING TELESCOPES

The Newtonian Design

Isaac Newton dealt with the problem of chromatic aberration by substituting a curved mirror for the objective lens. His telescope was a tiny instrument, the mirror only 5 cm across—a toy by our standards. However, Newton had built it not for its light-gathering power, but to overcome chromatic aberration, which it did. In telescopes of this design, called Newtonian reflectors, the image, formed in front of the main mirror, is shunted to the side by a small plane mirror set diagonally in front of the focal point (Fig. 5.11). Here it is magnified by a small lens—the eyepiece.

Although reflectors are free of chromatic aberration, they are subject to spherical aberration, which is overcome by giving the mirror a parabolic—rather than spherical—curvature (Fig. 5.12).

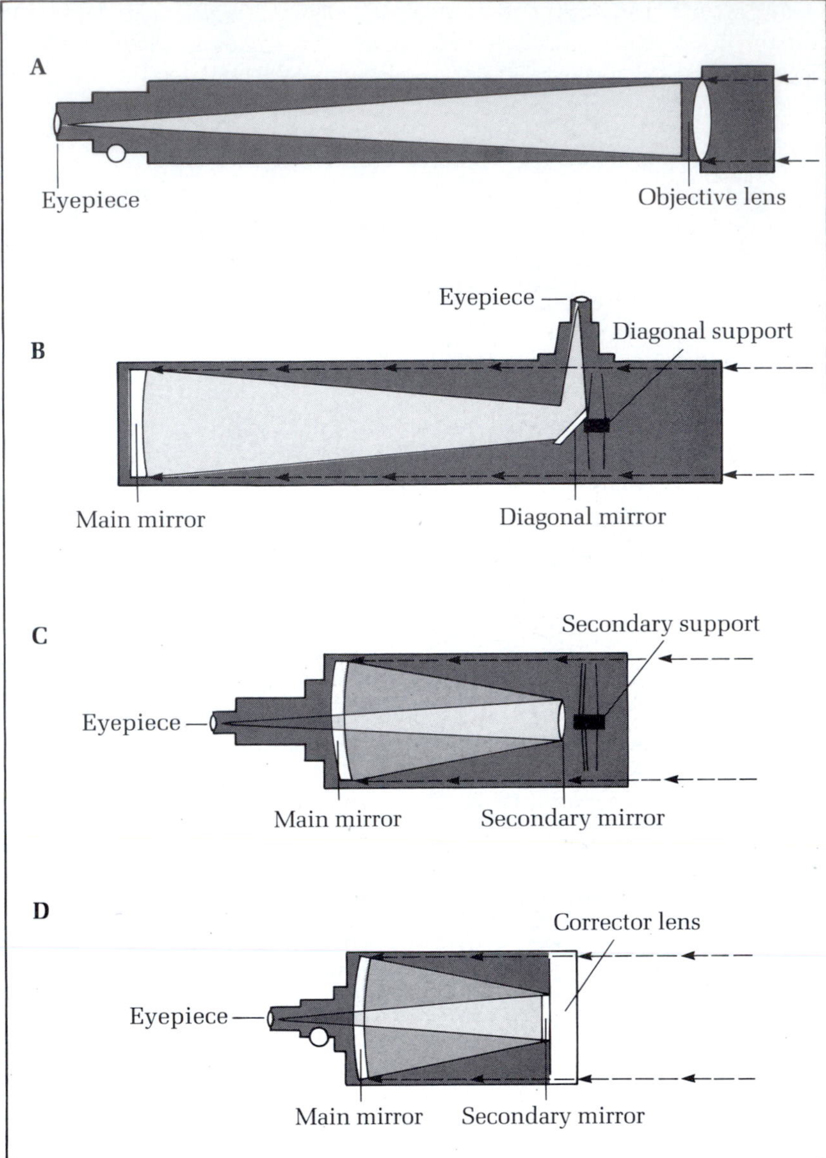

Figure 5.11 Four basic telescope designs: *A*, refractor; *B*, Newtonian reflector; *C*, Cassegrain reflector; and *D*, the Schmidt-Cassegrain reflector.

The Cassegrain Design

A few years after Newton presented his reflector, French astronomer N. Cassegrain introduced a modified model in which the plane diagonal mirror was replaced by a convex one placed parallel to the objective (the large mirror). The image from the convex mirror was reflected through a small, circular hole cut in the objective mirror to an eyepiece (Fig. 5.11).

The design was, without a doubt, ingenious and had a definite advantage over Newton's design, the eyepiece being at the lower (and therefore more accessible) end of the telescope. Because Newton dismissed the design as having no advantage over his, it was not widely adopted for some time. Today, most large telescopes are of this type.

As with refractors, the most important element in reflecting telescopes is again the size of the objective (in this case, a mirror), which determines the light-gathering power and resolution of the instrument.

Large Reflectors

The "father" of the large telescope is, without doubt, William Herschel, who developed most of the techniques needed for building them. His largest, a 1.22-m monster, was, unfortunately, too large to maneuver easily, and few of Herschel's many important discoveries were made with it. The largest telescope of this era was a 1.83-m giant built by Lord Rosse in Ireland.

Two important developments permitting the building of larger reflectors came in the 1930s; these were the development of Pyrex (a glass that expands very little with heating) and the discovery that aluminum coatings tarnish much less quickly than silver ones. Before this, plate glass was used for mirror blanks and silver for the coating. The 5-m reflector at Mt. Palomar (Fig. 5.13), until 1976 the world's largest optical telescope, incorporated both these developments. Many others have also: a 3-m reflector at Lick Observatory; a 2.1-m and a 4-m reflector at Kitt Park Observatory; a 4-m reflector at Cerro Tololo InterAmerican Observatory in Chile; and a 2.2-m reflector on Mauna Kea, Hawaii, where a 3.6-m reflector will also soon be placed.

The largest reflector is the 6-m one in the Caucasus in Russia, and it is not likely that larger ones (with a single mirror of conventional design) will be built in the near future. The major obstacle is cost: The 5-m telescope on Mt. Palomar, which cost $6.5 million, would cost many times that amount today. However, there are plans for large telescopes of *non*conventional design. The University of California has been considering building a 10-m reflector in which the mirror would be either a solid disk, much thinner than those normally used,

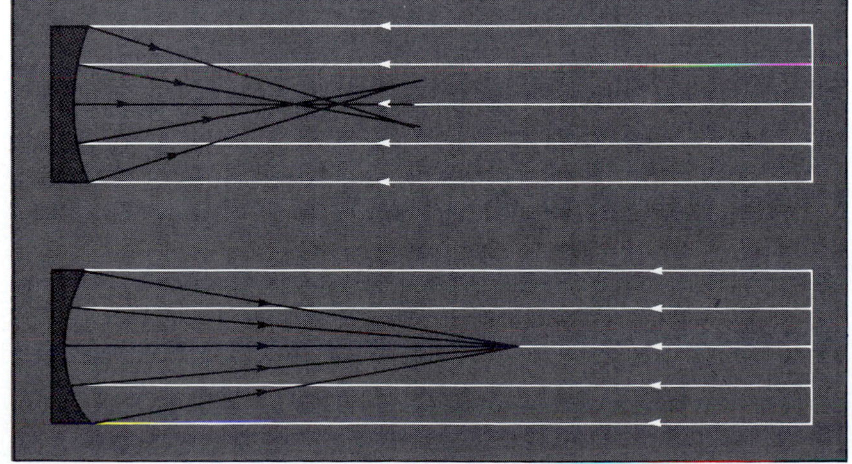

Figure 5.12 Rays reflected from a spherical mirror (top) do not come to a single focus; those reflected by the parabolic mirror (bottom) do.

Figure 5.13 The 5-m reflector on Mt. Palomar. Photograph was taken at the time of its dedication in 1949.

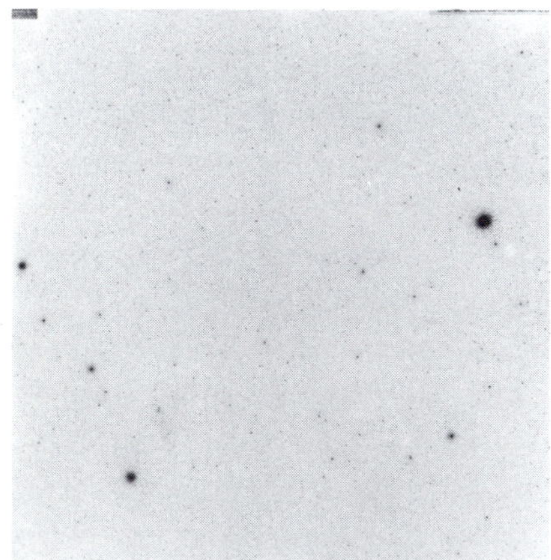

Figure 5.14 Typical wide-angle photograph taken with a Schmidt camera.

or a segmented disk (p. 91). Similar large telescopes are being considered by Kitt Peak Observatory in Arizona, by McDonald Observatory in Texas, and by the Soviet Union.

The Schmidt Design

In the early 1930s Bernhard Schmidt, a German optician, devised a correcting lens for reflectors that made parabolic mirrors unnecessary. Though it was soon hailed as the instrument of the future, interest in it faded for a time. In the United States the first Schmidt telescope was built by an amateur, H. Page.

The main advantage of this type of instrument, demonstrated by the photograph in Fig. 5.14, which was taken with a Schmidt Camera,[3] is that the area encompassed is much larger than that in most astronomical photographs; the Schmidt is a wide-angle instrument whose images, unlike those of other wide-angle telescopes, are sharply

[3]The Schmidt telescope is referred to as a camera because most are designed only to take photographs—one cannot look through them. All the larger instruments of this design are, in fact, of this type.

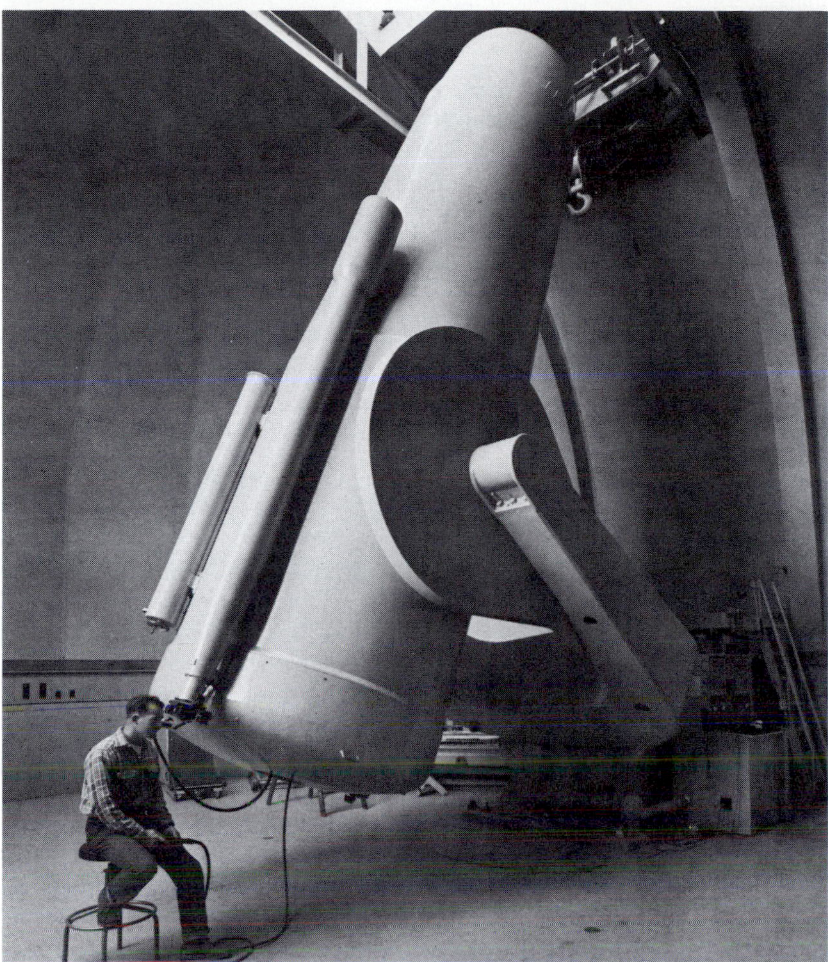

Figure 5.15 The 1.22-m Schmidt camera at Mt. Palomar.

focused in the whole field of view. This makes them ideal for routine surveys of the sky and for satellite tracking.

The 1.22-m Schmidt at Mt. Palomar (Fig. 5.15) is one of the world's largest. It was used to carry out a survey of the entire sky visible from Palomar through both red filters[4] and blue-sensitive photographic plates. (A similar survey has been done in the Southern Hemisphere.) The photographs from these surveys are currently available to astronomers around the world, and many significant discoveries have been made using them. (Each photograph covers over 30,000 times the area that the 5-m reflector at Palomar would.) Other large Schmidt cameras are in Germany, Australia, and England. Another telescope of similar design is the Maksutov, which was devised by Russian astronomer D. D. Maksutov in 1941. Maksutov replaced the Schmidt corrector plate with a simpler one that is thicker, with one side concave and one convex. The well-known "Questar" (a small, commercially available amateur telescope) is of this design.

[4]A filter allows only a small range of wavelengths of light to pass through it. A red filter, for example, is transparent to a band of wavelengths from about 6500 Å to about 7500 Å—the portion of the visible spectrum usually identified as *red*.

ATMOSPHERIC AND "SEEING" PROBLEMS

Without the atmosphere, our protective blanket of air, life could not exist on Earth. To the astronomer, however, the atmosphere is a nuisance. We have discussed how water vapor in it blocks infrared radiation; for the optical astronomer, even equipped with a telescope of tremendous light-gathering power and high resolution, it is also troublesome.

Astronomers are most concerned with the degree of stability of the atmosphere, which they call "the seeing." (Note that this refers not to transparency, but to the constancy of the air—its lack of turbulence.) A ray of light from a star is bent as it passes through the atmosphere. When the atmosphere is unstable, the angle of bending continually changes, so that the image constantly shifts. The result, if one is attempting to record the star's position in a photograph, is a blurred image. Thus a breathtaking night sky in which bright stars twinkle and glisten is not the astronomer's ideal; far from it; twinkling stars, as we have just seen, mean the seeing is poor. Astronomers look for nights when the stars shine steadily.

A recent technique for overcoming the problem of dancing images is called "speckle" photography. In it a series of short (approximately $\frac{1}{100}$ sec) exposures is made using a large telescope. Each exposure gives a pinpoint image (assuming one is photographing a star) and the result is a speckled pattern. By use of a computer in conjunction with a photoelectric photometer (p. 91) to measure light intensities, the pinpoints are reassembled into a sharp image of the star. This technique recently gave us our first view of the surface of a star (Betelgeuse) other than our sun.

As cities have expanded in recent years, bringing their unwanted light closer and closer to major observatories, light pollution has become an increasingly serious problem. The tremendous growth of both Los Angeles and Pasadena, for example, has brought the problem to Mt. Wilson Observatory, once almost entirely free of light pollution. The problem of light from cities is augmented by the slight glow caused by atomic interactions in the upper atmosphere. Extraneous light—light pollution—affects observations by lightening the backgrounds of astronomical photographs (of stars or galaxies) so that they are gray instead of black and, thus, reduced in contrast. This restricts the length of exposure (the longer the exposure the lighter the background will become) and, hence, limits what we can see.

Photoelectric Image Enhancement

Several means of compensating for atmospheric effects have been devised. One of these (which is also why large reflectors have not been built for over thirty years) is the image intensifier, which can extend the light-gathering power—and, hence, the effective diameter—of a telescope.

We see the image of a star on a photographic plate because a certain number of photons hit this region and sensitize the crystals that are here; these crystals darken when the plate is processed, or developed. But a minimum number of photons is required for this darkening; if less than this number hit, no change takes place on the plate. The image intensifier, in effect, increases the number to the level where the source of the light will register on the photographic plate. In actual operation, when a single photon enters one end of the image tube, hundreds (sometimes thousands) leave at the other end. Placed at the focus of a telescope, it produces hundreds of photons at the photographic plate for each photon that enters it. Thus, a plate made with the help of this device will show not only brighter images but a large number of apparently "new" stars (stars previously too faint to be recorded) as well.

Closely related is the charge-coupled device (CCD), which is becoming important in astronomy. The CCD allows astronomers to view the skies on a TV screen—usually in the comfort of a warmed room.[5] A CCD is an array of "pixels," or tiny photoelectric cells, each of which reacts to light; when light hits the array, a pattern of electrons is released according to the intensity of the light at the various pixels. These electrons impinge on a TV screen, where the pattern then becomes visible. Sometimes the pattern is stored in a computer for processing or later viewing.

Also widely used is the photoelectric photometer, in which light from a star (or other object) is also converted to an electric signal. Here, however, the signal is passed to a device in which an inked tracing of the signal's intensity (in effect, a graph) is recorded. The instrument is particularly useful for comparing brightnesses of stars.

Besides the use of larger objectives and image intensifiers to increase light-gathering power, there are other methods for improving the quality of images. In one of these, outputs of several small telescopes are connected, to give (by totaling the radiation received) the equivalent of a much larger one. This technique is used routinely with radio telescopes, but light waves, having much shorter wavelengths than radio waves, are harder to superimpose.

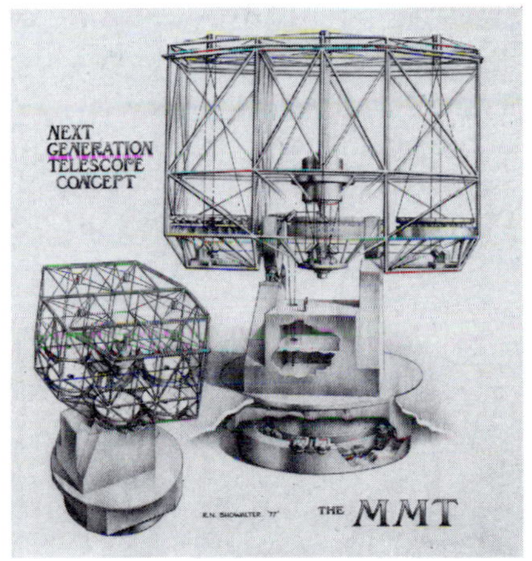

Figure 5.16 The multiple-mirror telescope (MMT), composed of six 1.8-m reflectors.

The first telescope of this type was built at Mt. Hopkins near Tuscon, Arizona. It is called the multiple mirror telescope (MMT) and is composed of a cluster of six 1.8-m reflectors (Fig. 5.16). Together they are equivalent to a 4.44-m reflector—at less than half the cost.

The MMT, already successful, may be the prototype for future telescopes. With instruments of this type astronomers may be able to measure the shapes and sizes of stars as never before.

Long-range plans are being made for telescopes—called NGTs (next-generation telescopes)—as large as 25 m, about six times the largest existing one. Two of the designs are, in effect, MMTs; one is composed of sixteen 6-m instruments, the other of six of 10 m each. There are also plans for a telescope quite different in design, called a "segmented mirror telescope," since the primary (main) mirror is

[5]Observatories cannot be heated because the seeing would be disrupted by the currents of heated air that would pour from the slit in the dome or roof through which the telescope passes.

made up of many sections placed carefully together (Fig. 5.17), each with its separate support. One of the designs features a spherical primary; in this case the surface of each segment is a portion of the sphere. But, of course, a spherical mirror will not focus light to a point, so the secondary must somehow compensate for this.

Telescope design is not the only technology of significance to astronomers. In recent years computers have become increasingly important; in fact, most astronomers now spend much more time with the computer than with the telescope, and it is used not only for data processing but also for pointing telescopes and for running observing programs.

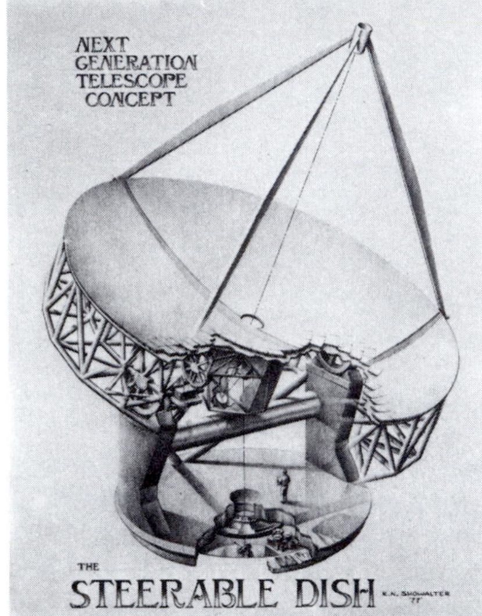

Figure 5.17 A segmented-mirror telescope.

RADIO TELESCOPES

In 1931 Karl Jansky, investigating why long-distance telephone calls were plagued by static and hissing, made a discovery that led to the opening of a radio window on the universe. With a large directional antenna assembled from odd parts (Fig. 5.18), which could be moved in a circle so as to cover all parts of the sky, Jansky began his search. Thunderstorms were quickly pinpointed as sources, but even with the storms accounted for there was still a low hiss, which seemed to be coming from a particular region of the sky. An important clue was the fact that the source appeared 4 minutes earlier each day, and so must somehow be connected with the stars. Jansky finally pinpointed the source as the constellation Sagittarius, which astronomers had just established as the direction of the center of our galaxy; the noise was coming from the core of our galaxy.

Figure 5.18 Karl Jansky's directional antenna, used to detect cosmic radio waves in 1931.

Although Jansky encouraged astronomers to explore the area further, there was little interest at first. However, an electrical engineer, Grote Reber, built an antenna in his back yard in Chicago and, night after night, plotted intensities from various regions of the sky. His report, submitted to the *Astrophysical Journal,* was dismissed by its review board as of no importance to astronomers; nevertheless, it was accepted and published.

Reber continued for almost five years (1938–42) as the world's only radio astronomer. It was not until after World War II that the English took an interest in the new device, and others followed.

The radio telescope differs little from an optical telescope. Though in most radio telescopes radio waves are collected by a large parabolic dish covered with a fine mesh,[6] the waves are focused by this dish just as light waves are in an optical telescope. The focused radio waves are then converted into an electrical current, which is amplified and passed to a chart recorder. Because the waves are so long (relative to light waves) the smoothness of the reflecting surface is not critical, as it is in the case of optical telescopes. Thus, centimeter for centimeter, radio telescopes are much cheaper than optical telescopes.

Radio telescopes are generally of three main types; movable, partially movable, and fixed, and they are constructed in arrays (which we discuss below). A good example of the first type is the 76-m Mark I telescope of Jodrell Bank, near Manchester, England. The largest of its type for years, its original mesh surface is now smooth, the change having been made so that shorter wavelengths could be received. The largest fully steerable radio telescope in the world, at the Max Planck Institute in Bonn, Germany (Fig. 5.19), has a diameter of 100 m.

The largest single radio telescope in the world—of any type—is the 308-m fixed dish at Arecibo, Puerto Rico (Fig. 5.20). Built into a natural bowl-shaped valley, it is not completely fixed; the position of the collector can be changed, allowing for a variation of about 15° on either side of the vertical. Of course, the earth's rotation also allows it to make a complete 360° survey each day.

Radio telescopes have to be much larger than their optical counterparts because of the long wavelengths involved. This makes it necessary to gather waves over a considerably larger area in order to get even moderate resolution (p. 83). As mentioned earlier, the use of arrays is a means of avoiding the high costs and other obstacles in the construction of larger optical telescopes. Radio astronomers succeeded in the use of arrays many years ago. In fact, their efforts have led to some rather strange figures—crosses, Ts, Ys. The largest array—the VLA (Very Large Array) in New Mexico—consists of 27 dishes, each 26 m in diameter, placed in the shape of a Y. Each leg of the Y is about 20 km long.

Even huge arrays like these suffer from resolution problems. To resolve stars and galaxies in the same detail as optical telescopes, a

Figure 5.19 The largest fully steerable radio telescope. Located at Bonn, Germany, it has a diameter of 100 m.

[6]The openings in the mesh are smaller than the wavelength (from a few millimeters to several meters) of the waves being collected.

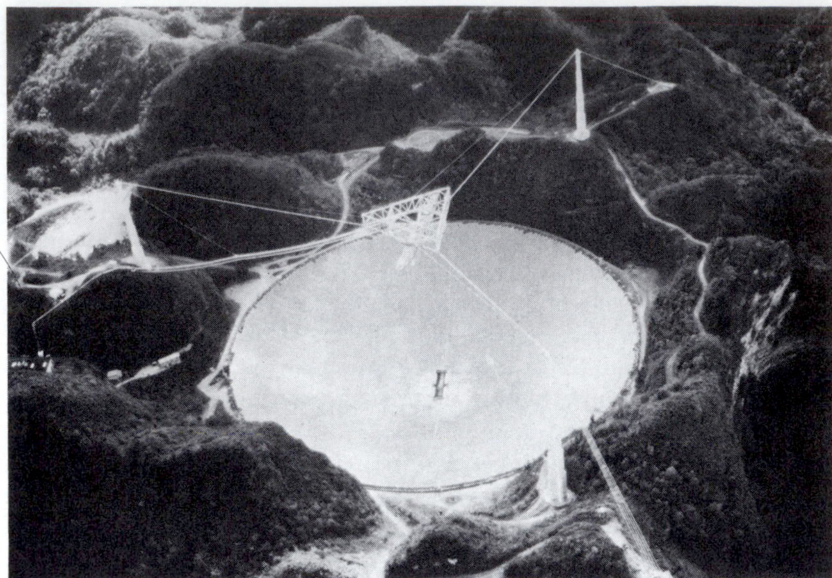

Figure 5.20 The largest nonsteerable radio telescope, a 308-m dish at Arecibo, Puerto Rico.

radio telescope would need to be as large as the earth. In a sense this has been accomplished.

Combined, the outputs of two radio telescopes are equivalent to a single telescope with a diameter equal to their separation. Early hookups of this type were between radio telescopes only a few miles apart, and consequently their resolution was still relatively low. But with the development of accurate atomic clocks and the use of computers, much longer baselines (distances between installations) became possible. One of the first of these long-distance hookups was between two instruments at opposite ends of Canada. With its success, baselines became even longer; finally a linkup was made between a radio telescope in England and one in Australia—spanning the world. In these long-distance hookups there is no direct link between the telescopes. Recordings are made at each, and these are combined by a computer. The process involved in long-distance setups is usually referred to as very-long-baseline interferometry (VLBI).

INFRARED, UV, AND X-RAY TELESCOPES

Besides the problems (mentioned in Chapter 4) in detecting infrared waves, ordinary photographic plates are not sensitive to them. In the mid-1960s, however, Robert Leighton built a specially equipped 1.52-m infrared telescope with which he made an extensive survey of the sky, finding over 20,000 infrared sources. A number of projects have followed Leighton's lead. Three large, earth-based infrared telescopes have now been completed, and, because of the obstacles presented by atmospheric water vapor (see Chapter 4), each has been placed at a high altitude. The first to go into operation (in 1977) was the 2.3-m University of Wyoming telescope on Mt. Jelm in Wyoming (Fig. 5.21; color). A similar 3.1-m instrument that began operations on Mt. Mauna Kea, Hawaii, in 1979 is designed for study of the planets. The largest of the three, also at Mauna Kea Observatory, was built by the United Kingdom and is 3.8 m across.

Major difficulties in the design of infrared telescopes are presented by the telescope itself (its motors and other devices being strong emitters of infrared radiation) and by the cosmic background, which gives off a considerable amount of infrared radiation. In the Wyoming instrument this is compensated for by first measuring radiation from the object (being viewed) *and* the background. The secondary mirror is then moved to another position, where background *alone* is measured. Computers then subtract the background "noise."

In infrared telescopes photographs are not taken as with an optical telescope. The sensing instrument in this case, called a bolometer, is little more than an extremely accurate thermometer. In practice, several bolometers operate in an array, each sensitive to a different range of infrared wavelengths. The entire array is shielded from stray radiation by liquid helium at −269°C.

In the 1960s, Dr. Frank Low made some important improvements in detection, and a short time later infrared observations were being made above our atmosphere (or at least above most of it) from balloons and airplanes.

Numerous discoveries were soon made. A number of stars—called "cocoon stars" because they appeared to have a cocoonlike cloud of dust and debris around them—were found to give off huge amounts of infrared. Two infrared sources were discovered (one in the Orion Nebula) that appeared to be newly forming stars. (These objects will be discussed in Chapter 16.) A number of galaxies also appeared to be strong infrared sources.

Another recent development is UV astronomy. In 1968 the planets, several galaxies, and even a quasar (Chapter 23) were scanned and studied from a satellite equipped with UV detectors. Of particular interest in UV astronomy are very hot stars, as they emit a considerable amount of their radiation in this region. The UV spectrum of our sun is also of considerable interest, and it has been studied extensively. Many UV sources have been found in galaxies and novae, and several comets have been studied in the UV region of the spectrum.

X-ray and gamma-ray astronomy are the most recent of the new astronomies. X-ray astronomy came into its own in 1970 with the launching of the X-ray satellite UHURU. As UHURU scanned the sky, numerous new sources were discovered. Objects of particular interest were supernova remnants such as the Crab Nebula, quasars, and radio galaxies. Figure 5.22 gives further details on the new astronomies.

Figure 5.22 Various astronomical means (including the "new astronomies") and their respective focuses of investigation are shown on the following page.

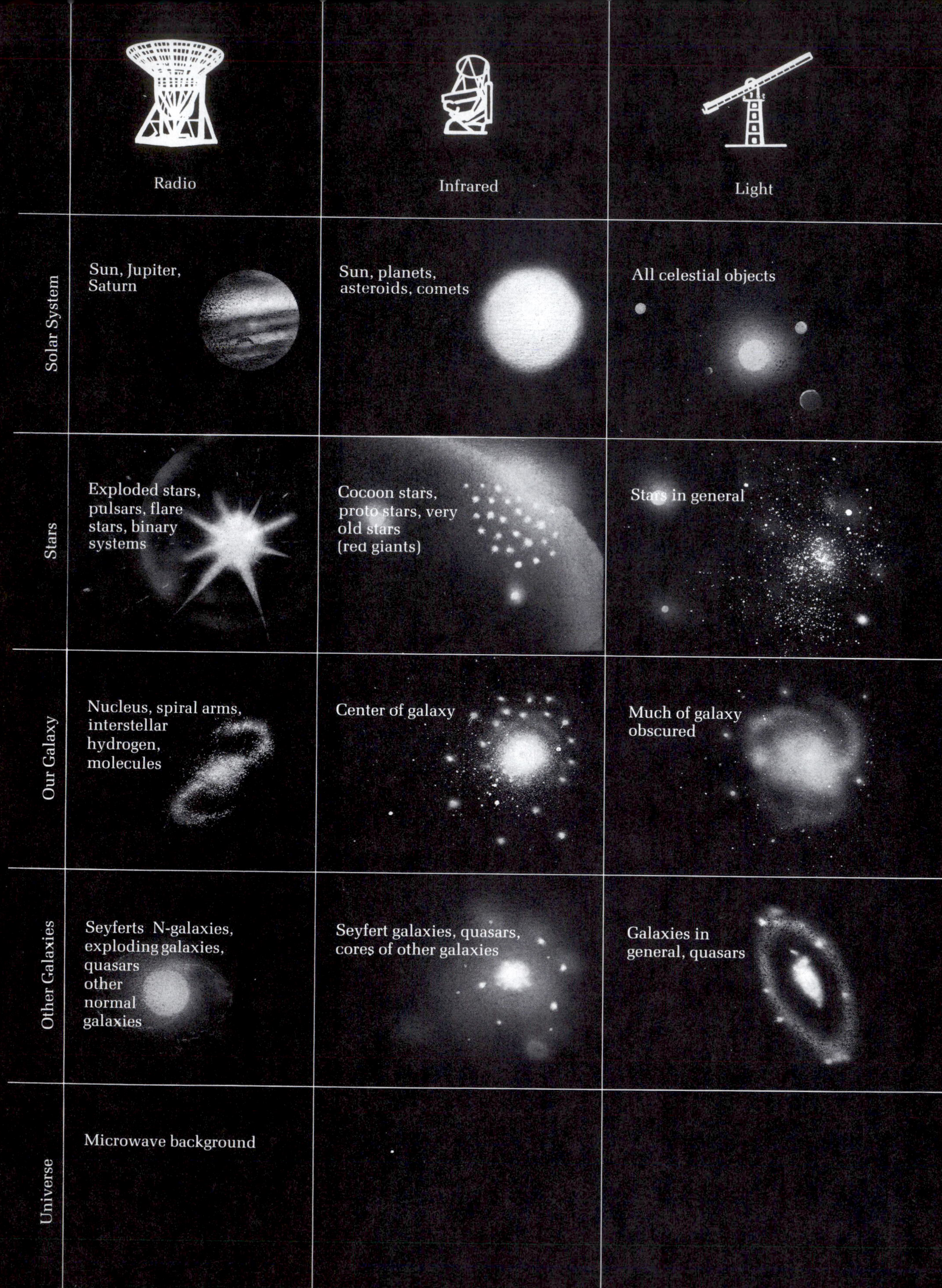

Radio
Infrared
Light
Solar System
Sun, Jupiter, Saturn
Sun, planets, asteroids, comets
All celestial objects
Stars
Exploded stars, pulsars, flare stars, binary systems
Cocoon stars, proto stars, very old stars (red giants)
Stars in general
Our Galaxy
Nucleus, spiral arms, interstellar hydrogen, molecules
Center of galaxy
Much of galaxy obscured
Other Galaxies
Seyferts N-galaxies, exploding galaxies, quasars other normal galaxies
Seyfert galaxies, quasars, cores of other galaxies
Galaxies in general, quasars
Universe
Microwave background

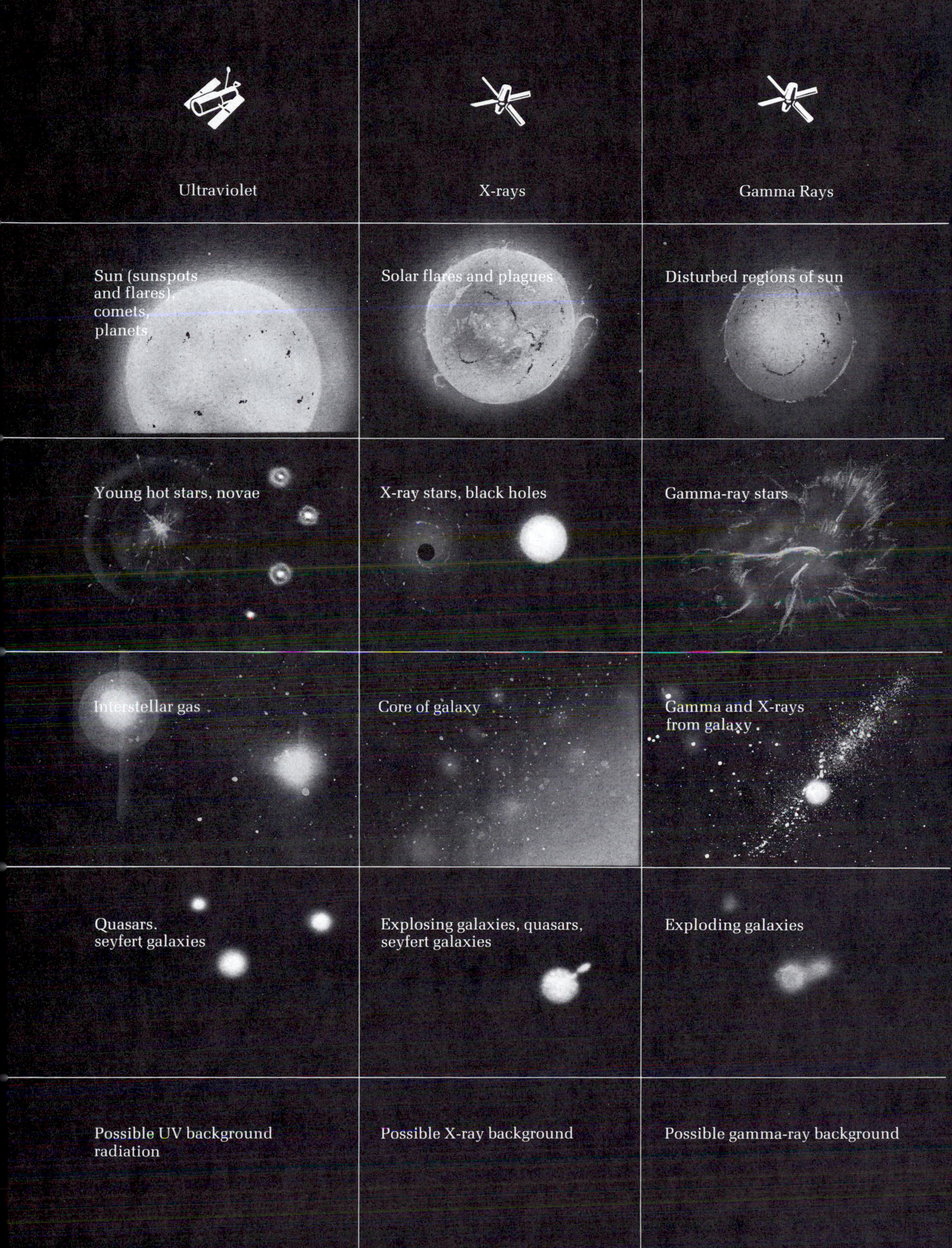
Ultraviolet
X-rays
Gamma Rays
Sun (sunspots and flares), comets, planets
Solar flares and plagues
Disturbed regions of sun
Young hot stars, novae
X-ray stars, black holes
Gamma-ray stars
Interstellar gas
Core of galaxy
Gamma and X-rays from galaxy
Quasars. seyfert galaxies
Explosing galaxies, quasars, seyfert galaxies
Exploding galaxies
Possible UV background radiation
Possible X-ray background
Possible gamma-ray background

Figure 5.23 UHURU, a major X-ray satellite (no longer operating).

TELESCOPES IN SPACE

In recent years satellites designed specifically for astronomical observations have been launched. One of the largest and best equipped of these was launched in 1968. Called "Copernicus," it commemorated the five-hundredth anniversary of his birth. Weighing over 4000 pounds, it contains 11 telescopes, the largest of which is 80 cm. Although it is primarily a UV satellite, there is some X-ray detection equipment aboard. The major X-ray satellite for years, the above-mentioned UHURU (Fig. 5.23), weighed 315 pounds and was about 2.4 m long. Numerous discoveries were made by UHURU after its launch in 1970, one of the most important of which was the discovery of the black hole (p. 366) candidate (CYG X-1) in the constellation Cygnus. In many ways Copernicus and UHURU operated as a team, UHURU scanning the sky rapidly for new sources, Copernicus carefully studying the object once found. In 1975 UHURU's useful life came to an end, and shortly thereafter a British X-ray satellite, Ariel 5, was put into orbit. The principal American X-ray satellites now in orbit are HEAO-1 and HEAO-2 (HEAO stands for high energy astronomical observatory; HEAO-2 is also known as the Einstein Observatory). Many significant discoveries were made by these two satellites, which also no longer function.

Skylab also functioned as an orbiting observatory. Several telescopes and a number of other instruments were aboard it, most designed for study of the sun. Some excellent X-ray photographs of the sun were obtained, as well as a number of UV photographs of the comet Kohoutek.

Doubtless, more satellites will be placed into orbit with various types of telescopes aboard. Indeed, plans are now under way for putting a 2.4-m optical telescope aboard a space shuttle in 1983. With an instrument this large above the atmosphere, many important discoveries will no doubt soon be forthcoming.

SUMMARY

1. When a ray of light enters glass (from air) at an angle to the surface, it is bent toward the normal.
2. A convex lens makes a beam of light converge or come to a focus; a concave lens makes it diverge or spread out.
3. The telescope was invented about 1604 by Hans Lippershey. The first to use it for astronomical observations was Galileo, who made several significant discoveries with it.
4. The main component of a telescope is the objective. The objective gathers and focuses light to form an image. This image is magnified by the eyepiece.
5. The resolving power of a telescope is a measure of its ability to resolve or separate two objects that subtend a small angle in the sky.
6. Lens aberrations, or faults, can be corrected by adding a second lens of different shape and refractive index.
7. The first reflecting telescope was built by Newton. The Newtonian style (eyepiece on side) and Cassegrain design (eyepiece at end) are the two most common types of reflectors.
8. The largest refractor in the world (100 cm in diameter) is at Yerkes Observatory. The largest reflector in the world (600 cm in diameter) is in the Caucasus Mountains of the Soviet Union.
9. Other special types of telescopes are the Schmidt, which incorporates both a mirror and a lens, and the infrared telescope.
10. The stability of the atmosphere is referred to as "the seeing." The more stars twinkle, the less stable the atmosphere and the poorer the seeing.
11. Image intensifiers are now used in conjunction with telescopes to enhance photographs. For each photon that enters one end of the image tube many leave at the other end.
12. A multiple mirror telescope (MMT) has been built at Mt. Hopkins in Arizona. Other designs are also being considered.
13. Cosmic radio waves were first detected by Karl Jansky in 1931. The first radio telescope was built by Grote Reber.
14. The largest fully steerable radio telescope in the world (100 m) is at the Max Planck Institute in Bonn, Germany. The largest fixed-antenna radio telescope (308 m) is at Arecibo, Puerto Rico.
15. Several orbiting observatories are now operating. They are concerned with the parts of the electromagnetic spectrum other than the visible and radio regions.

REVIEW QUESTIONS

1. Sketch a ray entering a piece of glass with parallel sides. (Assume it is incident at an angle to the normal.) Explain its path. Do the same for a prism.
2. Draw a convex lens. Show the path of light rays through it. Do the same for a concave lens.
3. What is the focal length of a lens? How is it found?
4. What discoveries did Galileo make with his telescope?
5. Sketch a simple refractor and label its parts.
6. What is light-gathering power?
7. What is resolving power? What determines it?
8. Describe two types of aberrations a lens can have. How can they be overcome?
9. Why is it unlikely that refractors larger than 100 cm will be built?
10. How does the Newtonian reflector differ from the Cassegrain reflector?
11. How do we overcome spherical aberration in reflectors?
12. How does the Schmidt camera differ from the reflector?
13. What sensing instrument is used in the infrared telescope? Explain.
14. What do we mean by "the seeing?" Is the seeing poor when it is foggy? What causes a plate to "fog"?
15. Explain briefly how an image intensifier works. What is a CCD?
16. What is the MMT, the NGT, the VLA, the VLBI?
17. What was Jansky looking for when he discovered cosmic radio waves?
18. Who was Grote Reber?
19. Explain the difference between a radio telescope and an optical telescope.

THOUGHT AND DISCUSSION QUESTIONS AND PROJECTS

1. Draw a concave lens slightly to the right of a convex lens. Assume that parallel rays enter the convex lens at the left. Sketch the path of these rays through both lenses. (Note: Rays will not be parallel when they enter the concave lens.)
2. List and discuss the factors that limit detail in a telescope.
3. In the Newtonian and Cassegrain telescopes there is a small mirror in the way of the entering light, yet when you look through these telescopes the mirror cannot be seen. Why?
4. List the advantages and disadvantages of the (a) Newtonian reflector, (b) refractor, (c) Cassegrain reflector, (d) Schmidt camera.
5. What is the magnification of a telescope with a focal length of 120 cms and an eyepiece of focal length $\frac{3}{4}$ cm? What focal-length eyepiece would give a magnification of 300?
6. Discuss the advantages of a telescope in space.
7. *Project:* Obtain several convex and concave lenses of different focal lengths. Observe a given object through each. Describe the position and size of the image you see. Explain why.
8. *Project:* Observe the moon and planets through different types of telescopes (reflector, refractor) of various objective sizes. Describe the differences in brightness and clarity of the images.

FURTHER READING

Asimov, I. *Eyes on the Universe: A History of the Telescope.* Houghton Mifflin, Boston (1975).

Heys, J. S. *The Evolution of Radio Astronomy.* Neale Watson Acad. Pub. (1973).

Page, T. and Page, L. W. *Space Science and Astronomy.* Macmillan, New York (1976).

Pike, R. "Schmidt Cameras," *Astronomy,* p. 50 (Nov. 1976).

Verschuur, G. L. *Starscapes.* Little, Brown and Co., Boston (1977).

Woodbury, D. O. *The Giant Glass of Palomar.* Dodd, Mead and Co., New York (1970).

Aim of the Chapter

To explain the stellar spectrum and show how it is used by astronomers.

chapter 6

Radiation Laws and Spectra

All objects emit radiation that is characteristic of their composition and state. Radiation in the visible part of the electromagnetic spectrum can be analyzed by passing it through a spectroscope (see box), which produces a pattern (spectrum) of colored bands and lines that can tell us much about the composition and other characteristics of the source—as, for example, a star—of the light being analyzed.

SPECTRA AND THE SPECTROSCOPE

Discovery and Development

Newton was the first to explain how a prism separates light into its spectrum of component colors. Passing a beam of light from a small circular hole through a prism, he noticed that the once circular beam was oblong when it emerged—almost as if it had been stretched. Since red light was at one end, violet at the other, and the rest between, Newton concluded that white light must be composed of light of all colors, which the prism simply separated. By way of confirming this hypothesis he passed the colored rays through a second prism just beyond the first, and found that, indeed, it brought the rays together to form white light again.

Although little progress was made in optics for many years after Newton's experiment, lens-grinding techniques gradually improved and, by the early 1800s, prisms of high quality were being produced. In 1802 William Wollaston, using a prism to examine the spectrum of the sun, noticed several dark lines crossing it but came to no conclusions concerning their meaning. Twelve years later, German physicist Joseph Fraunhofer directed a beam of sunlight into a prism from a slit, so that the light entered in a narrow line. The resulting spectrum showed hundreds of narrow dark lines.[1] Over a period of years

*Narrowing the opening provides better delineation of individual lines, since each line is, in effect, an image of the opening through which the light passes.

he counted about 600 and determined the wavelength of many of them. These lines, named in his honor, are also called absorption lines.

In 1859 Robert Bunsen and Gustav Kirchhoff developed the spectroscope and performed the first of a series of important experiments. They found that a substance heated until it was a glowing gas had a spectrum completely different from that of the sun. They examined the spectra of other substances, and in each case the lines were different. Each substance apparently had its unique set of lines—as specific as a fingerprint. Kirchhoff and Bunsen soon found that they could easily identify a substance from its spectrum. They also found that there were three types of spectra (Fig. 6.1, color): continuous (like that of light from a white-hot metal), bright-line (like that from a heated gas), and dark-line (as produced from sunlight). Kirchhoff, in trying to determine their relations, examined two of the darkest Fraunhofer lines, which appeared to be in the same position (i.e., at the same wavelength) as two bright lines that occurred in the spectrum of sodium. When he passed a beam of sunlight through glowing sodium, he found that the dark lines became more intense. This caused Kirchhoff to realize that there was a relationship between the dark and bright lines: glowing sodium produced bright lines, but in the case of white light (with its continuous spectrum) passed through the gaseous sodium, the wavelengths corresponding to these lines were somehow *absorbed* by the gas. Thus they would appear as dark lines against the bright, continuous background spectrum.

Using this phenomenon, Kirchhoff soon identified hydrogen, iron, calcium, and nickel as components of the sun. He summarized his work on spectroscopy in three rules, now called Kirchhoff's rules of spectral analysis:

1. Hot, dense gases, liquids, and solids emit a continuous spectrum.
2. Hot, rarefied gases have bright *emission* lines that are characteristic (in position and number) of the gas.
3. Light from a continuous source passed through a cool gas shows dark *absorption* lines characteristic of the gas. (Thus, a spectro-

The Spectroscope

Starlight carries a hidden message, which can be decoded with an instrument called a spectroscope, a simple version of which is shown at right. Light from a star enters through the slit, passes through a collimating lens (which makes the rays parallel), and then through a prism. The light from the prism (dispersed into its various colors) is focused onto a photographic plate. The resulting photograph, called a spectrogram, shows a series of spectral lines. The spectrum can also be examined directly, and the prism is now generally replaced by a grating—a small block of material on which closely spaced lines have been ruled (p. 72).

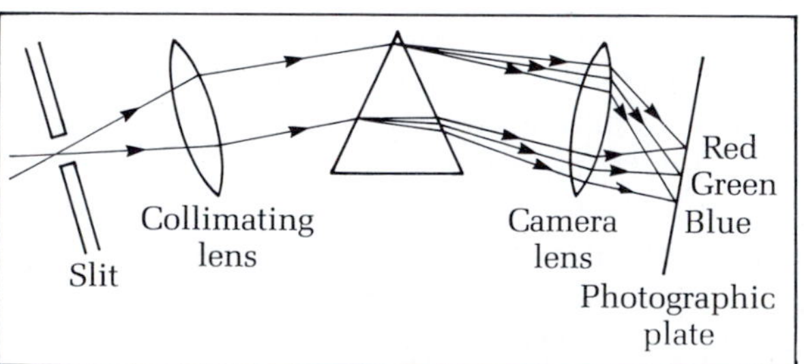

gram of the light that is emitted at the sun's "surface" and then *passes through the relatively cooler gas composing its outer regions*, shows absorption lines.)

Only a few years before Kirchhoff's experiment, prominent scientists had publicly stated that the composition of the stars was unknowable. Spectra makes such knowledge possible.

RADIATION LAWS AND THEIR APPLICATION TO ASTRONOMY

If an object—say an iron ball—is heated sufficiently, it will glow, according to the temperature, in various colors. In the past hundred years or so, physicists have formulated laws governing such radiation (and absorption), and these have been applied by astronomers in dealing with their objects of study, which can be observed mainly on the basis of the light (or other radiation) they emit. What these laws are and how they are applied are discussed below.

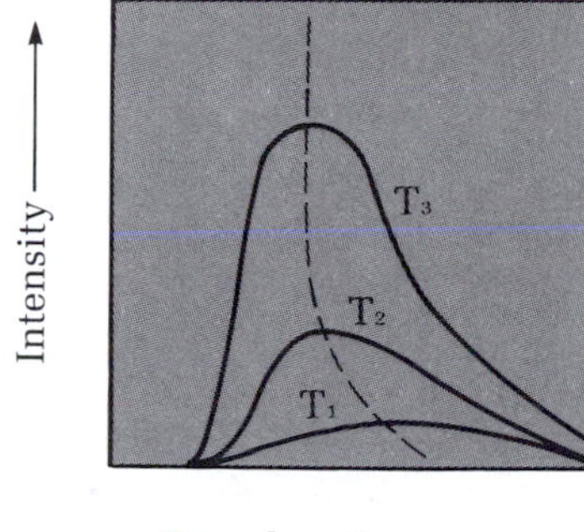

Figure 6.2 Energy radiated in relation to wavelength. *T* indicates temperature of the body, T_3 being the hottest. Note that wavelength of the peak decreases as temperature increases (dashed line).

By the late nineteenth century, scientists had noted that the increase in radiation (with increased temperature) from an object—say an iron ball—occurs at all wavelengths but that *most* radiation for a specific temperature occurs at a specific wavelength (λ_{max}). (Indeed, it had even been noticed that for a brass ball the curves—for the same temperatures—were slightly different.) They also noted that the curve of intensity (relative to wavelength) has a characteristic shape, like a skewed "normal" curve, falling off on both sides of the λ_{max} high point, but with the dropoff generally steeper to the left—toward shorter wavelengths. Figure 6.2 summarizes these points.

To help standardize these absorption–emission phenomena, scientists introduced the concept of an ideal body—one that could absorb all (i.e., reflect none) or emit all of the radiation that fell on it, called a *black body*. German physicist Wilhelm Wien devised a formula that predicted (but did not explain) the shift in wavelength of the peak of radiation (λ_{max}) with temperature. In 1879 Joseph Stefan showed that the radiant output increased as the fourth power (T^4, where T is temperature). This is now usually referred to as the Stefan–Boltzman Law. According to it, doubling the temperature, say, of a candle flame would increase radiant output by $(2)^4 = 2 \times 2 \times 2 \times 2 = 16$ times. (This is one reason that a hotter flame requires much more fuel proportional to the increase in temperature: It is losing radiant energy at a vastly increased rate.)

Even with the Stefan–Boltzmann and Wien laws the radiation curve itself remained unexplained. Just before the turn of the century German scientist Max Planck (Fig. 6.3) attacked the problem from a radically different point of view—so radically different, in fact, that for a time he disbelieved it himself. Planck assumed that the conversion of heat to light could occur not in units of any random size, but in units *(quanta)* whose size depended on the frequency (and hence the wavelength) of the emitted light. These quanta of radiation energy are now called photons (p. 74). Using this approach, Planck derived a formula that fitted the black-body radiation curves perfectly, and in so doing he explained both Wien's and the Stefan–

Figure 6.3 Max Planck.

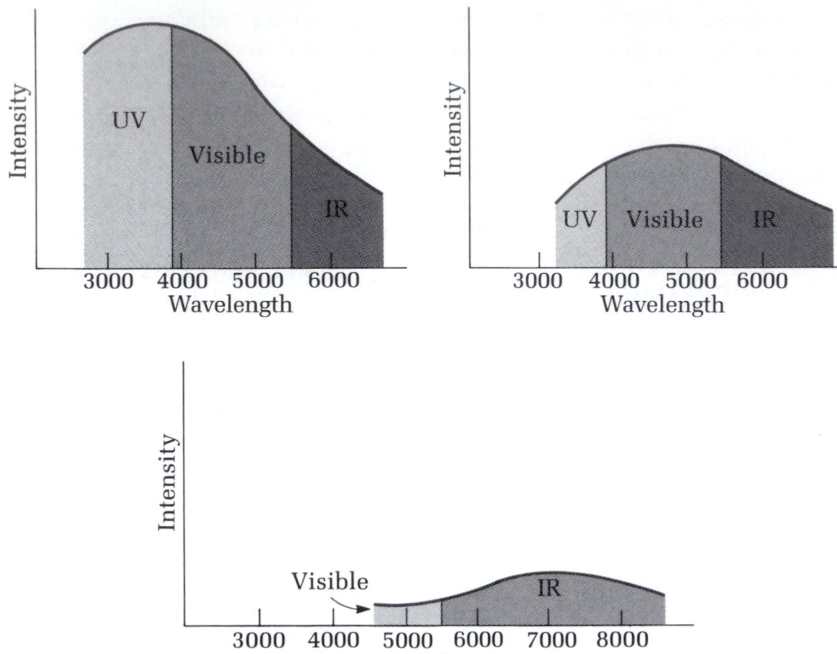

Figure 6.4 Radiation (intensity) curves for three hypothetical stars with surface temperatures of 8000°K (left), 6000°K (right), and 4000°K (lower center).

Boltzmann formulas. Although Planck did not realize it at the time, his formula would have a tremendous impact on the future of physics, ushering in a new era in science.

Stars emit radiation—in fact, in ways very similar to those of our iron ball. This means, then, that their radiation curves are similar to the ones we have just been considering. The radiation curves of three stars (Fig. 6.4) of different surface temperatures—say, 4000°K, 6000°K, and 8000°K—are similar to those shown in Fig. 6.2. From Fig. 6.4 we see that stars radiate over a large region of the electromagnetic spectrum, the hottest star radiating mostly in the ultraviolet, though also in the visible and infrared regions. The star with a surface temperature of 6000°K is radiating mostly in the visible region, and the 4000°K star is radiating mostly in the infrared, though again, in both cases there is emission over the entire spectrum.

Figure 6.5 Ernest Rutherford.

The Atom and Absorption–Emission Phenomena

In the early twentieth century, British physicist Ernest Rutherford (Fig. 6.5) postulated that the atom had a structure similar to that of the solar system, with a nucleus (having a positive charge) orbited by negatively charged electrons. Difficulties with Rutherford's model were resolved by Niels Bohr (Fig. 6.6), who proposed that electrons—unlike the case with Rutherford's model—are normally restricted to specific orbits. When the electrons move to and from different orbits, they absorb or emit energy (the state of lowest energy of the atom is called the "ground state"), depending on whether they move outward or inward relative to the nucleus. Energy is emitted (radiated) in the form of a photon (p. 74), whose frequency is in proportion to the energy lost in the move. There are many possible orbits, or energy levels as they are usually called. Thus, if you were to heat a sample of hydrogen (one way of applying energy to it), the electrons in the various atoms making up the sample would move rapidly to many different outer levels. (This phenomenon is illustrated in the left-hand diagram in Fig. 6.7). When they drop back to various lower levels (right-hand diagram), they emit photons of various wavelengths. All electrons dropping to the same energy level will emit photons of the same wavelength. It is the effect of this grouping of photons of the same wavelength that we see as bright colored lines—the emission lines referred to earlier p. 104)—at various positions in a spectrogram. Figure 6.8 illustrates these phenomena as they occur in hydrogen. Electrons dropping from any outer orbit to Orbit 1 give rise to the spectral lines called the Lyman series; similarly, those dropping to Orbit 2 give the Balmer series, and those dropping to Orbit 3, the Paschen series. All these series have been observed in hydrogen, and the positions of their lines in a spectrogram correspond closely to those predicted by Bohr's picture of the atom. The cause of the absorption lines for hydrogen can also be seen in this model: White light passed through a cloud of hydrogen will be absorbed at the frequencies of the photons emitted by the hydrogen electrons, with the result that the dark lines are seen at those frequencies in the spectrogram.

Figure 6.6 Niels Bohr.

Although heavier atoms differ from hydrogen in that they have *two* types of particles in the nucleus—protons and neutrons (the neutron uncharged)—the absorption–emission phenomenon is the same as in hydrogen. When energy is absorbed, the electrons move outward; when the electrons fall inward, energy (radiation) is emitted. However, it is important to remember that, in these heavier atoms, each orbit has a maximum number of electrons it can hold; if the orbit below a given electron is filled, the electron cannot drop to that level.

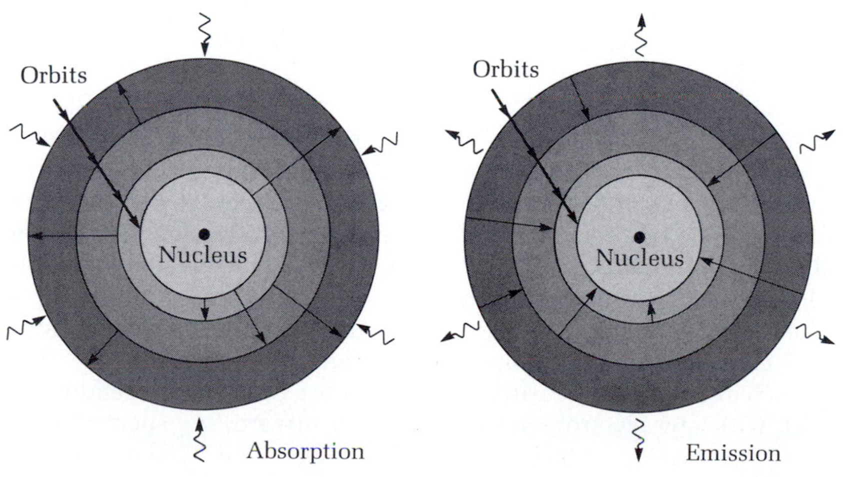

Figure 6.7 Schematic representation of absorption (left) and emission (right) of radiation by an atom. Straight unbroken arrows indicate movement of electrons; wavy arrows, energy being applied or radiated.

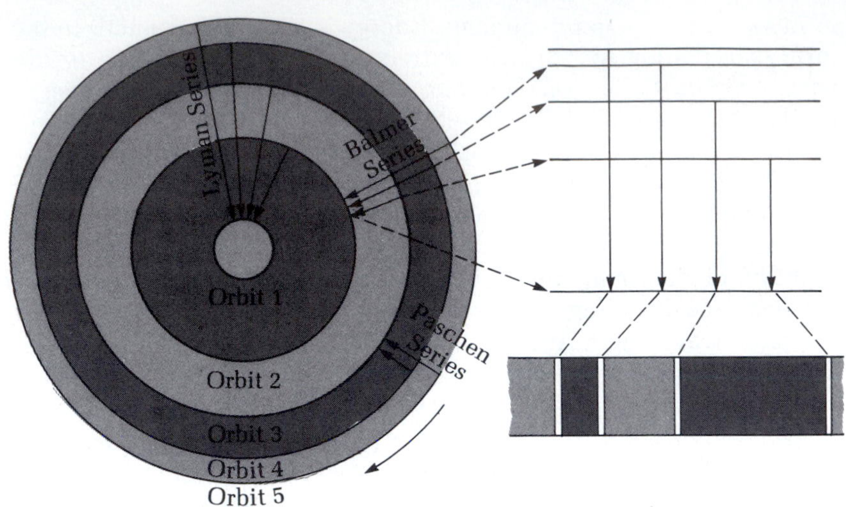

Figure 6.8 Process illustrated in Fig. 6.7 as it occurs in the hydrogen atom. Electrons dropping to Orbits 1, 2, and 3 give rise to the Lyman Balmer, and Paschen series, respectively. At right are shown the spectral lines of the Balmer series.

STELLAR SPECTRA

With these developments, beginning with Kirchhoff and Bunsen and given impetus and resolution by theorists like Bohr, a new door was open to astronomers. It was now possible to find out what stars were made of by identifying the lines in their spectra. One of the first to do this was Sir William Huggins, who looked at the spectrum of Aldebaran, the brightest star in the constellation Taurus. Soon astronomers were examining the spectra of all the brighter stars; most contained lines corresponding to hydrogen, but lines of other elements, such as calcium, iron, silicon, magnesium, and zinc, were also seen. (Figure 6.9 shows these lines in the spectrum of the sun.)

Although, as mentioned earlier, when an element is present in abundance its spectral lines are generally prominent, in the case of stars, other things affect the intensity of the lines. In the case of hydrogen lines, for example, sharper and more intense lines do not mean more hydrogen.

The absorption spectrum—in particular, the lines of the Balmer series near the visible region of the spectrum—of a sample of hydrogen depends on the number of electrons in the outer levels of the various atoms—and the lower the temperature, the fewer electrons there will be out here; most will be in the ground state. This means that a star with a cool surface, say about 3000°K, will show relatively weak hydrogen lines in its visual spectrum. If the temperature is about 10,000°K, however, these outer levels will be well populated and the lines will be intense. At still higher temperatures the electrons leave the atoms entirely (the atoms are then said to be ionized) and, with few electrons available to fall inward, the spectral lines

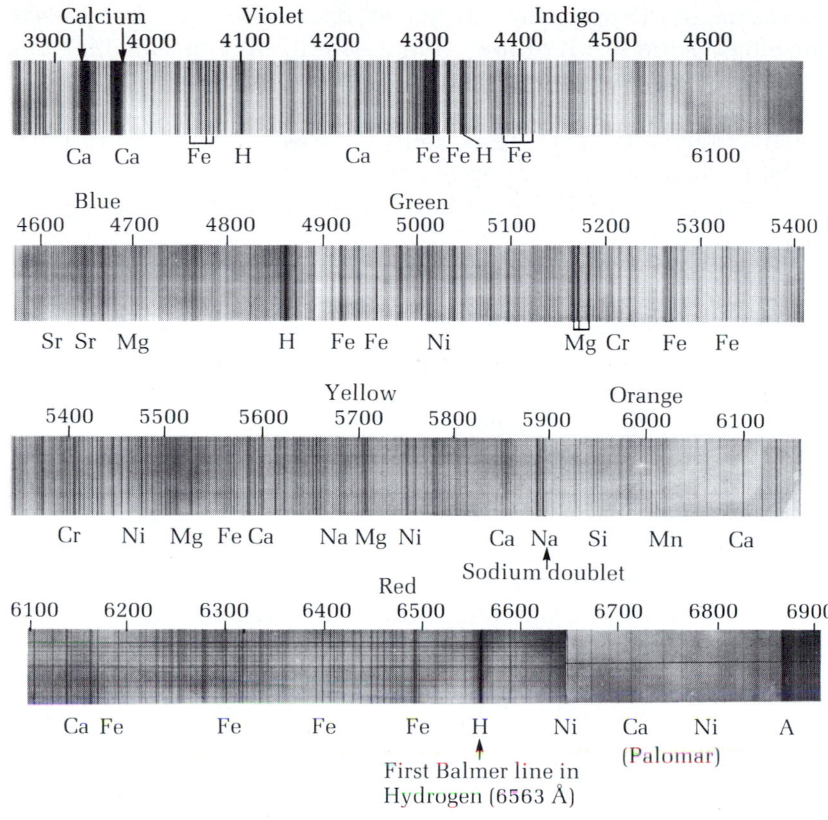

Figure 6.9 Spectrum of the sun, with dark (absorption) lines and, at bottom edge, the elements to which they correspond. Note also the first Balmer line in hydrogen (Fig. 6.8).

will again be weak. Thus the hydrogen lines are weak in the spectra of both cool stars and of exceedingly hot ones.

It turns out, however, that all stars have about the same general composition: They are composed mostly of hydrogen with some helium and traces of other elements, and consequently the astronomer is not usually concerned with composition when taking a spectrum. Other things that spectra tell (and that are discussed below) are of much more interest: whether the star is spinning, if it is approaching or receding from us, whether it has a magnetic field (and how intense), and many other things. Information from spectra is important even in relation to the planets; this subject is taken up in Part III.

The Doppler Effect

Another influence on spectra is the Doppler effect, which produces a shift (Fig. 6.10) of the lines in the spectrogram of a star that is approaching or receding from us. This effect, often associated with sound waves, is applicable to light waves as well. For example, the pitch (or frequency) of the whistle of a rapidly approaching train

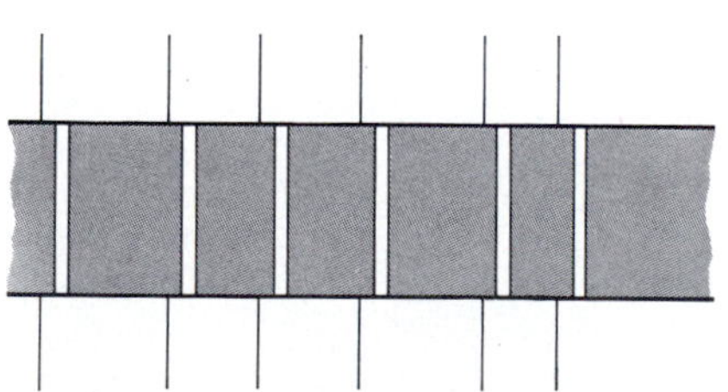

Figure 6.10 Spectrum with shifted lines. The positions of the spectral lines for the same source when stationary are shown above and below.

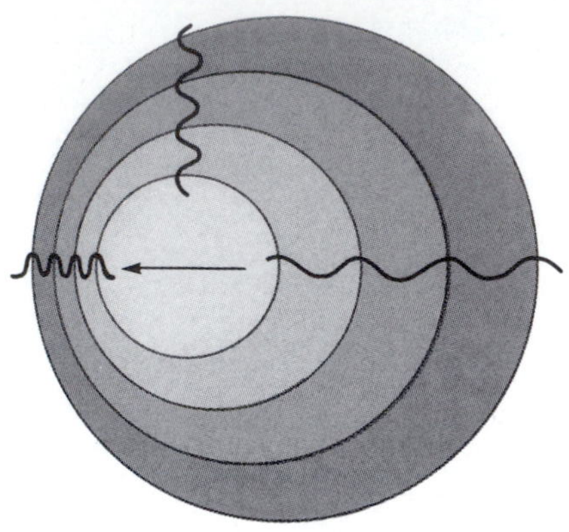

Figure 6.11 Illustration of the Doppler effect. Note change in wavelength in direction of motion. The source is considered to be at the center of the inner circle.

rises as the train approaches the hearer, then falls once it has passed. The effect is the result of the compression (train approaching) and, then, stretching out (train receding) of the sound waves (Fig. 6.11). In the case of light waves a much greater velocity is necessary if the effect is to be noticeable. Thus, since the spectral lines are at a particular frequency, and the Doppler effect changes this frequency, we would expect a shift of the lines. In fact, the greater the velocity, the greater the shift.

Note that the frequency is increased (wavelength decreased) if the source is approaching and is decreased (wavelength increased) if it is receding. As far as the spectral lines go, this means a shift toward the blue end of the spectrum for an approaching source and a shift toward the red end for a receding one. In the case of the spectrogram shown in Fig. 6.10, the star was receding.

Line Shape

We have seen that a star's spectrum can indicate the presence of certain elements, the surface temperature, and whether a star is approaching or receding from us. The widths of the spectral lines—some sharp and fine, others wide and diffuse—can also give us information. Since line width is a relatively crude measure, the astronomer scans the spectrum with an instrument called a microdensitometer, in which a fine beam of light is shone on the spectral photographic plate. Beneath the plate a photometer measures the amount of light (depending on the darkness of the line) that gets through. This is recorded as a tracing, giving the pattern of density of the spectral line.

A typical tracing shows the greatest intensity near the center of the line, with a gradual decrease at both sides. All lines have a *natural width,* which results because the orbital electrons do not have precise positions in their orbit. It is impossible for the line to be narrower than this natural width.

But a line is frequently much broader than its natural width. Astronomers distinguish three kinds of broadening. One, called *pressure broadening,* is a result of the fact that, though the atoms in a star have exceedingly high velocities, they usually do not go very far before they collide with other atoms. These collisions perturb (change slightly) the energy levels in the atom, producing photons that cover a range of frequencies, thereby broadening the spectral line. Since the collision rate depends on the density and temperature of the star, which in turn determines the "pressure," this broadening gives us a measure of the atmospheric pressure of the star.

A second type of broadening, *Doppler broadening,* depends primarily on the temperature of the star. One element in this phenomenon is the just-mentioned high velocities of atoms in a star. Some of these atoms, in their rapid random movement, are approaching us and some receding from us. (Although in most cases this will be neither *directly* away from us nor *directly* toward us, a component of the velocity will be.) According to the Doppler effect, just discussed, those that are approaching will undergo a slight blue shift, and those that are receding will undergo a slight red shift. The overall effect, because of the increase in the range of frequencies, will again be a broadening of the line.

A third type of broadening is the result of rotation. If the axis of rotation of a star is perpendicular to us (as in Fig. 6.12), the atoms to the left of the axis are approaching us, and those to the right are receding. This means that the spectral lines of the approaching atoms will be blue-shifted, and those of the receding atoms will be red-shifted. The maximum shift will occur for those on the equator; other positions above and below will contribute to the resulting range of frequencies. As in the previous two cases, the overall effect is a broadening of the spectral line.

Distinguishing among these broadening effects is often difficult; the key is the profile or shape of the line, which is characteristically different for the different types of broadening.

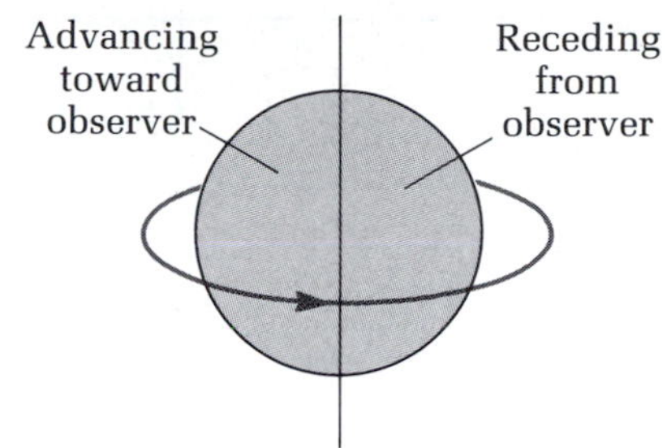

Figure 6.12 Rotation of star results in some of its matter advancing toward and some receding from observer, which produces (by the Doppler effect) a broadening of spectral lines.

The Zeeman Effect

The presence and magnitude of magnetic fields can also be determined from spectra. This was discovered in 1896 by Dutch physicist P. Zeeman, who showed that the spectral lines from atoms in a magnetic field split into a number of separate components (Fig. 6.13). Examining the effect further, he found he could increase the splitting by increasing the strength of the magnetic field. Thus, one could determine the magnitude of a magnetic field by measuring the splitting; this is now known as the *Zeeman effect*. Astronomers using it have determined the magnitude of magnetic fields associated with stars—in particular, our sun. The intense magnetic fields of sunspots (p. 144), for example, have been determined in this fashion.

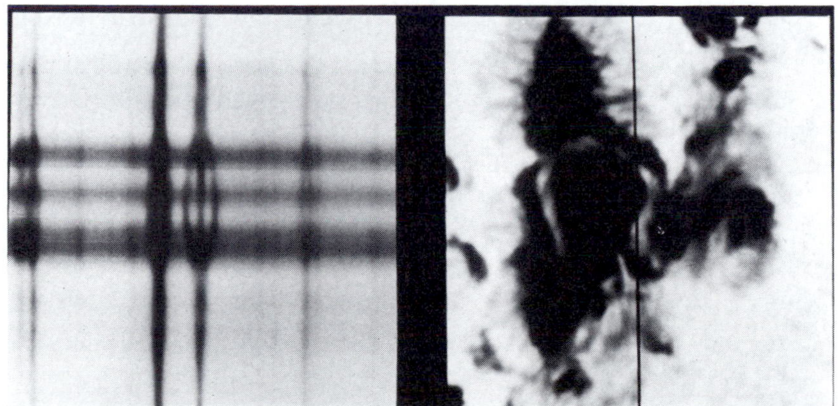

Figure 6.13 Splitting of spectral lines by magnetic field in the sun (Zeeman effect). Panel at right shows the position of the spectrograph slit (straight line) in relation to the sunspot whose magnetic field produced the effect. Note that splitting occurs where slit covers sunspot; otherwise the line is single.

SUMMARY

1. The continuous spectrum was first explained correctly by Newton: White light is composed of light of all colors and can be separated into those colors by use of a prism.
2. Fraunhofer, using a slit source, discovered dark lines in the continuous spectrum of the sun.
3. Bunsen and Kirchhoff discovered that each substance has its own characteristic spectrum.
4. There are three known types of spectra: continuous, bright-line, and dark-line.
5. Kirchhoff formulated three laws of spectral analysis: (a) Hot, dense gases, liquids, and solids emit continuous spectra. (b) Hot, rarefied gases have bright emission-line spectra characteristic of the gas. (c) Continuous-spectrum light passed through a cool gas shows absorption lines characteristic of the gas.
6. When a substance is heated, the intensity of radiation emitted depends on the wavelength, on the temperature, and to some degree on the type of material in the substance. To get around the problems associated with actual substances, scientists introduced the idea of the "black body," a perfect radiator and absorber.
7. In 1900 Planck derived a formula that explained the radiation curve (intensity of radiation versus wavelength) of a black body.
8. One of the earliest models of the atom was devised by Rutherford. At the center was a small, but heavy, positively charged nucleus and around it, in various orbits, negatively charged electrons.
9. Bohr later refined the model by restricting the orbits and allowing radiation emission and absorption to occur only when electrons moved between orbits. Using this model, he explained the spectrum of the hydrogen atom. Emission lines occur when electrons fall inward; absorption lines occur when they are moved outward.
10. Spectra give us considerable information about stars: atmospheric composition and pressure, magnetic characteristics, surface temperature and spin, and whether the star is approaching (blue-shifted) or receding (red-shifted).

REVIEW QUESTIONS

1. What is a Fraunhofer line?
2. What is the difference between continuous, bright-line, and dark-line spectra? Explain how each is produced.
3. Write Kirchhoff's rules in your own words.
4. Describe a representative spectroscope.
5. What is a black body?
6. Describe the form of a black-body radiation curve. What are plotted on the axes?
7. What was Planck's major contribution to the theory of radiation?
8. What is the major difference between Rutherford's and Bohr's model of the atom?
9. Describe the Balmer series in the spectrum of hydrogen and explain how it arises.

10. How do atoms differ in general? Compare a heavy one to a light one.
11. How can we determine the surface temperature of a star from its spectrum?
12. List some of the things that stellar spectra tell us.
13. Explain the Doppler effect.
14. Assume a star is spinning and its axis of rotation is perpendicular to us. Describe its spectral lines. What would happen to them were the axis pointed toward us?
15. Can we determine if a star has a magnetic field? If yes, how?

THOUGHT AND DISCUSSION QUESTIONS AND PROJECTS

1. At what point on the centigrade scale is the reading the same as on the Fahrenheit scale? (Use scale in Appendix.)
2. One star has the peak of its radiation curve at 3000 Å; another has its peak at 4000 Å. Which emits the most radiation in the visible region? Explain. (See Fig. 6.4.)
3. Ordinary stars generally exhibit absorption spectra, and nebulae exhibit bright-line emission spectra. What does this tell about stars and nebulae? Objects called quasars exhibit both types of spectra simultaneously. What does this tell about them?
4. *Project:* Observe the spectra of several different elements (e.g., mercury, oxygen, hydrogen). Note the colors of the lines. Compare the observed lines to those on a standard chart. Note the wavelengths.

FURTHER READING

Asimov, I. *Inside the Atom.* Abelard-Schuman, New York (1966).

Boorse, H. and Motz, L. *The World of the Atom.* Basic Books, New York (1966).

Feinberg, G. *What Is the World Made of?* Anchor/Doubleday, New York (1977).

Hoffman, B. *The Strange Story of the Atom.* Dover, New York (1959).

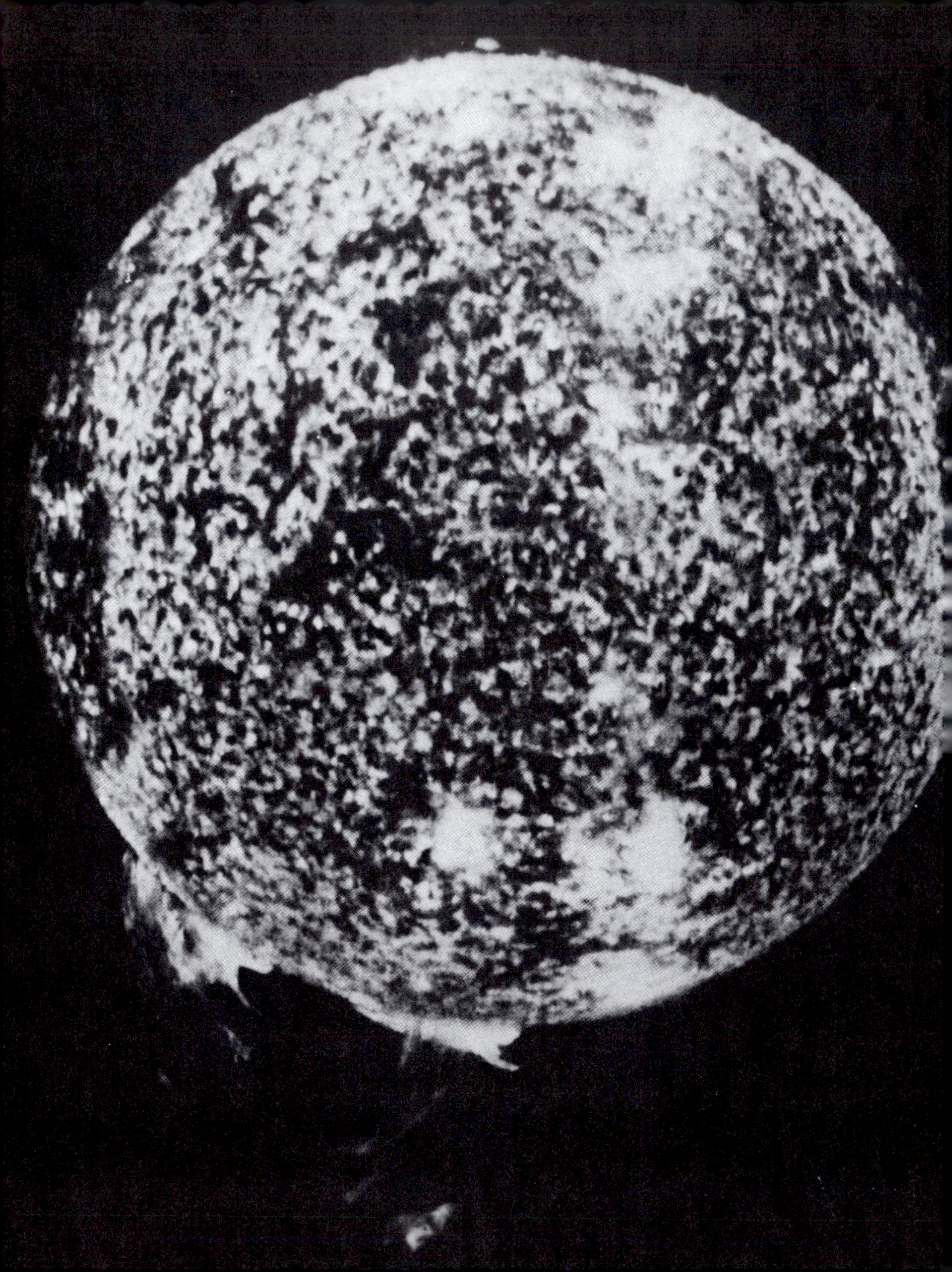

part 3

The Solar System

In the last few years our knowledge of the solar system has increased significantly. New things are being discovered almost daily. This knowledge explosion began with the dawn of the space age.

The launching of Sputnik was, of course, just the beginning. Flights to the moon and to many of the planets soon followed. The two major highlights of the last decade or so have been the first manned landing on the moon and the landings on Mars. Much was learned as a result of these truly significant and inspiring events. But, of course, there has been much more: trips to Mercury, Venus, and Jupiter, and beyond. Each has contributed to our knowledge.

In this part we will be looking at both what we have learned as a result of the space program, and what we have learned of the solar system using earth-based equipment. The amount of detail—names, physical features, distances, sizes, orbital periods—can be staggering, but most important is to extract from all this information a familiarity with major features and an appreciation of them.

Aim of the Chapter

To describe the size and composition of the solar system, and to show how it is likely to have formed.

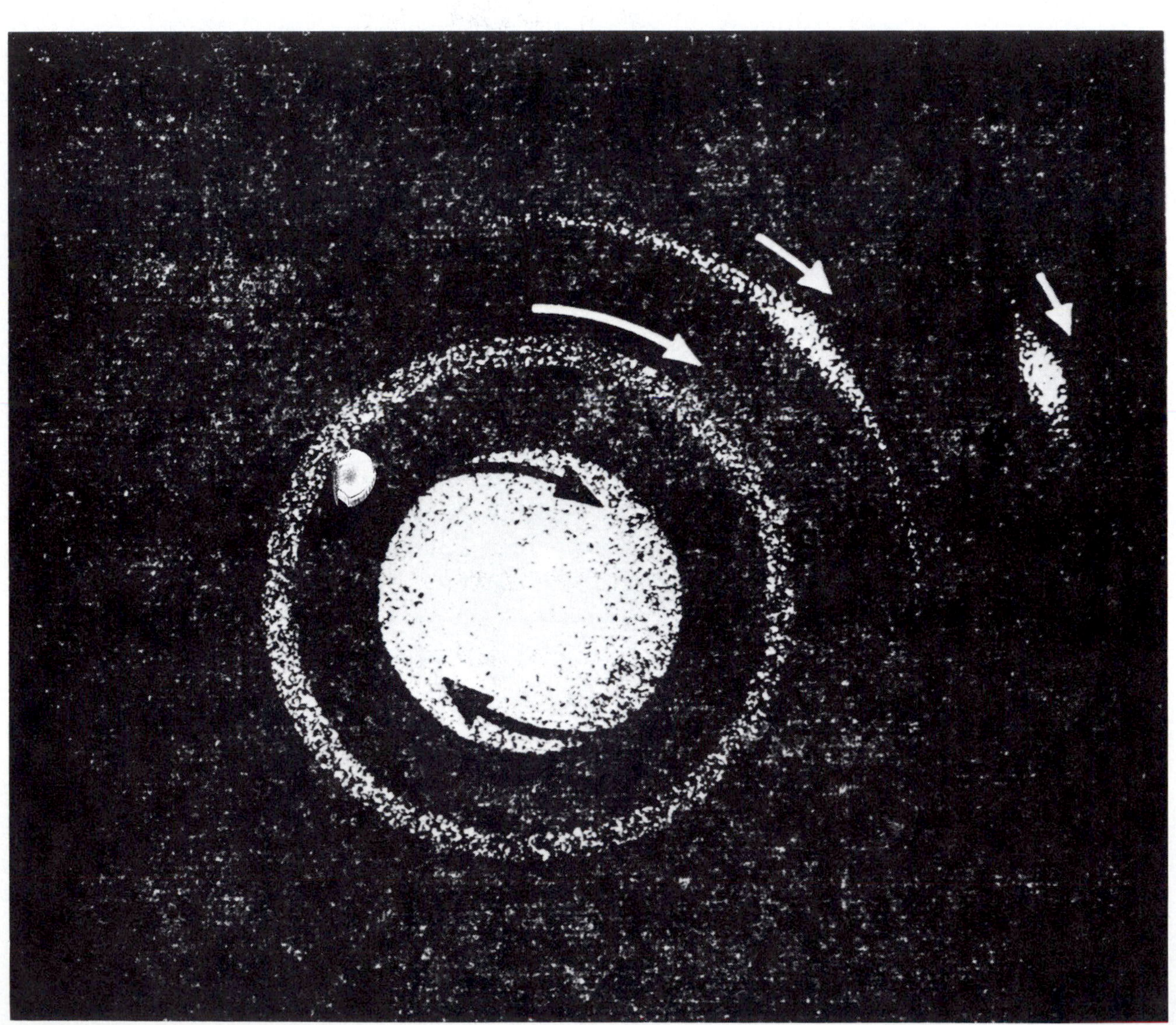

chapter 7

The Solar System

Our solar system is surprisingly regular. We know, for example, that all of the planets lie approximately in the plane of the sun's equator. We also know that all the planets revolve in nearly circular orbits and that all orbit in the same direction. Indeed, with the exceptions of Uranus and Venus, they even rotate in the same direction. Furthermore, their satellites (with a few exceptions) revolve in this same direction.

Is this a coincidence? It hardly seems likely. A more reasonable assumption would be that there was at one time some sort of connection between the planets and the sun, that perhaps they formed out of the same rotating gas cloud.

There are more regularities: The four inner planets are small, rocky (terrestrial-like), and fairly close to the sun; the four outer ones are large and composed primarily of light gases. (We are neglecting Pluto here because it obviously does not fit into the overall picture; many astronomers believe, in fact, that it is just an escaped satellite.) The orbits of the planets (Fig. 1.4) are spaced roughly according to a definite sequence, obtainable as follows: To each of the numbers 0, 3, 6, 12, 24, . . . (note that each number after 3 is double its predecessor) add 4 and divide by 10. The resulting series (.4, .7, 1.0, 1.6, 2.8, . . .) is the distance to each of the planets from the sun in astronomical units (A.U., the average distance from the sun to the earth). (Table 7–1 shows predicted and actual distances.) This was first noticed by Titius of Wittenberg, Germany, in 1766. J. E. Bode, director of the Berlin Observatory in the early nineteenth century, popularized the set of numbers, and it eventually became known as the Bode–Titius Law or simply Bode's Law.

Many astronomers believe that this arrangement is just a coincidence, but M. J. Ovenden of the University of British Columbia believes otherwise; he has shown that small perturbations acting over billions of years can change any original distribution of nature into a Bode's Law distribution.

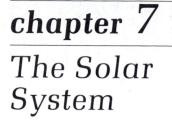

In the last few years our knowledge of the solar system has increased tremendously. This is mainly a result of space exploration

Table 7–1. Actual and Predicted Distances from the Sun to the Planets

Planet	Distance (A.U.) Bode–Titius Prediction	Observed
Mercury	0.4	0.39
Venus	0.7	0.72
Earth	1.0	1.00
Mars	1.6	1.52
Asteroid (Ceres)	2.8	2.65
Jupiter	5.2	5.20
Saturn	10.0	9.54
Uranus	19.6	19.19
Neptune	38.8	30.06
Pluto	77.2	39.44

programs, but significant results have also been obtained from earth-based studies. A summary of the basic statistics of the system is given in Table 7–2, and data on planetary satellites are summarized in Table 7–3.

One of the main objectives of the space program was to collect enough data to allow us to arrive at a theory describing the origin of the solar system—a *cosmogony*. Such a cosmogony would explain the regularities just discussed, but it would also explain differences in the general makeup of the planets, the origins of the various minor components of the solar system, (such as comets and meteorites), and many other things.

Table 7–2. Basic Planetary Data

	Mercury	Venus	Earth	Mars	Jupiter	Saturn	Uranus	Neptune	Pluto
Distance from sun									
Astronomical units (A.U.)	.39	.72	1.00	1.52	5.20	9.54	19.19	30.06	39.44
Millions of kilometers	58	108	150	228	778	1427	2869	4497	5900
Millions of miles	36	67	93	142	483	887	1783	2795	3670
Orbital period (yrs)	.24	.62	1.00	1.88	11.9	29.5	84.0	164.8	247.7
Average orbital speed (km/sec)	47.9	35.0	29.8	24.1	13.1	9.6	6.8	5.4	4.7
Diameter									
Earth = 1	.38	.95	1.00	.53	11.19	9.47	4.15	3.88	.20
Kilometers	4878	12,100	12,756	6795	142,800	120,800	52,900	49,500	3500
Mass (Earth = 1)	.055	.815	1.00	.11	317.8	95.2	14.6	17.2	.0025
Density or mass per unit volume (gm/cm^3)	5.4	5.2	5.5	3.9	1.3	.7	1.2	1.6	.7
Surface gravity (Earth = 1)	.37	.88	1.00	.38	2.64	1.15	1.06	1.41	.2

Table 7–3. Data on Planetary Satellites in the Solar System

Planet	Satellite	Approximate Diameter (km)	Distance from Planet (km)	Orbital Period (days)
Earth	Moon	3,475	384,410	27.3
Mars	Phobos	25	9,350	0.32
	Deimos	13	23,500	1.26
Jupiter*	Amalthea	150	181,000	0.42
	Io	3,680	422,000	1.77
	Europa	3,100	671,000	3.55
	Ganymede	5,300	1,071,000	7.16
	Callisto	4,840	1,884,000	16.69
	Leda	very small	11,090,000	239
	Himalia	very small	11,500,000	251
	Elara	very small	11,750,000	260
	Lysithea	very small	11,800,000	264
	Ananke	very small	21,000,000	625
	Carme	very small	23,000,000	714
	Pasiphae	very small	23,500,000	735
	Sinope	very small	23,700,000	758
Saturn**	Mimas	400	185,400	0.94
	Enceladus	600	238,200	1.37
	Tethys	1,000	294,800	1.89
	Dione	800	377,700	2.74
	Rhea	1,500	527,500	4.52
	Titan	5,118	1,223,000	15.94
	Hyperion	500	1,484,000	21.26
	Iapetus	1,300	3,563,000	79.32
	Phoebe	300	12,950,000	550.37
Uranus	Miranda	550	130,000	1.42
	Ariel	1,500	191,800	2.52
	Umbriel	1,000	267,300	4.14
	Titania	1,800	438,700	8.71
	Oberon	1,600	586,600	13.47
Neptune***	Triton	4,500	353,600	5.88
	Nereid	500	5,600,000	359.4
Pluto	Charon	1,500	18,000	6.4

*Recently discovered satellites of Jupiter have been designated as 1979J1, 1979J2, and 1979J3.

**Eight recently discovered satellites of Saturn have been designated as S10 through S17.

***A third moon of Neptune was discovered in 1981.

EARLY THEORIES OF ORIGIN

The first scientific theory of the origin of the solar system was proposed by French philosopher René Descartes in 1644. According to Descartes, space was initially filled with gas. In this gas there were gigantic vortices, or circular eddy currents, like the small ones seen in a swiftly moving stream. Over a long period these vortices evolved into the sun and the planets.

By present-day standards Descartes's theory was crude, to say the least; however, it was formulated before Newton introduced his

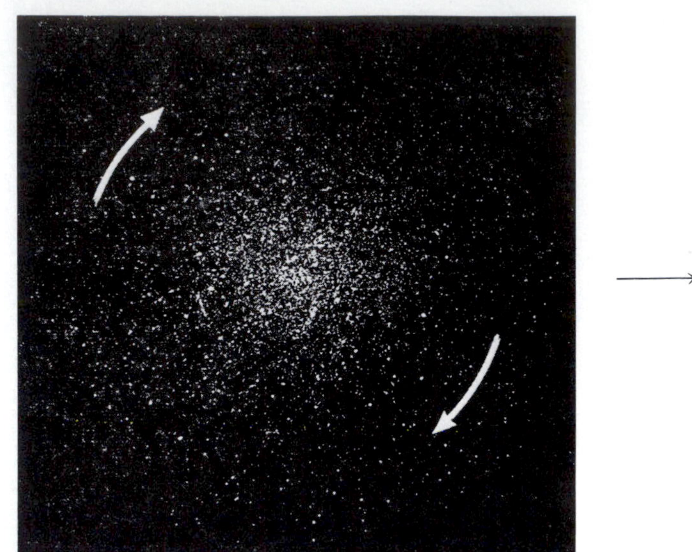

Figure 7.1 The Laplace cosmogony of 1796 (above) and the Chamberlin–Moulton cosmogony of 1901–5 (top of p. 121). According to the Laplace cosmogony, as the solar nebula rotated, rings of matter were formed that eventually condensed into the planets. The Chamberlin–Moulton cosmogony postulated that tongues of matter pulled from the sun by a passing star condensed into the planets.

theory of gravity, and so Descartes could only guess at how and why the evolution would occur.

Descartes's was the first of a type now referred to as *evolutionary theories,* which postulate a system that developed (or evolved). About a hundred years later French scientist G. L. de Buffon proposed a completely different cosmogony, which was of a type now referred to as a *catastrophe theory.* In theories of this type the solar system is assumed to have begun as a result of an unusual event, generally a collision. Buffon visualized a collision between the sun and a comet; presumably the comet hit the sun with such force that hot streams of gas were thrown out from it, and these cooled and condensed into the planets. Of course, we now know that the nucleus of a comet is only a few miles across and contains little mass, so that the collision would not be violent enough to spew out material. In Buffon's day, however, it was believed that comets had a mass about 10% that of the sun.

The next major contribution to cosmogony came in 1755, when German philosopher Immanuel Kant applied Newton's theory of gravity to the gas cloud theory. He argued that, as the gas cloud spun, its outer regions would gradually flatten into a disk, and "knots" in the gas would then presumably condense into planets. French mathematician P. S. Laplace was convinced that condensation would not occur from the "knots" and introduced his "nebular hypothesis" in 1796. Like Kant, he started with a spinning cloud of gas (Fig. 7.1). He showed that this cloud would not only flatten but also gradually contract under the force of gravity, and he believed that as it contracted it would leave rings of matter that would eventually condense into planets. Laplace was able to explain many of the properties of the solar system with his model, and, though now nearly 200 years old, his theory is quite similar to the one we accept today.

One major difficulty with the Laplace model is its failure to explain why, if most of the nebula condensed into the sun, it is not

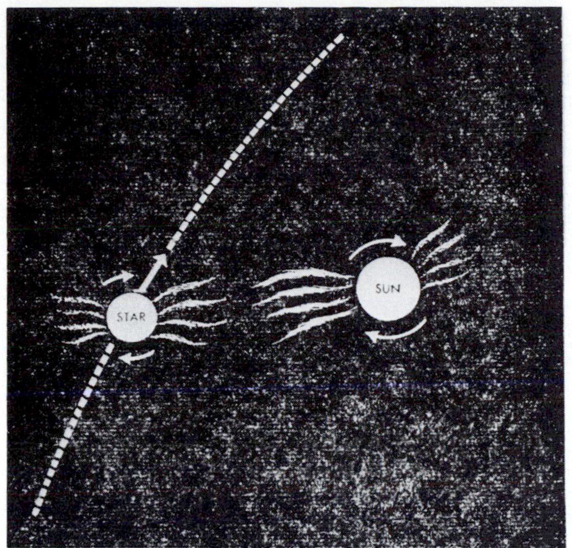

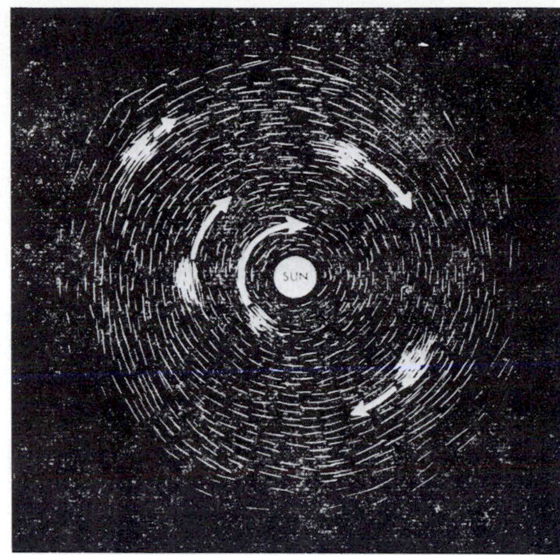

spinning much faster.[1] Moreover, C. Maxwell showed that, according to accepted physical principles, rings such as those visualized by Laplace would not condense into planets. Because of problems such as these, many astronomers began to turn back to catastrophe theories.

Shortly after the turn of the twentieth century, T. C. Chamberlin and F. R. Moulton proposed a variation on Buffon's theory—that a passing star pulled tongues of molten matter from the sun, and that these and similar ones drawn from the passing star collided, cooled, and condensed to form the planets (Fig. 7.1). But H. Jeffreys showed that colliding tongues of hot matter would not condense into planets; they would be too hot to condense and would dissipate instead. Jeffreys, along with Sir James Jeans, then tried to patch up the theory. They began by showing that a collision was not necessary: Tidal forces would cause a large bulge to form in the star, and the material from this bulge would be pulled into a large, ellipsoidal (cigar-shaped) filament in the region between the stars. This filament would then condense to produce the planets (Fig. 7.2). They pointed out that, as the stars passed one another, the filament would be given considerable angular momentum, which would account for the angular momentum of the planets. The shape of the filament could also account for the size of the planets—the largest would tend to be near the center of the filament.

But it soon became obvious that catastrophe theories were not the answer after all. It was shown that the probability of collision of two isolated stars is extremely small; there have probably been less than a dozen such stellar collisions in our galaxy since its birth.

[1] Momentum of spinning or revolving objects is called *angular* momentum and, like momentum along a straight line, is a product of mass and velocity, but also of the radius of the object. Thus, the (circular) velocity of a spinning ball having a 4-in. radius would double were the radius reduced to 2 in. (and the mass kept constant). In the case of the Laplace model, since *most* of the spinning matter composing the nebula *condensed* to form the sun, its angular momentum should be much greater than it is.

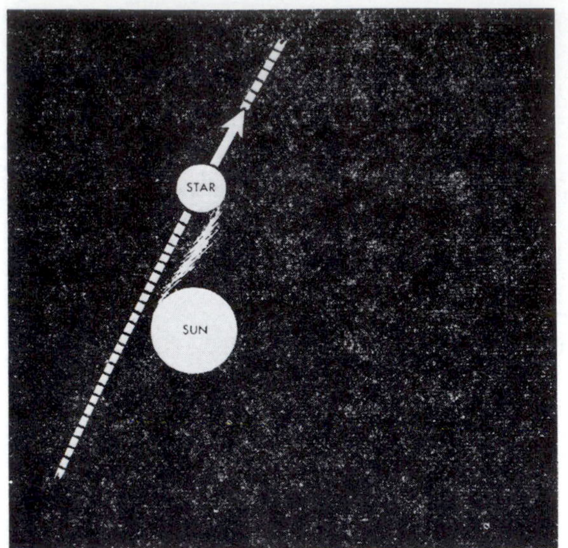

Figure 7.2 The Jeffreys–Jeans cosmogony of 1917 theorized that the sun was tidally disrupted by a passing star, and a long, cigar-shaped filament pulled out. The planets formed from the filament.

Thus, if systems like ours occurred only as a result of collisions, they would be exceedingly rare. Furthermore, such a theory could not explain the observed differences in the properties of the planets.

EARLY CONTEMPORARY MODELS

By the mid-1940s astronomers had once again returned to evolutionary theories. One of the major remaining shortcomings of these theories was their inability to explain how uniform rings of matter would condense to planets. C. F. von Weizsäcker got around this difficulty in 1944 by postulating considerable turbulence in the outer regions of the solar nebula and a resulting series of regularly spaced vortices, along whose boundaries the planets formed (Fig. 7.3).

G. Kuiper presented a variation on von Weizsäcker's cosmogony in 1951. Like von Weizsäcker, he assumed that vortices led to the planets, but his distribution of vortices was different: It was random (Fig. 7.4). He also introduced the idea that accretion (accumulation of matter) occurred around grains of solid matter that condensed out of the cloud.

The theory that we now accept is quite similar to Kuiper's, but it contains several additional features. For example, Kuiper's theory did not explain the difference in composition of the planets, nor did it resolve the abovementioned angular-momentum problem: Most of the angular momentum of the solar system is tied up in the planets. The key to the angular-momentum problem was found by H. Alfvèn. Realizing that the protosun (the sun in its early formative stages) probably had a strong magnetic field, he showed that, as the ions of the hot, ionized (charged) gas in the solar nebula moved outward, they would be forced to travel along the magnetic field lines. This

Figure 7.3 Von Weizsäcker's cosmogony of 1946, according to which vortices, or eddy currents, formed in a regular pattern throughout the solar nebula. Accretion into planets then took place along the boundary of the eddy zones.

would exert a dragging force ("magnetic breaking") on the protosun that would gradually slow its spin, at the same time speeding the nebula's rotation. In effect, there ould be a transfer of momentum from the protosun to the nebula surrounding it.

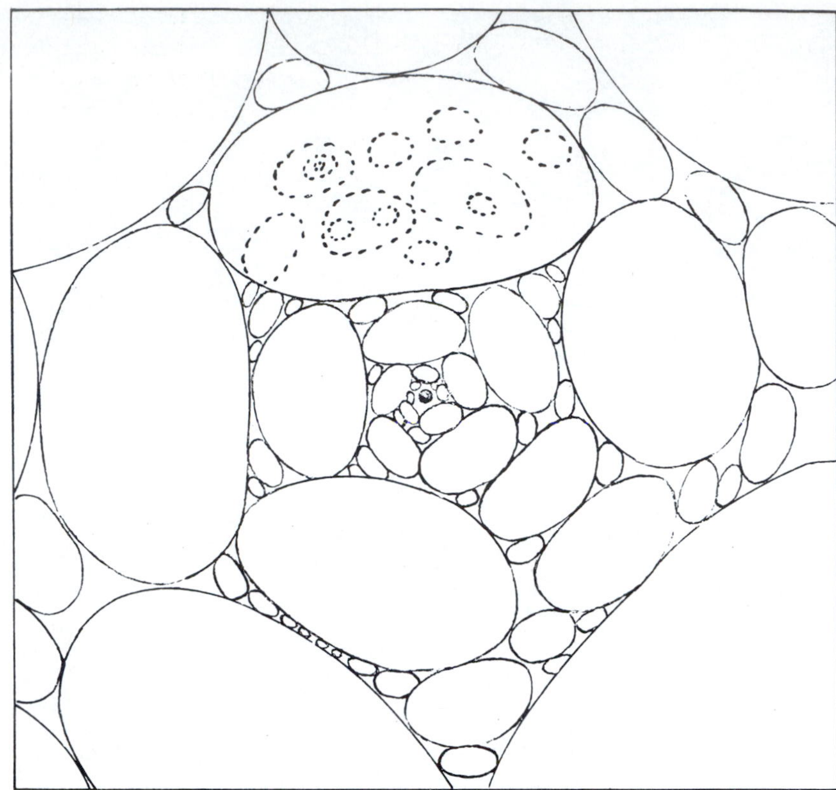

Figure 7.4 Kuiper's cosmogony of 1951 is similar to von Weizsäcker's (Fig. 7.3) but postulates that the vortices formed at random.

CONTEMPORARY MODELS

As mentioned earlier, the cosmogony that we now accept assumes that the solar system condensed from a large gas cloud. In most of the early models it was assumed that the gas cloud had a mass only slightly greater than that of the sun, and that most of the cloud went into the making of the sun. It is difficult, however, to see how the abovementioned transfer of angular momentum could take place in the early stages of the solar system if this was the case. To get around this difficulty, we now assume that the original gas cloud had a mass of about 3 $M_\odot$ (where $M_\odot$ is the mass of the sun), and that two thirds of this mass went into the making of the planets and about one third into the sun. (About 97% of the present mass of the solar system resides in the sun. How most of the planet mass was lost along the way will be discussed later.)

As the gas cloud condensed and, therefore, spun faster and faster, it flattened and spread. (The spin prevented collapse perpendicular to the rotation axis.) Temperatures in the center of the cloud began to increase as pressure built up in this region.

The differences in composition of the planets are attributable mainly to the differences in temperature in the different parts of the solar nebula. Soon after the system formed, temperatures through most of the solar nebula were high and everything was in a gaseous state. But in time the outer regions began to cool. Near the protosun they were perhaps 2000°K at this stage, but they dropped off rapidly from here outward, approaching absolute zero near the edge of the nebula. One of the major reasons for this was that the nebula itself blocked radiation to the outer parts from the protosun. There was, then, a *gradient* of temperatures throughout the nebula. This gradient, and the fact that most materials were originally in a gaseous state, can account for the distribution of materials that condensed out of the nebula at various distances from the protosun.

The temperatures in the inner region, where the planet Mercury now is, were so high that only elements such as iron, nickel, aluminum, and silicon compounds (of magnesium) could condense. Further out, near the present position of Earth, temperatures were cooler, and silicates of iron and magnesium, along with various oxides, condensed. Still further outward, in the region near Jupiter, carbon, nitrogen, oxygen, and water condensed and formed various kinds of ice. Methane and ammonia also condensed in this region and beyond.

The various elements condensed from the nebula as small grains (Fig. 7.5). Eventually, as a result of the forces within the nebula, these grains formed a sheet along the midplane of the nebula, no doubt like a larger version of the rings of Saturn, but with the difference that this ring was immersed in a dense cloud of hydrogen and helium.

As the grains whirled around the protosun, they occasionally collided and stuck together, forming even larger grains. Gravitational instabilities (slight fluctuations in density) soon developed in the disk, however, and it began to break up. Clumping and aggregation of the small particles continued to increase their size. Eventually some were a few kilometers across, then a few hundred kilometers. Formations at this stage, when the solar system was about 80 million years old, are referred to as *planetesimals*. (Fig. 7.5).

The planetesimals then began to collide and coalesce. Although they had high orbital velocities, their velocities relative to each other were low, and they tended to coalesce when they collided (rather than smash apart). Within a million years or so the first *protoplanets* appeared. Because the composition of the planetesimals varied with distance from the protosun, the composition of the protoplanets varied correspondingly.

The four inner protoplanets were relatively small and had rocky surfaces. Beyond them were four giants, composed mostly of ices of various types, though some may also have had small rocky cores. Because they were so massive, they attracted large atmospheres of hydrogen and helium.

At this stage water ice condensed in a region well beyond the present position of the earth—out near Jupiter. How, then, did the earth end up with its abundant water? It is likely that some was trapped in the minerals that condensed to form the earth, but the bulk of the water must have been transferred in from the region around

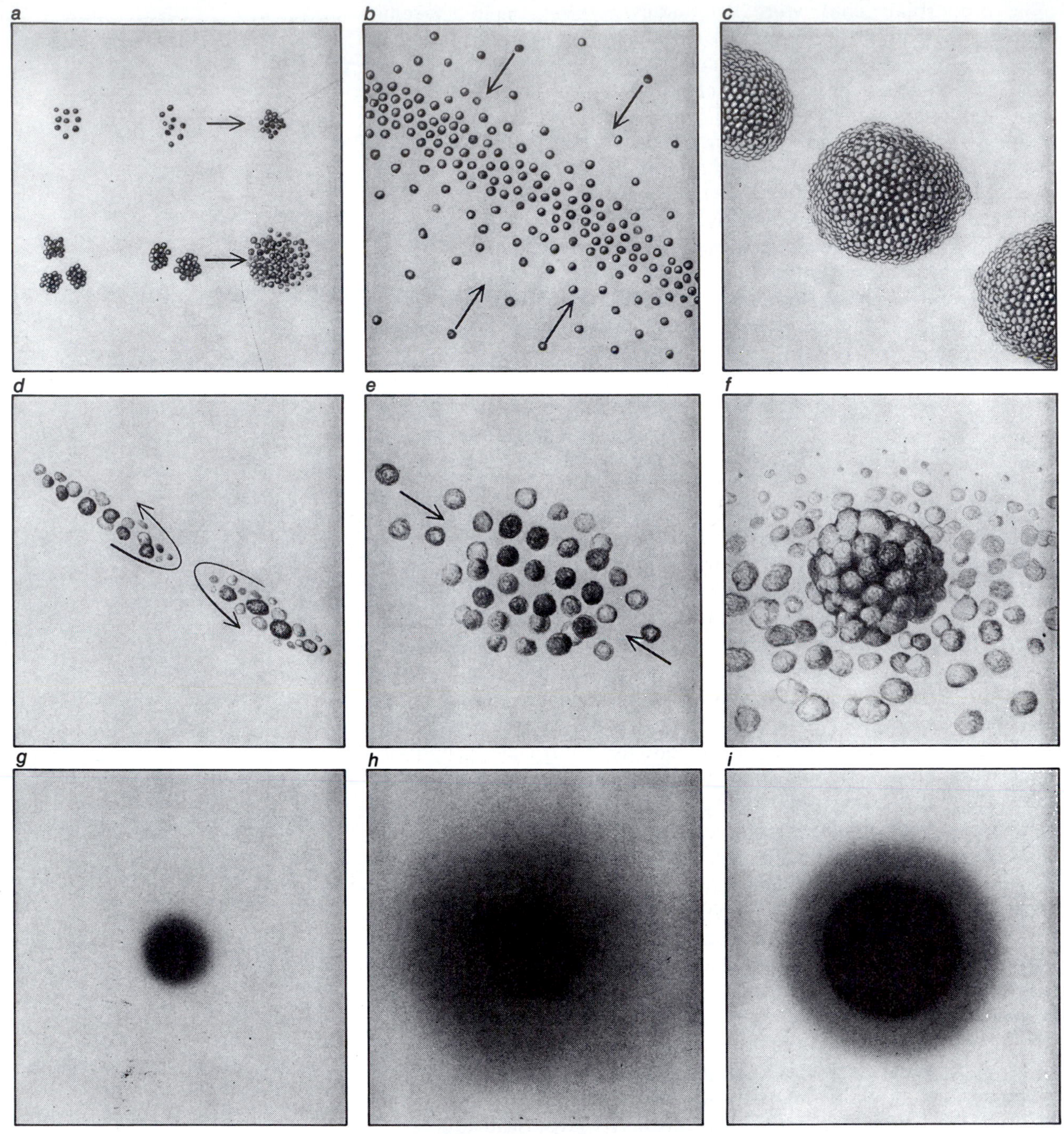

Figure 7.5 Formation of the planets according to current cosmogony: *(A)* Interstellar grains collide and stick together; *(B)* clumps of grains fall to midplane of the nebula; *(C)* further collisions generate planetesimals; *(D–F)* planetesimals collide and coalesce to form protoplanets; and *(G–I)* further evolution of the protoplanets occurs.

Jupiter. The mechanism of this appears to have been as follows: The gravitational forces in this region seem to have been different from those in most other regions of the solar system. The strong gravitational forces of the sun and that of massive Jupiter coupled to create forces that tended to deflect many of the planetesimals in this region into other parts of the solar system. Many of these planetesimals struck the newly formed surface of the earth, bringing water to it.

In 1978 the Pioneer Venus probe revealed approximately one hundred times greater concentrations of argon in the Venusian atmosphere than in the earth's. But according to the cosmogony we are discussing, there should be—because of the sharp temperature gradient outward from the sun—more argon in the earth's atmosphere than in that of Venus. The explanation may be that the incorporation of gases into the nebula occurred in more than one step. In particular, the temperature gradient may not have been the only thing that controlled the amount of gas incorporated at various distances from the protosun. It has been suggested that the higher pressures near the orbit of Venus may have forced more argon into the dust grains—those that eventually coalesced to produce Venus—in this region. Were this the case, Mars would have less argon than Earth—and it does.

The atmosphere of the earth at this stage was composed mainly of hydrogen and helium, as was the entire region around the earth. Nuclear reactions were initiated at this time in the protosun, changing it into a star (our sun) and at the same time releasing a violent wave that, in the case of the sun, is called the solar gale or the T Tauri wind (Chapter 16). As this gale rushed through the inner solar system, it blew the helium and most of the hydrogen from the atmospheres of most of the inner planets and from the region around them. However, it was not strong enough to blow the hydrogen and helium from the outer planets; more massive (and consequently with stronger gravitational fields) and further away, they still retain their large atmospheres today.

Looking back over our model, we see that it does indeed explain most of the features of our solar system: shape, distribution of angular momentum, and so on. It also explains the two basic groups of planets and, in particular, the differences within the groups. According to our model, for example, iron and nickel condensed mainly in the inner region of the solar system, silicates generally further out, which means that the innermost planets would have the largest (iron–nickel) cores (these heavier elements tending to sink toward the center), and those farther out would have more silicates and lighter elements, and thus in general larger mantles. And indeed this is the case: Mercury has a larger percentage of its volume taken up by its core than Earth does; Mars, on the other hand, has a smaller percentage of its volume taken up by its core. In general, the metallic cores do decrease in size as we move outward through the planets, and the mantles do increase in size relative to the planetary volume.

Figure 7.6 Stages in the development of the solar system: *Stage 1 (A, B)*: The solar nebula gradually flattens as a result of its spin. *Stage 2 (C)*: A ring system forms; grains begin to condense. *Stage 3 (D)*: Grains collide and planetisimals form. Collisions of planetisimals give protoplanets. The protosun develops. *Stage 4 (E and F)*: Protoplanets evolve into planets; protosun evolves into sun.

Summary of Stages (Fig. 7.6)

Stage 1 (Almost 5 billion years ago)

A giant cloud of hydrogen and helium, with small amounts of other elements present, rotates slowly and condenses. As it condenses, its central region becomes hotter and its spin rate increases.

Stage 2 (From 50–60 million years after Stage 1 to approximately 80 million years after)

As the system begins to take shape, most of the angular momentum is transferred outward via magnetic breaking. Grains then begin to condense as the nebula cools. The grains collide and coalesce, and a gigantic ring system forms in the outer parts of the gas cloud.

Stage 3 (A few million years duration)

Grains continue to collide and grow until, finally, planetesimals form. Collisions of the planetesimals then create protoplanets, which, in turn, evolve into planets. A protosun forms (at the center), which eventually evolves into the sun.

Stage 4 (About 4.6 billion years ago)

Nuclear reactions are triggered in the protosun, causing it to become a star (our sun), and the resulting solar gale blows much of the atmospheres from most of the inner planets but leaves the outer ones unaffected.

SUMMARY

1. There are many regularities in the solar system. One, the spacing of the planets, is formulated in the Bode–Titius Law.
2. A cosmogony is a theory of the origin of the universe or of its parts as, in this chapter, the solar system. There are two basic types of cosmogonies: *evolutionary* and *catastrophic*.
3. The earliest cosmogony, that proposed by Descartes, was of the evolutionary type. He assumed that the solar system (and universe) began as a condensation of a large gas cloud. Kant later applied Newton's theory to Descartes's ideas.
4. Laplace, in his nebular hypothesis, introduced in 1796, assumed a rotating nebula that left rings of matter, which condensed into the planets.

5. The first catastrophe theory, Buffon's, assumed a comet collision with the sun. Chamberlin and Moulton later introduced a second theory of this type. Theirs assumed that tongues of matter, pulled from the sun and another star when a near-collision occurred, condensed to form the planets. Jeffreys and Jeans modified the theory.
6. Von Weizsäcker, in a return to evolutionary theories in the late 1940s, introduced vortices. Kuiper presented a more successful variation on von Weizsäcker's theory in 1951.
7. H. Alfvèn showed that the angular momentum of the protosun was probably transferred outward via its magnetic field lines.
8. The difference in composition of the planets is due to a gradient of temperatures throughout the solar nebula. Different elements condensed at different distances from the sun.
9. Planet formation occurred as follows: Grains collided and stuck together to form planetesimals. Planetesimals collided and coalesced to form protoplanets. Protoplanets condensed to planets.
10. A solar gale, triggered when nuclear reactions began in the core of the sun, swept away the hydrogen–helium atmospheres of the planets of the inner solar system.

REVIEW QUESTIONS

1. List the various regular features of the solar system. Are there exceptions?
2. What is the Bode–Titius Law? Can it be derived from physical principles?
3. What sorts of things should a satisfactory cosmogony explain?
4. What was the major difficulty with Descartes's cosmogony?
5. Explain the difference between a catastrophic cosmogony and an evolutionary one.
6. What was the major difference between the Moulton–Chamberlin theory and the Jeffrey–Jeans theory?
7. Briefly outline Kuiper's cosmogony. Could his theory explain the difference in composition among the planets?
8. What was the significance of the gradient of temperatures in the solar nebula? What happened as a result of it?
9. What is the difference between a planetesimal and a protoplanet?
10. Briefly outline how the planets formed from grains.
11. Where in the solar nebula did water condense? What is the significance of this in relation to the earth?
12. What was the importance of the solar gale?

THOUGHT AND DISCUSSION QUESTIONS

1. What features of the current model of the solar system's origin would have to be modified if the planets did not orbit in a single plane?
2. On the basis of what we know about the formation of our solar system, what can we say about the prevalence of systems of this type throughout the universe? How would this view be modified if we accepted the Jeffrey–Jeans theory?
3. Extend the arguments in the present chapter to the formation of a binary system (two stars) plus an array of planets.

FURTHER READING

Cameron, A. G. W. "The Origin and Evolution of the Solar System," *Scientific American,* p. 32 (September 1975).

Hartmann, W. K. "In the Beginning," *Astronomy,* p. 6 (January 1976).

Williams, I. P. *The Origins of the Planets.* Crane-Russak, New York (1977).

Aim of the Chapter

To describe the sun and its importance and influences.

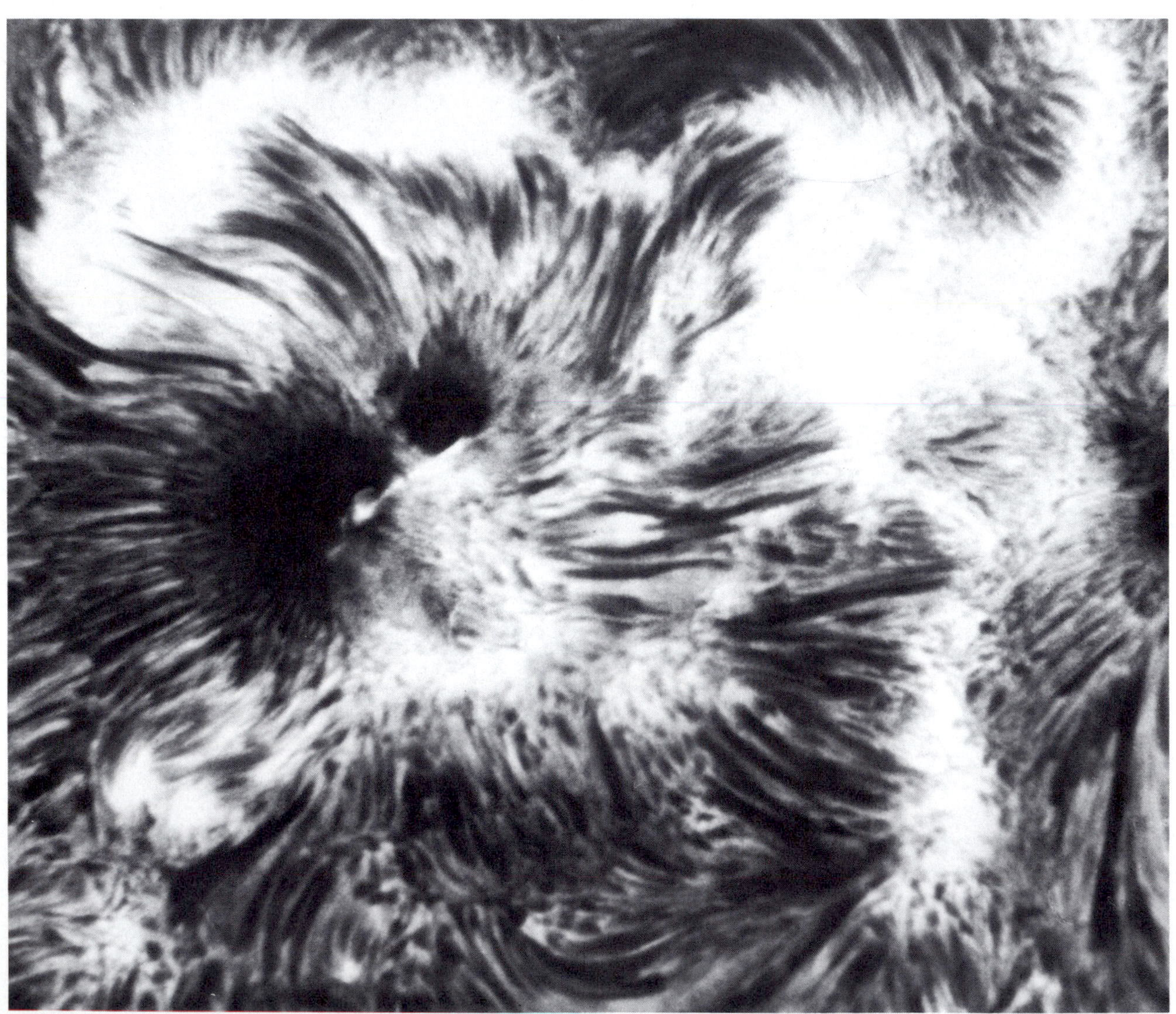

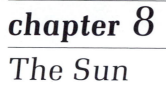

chapter 8
The Sun

Of all the objects beyond the earth none is more important to us than our sun. It supplies us with warmth and light and is indirectly responsible for meeting most of our other needs. Although it is an ordinary, run-of-the-mill star, similar to millions of others in the universe, to life on Earth it is indeed the perfect star—its size and brightness, and even its color, critical to our existence. A change in any of these features or a slight change in our average distance from it would be disastrous.

THE SUN: DIMENSIONS AND COMPOSITION

Average distance from Earth to the sun is 150 million km (93 million mi). There is a slight variation in this distance because of the ellipticity of the earth's orbit (Chapter 2), so that at certain times of the year it is 300,000 km closer than its average distance; at other times it is 200,000 km more distant. It has a diameter of approximately 1,400,000 km (864,000 mi) and a volume 1 million times that of the earth. However, because its average density (about 1.4 gm/cm^3) is much less than that of Earth, its mass is considerably less than 1 million $M_{\oplus}$ ($\oplus$ is the symbol for Earth)—about 330,000 $M_{\oplus}$.

The sun is made up mostly of hydrogen and helium along with small amounts of other elements (Table 8–1). In its outer regions it is about 75% hydrogen and 24% helium by weight; the rest of the elements compose approximately 1%. Although the atoms of the heavy elements are few in number, they contribute significant mass per atom compared to hydrogen. If instead of calculating mass we count atoms, we find more than 90% of the atoms are hydrogen, more than 8% helium, and less than 1% heavier elements. The outer region of the sun has changed very little since it was formed. In essence, it is composed of the same mixture of gases that existed in the solar nebula

Table 8–1. Chemical Composition of the Sun

Percentage of the sun's mass							
Hydrogen	Helium	Oxygen	Carbon	Nitrogen	Neon	Nickel	Silicon, Sulfur, Iron
78.4	19.8	0.8	0.3	0.2	0.2	0.2	0.12

(Chapter 7). Deep beneath the surface, on the other hand, near the center, hydrogen is being changed into helium and the hydrogen–helium ratio is quite different. At present this region consists of about 35% hydrogen and 65% helium by mass.

To the unaided eye the sun looks like a shiny, smooth, tranquil disk; it is in reality an object of tremendous turmoil, its surface a seething inferno of exploding gases. Its core temperature is so high (about 15,000,000°K) that the entire body is gaseous. The luminous surface of the sun is a thin layer (several hundred miles thick) called the photosphere (Fig. 8.1). Above the photosphere is the atmosphere of the sun. (More properly, the photosphere is the bottom layer of the atmosphere.) The region directly above the photosphere is a transparent one called the chromosphere, and above it, extending outward for millions of kilometers, is a pearly-white, tenuous halo called the corona (Fig. 8.2, color).

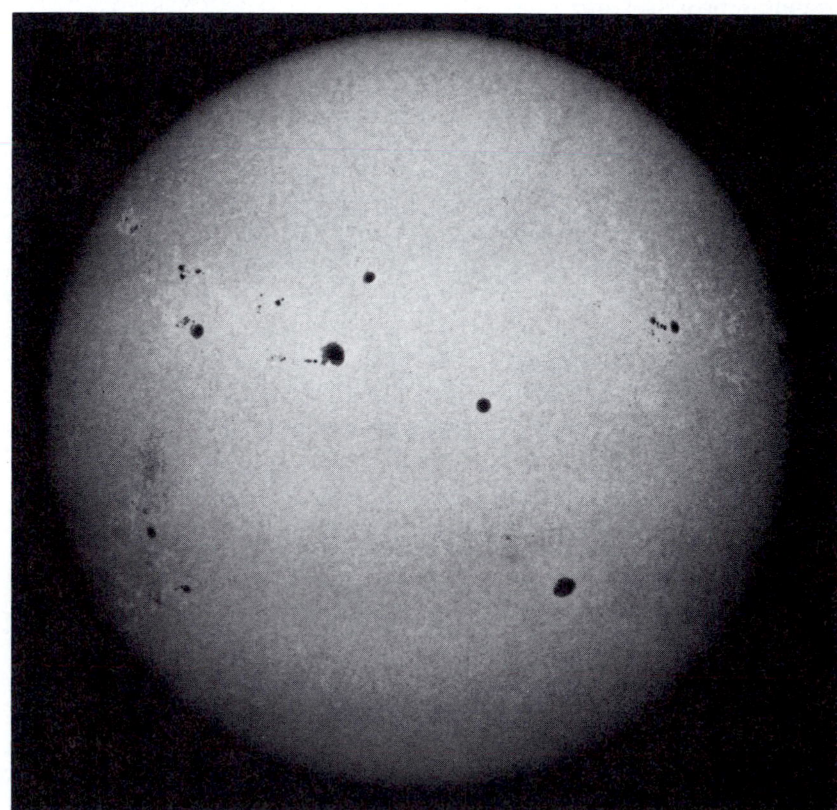

Figure 8.1 Photosphere of the sun; the dark regions are sunspots.

From the photosphere inward temperatures and pressures increase rapidly, as do densities. In the outer region of the sun the density is exceedingly low (about 10^{-7} gm/cm^3); in the core it is approximately 158 gm/cm^3. The sun's energy is generated in the core via thermonuclear reactions. The radius of the core is approximately 200,000 km, just over one quarter of the sun's overall radius.

ENERGY OF THE SUN

A fuller discussion of energy generation in stars (including our sun) will be given in Chapter 15. However, it is useful at this point to give a brief overview.

One of the major scientific dilemmas around the turn of the century was the source of the sun's energy. How could it produce so much energy, day after day, year after year, and still go on shining at the same rate? One of the first to consider this difficulty was H. von Helmholtz, who showed that the energy could not possibly be generated by chemical combustion; given its known mass, the sun would have burnt itself out long ago had this been the case. Helmholtz and Lord Kelvin finally came to the conclusion that the energy had to be generated as a result of the sun's contraction. According to their hypothesis, as the material of the sun was slowly squeezed to a smaller volume as a result of self-gravity, it heated; in other words gravitational energy changed to heat energy. But calculations showed that even if this was the case, the sun could have lasted no more than 30 million years; yet, according to geological evidence, the earth was much older than this. The first step toward a solution was the formulation by Einstein in 1905 of his theory of relativity (Chapter 19), which included his now-famous equation of mass with energy ($E = mc^2$, where E represents energy, m, mass, and c, the speed of light). However, because a clear understanding of nuclear reactions was lacking, it was not until the 1930s that physicists understood that the sun's energy was produced via a cycle of nuclear reactions in which a small amount of matter was converted into a huge amount of energy. In these reactions hydrogen is converted into helium in a process we now refer to as nuclear fusion—the same process that goes on in the explosion of a hydrogen bomb.

In the billions of years since the formation of our solar system, only about 5% of the total mass of the hydrogen in the sun has been converted to helium. The conversion takes place in the core, a region sometimes referred to as the thermonuclear furnace. Most of the energy released in this region is in the form of high-energy gamma rays, which gradually make their way to the surface. (Gradually may, in fact, be an understatement; it may take over a million years for the energy associated with a photon released near the core to be delivered to the surface.) At the surface most of this energy will be in the visible and infrared regions of the spectrum. (The radiation spectrum of the sun peaks in the visible region.)

THE PHOTOSPHERE

We have noted that the sun is completely gaseous and that what we see with the naked eye[1] is the photosphere. (All the light we see comes from this very thin layer, which is only about 400 km thick.) That the edge of the sun is so sharply defined even though it is a ball of gas is explainable partly on the basis of the opacity of a gas. All gases, including that which makes up the atmosphere of the sun, diminish light passing through them, no matter how slightly; hence a sufficiently large volume of gas will be opaque. Thus there is an amount of any gas beyond which we cannot see. In the case of the sun we see through the outer layers but eventually reach a point where we can see no farther, if we are looking at the center of the sun (see below). (This region is defined as the *base* of the photosphere—its hottest, and therefore brightest, part.)

When we look at a point very close to the edge of the sun's disk, we can see completely through the atmosphere (Fig. 8.3); a very slight inward shift of our line of sight, and we cannot. The angle subtended by the transition is only about one second of arc, which is beyond the ability of the eye to detect. Thus, while there *is* a transition from transparency to opaqueness, the unaided eye cannot see it; hence the seemingly sharply defined edge.

However, there is a definite change in appearance—a darkening—as we move from the center toward the edge, or limb, of the sun. This limb darkening can also be understood by examining Fig. 8.3: the closer toward the edge one looks, the less directly one sees the base of the photosphere. Near the edge, since we are looking in at an angle, we are seeing layers that lie above the base of the photosphere. These layers are cooler than the base layer and, hence, give off less radiation and appear slightly darker to us. Limb darkening can, in fact, give us a measure of the temperature of these layers.

The temperature of the photosphere ranges from about 4000°K to 8000°K, with an average of about 5700°K. With the exception of an occasional visible sunspot it appears uniform and structureless to the naked eye. However, a close examination with a telescope reveals that it is actually covered with small granules (Fig. 8.4). These granules, approximately 1000 km across, persist for only a few minutes, but as one disappears it is replaced by another.

The spectra of these granules tell us that they are in continual motion. The central bright region is rising and the outer dark regions descending; they are, in many ways, similar to the small regions that rise and fall in boiling porridge or oatmeal. A similar phenomenon is also frequently seen in the clouds—particularly when viewed from above. All these phenomena (including the granules) have the same cause—convection currents, or movements in gases or fluids that result from differences in temperature or density in different parts of the substance. Such currents are a factor in weather: When masses of air become heated, they rise and are replaced by cooler air; the currents caused by this are convection currents.

[1] A reminder: When viewing the sun, *always use an adequate filter!*

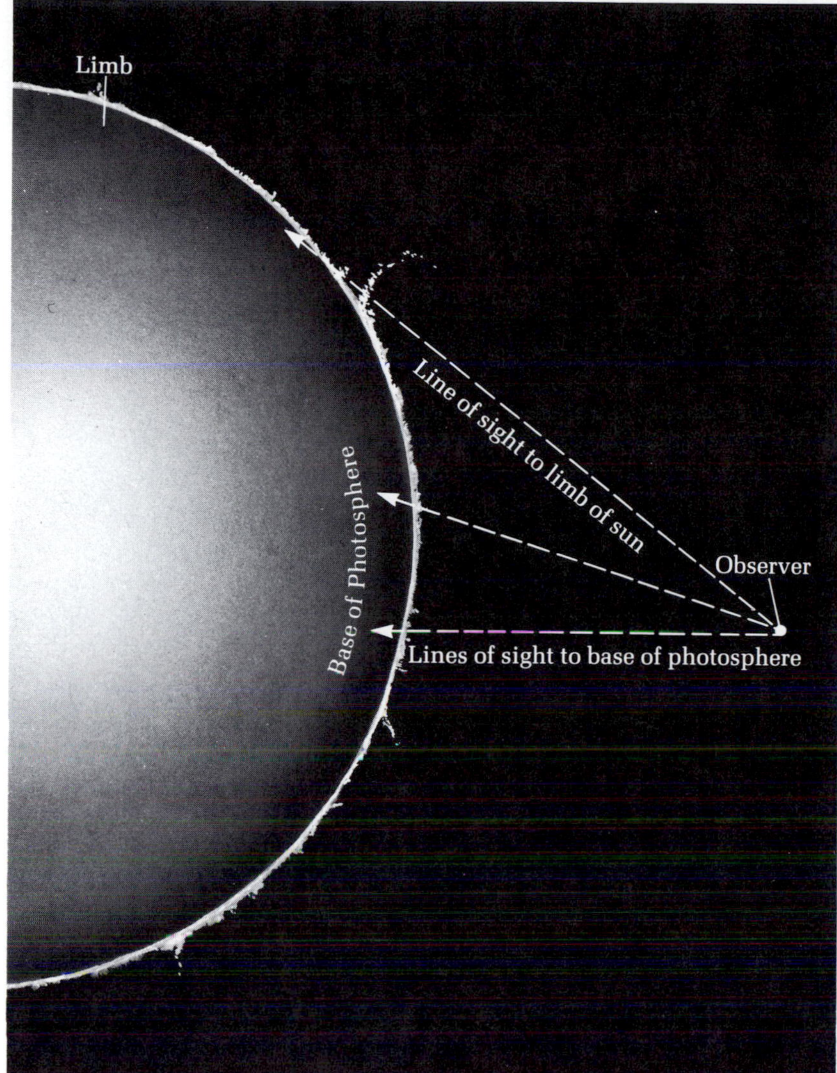

Figure 8.3 Mechanism of limb darkening. Base of photosphere is hotter than upper layers, hence is brighter.

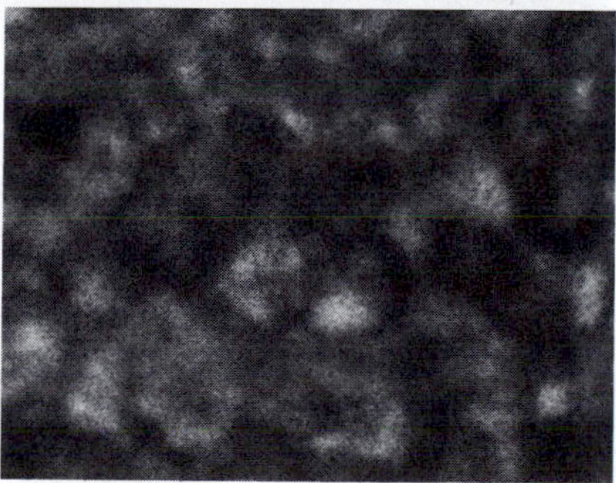

Figure 8.4 Granulation of the photosphere.

In the case of the sun, the gases near the core are extremely hot and the atoms composing them, as a result, are extensively ionized; thus, energy transfer from this region is almost entirely via radiation. With movement outward, toward the surface, however, the gases

Figure 8.5 Mechanism of granulation (top), and cross section of sun showing principal regions (bottom). Note zone of convection.

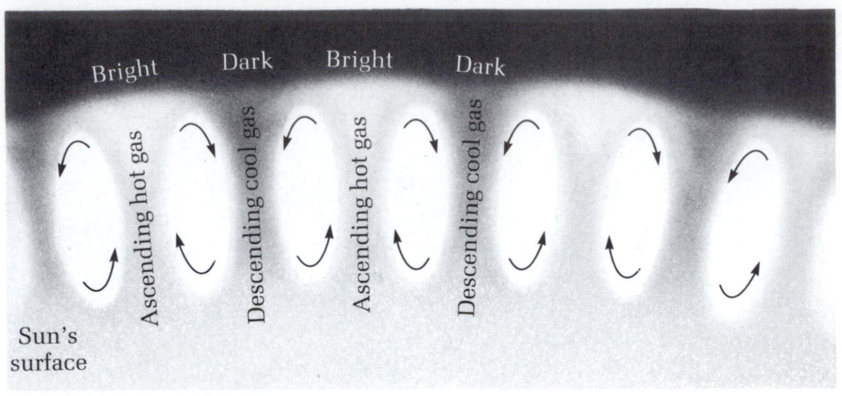

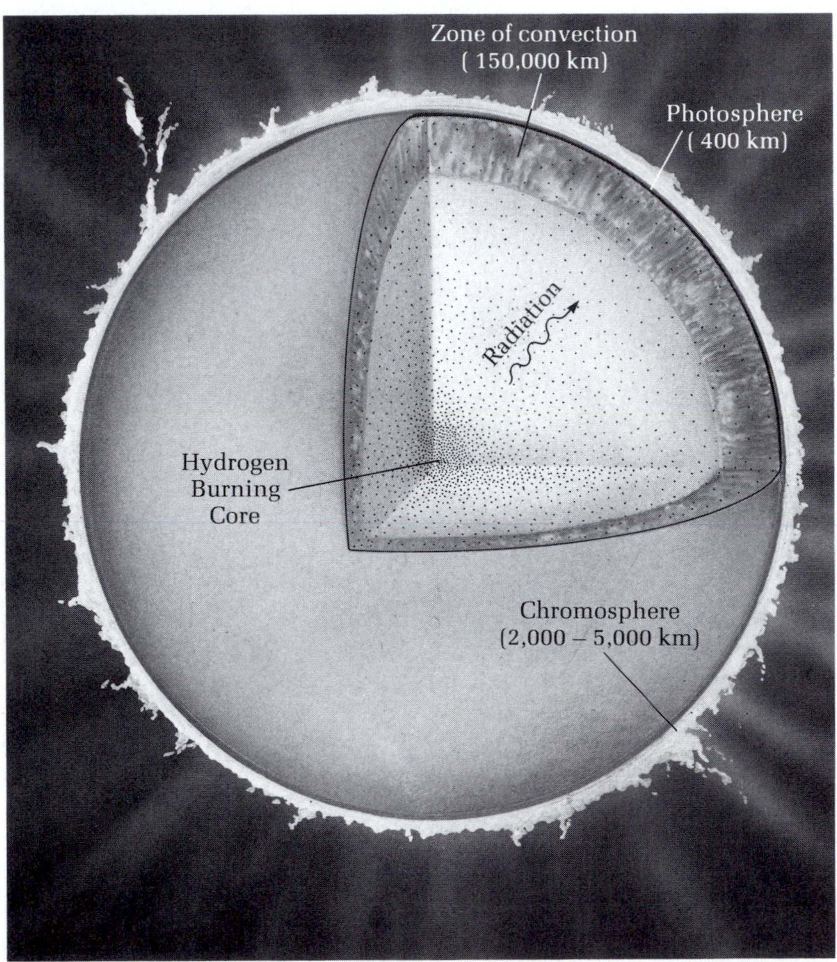

grow cooler and ionization decreases. In a relatively cool region of the outer 150,000 km or so of the sun, radiation is no longer the major mode of transfer, and most of the heat is transferred via convection; this is the convective zone of the sun. As Fig. 8.5 shows, transfer of hot gas to the surface occurs in small cells (the granules), with the hot, rising gas constituting the bright region, and the cooler (once energy has been radiated), descending gas the dark regions. Obviously the outer convective layer of the sun plays an important role in releasing the energy generated in the core.

In addition to the granules just discussed, there are supergranules—much larger (about 30,000 km across) and much longer-lived. These contain about 300 of the smaller granules. Again, as in the case of the smaller, there is heat transfer, though generally the flow in this case is horizontal to the outer edge.

Besides the percolating motions of the granules and supergranules, the overall photosphere seems to have a radial oscillation with a period of about 5 minutes—hence its name, the 5-minute oscillation. (There is also a component with a longer period of approximately 160 minutes.)

THE CHROMOSPHERE

Through a telescope tiny structures shaped like flames or blades of grass can be seen shooting out from the supergranules. These "spicules" (Fig. 8.6) are huge columns of gas, some over 5000 km long, each lasting 5–15 minutes. They emanate from the bottom of the chromosphere.

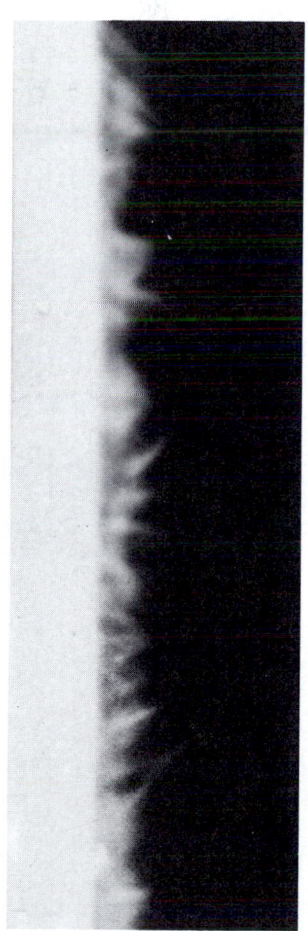

Figure 8.6 Spicules, some over 5000 km long.

The chromosphere is much less dense than the photosphere (approximately $\frac{1}{100}$) and considerably hotter. One of the best times to study it is during an eclipse, when it can be seen both at the beginning and end of the eclipse as a brief red flash. Its spectrum, usually called the flash spectrum, can be photographed during the one or two seconds it is visible (Fig. 8.7).

The flash spectrum, unlike the normal solar spectrum, consists entirely of emission lines, rather than absorption lines. However, with the exception of a few new lines, the two spectra are generally the same. The new lines are a result of increased ionization, which is caused by the high temperatures—considerably higher than those of the photosphere—in this region. A graph of the chromosphere's temperatures (relative to distance above the photosphere) shows a slight decline with increasing altitude, then a sudden increase at about 2000 km above the surface (Fig. 8.8).

Most astronomers believe this increase to be due to shock waves that originate in the convective zones. These waves reach supersonic speeds and, when they crash into the chromosphere, act like the crack of a whip, compressing (and heating) the gases of the region.

Astronomers usually study the chromosphere with an instrument called a spectroheliograph, which spreads waves so that a single wavelength can easily be selected for detailed study. Restricting study to one wavelength allows us to see the sun only in the light of a particular gas in a specific atomic state. The temperature and pressure at the surface of the sun are such that the α line of the Balmer series of

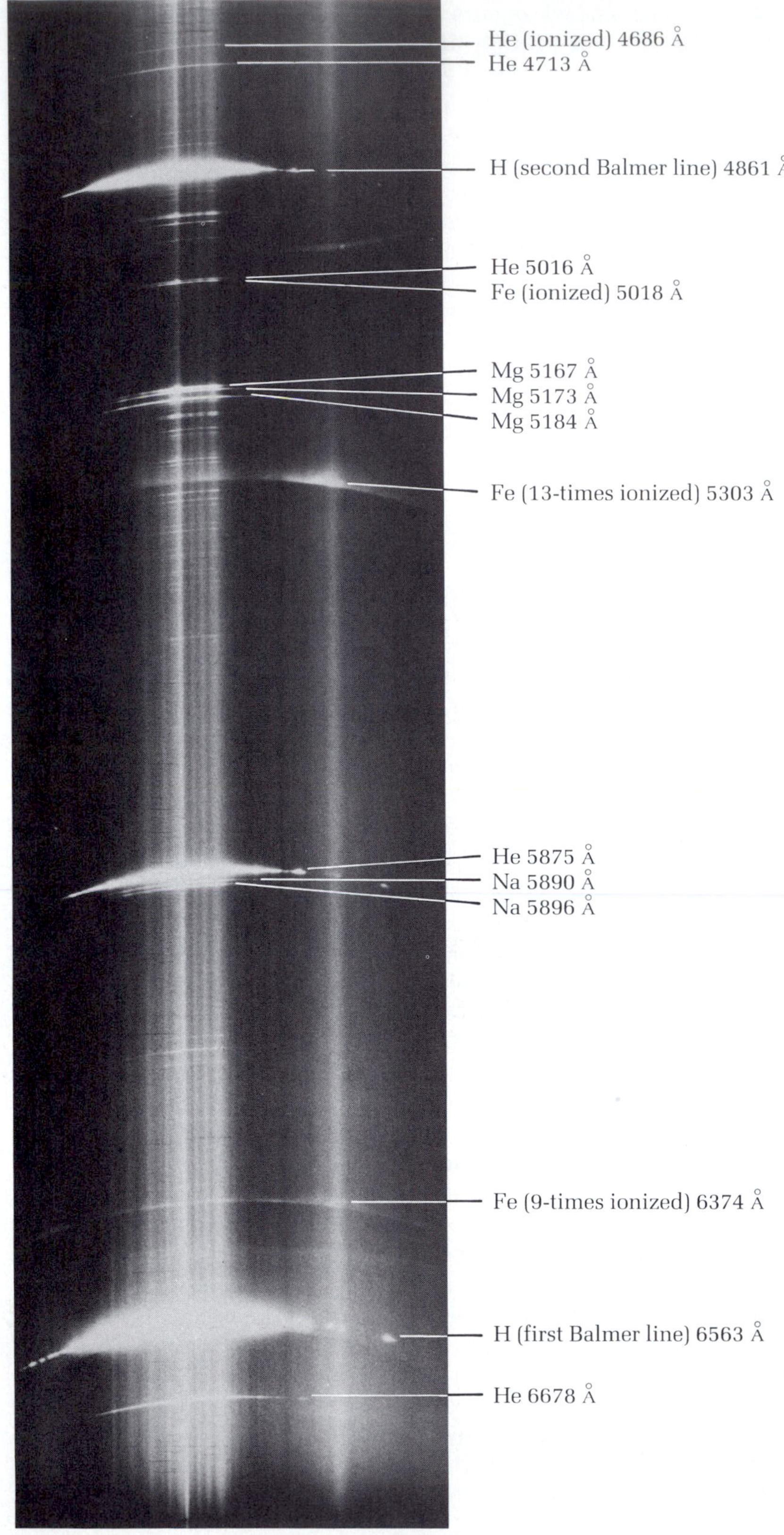

Figure 8.7 Flash spectrum of the chromosphere taken during the 1973 eclipse at Mauretania.

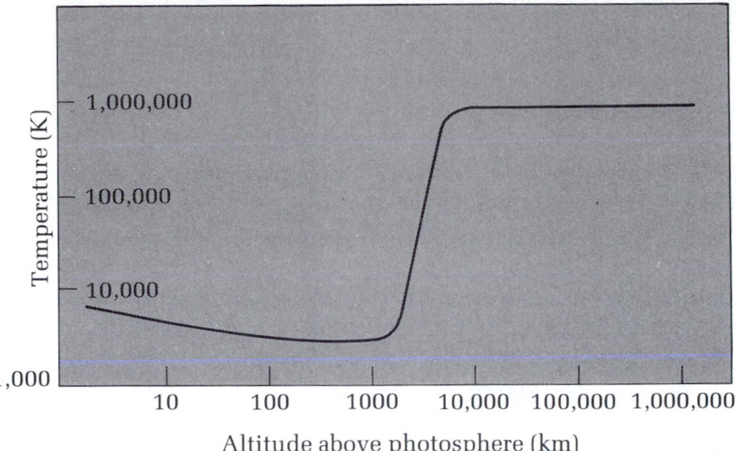

Figure 8.8 Temperature of the chromosphere relative to altitude (distance above photosphere).

hydrogen is of particular interest, but we could just as easily look at the light emitted by the helium or the calcium in the sun by narrowing in on the appropriate wavelength. Hα light gives us an excellent view of the tremendous turbulence of the surface (Fig. 8.9).

THE CORONA

The soft, white glow that surrounds the eclipsed sun at totality is the corona. Because its form depends on the magnetic field of the sun (which changes), it varies slightly in shape from eclipse to eclipse.

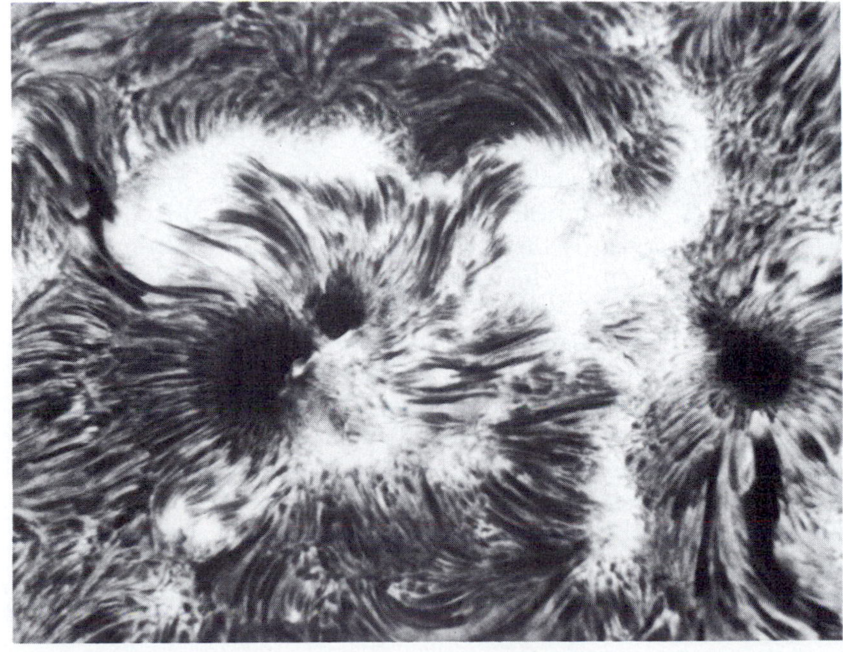

Figure 8.9 Turbulence of the surface of the sun. The photograph is of radiation emitted by hydrogen in the sun (the Hα absorption line, which is in the red region of the spectrum).

Brilliant streamers (Fig. 8.2; color) can frequently be seen extending out from the equatorial regions.

When astronomers first looked at the spectrum of the corona, several lines were found that did not correspond to known elements. Many were convinced that there was a new element in the sun (which they called *coronium*). It soon became evident, however, that it must be a known element in disguise.

In 1940, Swedish physicist B. Edlen noticed that some of the abovementioned lines in the coronal spectrum correspond to ionized[2] iron. He showed, further, that some of these lines correspond to iron atoms that have lost as many as 12 electrons. Such ionization was unheard of on earth, and meant that the temperature of this region was over 1,000,000°K, and perhaps as high as 2,000,000°K.

As in the case of the chromosphere, the cause again appears to be the shock waves emitted by the convective layer. These waves pass through the chromosphere and out into the corona, where they severely agitate the coronal gases. Because of the region's exceedingly low density (less than one billionth that of our atmosphere), the actual heat content of this region is small.

At one time we could see the corona only during an eclipse. In 1930, however, French astronomer B. Lyot invented the coronagraph, which allowed him to produce an artificial eclipse on any clear day. (This is more difficult than it might seem, mainly because of light scattering by our atmosphere.) Coronagraphs are set up at sites where the atmosphere is thin and scattering is minimal, such as Climax in Colorado, Haleakala Crater in Hawaii, and Pic du Midi in the French Pyrenees. More recently, coronagraphs have been employed in orbiting solar observatories; there was one aboard most of the OSO series, and there was also one aboard Skylab.

SURFACE FEATURES

Sunspots

Sunspots (Fig. 8.10), dark regions on the solar surface, have been seen for centuries. The Chinese noted them as early as 20 B.C., and they were also observed by the early Greek philosophers. In most cases they were considered to be transits (across the face of the sun) of the planets Mercury and Venus. The telescope was first used to study them in 1610, when Galileo and several other astronomers observed them. Galileo noted that they moved across the face of the sun, and assumed (correctly) that this was due to the sun's rotation.

Sunspots are gigantic by earthly standards—most about 10,000 km across, a few are as large as 150,000 km. They appear black because they are 1500–2000°K cooler than the surrounding photosphere (p. 136); against the sky as background they would appear brilliant.

[2] An ion is an atom that has become electrically charged (+ or −) through the loss or gain of one or more electrons.

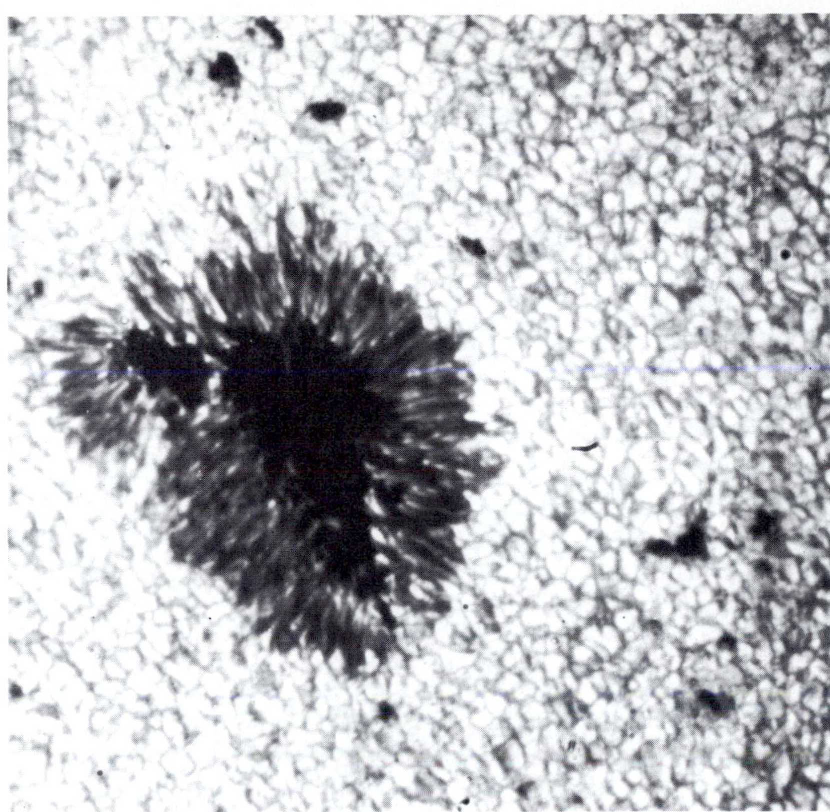

Figure 8.10 Closeup of a sunspot, showing the umbra and penumbra.

The central region of a sunspot is much darker than the surrounding area, seeming almost a hole in the surface (Fig. 8.10). This region is called the umbra, and the grayer region around, the penumbra (both terms used also in relation to eclipses).

About 1826 an amateur astronomer in Germany, Heinrich Schwabe, began counting sunspots as a hobby. For many years he kept records of the number present at any time, and it became evident to him that their total varied cyclically—from one minimum (or maximum) to the next—approximately every 11 years. Subsequently it was found that many climatic effects were apparently associated with sunspots: cycles of drought, depths of lakes each spring, and so on, initially appeared to be related to the sun's cycle. Within a few years, however, it was clear that there was no definite pattern linking the sun's 11-year cycle and the earth's climate.

But there were other ways of looking at the cycles. In the late 1800s German astronomer Gustav Spörer, after examining old records of sunspots, reported that there had apparently been very few (if any) sunspots during certain periods (Fig. 8.11). British astronomer Walter Maunder verified Spörer's discovery of inactive periods and noted that early in the sunspot cycle (sunspot minimum) most sunspots formed at a latitude of about 30°—well away from the sun's equator. As the sunspot cycle progressed, however, new ones formed in regions closer and closer to the equator. Finally, near the end of

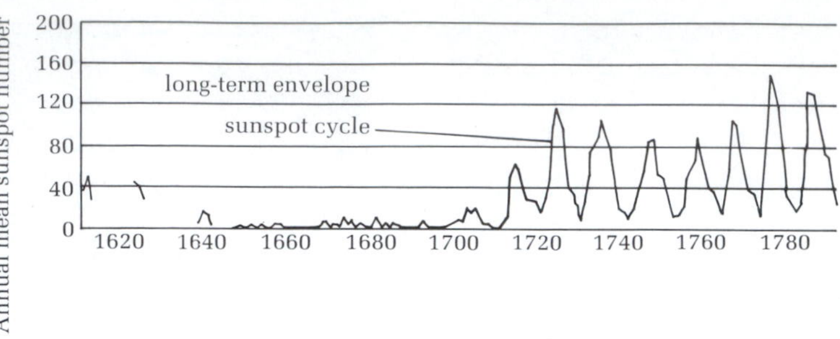

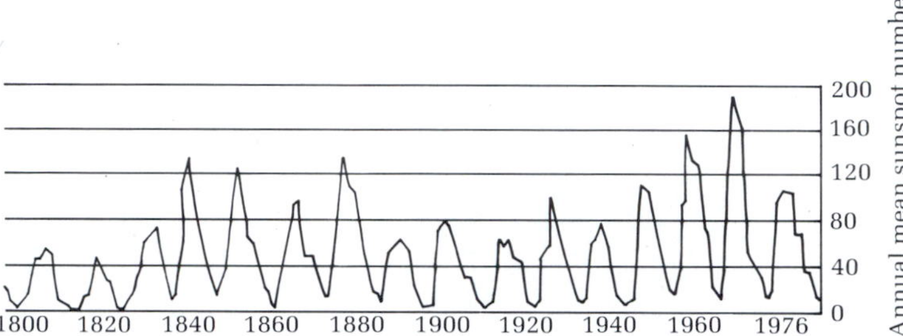

Figure 8.11 The sunspot cycle from 1620 to 1976.

the cycle, all sunspots were near the equator. Then a few would appear at much higher latitudes, and the cycle would start again.

Astronomers believe sunspots are intimately connected to the sun's magnetic field. The earth has a magnetic field much as a conventional bar magnet (Fig. 8.12) does, which is why a compass indicates magnetic north. (The metal pointer of the compass aligns itself with the lines of force of the earth's magnetic field.) Like the earth, the sun has a magnetic field, and so do sunspots.

The first clue to the nature of sunspots came in 1908, when American astronomer E. Hale found that the spectral lines of the light emitted by sunspots were split, indicating that there were strong magnetic fields in the vicinity. Only a few years earlier (1896) P. Zeeman had shown that when light is passed through a magnetic field, spectral lines separate into two or more lines, and that the amount of separation depended on the strength of the field (p. 111). Hale measured the separation for the lines associated with sunspots and determined that the magnetic fields were exceedingly strong—in some cases thousands of times greater than that of the overall field of the sun.

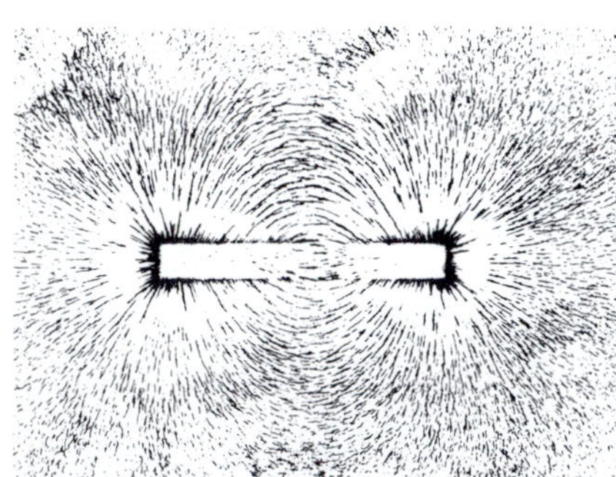

Figure 8.12 Magnetic field of a conventional bar magnet. Iron filings sprinkled in a sheet of paper lying on the magnet align themselves along the lines of the field.

The lower temperature of sunspots, then, can be explained as follows. Charged particles move easily along magnetic field lines but cannot easily cross them. In essence the charged particles in sunspots are constrained by the lines of force of the powerful magnetic fields associated with them. This, in turn, impedes the transfer of heat through the convective zone, with the relatively cooler sunspots the result.

A few years later Hale and his associates noted that when sunspots occurred in east–west pairs, as they frequently did, the polarity of the two spots was opposite. In other words, if one exhibited a north (N) pole, the other exhibited a south (S) pole. He also found that the polarities of pairs changed with every cycle. Furthermore, polarities of lead (easternmost) spots in the southern and in the northern hemisphere were opposite. This meant that the sunspot cycle was, in reality, 22 years long rather than the 11 years determined by Schwabe. The reason for this has recently been discovered: The polarity of the sun changes every 11 years. Thus 11 years of one polarity plus 11 years of opposite polarity give 22 years for a complete cycle. (Earth's polarity also reverses—every few hundred thousand to a million years.)

The most generally accepted model of sunspots is one proposed in 1961 by Horace W. Babcock and modified later by Robert Leighton, Eugene Parker, and others. Babcock's model is centered on the fact that the sun's relatively weak overall magnetic field (which moves along with the sun as it spins) has differential motion—that is, its rotational period is different depending on latitude: Near the equator it is about 25 days, at 45° latitude it is approximately 29 days, and near the poles it is about 35 days.

The effect of this differential motion on the lines of the magnetic field (which are assumed to lie just below the surface) is to stretch out and pull them in an east-west direction around the equator, as shown in Fig. 8.13. Here, at the beginning of the cycle the field lines are "normal" (run north–south). After a few turns of the sun, however, they develop a distinct forward displacement in the vicinity of the equator and, eventually, take on a generally east–west direction. In time the lines are pushed closer and closer together, with the result that the magnetic field becomes stronger and stronger—first in a region about 30–35° above and below the equator. As rotation continues, they strengthen in regions closer to the equator. However, magnetic field lines of the same polarity lying close together will repel one another. Furthermore, because the magnetic field deflects particles, the density in this region is lower. Together, these phe-

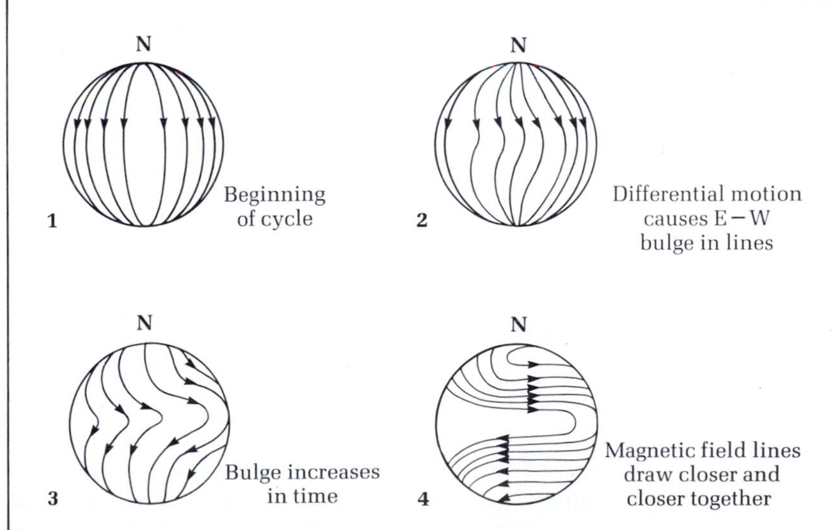

Figure 8.13 Babcock's model of the origin of sunspots. The lines of the sun's magnetic field, distorted by the nonuniform rotation of the sun, tend to wrap around it.

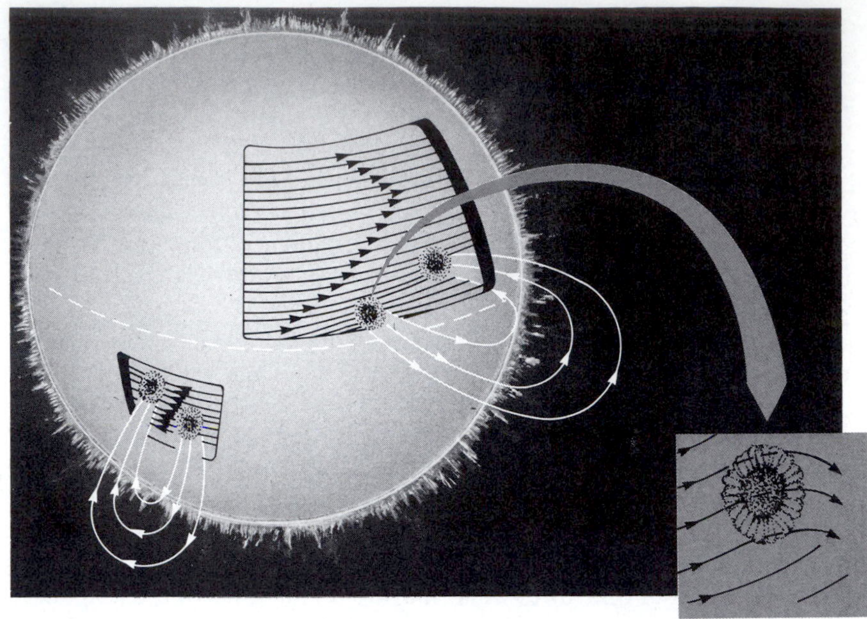

Figure 8.14 Archlike fields of magnetic force erupting from just below the surface of the sun.

nomena create a buoyant force that tends to lift the lines of force upward out of the surface to create an arch of magnetic field lines (Fig. 8.14). As seen from above, one end of this arch will exhibit a north magnetic pole, and the other end a south magnetic pole.

The Babcock model, then, explains most of the observed phenomena associated with sunspots; it explains:

1. Why sunspots occur in pairs of opposite polarity, and why these pairs are generally aligned in east–west directions.
2. Why sunspots appear at some distance from the equator at the beginning of the cycle, and why they appear closer to the equator as the cycle progresses.
3. Why lead-spot polarity during a cycle is always opposite that of lead spots in the other hemisphere. (Note the direction of the field lines in Fig. 8.14.)
4. Why lead-spot polarity changes with every cycle.
5. Why sunspots are cooler than the surrounding area. (The magnetic field lines below the surface constrain the motion of the charged particles, and this, in turn, impedes the transfer of heat through the convection region.)

Sunspots and Climate

Earlier we described the efforts to connect the earth's weather and the sun's 11-year cycle, and, in the end, the concession that there was no apparent relationship. Why, however, did it take so long to discover the 11-year cycle? Galileo and several other astronomers saw sunspots through a telescope as early as 1610, but the cycle was not discovered until almost 200 years later. This was probably due in large measure to absence of sunspots over a considerable part of this period. Maunder and Spörer both studied this period of the sun's history and concluded that the absence of sunspots did not result from lack of observations. Today we have further evidence that this was the case. Records indicate that there were few Aurora Borealis, as well, and that the corona was much less impressive. This period (1645–1715), which became known as the Maunder minimum (Fig. 8.15), also coincided with a period of extensive cold (called the "little ice age") throughout Europe.

In recent years John Eddy and others have restudied all available early records of sunspots for sun–earth climate correlation. They looked, not at the 11-year cycle, but rather at what is called the "envelope" (Fig. 8.15), and found relationships that have been confirmed by recent studies of rings (growth patterns) in exceedingly old trees. The Maunder minimum is, apparently, not the only period of this type that the sun has undergone. Between 1460 and 1550 there was an earlier minimum, now called the Spörer minimum, and it is likely that, before that, there were others. In the twelfth and thirteenth centuries there was also a peak of sunspot activity similar to what we are now in. (We are, incidentally, now on our way out of this era.) Furthermore, in addition to the two conspicuous minima, there appear to be lesser minima in the envelope every 80–100 years, and

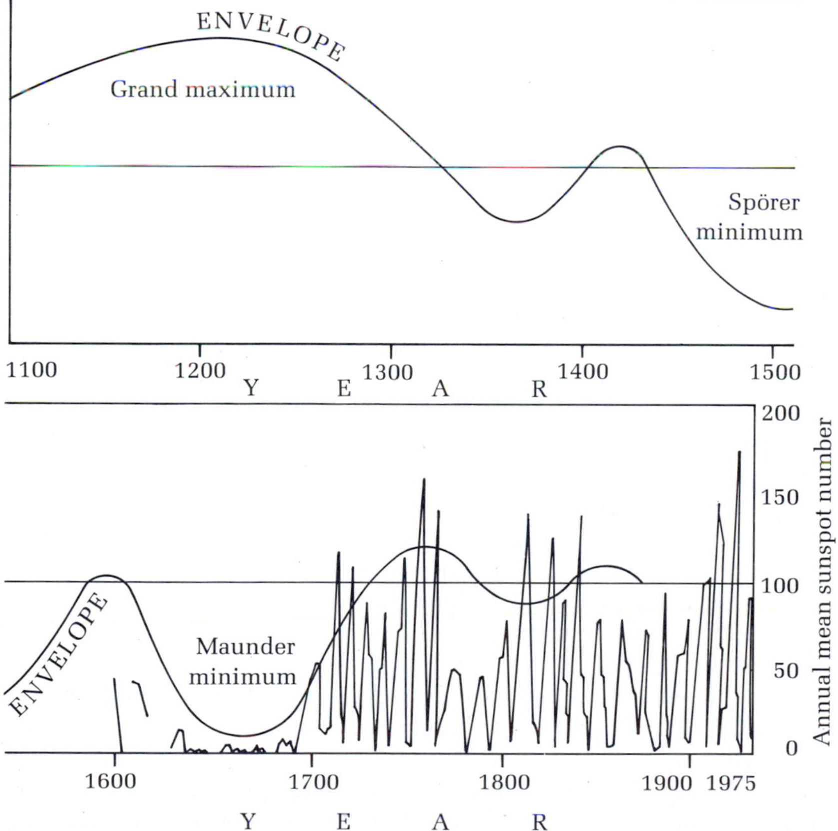

Figure 8.15 The sunspot cycle for a period of almost 1000 years. The curve of the "envelope" before about 1600 was obtained by other than astronomical means.

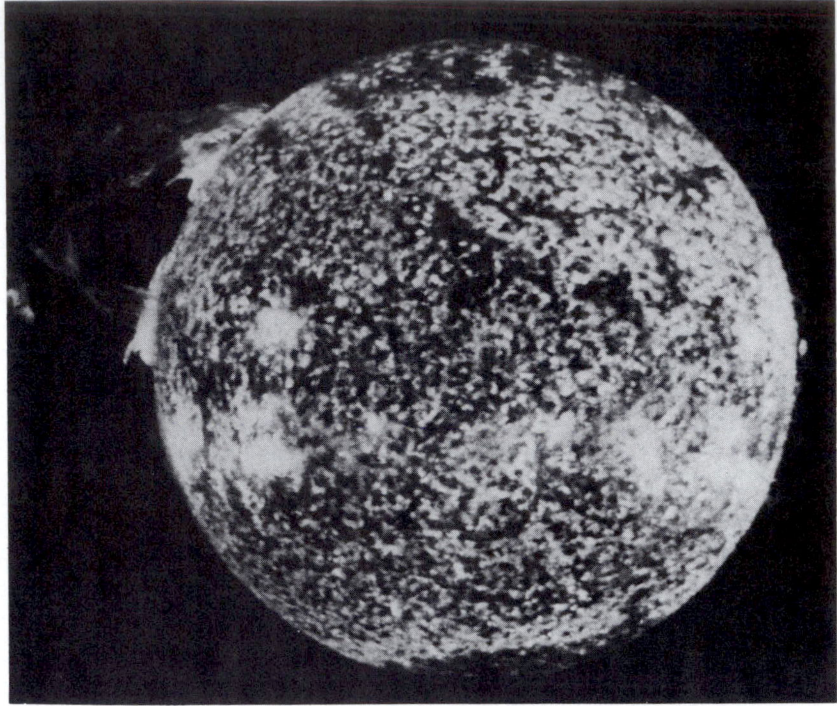

Figure 8.16 Plages (large bright areas) on the surface of the sun. The photograph is of light emitted by ionized helium in the sun.

recent studies show that droughts in the United States and elsewhere seem to have a period of 20–22 years, which coincides with the true 22-year cycle of the sun.

PLAGES AND PROMINENCES

Bright areas seen with a spectroheliograph in the regions near sunspots (known as active regions) are called *plages* (Fig. 8.16). Like sunspots, they are a result of strong localized magnetic fields on the surface, but unlike sunspots, they do not require exceedingly strong fields to maintain them and are therefore seen long after the sunspots disappear. They last for weeks or months, compared to days or weeks for sunspots.

When active regions of the surface occur on the limb of the sun, huge flamelike structures are frequently seen projecting into space above them. They are called prominences (Fig. 8.17, color). Because the gases of the photosphere have a high temperature, they are strongly ionized and are therefore more strongly influenced by magnetic fields. Since the particles move easily along the magnetic field lines but find it difficult to cross them, they take on the shape of the field (Fig. 8.18)—usually arches or loops (Fig. 8.19). Sometimes the gas seems to rise upward in great surges; at other times it appears to rain downward.

Prominences can be divided into two types: quiescent and eruptive. The quiescent type, which change much more slowly than

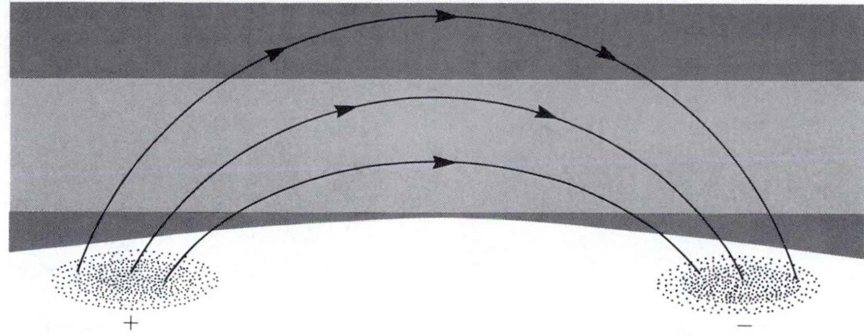

Figure 8.18 Shape of the magnetic field of a pair of sunspots.

the eruptive type, can last for days or even weeks and occasionally extend to heights of 1 million kilometers above the surface. Eruptive prominences move at high speeds (1000 km/sec) and usually last for much shorter periods.

The sun in Hα light often shows long, dark, snakelike filaments on its surface. These are also prominences but, because they are being seen against the disk of the sun, appear different. They are dark because they are cooler than the coronal gas that surrounds them.

FLARES

In early 1942 the entire network of radio communications and radar, which was an integral part of Britain's defense during World War II, appeared to have been jammed. Scientists, called in to see whether the Germans had actually developed a system powerful enough to do this, found that the disruption was being caused by the sun. The sun

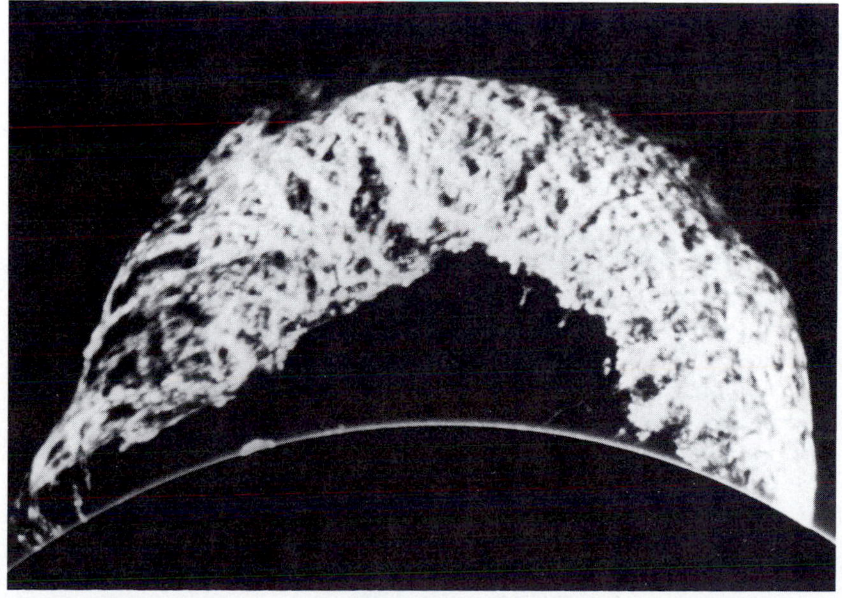

Figure 8.19 One of the largest arch prominences recorded. Note correspondence with field shown in Fig. 8.18.

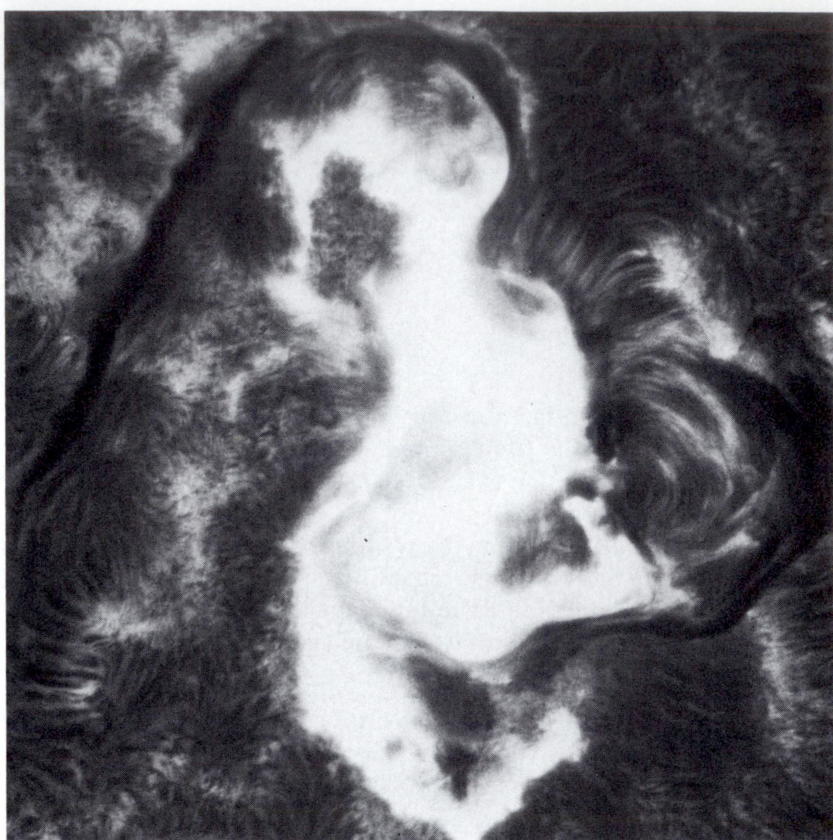

Figure 8.20 Solar flare.

was near the maximum in its cycle, and active regions on its surface were producing radio waves that interfered with the British radar and radio transmissions. Short-wave radio is particularly susceptible to these disturbances, as the ionosphere (Chapter 10) is used to reflect radio beams. UV and X-rays from the sun tend to thicken the lower region of the ionosphere and cause it to absorb, rather than reflect, these waves.

The regions of the sun responsible for these radiations are regions of intense turmoil called flares (Fig. 8.20). These occur when several sunspots come together and the magnetic fields interact and become exceedingly complex. The entire area may explode, releasing a tremendous amount of energy in the form of high-speed particles and radiation. Temperatures may reach 5,000,000°K. The fastest-moving particles will reach the earth in a few hours; others may take a few days. The radiation, on the other hand, arrives in only 8 minutes.

What triggers a flare is not known, and they cannot be predicted. They occur most frequently at sunspot maximum, when small ones occur about once an hour, and large ones about once a month.

In addition to affecting radio communications, flares can cause surges in power lines, and pose a considerable hazard to passengers in high-flying aircraft and, of course, to astronauts. Because of this the sun is carefully monitored for them.

SOLAR WIND

Apart from the tremendous number of charged particles given off during a flare, the sun is continually throwing particles into space via the solar wind, which consists mostly of electrons and protons. Near Earth the velocity of these particles is approximately 600 km/sec. Most of the solar system, perhaps even out to Pluto, is bathed in this wind, whose temperature is high but whose density is low. We are protected from it by the magnetic field of the earth (magnetosphere), which deflects charged particles. Moreover, the solar wind itself acts to shield us from high-speed nuclei (cosmic rays) from space. When the sun's activity is greatest, the cosmic ray intensity is at its lowest, and vice versa. The mechanism seems to be as follows: During maximum solar activity the intensity of the solar wind is greatest, and the solar wind carries the sun's magnetic field with it; this magnetic field deflects the cosmic rays.

One of the most important recent discoveries in relation to the solar wind was made by Skylab astronomers, who noticed that the plasma density was particularly low in certain regions around the sun. These regions occur mainly near the poles (but can occur elsewhere). They are now referred to as *coronal holes* (Fig. 8.2), and there are strong indications that the solar wind may originate in these areas.

SOLAR ENERGY

Most of our energy comes to us, in one way or another, from the sun. Some of it comes in the form of plant energy stored via photosynthesis, some as older plant energy—namely, fossil fuels—and some as hydroelectric power.

Since all three of these energy forms are produced indirectly by the sun, it is perhaps reasonable to look to the sun for an alternative source. Sunlight itself represents a considerable amount of energy. Large-scale conversion of it directly to electrical energy by solar cells is not now feasible. Even were we able to construct the 6000 square miles of collecting surface that would be necessary to supply the needs of the United States, these panels would represent a large heat sink and might, in time, actually change the climate of the area. Perhaps the easiest and best way at the present time to use solar energy is to employ it in numerous small projects—systems that can easily be mounted on individual houses.

There is an indirect way we can use the sun—by gaining further knowledge of it. In its core it is converting hydrogen to helium via nuclear fusion. The process is similar to the one that goes on in the explosion of a hydrogen bomb. We can produce similar reactions here on earth. Slowed and controlled, this could supply enormous energy.

SUMMARY

1. The visible disk of the sun is the photosphere; its diameter is 1,400,000 km. The volume of the sun is approximately 1 million times that of the earth; its mass is 330,000 times as great.
2. The sun is composed mostly of hydrogen and helium. Its energy is generated by a fusion process in which hydrogen is converted into helium.
3. The average temperature of the photosphere is 5700°K. A close examination shows it to be composed of convective granules. Supergranules (each containing about 300 granules) also exist at the surface of the sun.
4. Spicules emanate from the bottom of the atmospheric level called the chromosphere. The chromosphere is generally much hotter than the photosphere. High temperatures are likely the result of shock waves.
5. The outermost layer of the solar atmosphere is the corona. Temperatures in the corona are at least 1,000,000°K.
6. Sunspots were observed by many early civilizations. Galileo was one of the first to observe them with a telescope. He used them to determine the rotational period of the sun and noted that they had a definite structure: an outer penumbra and an inner umbra.
7. The 11-year sunspot cycle was discovered by Schwabe in the 1830s. Maunder's study of the cycle led to the recognition that the first sunspots of each new cycle form well away from the equator, and later ones form near the equator.
8. Using the Zeeman effect, Hale showed that there are strong magnetic fields associated with sunspots.
9. Babcock's theory of sunspots—now generally accepted—assumes that magnetic field lines are wound around the equatorial regions of the sun as a result of its differential rotation. Spots occur when field lines break through the surface.
10. There is no distinct relationship between the earth's climate and the sun's 11-year cycle; however, there is some correlation with the 22-year cycle. There are also two minima of interest in the envelope of the cycles: the Maunder minimum and the Spörer minimum.
11. Plages are visible only at certain wavelengths. They are active regions of the surface; sunspots are usually seen as the region develops.
12. Prominences, best seen on the limb of the sun, are radiating gas clouds that extend from the solar surface into the corona. They are of two types: quiescent and eruptive.
13. Flares are highly transient regions of intense activity. Charged particles and energetic radiations projected into space from these regions affect radio communications on Earth and create magnetic storms that interfere with navigation and radio transmission.
14. The sun continually projects a low density of particles into space. Most of the solar system is bathed in this solar wind.

REVIEW QUESTIONS

1. What are some of the properties of the sun that are critical to our existence?
2. Give the size, mass, volume, and diameter of the sun.
3. Why is the composition of the outer regions of the sun different from the composition near the center?
4. Why is the density of the sun so much higher near the core than in the outer regions?

5. How is the energy of the sun generated?
6. Explain why the sun appears darker near its limbs.
7. Describe a granule. How are convective currents important in relation to granules?
8. What is the difference between convection and radiation?
9. What is the flash spectrum? Describe it. Why is it so named?
10. Why is the chromosphere hotter than the photosphere?
11. What surprised astronomers first about the spectrum of the corona? How was the phenomenon explained?
12. What is a coronagraph?
13. Describe a typical sunspot (temperature, size, structure, etc.).
14. What is the 11-year cycle? Who discovered it?
15. Describe the evolution of sunspot pairs as the sun goes through its cycle. What does this tell us?
16. Briefly outline Babcock's model of sunspots. Show that it explains most of the observed features.
17. Is there any relationship between the sunspot cycles and the earth's climate? Explain.
18. What are plages? How do they relate to sunspots?
19. What creates prominences? Are they seen only on the limb of the sun? Explain.
20. How do sunspot maxima sometimes disrupt radio communications?
21. What is the difference between the solar wind and the wind on Earth? How do coronal holes relate to the solar wind?
22. Explain some of the problems that would be encountered in large-scale solar energy conversion projects.

THOUGHT AND DISCUSSION QUESTIONS AND PROJECTS

1. Give some convincing arguments against the following: (a) a solid sun, (b) a sun with a hollow core, (c) a sun with a cool interior.
2. If you observed several sunspots near the ends of two consecutive cycles, how could you determine which cycle they were from?
3. What does the lack of sunspots during the Maunder minimum tell us about the sun's magnetic field at that time?
4. Discuss the advantages and disadvantages of solar energy as compared to other sources of energy.
5. *Project:* Project the image of the sun onto a white sheet of paper. Count the number of sunspots you can see. Do this on several days and see if the number varies. Also note the changing position of given spots. Try to determine the rotational period of the sun from them.
6. *Project:* Determine the magnetic field of a bar magnet, using iron filings and a sheet of paper. Sketch the pattern you see. What does it tell you?

FURTHER READING

Eddy, J. A. "The Case of the Missing Sunspots," *Scientific American* p. 80 (May 1977).

Menzel, D. H. *Our Sun,* Revised Edition, Harvard Univ. Press, Cambridge, Mass. (1959).

Parker, E. N. "The Sun," *Scientific American,* p. 42 (September 1975).

Parker, E. N. "The Solar Wind," *Scientific American,* p. 66 (April 1964).

Pasacoff, J. M. "Our Sun," *Astronomy,* p. 6 (January 1978).

Pasacoff, J. M. "The Solar Corona," *Scientific American,* p. 68 (October 1973).

Aim of the Chapter

To introduce the various features of the moon, and to show you how much our knowledge of it has increased in recent years.

chapter 9

The Moon

The moon when full is indeed a majestic sight: A pristine silver ball, it stares down out of the sky, lighting the night. Countless lovers have sighed over it, and songs have celebrated its beauty and powers; but in reality, we now know that, compared to Earth, it is a dusty, harsh, dead world.

When Galileo turned his telescope to the moon in 1609, he saw a strange sight: craters—the whole surface covered with them—and mountains and dark areas that looked like seas. He called these dark regions *maria* (Latin for seas), a name we still use, though we now know there is not a drop of water on the moon. Galileo noticed that the mountains cast shadows, and from the lengths of them he calculated (actually overestimated) the heights of the mountains. In 1610 he published his results, including the first map of the moon.

A much more extensive map, with many of the seas and other features named, was published by Hevelius in 1649. Some of the names he selected are still in use today. Hevelius also estimated the heights of some of the mountains, arriving at figures very close to those we accept today. A few years later his friend Riccioli published a two-volume work on the moon in which most of the main features were named in the included maps.

Serious investigation of the moon had hardly begun, though, when it ended. Almost a hundred years elapsed between Riccioli's volumes and the next important work on the moon. In 1775 Tobias Mayer published a book that included extremely accurate positions and sizes for the craters and other markings; he obtained these with a fine grid in the eyepiece of his telescope.

Maps continued to improve over the next few years, and more and more features were named. During the latter part of the nineteenth century photography came into use, and there was little further need for free-hand drawings. Many of the first good photographs were taken at Lick Observatory.

The moon's average distance from the earth is 380,000 km (236,000 mi), and it has a diameter of 3476 km (2160 mi), which is approximately one quarter that of the earth. Compared to some of the moons of the solar system, it is rather small (it ranks fifth), but

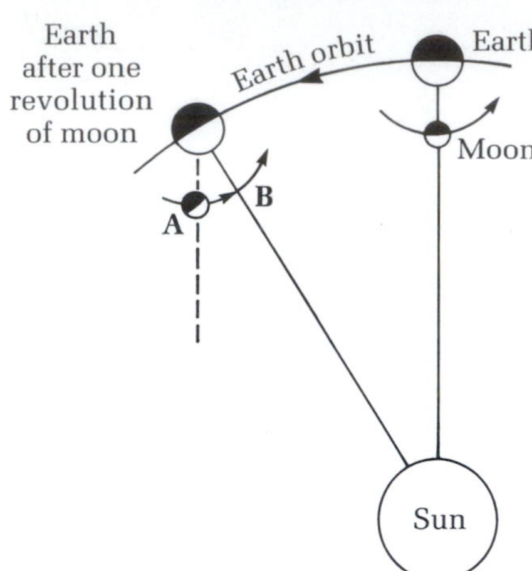

Figure 9.1 Causes of the difference between the sidereal and synodic periods of the moon. *AB* is the additional distance the moon must travel to reach point where it is in same relation to the earth. A similar mechanism governs the difference between the sidereal and solar days, illustrated in Fig. 3.24.

it has the distinction of being one of the largest relative to its parent planet (i.e., the moon–earth size ratio is large). The mass of the moon is $\frac{1}{81}$ and its surface gravity $\frac{1}{6}$ that of the earth.

As is the case with the earth (p. 58), when we talk about the revolution of the moon we have to distinguish two periods—the sidereal and the synodic. The first is the time required to get back to the same position relative to the stars ($27\frac{1}{3}$ days). But while the moon is in motion, the earth also moves (Fig. 9.1), so that, when the moon has returned to the same position relative to the stars, it is not in the same position relative to the earth. To make up this additional distance, the moon has to continue moving for approximately two more days—a total of $29\frac{1}{2}$ days, the time it takes the moon to go through its complete cycle of phases (p. 46).

The moon always keeps the same face (Fig. 9.2) toward us; therefore, according to popular (but incorrect) belief, it does not rotate on its axis. That it *does* rotate can easily be seen with a simple demonstration: move any object, a textbook for example, in a circle around you, keeping the cover always facing you. In order to do so you obviously have to rotate the book relative to the walls and surrounding objects of the room; in fact, the book's period of revolution (as

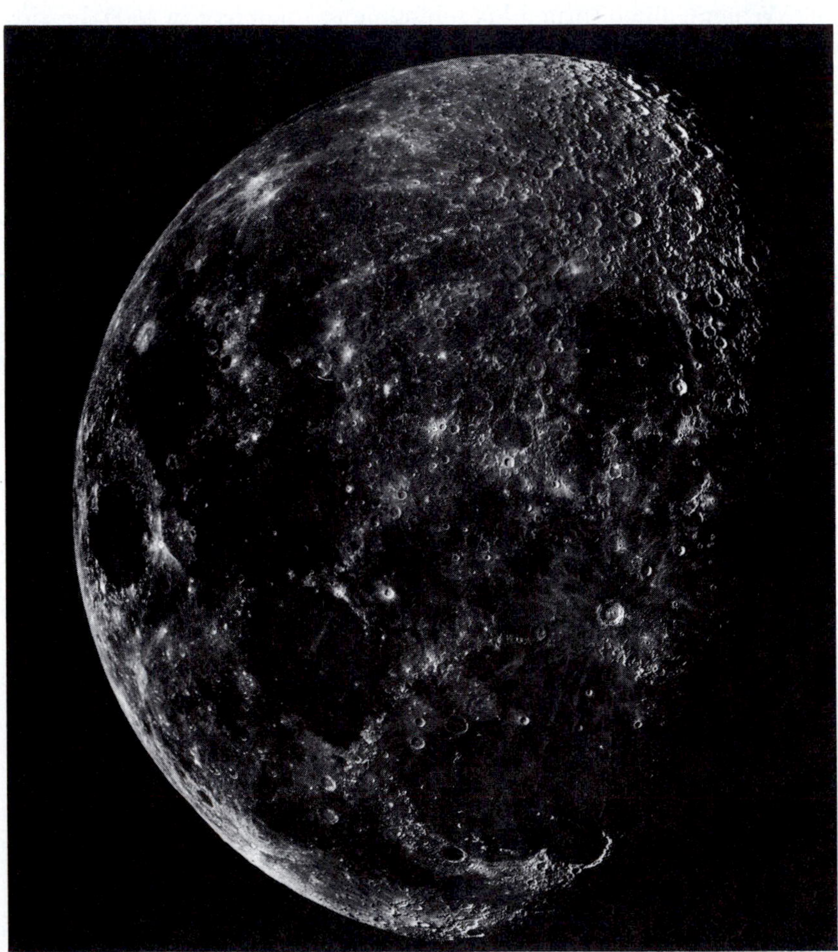

Figure 9.2 Portion of the Earth-facing side of the moon. The two most conspicuous craters are Copernicus (just below center on the right) and Tycho (far upper right).

with the moon in respect to the earth) will be equal to its period of rotation.

Millions of years ago the moon rotated much faster than it does today. Gradually, tidal forces generated by the earth's gravitation exerted a frictional force on the moon that slowed it down. Today it is "tide-locked" to the earth; we say it has *synchronous motion*.

The motion of the moon was, at one time, so misunderstood by the general public that the side facing away from Earth, was sometimes referred to as the "perpetually dark side." We know, of course, that this is not the case; the other side has night and day just as the facing side does. Actually, we do see some of the other side—about 9 percent—for several reasons: First, because the rotational axis of the moon is not exactly perpendicular to the plane of its orbit, we are able to see slightly beyond its north and south poles when it is at certain parts of its orbit. In addition, although the moon's rotational period is constant, its orbital speed varies (Kepler's second law), so that its rotation and revolution are not always exactly in step. Finally, the rotation of the earth carries us thousands of miles during the twelve-hour period in which the moon is visible; this allows us to see it from different vantage points. Collectively, these phenomena—called librations—give the moon a slightly wobbly motion as seen from the earth. Of course, with the recent flight to the moon we have not only seen the other side but have photographed and mapped it extensively and named most of the craters there. The other side is, interestingly, quite different from the side facing us (Fig. 9.8), the major difference being the small number of seas. This is probably due to the varying thickness of the lunar crust (p. 166).

The abovementioned tidal forces have another important effect on the moon. As Fig. 9.3 shows, the axis of the earth's tidal bulge is

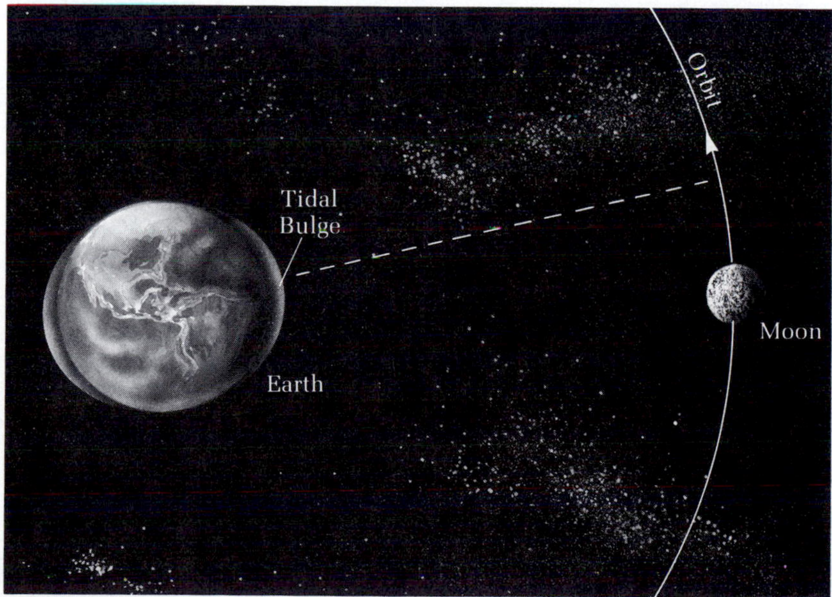

Figure 9.3 The tidal bulge of the earth produced by the moon. Note that the axis of the bulge points slightly ahead of the moon.

not perpendicular to the moon's course but points slightly ahead of it. (This is because the earth rotates much faster than the moon revolves, so that the bulge is carried ahead of the earth–moon line before it can subside.) The gravitational pull of the bulge tends to accelerate the moon forward and outward, causing it to spiral slowly away from the earth at about 3 cm per year—a tiny amount, but significant over millions of years.

Because of this outward movement of the moon astronomers believe that millions of years ago it may have been as close as 18,000 km (11,180 mi); it would have revolved around the earth in less than 7 hours at that time.

We will not, however, eventually lose our moon, for the sun also—indirectly—affects the moon, by producing tides on the earth along with the ones produced by the moon. These tides are gradually slowing the earth's rotation. In the distant future our month and our day will be of equal length (approximately 55 of our present days), and eventually, because of this, the moon will begin to move back toward the earth. As it approaches closer and closer, though, the tidal forces will increase, and the moon will become more and more elongated (egg-shaped). Finally, the forces pulling it apart will overwhelm the ones holding it together, and it will disintegrate into a cloud of particles.

A star occulted (eclipsed) by the moon blinks out of sight rapidly—one moment it is there, the next it is gone. Since, if the moon had an atmosphere, the star would dim slightly and change in color (as its light passed through the atmosphere before it disappeared), this confirms that the moon does not have an atmosphere. Basically, as James Clerk Maxwell told us long ago, the moon does not have enough gravitational pull to retain an atmosphere for any time. Most of its atmosphere, if indeed it was formed with one, was lost to space in the first few thousand years of its existence. Actually, we may have evidence that the moon had an atmosphere at one time: Outgassings from beneath the surface have been seen; this may be gas that was trapped at the time atmosphere was formed, or it may be volcanic gas.

Because the moon has no atmosphere, its temperatures are extreme. When the sun is directly overhead, surface temperature is a searing 132°C (270°F); but in the moon's night it drops to −168°C (−270°F).

SURFACE FEATURES

A pair of binoculars or small telescope will reveal many types of features on the moon. Craters are seen across its entire face; large dark areas (seas), mountain ranges, valleys, rays, and long, cracklike fissures are also visible. When full, the moon is exceedingly bright, and there are no shadows to give a feeling of depth. The best time to observe the moon is when it is in partial phase, when craters,

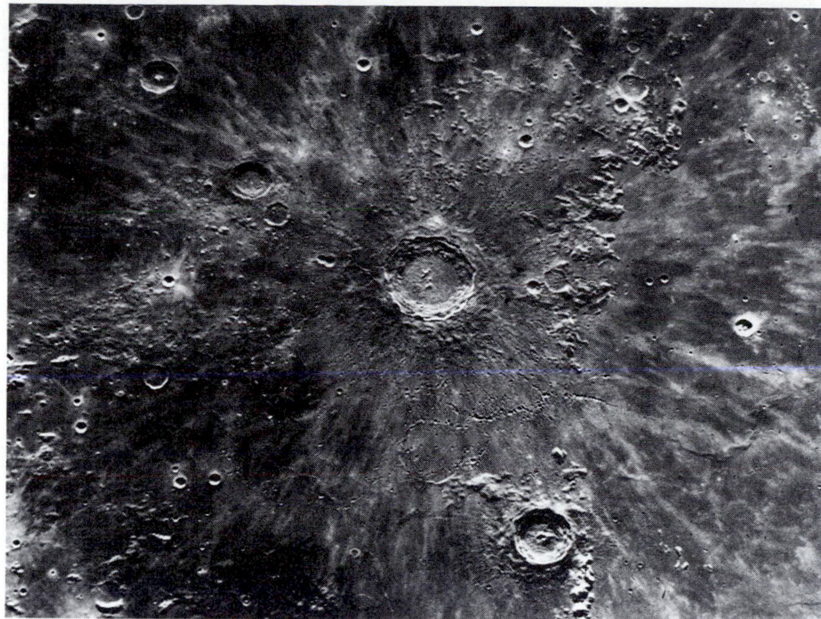

Figure 9.4 The moon crater Copernicus.

mountains, and so on stand out most clearly near the terminator line (dividing line between day and night).

The most striking feature of the moon's surface is the numerousness of craters. There are literally thousands; the moon has obviously had a violent past. In many ways the craters look like collision sites, but for years it was argued that they might be extinct volcanoes. The controversy was finally resolved with the first flights to the moon, and we now know that about 99% of the craters are due to meteorite collision. The moon did once have active volcanoes, however, and the remnants of a few are still visible.

The craters are generally round and up to 240 km (150 mi) across, with walls that rise a few thousand meters above the crater floor. The inner wall is generally steeper than the outer, and some craters have a sharp peak at the center. Such a peak may be seen in one of the most prominent craters on the moon, Copernicus (Fig. 9.4), which has a diameter of approximately 93 km (58 mi) and walls that rise 3700 m (12,140 ft) above the crater floor. Even a casual glance shows that it has a spectacular ray system (made up of material thrown out during its explosive origin). It is, interestingly, only a few centimeters deep.

Another crater with a large ray system is Tycho. Seen in the proper light, its rays appear to stretch across the entire face of the moon (Fig. 9.2 and 9.6). The largest crater on our side of the moon, Clavius (Fig. 9.5), has a diameter of 240 km (150 mi); the numerous smaller craters superimposed on it indicate that it is relatively old. This is, in fact, a good way of comparing the ages of various features; we know, for example, that the seas are younger than the mountain areas because they are relatively free of craters (p. 162).

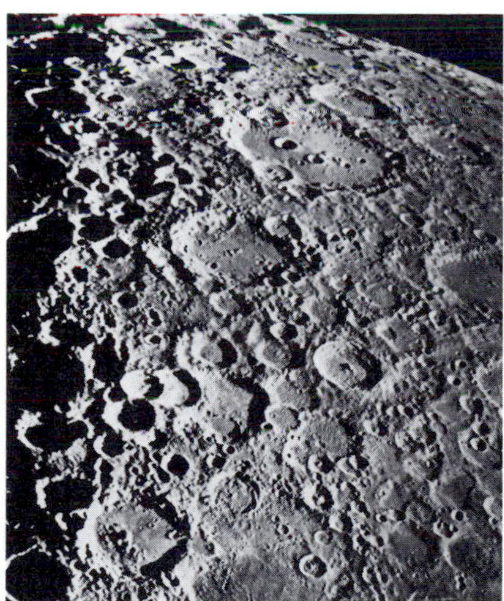

Figure 9.5 Clavius (just right of center, in top quarter of photograph) is the largest crater on the Earth-facing side. Visible in the photograph are several of the smaller craters within Clavius.

The seas of the moon (Fig. 9.6) (generally referred to as the lunar lowlands) are at a lower elevation than the mountain regions (lunar

Figure 9.6 Some major seas (*Mare*) and other surface features of the moon.

1. Clavius (crater)
2. Tycho (crater)
3. Mare Nectaris
4. Mare Nubium
5. Mare Humorum
6. Mare Foecunditatis
7. Mare Tranquillitatis
8. Mare Vaporum
9. Copernicus (crater)
10. Kepler (crater)
11. Oceanus Procellarum
12. Mare Crisium
13. Mare Serenitatis
14. Apennine Mountains
15. Archimedes
16. Mare Imbrium
17. Alps Mountains
18. Aristoteles (crater)
19. Plato (crater)
20. Mare Frigorous

highlands). There are several mountain ranges on the moon; interestingly, most surround Mare Imbrium (Fig. 9.6). They are comparable in height to those of Earth but in most other respects are quite different. Although there are a few jagged peaks here and there, lunar mountains are generally much more rounded than Earth's mountains. To us they would look like gigantic hills. They could have become rounded only as a result of erosion—on earth, the result of the action of wind or rain. On the moon erosion is a much slower process and is brought about by continual bombardment of the moon's surface by tiny grains from space called micrometeorites which very gradually wear away the surface. So slow is the process that the footprints left by our astronauts will not disappear for at least 20,000 years; but, of course, on an astronomical scale 20,000 years is but the blink of an eye.

Two of the many other features on the moon—too many to consider in detail—are the long cliff in the Sea of Clouds, known as the Straight Wall, and the famous Alpine Valley. The Straight Wall is about 130 km (80 mi) long and perhaps 1500 m (4921 ft) high. For unknown reasons the lunar floor collapsed along this line, creating the wall. The Alpine Valley (Fig. 9.7) is approximately 120 km (75 mi) long and has a width of 6.5–10 km (4–6 mi). Although there has been considerable speculation, astronomers are not certain what caused it.

Figure 9.7 Alpine Valley on the moon. It is about 120 km long.

Figure 9.8 Earth-facing (top) and far sides of the moon. The latter has few seas.

In 1959 the Soviet spacecraft Luna 3 radioed to Earth the first photographs of the moon's far side. Though the image was fuzzy, one thing was clear: the other side was quite different from the one that faces us (Fig. 9.8). Few seas were visible, and it appeared to be more heavily cratered than our side. The largest sea—small by comparison with most on our side—was named the Sea of Moscow. The first American series of space flights (Ranger) was designed only to crash into the lunar surface. One of the more interesting of the series was one that crashed into the center of the crater Alphonsus, where there had been some indication that the peak near the center might be a still active volcano; none was found. The first Surveyor craft resolved the controversy of whether the moon was covered with a thick layer of dust or fine sand. The soil was indeed found to be powdery, but capable of easily supporting the space craft.

During a series of flights called Orbiter, in which the entire surface of the moon was carefully photographed, it was noticed that the orbiter showed sudden accelerations at various points as it circled the moon. These were found to be caused by regions of peculiarly high density directly below the orbiter. These areas, where mass was highly concentrated, are now called mascons. Astronomers believe they may be the remnants of large asteroids (which created the seas), still buried beneath the surface.

HISTORY

In 1969 the first manned landing (Apollo 11) was made on the moon. Hundreds of pounds of rocks and soil samples were brought back on that flight and on the various subsequent Apollo flights. Thousands of photographs were also taken, and numerous experiments performed. As a result, we now have a much better understanding of the history of the moon and the conditions there today. The history of the moon is particularly important because of what it tells us about the history of the solar system. If we can reach back and determine what events occurred on the moon (fortunately, the history of the moon has been well preserved on its surface), we can better understand the overall picture. In the case of the earth we cannot reach back very far, for erosion has erased much of its history.

The rock samples brought back from the moon are quite similar to rocks found near volcanic and lava areas here on Earth. No new elements were found. Despite the similarities, however, there are distinct differences. First, there is no water whatever in moon rocks or moon soil, and no indication that there has ever been any. This means that the moon is devoid of life and likely always has been. No hydrocarbons or other basic molecules of life were found.

Moon rocks have also generally been depleted of volatile elements such as hydrogen, helium, and mercury. Since these elements have a low boiling point, they would have evaporated if the surface of the moon was strongly heated at any time. Relative to rocks on Earth, moon rocks have a high concentration of elements with high boiling points (e.g., aluminum and titanium). This also suggests that temperatures were exceedingly high at one time—these elements being left behind as others evaporated.

Moon rocks are, on the average, considerably older than rocks on Earth, highland rocks ranging from 3.9 to 4.4 billion years old and lowland rocks from 3.1 to 3.8 billion years old; thus the highlands are generally older than the lowlands. The oldest rocks found on Earth are 3.7 billion years old. (The technique used for dating the rocks, radioactive dating [see Box, "Radioactivity and Radioactive Dating"] tells us only the date since the material was last molten.)

The three main types of rock found on the moon are basalt (a volcanic rock), anorthosite (which contains mainly aluminum/calcium silicates), and KREEP. KREEP rocks are made up mostly of potassium (K), rare-earth elements (REE), and phosphorus (P). The basalt rocks were gathered from the lowlands and the anorthosite and KREEP from the highlands. Scientists believe the moon was entirely covered with anorthosite shortly after it formed. The basalt and KREEP flooded onto the surface from the interior later. The seas are almost entirely of basalt.

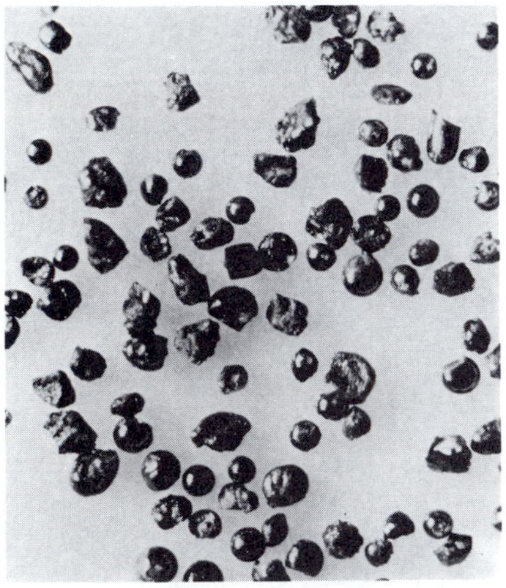

Figure 9.9 Glass spherules (diameter about .015 mm) found in the rocks and "soil" of the moon.

A close examination of the soil and rock samples showed that they contained numerous small, glass "spherules," which are a few millimeters across and usually brownish (Fig. 9.9). Some of them were probably generated long ago when the surface was extensively bombarded with meteorites, but it is now believed that most are of

Radioactivity and Radioactive Dating

The nuclei of the atoms of some heavier elements, such as uranium and radium, are unstable; that is, they spontaneously break up or "decay" into simpler nuclei. The result of this phenomenon, in which energetic rays and particles are emitted, is called radioactivity. In the early earth these rays and particles imparted considerable energy to the region around them, and this eventually caused the core of the earth to melt.

The half-life of such elements is the time it takes for half the nuclei present in a sample to decay to other nuclei. For example, ^{238}U (U stands for uranium, and 238 is the total number of particles in its nucleus) decays to nonradioactive lead; it has a half-life of about 4.5 billion years. This means that, of a certain number of uranium atoms, one half will have become lead in 4.5 billion years. Rocks have many different elements in them, some radioactive. Thus, by measuring the percentage of each type of element present—radioactive and nonradioactive—we can determine the age of the rocks (since they solidified).

more recent origin, a result of the abovementioned continual bombardment by micrometeorites. Although these micrometeorites are small, they travel at tremendous speeds and the impact, when they strike a rock, melts a tiny region of it, which crystallizes into a spherule.

Stages of Formation

It is generally believed that the moon formed about 4.6 billion years ago. This is the age of the earth (according to our best estimates), and it is assumed that both bodies were formed at the same time. The age of the moon rocks seems to confirm this: The oldest ones (fragments of rocks) are, as we have said, about 4.4 billion years old. It is not known whether the moon was hot or cold when first formed, but because some of its rocks are as old as 4.4 billion years, we can conclude that its surface had become molten within a short time after it was formed. There are two possible sources of the heat that brought this about: meteorite bombardment and radioactive heating of the interior of the moon. Scientists are now relatively certain that both of these played an important role but that the early molten stage was likely a result of intense meteorite bombardment, radioactive heating coming later. The intensity of the bombardment may have been 1000 times what it is now. It is convenient to subdivide the history of the moon into six stages, as follows.

Stage 1: Origin (See Fig. 9.10*(A)* and *(B)*.) The exact origin of the moon is still being debated and is discussed further later. Most likely, however, is the view that it formed along with the earth about 4.6 billion years ago as a result of accretion of particles in the solar nebula. By the time it was 100 million years old, meteorite bombardment had turned its surface into a sea of white-hot lava.

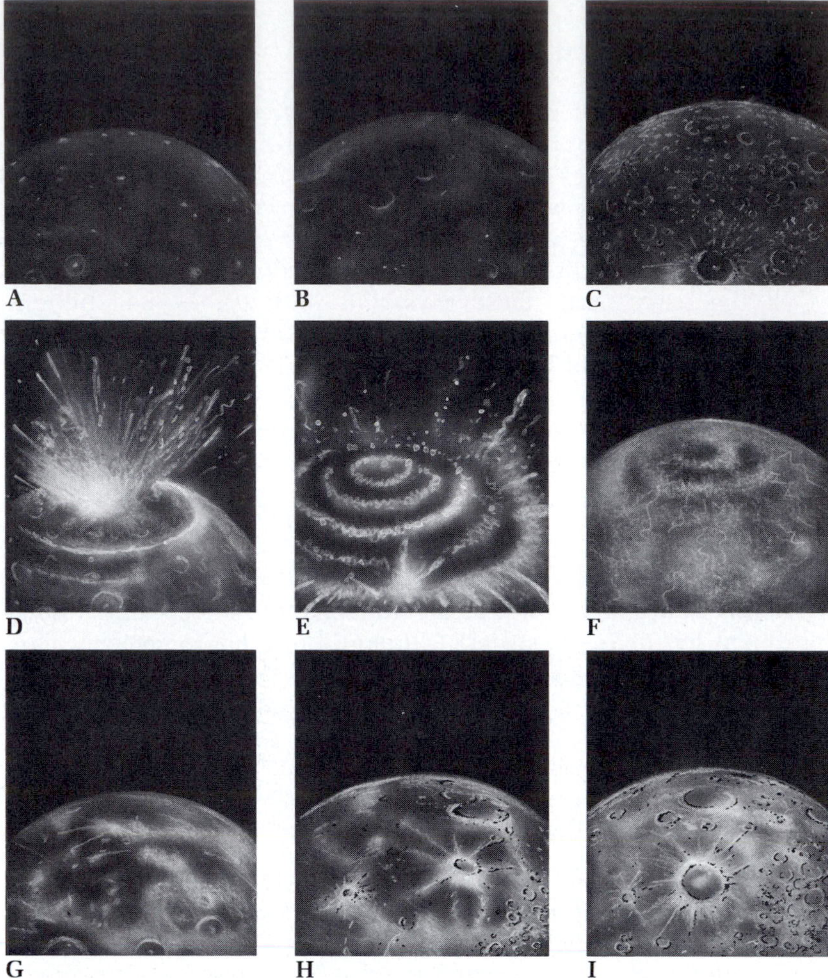

Figure 9.10 Stages in the development of the moon from shortly after its formation to the present. See text for description.

Stage 2: Separation of Crust (See Fig. 9.10*(C)*.) The meteorite bombardment ended about 4.4 billion years ago. The subsequent cooling period lasted about 300 million years, by the end of which the moon had a crust 50–100 km deep. The major component of this crust was anorthosite rock.

Stage 3: Early Epoch of Vulcanism Below the crust the moon was still molten, and lava soon began to make its way to the surface. With it came high concentrations of potassium, rare-earth elements, and phosphorus (components of the KREEP referred to earlier). This early era of vulcanism was much less extensive than that of Stage 5.

Stage 4: Mass Bombardment (See Fig. 9.10*(D–F)*.) About 4 billion years ago a new and larger population of meteorites—some as large as asteroids—were suddenly set loose in the solar system (Chapter 7). These huge flying rocks—many as much as 100 km across—crashed into the lunar surface with devastating force. The largest left gigantic depressions, which were to become basins for the seas; smaller ones produced most of the craters we see today.

Stage 5: Vulcanism (See Fig. 9.10*(G)* and *(H)*.) Radioactive heating had now melted the interior of the moon. Vast floods of lava soon began to pour onto the surface and into the basins left by the asteroids. (This was the first appearance of the abovementioned basalt material.) The lava eventually solidified, giving us the lunar seas. This era ended about 3 billion years ago.

Stage 6: Quiescence (See Fig. 9.10*(I)*.) For the past 3 billion years very little has happened to the moon; it has been in a quiescent stage. Occasionally—every million years or so—it is hit by a large meteorite, but in general its surface has changed little since the beginning of this stage.

ORIGIN

Despite hopes that the Apollo space exploration would provide one, we still have no completely satisfactory theory of the origin of the moon.

Of the three traditional theories put forward so far, one was advanced many years ago by George Darwin, the son of Charles Darwin. He assumed that the moon was once part of the earth and was formed from it by a fission process (Fig. 9.11), as follows. The earth spun much faster millions of years ago than it does today. Eventually an unstable condition developed and the earth became pear-shaped; finally, part of the earth's mass broke away and formed the moon. At one time it was suggested that it broke away from the area that is now the Pacific Ocean, but with the recent discovery of continental drift (p. 174) this is no longer seriously considered.

There are many problems associated with this theory: What happened to the spin of the earth? Why is it so much slower now? Is it, in fact, possible that the spin was great enough in the past that the moon could actually be thrown out from the earth?

One point in favor of the theory is that, according to it, the moon would have been ejected from the outer region of the earth (the mantle), and the average density of the mantle is about equal to that of the moon. However, the other difficulties cannot be explained at present, and consequently few astronomers take this theory seriously.

The second theory—the most acceptable of the three—assumes that the moon was formed in much the same way the earth was: from a gradual condensation of the materials of the solar nebula. Its main difficulty is the differences in composition and density between the earth and the moon. This can be overcome to some degree by assuming that the earth formed from the core of our local condensation of

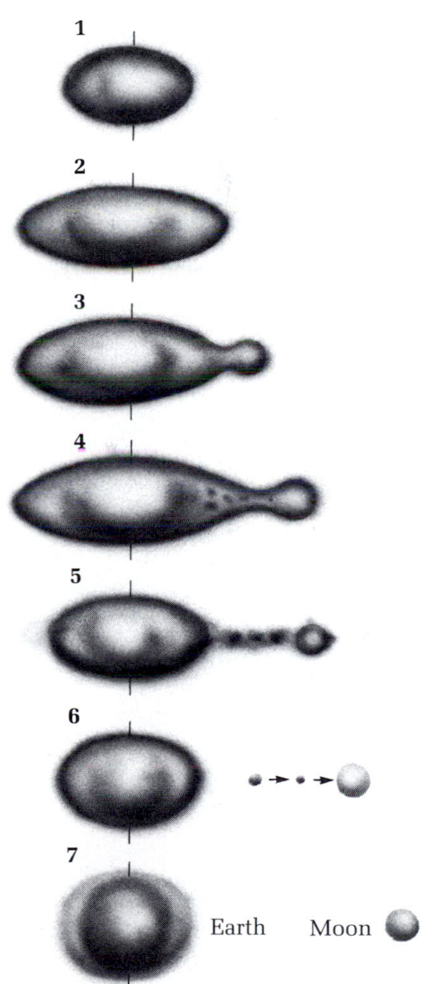

Figure 9.11 Formation of the moon by fission from the earth, a theory put forward by the son of Charles Darwin.

the solar nebula, and the moon from its outer regions. The heavier elements would have fallen toward the core. Strongly favoring this theory is evidence (found during the Apollo program) that the earth and the moon formed in the same general region of the solar nebula.

The third theory, usually referred to as the capture theory, assumes that the moon was originally a planet orbiting the sun. At some time during its life it passed close enough to the earth to be captured by it. Tidal forces may have been responsible for the capture. According to one version the moon was in a highly eccentric (elongated) orbit and moving in a direction opposite to the earth's rotation. It was captured as a result of tidal friction at its closest approach. Its orbit was then tipped to its present direction and gradually increased in size under the continuing influence of the tidal bulge. The major difficulty with this theory is the complexity of the mechanism of capture. Of the three theories it is the least acceptable.

INTERIOR

One of the major surprises of the Apollo program was the discovery that, like the planets, the moon has a differentiated interior (Fig. 9.12): a crust, a mantle, and a core. It was generally expected that the moon would have a simpler internal structure. A further surprise was the varying thickness of the crust: about 65 km on the side facing us, but about 130 km on the other side (Fig. 9.12). This is no doubt the result of the earth's tidal forces acting on the early moon.

Most of the information about the interior of the moon has come from seismic tests (Chapter 10). Like the earth, the moon also undergoes quakes, though they are much less intense (the surface barely trembles) than earthquakes. One of the major differences between lunar and earthly quakes is the depths at which they occur. Lunar quakes originate at depths of 700–1200 km; most earthquakes occur at depths of about 60 km. Lunar quakes are likely to be a result of the flexing of the surface that occurs because of the continually varying distance (and gravitational pull) of the earth; most occur when the moon is at perigee (closest to the earth).

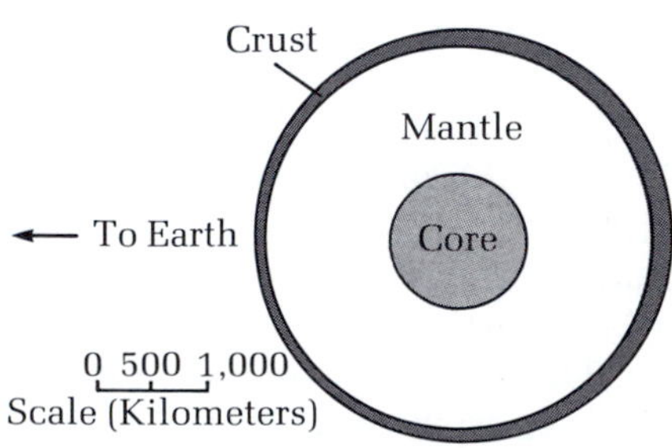

Figure 9.12 Cross section of the moon.

By studying seismic waves as they pass through the interior of the moon we can determine its structure. We know, for example, that there is a discontinuity (a change in density) at about 65 km, below which the waves travel slightly faster. This region, which extends to a depth of about 1100 km, is believed to be the mantle of the moon; it is somewhat denser than the material above it. There are also some indications that the core may be molten. Certain types of waves, which we know are unable to pass through liquid, do not pass through this region. Yet there are other indications that the core is not molten: The moon has no overall magnetic field. (If the core were molten, circulating charges in this region would create a magnetic field.) However, moon rocks have shown some small magnetism, which is considered to be residual from a much earlier period; the field that produced it must have had about 4% the force of that of the earth. These results seem to indicate that when the moon was first formed it had a molten iron core but that the molten part is now either very small (20% of the radius) or nonexistent.

SUMMARY

1. The average distance to the moon is 380,000 km. It has a diameter of 3476 km. Its mass is $\frac{1}{81}$ that of the earth.
2. There are two distinct periods of revolution associated with the moon: the sidereal (time to return to the same position relative to the stars—$27\frac{1}{3}$ days) and the synodic (time for a complete cycle of phases—$29\frac{1}{2}$ days).
3. The moon is tide-locked to the earth, and we therefore see only one side of it. Librations allows us to see 59% of its total surface.
4. At present the moon is moving slowly outward as a result of tidal forces. Eventually, forces associated with the sun will stop its outward motion, and it will begin to move inward.
5. The moon has no atmosphere. Although it was formed with one, the atmosphere was lost in a few thousand years because of the moon's weak gravitational field.
6. Major lunar features are craters, seas, mountains, valleys, and rilles. Some of the larger, more spectacular craters are Copernicus, Tycho, and Clavius.
7. The far side of the moon was photographed from the spacecraft Luna 3 in 1959.
8. The first manned landing on the moon, Apollo 11, took place in the Sea of Tranquility on July 20, 1969.
9. Soil and rock samples were brought back to the earth by Apollo 11 and subsequent Apollo flights.
10. The three main types of lunar rock samples found were basalt, anorthosite, and KREEP. Numerous small, glass *spherules* were also found in the soil and rock samples.

11. Mascons, or mass concentrations, were discovered during the lunar orbital missions. They are believed to be the remnants of the large asteroids that created the seas.
12. There are three traditional theories of the origin of the moon: the fission theory, the condensation theory, and the capture theory.
13. The moon has a differentiated interior structure. Like the earth, it has a core, a mantle, and a crust.

REVIEW QUESTIONS

1. What did Galileo use in his attempt to determine the height of the mountains on the moon?
2. What is the distance to the moon? What are its diameter and mass (relative to Earth)?
3. Explain the difference between the moon's sidereal period and its synodic period.
4. Why is the moon tide-locked to the earth?
5. What are librations? What do they allow us to do?
6. Describe the long-term motion of the moon.
7. What is the largest crater on the moon? Is it old or young? Explain.
8. Why does the moon not have an atmosphere?
9. Describe Alpine Valley. Do we know how it was caused? Explain.
10. Is there erosion on the moon? Explain.
11. What was one of the main purposes of the Surveyor landings? What was discovered when the first Surveyor craft landed?
12. How does lunar soil differ from earthly soil?
13. What are the approximate ages of lunar rocks? Does their age depend on where they were found? Explain.
14. Describe the three main types of rocks found on the moon.
15. What is a mascon? How were they discovered?
16. Summarize the history of the moon.
17. Briefly describe the three traditional theories of the moon's origin.
18. Describe the moon's interior.

THOUGHT AND DISCUSSION QUESTIONS AND PROJECTS

1. Galileo used the shadows projected by various mountains on the moon to determine their height. Explain in detail how you would go about doing this.
2. Using a good photograph of the moon, select several points on the surface. Assume you are an astronaut who has landed at these points. Describe what you would see.
3. Give several arguments suggesting that most of the craters on the moon are of impact origin rather than volcanic origin. Also give arguments that they occurred early in the moon's history.
4. How has the Apollo program helped us understand the moon? Discuss. Did we learn about the origin of the moon as a result of it? Discuss.
5. *Project:* Using a pair of binoculars or a small telescope, sketch the seas of the moon. Compare with a standard map.
6. *Project:* Using a large map of the moon with a scale on it, determine the sizes of Copernicus, Alpine Valley, and the ray system around Tycho.

FURTHER READING

Alter, D. *Pictorial Guide to the Moon.* Crowell, New York (1973).

Cooper, H. S. F. *Moon Rocks.* Dial, New York (1970).

French, B. M. *The Moon Book.* Penguin, New York (1977).

Hartmann, W. K. "The Moon's Early History," *Astronomy,* p 6 (September 1976).

Lewis, R. S. *The Voyages of Apollo.* Quadrangle, New York (1974).

Wood, J. "The Moon," *Scientific American,* p. 92 (September 1975).

Aim of the Chapter

To describe the evolution of the earth and its present makeup so that other planets can be compared to it.

chapter 10

The Earth: Our Standard of Comparison

The study of the earth, its atmosphere, and its magnetic field is usually considered to be in the realm of the science known as geophysics. Study of the earth to the detailed extent involved in geophysics is outside our scope; our examination will treat it as "just another planet," (Fig. 10.1; color) one we know much about, and can, therefore, use as a standard of comparison with others.

In this chapter we will look at some of the important terrestrial discoveries made in the last few years, such as continental drift, plate tectonics, and the Van Allen belts. We will also look at aspects of the earth's interior structure, its atmosphere, and its magnetic field. In this and later chapters we will answer questions such as: Why is the internal structure of the earth as it is? What causes its magnetic field and its radiation belts? Do other planets also have them? Are they generated by the same mechanisms?

INTERNAL STRUCTURE

With the exception of a small amount that has oozed out of deep fissures, we have never seen material from the interior of the earth. Although we have no direct way of determining the earth's internal structure, we *do* have indirect methods.

Using our knowledge of its size and mass, for example, we can calculate its average density (mass per unit volume). In the same way our knowledge of the earth's oblateness indirectly tells us something: As the earth spins on its axis, its equatorial regions bulge (Chapter 3), and the magnitude of this bulge gives us a general indication of its internal makeup.

Of much more value, however, is the study of the waves that are generated by earthquakes: seismic waves. These tell us that the interior is layered (differentiated), how thick the layers are, and whether they are molten or solid.

Earthquakes occur worldwide; most take place 50–100 km below the surface, but a few occur at depths as low as 700 km. The waves that are generated by these quakes are picked up by seismic stations at many different locations, usually several thousand kilometers apart. That each of these stations is "seeing" the quake from a different point of view allows us to pinpoint the quake's origin.

Figure 10.2 diagrams a typical earthquake. The region in which the quake occurs is the focus; the point on the surface directly above its center, the epicenter. The quake sends out two types of waves into the surrounding medium; P- and S-waves. (A third wave, which travels along the surface, will not be discussed.) The P or primary wave, is a longitudinal one; like a sound wave it creates a series of compressions and rarefactions (Fig. 10.3). The S or secondary wave, is a transverse one, like light: its vibrational motion is perpendicular to its direction of propagation. Both waves travel more rapidly the denser the material through which they are passing. They also travel at different speeds (the S-wave slower) regardless of the medium. S- and P-waves also differ in another respect: The former cannot pass through liquids; P-waves can. Furthermore, both waves are refracted or bent—but not by the same amount—as they pass into a medium of different density, much as are light waves (pp. 69–70).

Early in the study of seismic waves it was noticed that P-waves passed through the core of the earth but S-waves did not. This implied that the core was liquid. It was later shown, however, that there is a small solid core inside the liquid one (Fig. 10.4). (Although temperatures are obviously higher in the center, so is the pressure, which raises the melting point. Thus, despite the high temperature, the region remains solid.)

Seismic waves have been particularly valuable tools in our study of the earth's interior. Because of them we now know that the earth consists of three main layers: a core (of two parts), a mantle (which surrounds it), and at the surface, a crust. We will discuss each of these in turn.

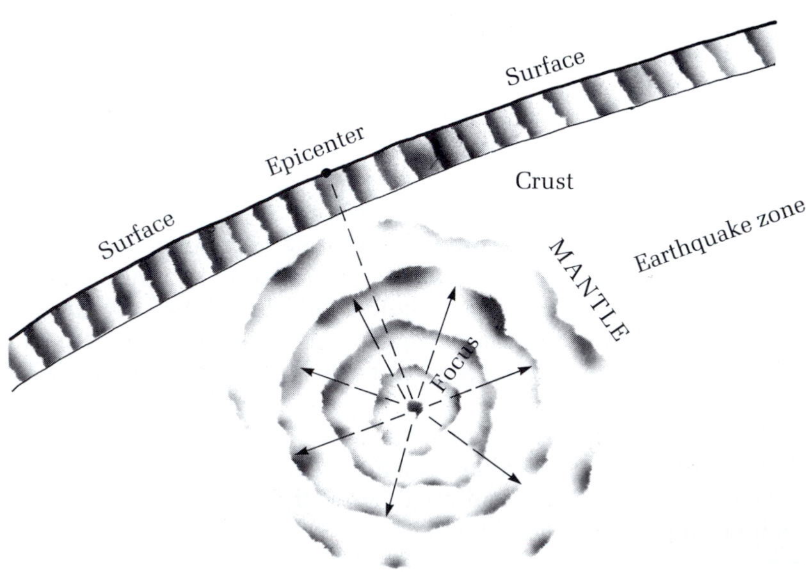

Figure 10.2 Focus and epicenter of an earthquake. Arrows indicate waves sent out from focus.

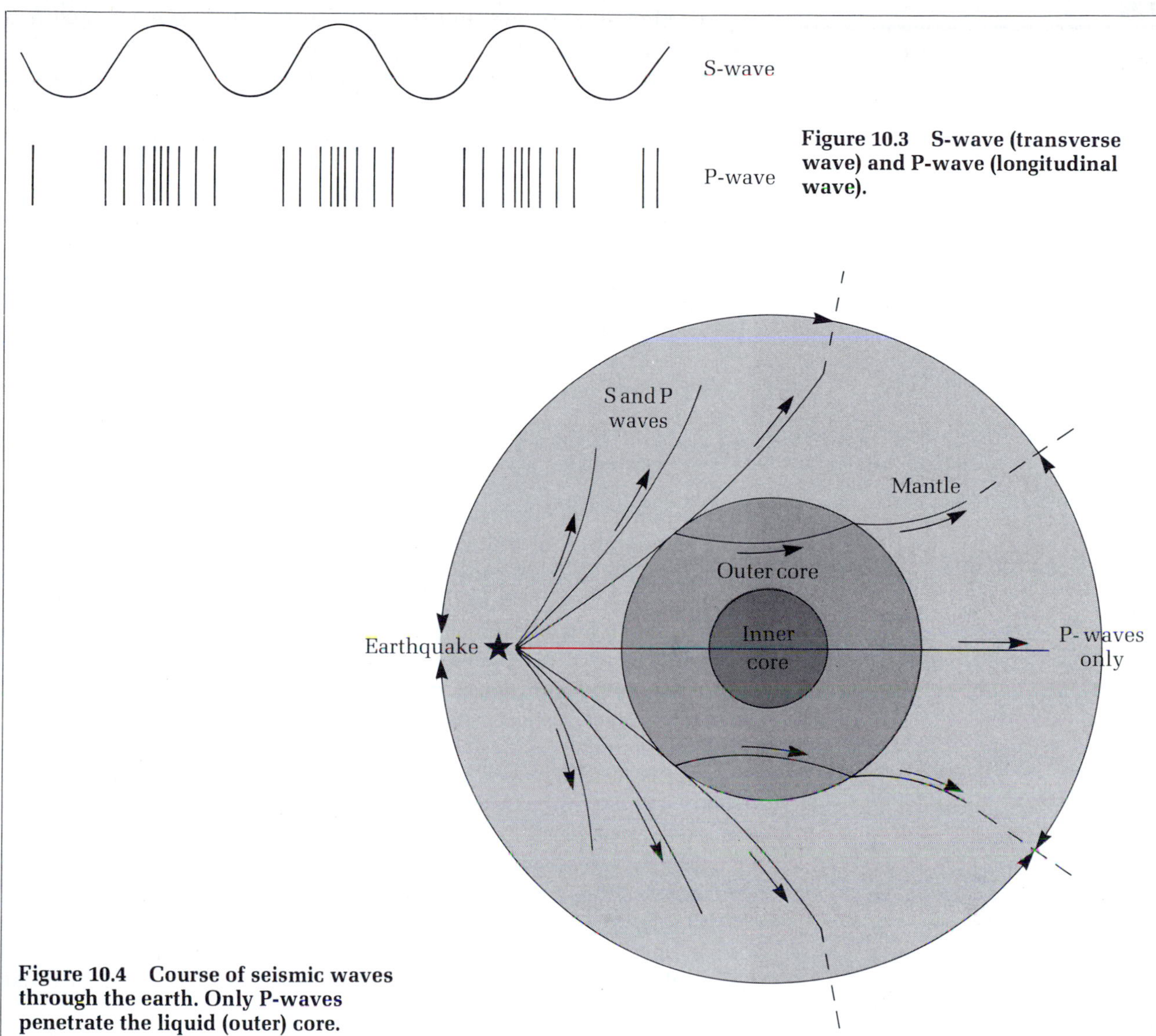

Figure 10.3 S-wave (transverse wave) and P-wave (longitudinal wave).

Figure 10.4 Course of seismic waves through the earth. Only P-waves penetrate the liquid (outer) core.

MANTLE AND CORE

The diameter of the earth's outer core is approximately 6950 km (4320 mi), about half the diameter of the earth. The inner, solid core has a diameter of about 2500 km and a density of approximately 10–12 gm/cm³. This compares to 2.5 gm/cm³ for the material at the surface of the earth and an overall average throughout the earth of about 5.5 gm/cm³. Because the density near the center is much higher than the average, it is obviously composed of heavier elements. Geologists believe that it is composed mostly of iron and nickel with perhaps some sulphur and silicon. This is reasonable: Heavy elements such as iron and nickel would have fallen to the core soon after this region became molten.

Surrounding the core is the mantle. It contains little iron and nickel, hence its much lower density than the core. It is composed primarily of magnesium-rich silicate minerals. Its inner regions are molten, but its outer layers exhibit plastic flow (they flow slowly) and are therefore considered to be a quasi-fluid. They are referred to as the asthenosphere.

With the goal of examining the mantle directly, Project Mohole was begun in the 1960s. Its object, to drill through the earth's crust to the mantle, was eventually abandoned because of technical difficulties.

CRUST

Surrounding the mantle is the outermost shell of the earth, the crust. This is composed of the lightest material of the earth—in effect, the slag that floated to the top when the earth was molten. Under the oceans it is relatively thin (approximately 5 km), but under the continents, much heavier than the oceans and therefore needing a much thicker layer to support them, its thickness averages about 30 km.

Though the continents are largely of granite, most of the crust is made up of basalt, a volcanic rock (Chapter 9) consisting of elements such as silicon, oxygen, aluminum, magnesium, and iron.

In the last few years we have learned to date rocks quite accurately. The techniques have been applied to both earth and moon rocks. We saw earlier that moon rocks do not differ significantly in age: Most are relatively old. This is not the case, though, with earth rocks. Many are quite young. Some, from the Rocky Mountains of North America, for example, are as young as 60 million years. The oldest known Earth rocks (found in Greenland) are about 3.7 billion years old. The reasons for this difference in the ages of Earth rocks are considered in the next section.

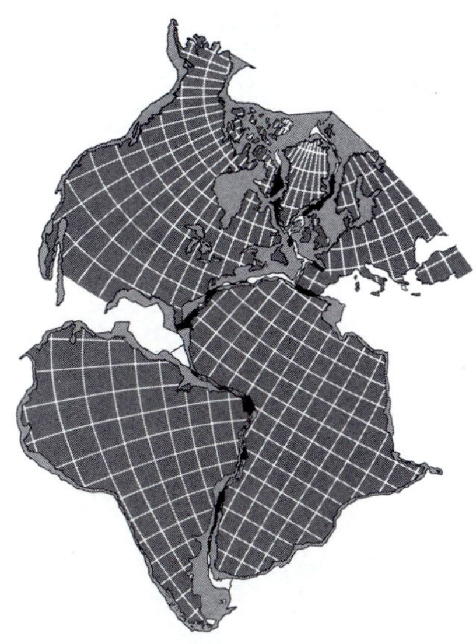

Figure 10.5 When the continental shelves are included, the degree of fit of North and South America, Europe, and Africa is striking.

PLATES AND CONTINENTAL DRIFT

Shortly after worldwide maps became available, in the early 1600s, Francis Bacon noticed the striking resemblance between the eastern coastline of South America and the western coastline of Africa (Fig. 10.5). They looked like two adjacent parts of a jigsaw puzzle. Bacon suggested, on the basis of this, that they might at one time have been part of a single large land mass. How such a large land mass could break apart, though, was beyond understanding.

The idea was resurrected in 1910 by German meteorologist Alfred Wegener. In addition to the resemblance between the shorelines of Africa and South America, he noticed similar resemblances between the eastern shores of North America and the western shoreline of Europe. He also found striking matches in the types of rocks along regions where the continents presumably had once abutted. Furthermore, fossil types were similar. On the basis of these findings Wegener formulated a theory according to which all of the present continents of the world were, at one time, together in a single supercontinent that later broke apart. But what force could push continents apart? There was no known mechanism, and, as late as 1960, Wegener's ideas were still looked on with considerable skepticism. But more data were uncovered. First, a huge underwater chain of mountains was discovered in the middle of the Atlantic Ocean. Almost midway between the continents and extending their entire length (Fig. 10.6), it is now referred to as the Mid-Atlantic Ridge. It was also discovered that earthquakes were common along its entire length and that considerable heat was being released along it.

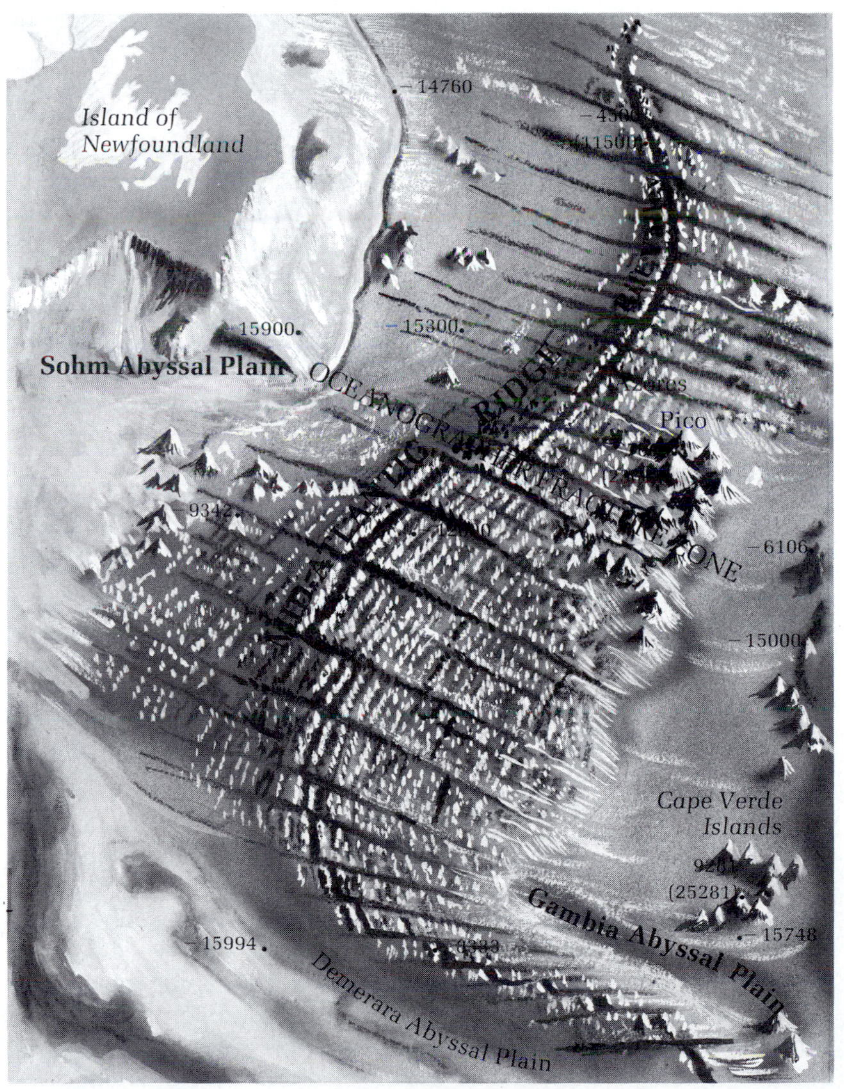

Figure 10.6 The Mid-Atlantic Ridge, from which the sea floor is spreading outward. Figures preceded by minus sign are feet below sea level.

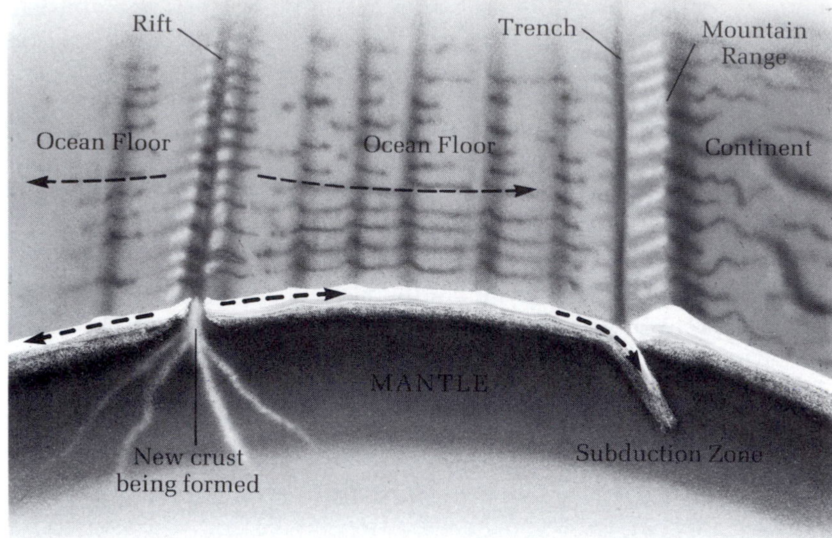

Figure 10.7 Mechanism of ocean-floor spreading. Dashed arrows indicate direction of movement of ocean floor.

H. Hess and R. S. Dietz proposed that magma (liquid rock from the interior of the earth) was flowing out of the ridge, creating in the process a new section of ocean floor (Fig. 10.7). And, as this section of floor moved away from the ridge, the continents were pushed apart. Further evidence for this point of view came when the ages of the rocks on the floor were determined. Near the ridge they were exceedingly young (age here is the time since last molten), and they increased in age with distance from the ridge. The oldest rocks—about 150 million years old—were found near the edges of the continents.

Another recent piece of evidence has convinced even the most confirmed skeptics. The magnetic polarity of the earth has reversed many times (p. 181). Because iron atoms in molten material orient themselves along magnetic field lines (p. 144), reversals of polarity can be detected in iron-containing materials that were once molten. Examination of the floor of the ocean showed evidence of several reversals, and, of particular interest, the patterns on the two sides of the ridge were identical: Iron at equal distances from and on either side of the ridge always indicated the same polarity (Fig. 10.8).

Scientists are now convinced that magma pushing up through the Mid-Atlantic Ridge is indeed pushing the continents apart. The rate of separation (continental drift) is small, only about 2–4 cm per year, but a simple calculation shows that even with this low rate the separation would be substantial in a few million years. Indeed, it would have taken about 150 million years to move South America as far from Africa as it is today.

According to present theories, about 250 million years ago all the continents of the earth were part of one gigantic continent, called Pangaea (Fig. 10.9). About 8300 km (5000 mi) wide and 16,600 km (10,000 mi) long, it stretched across the earth in a north–south direc-

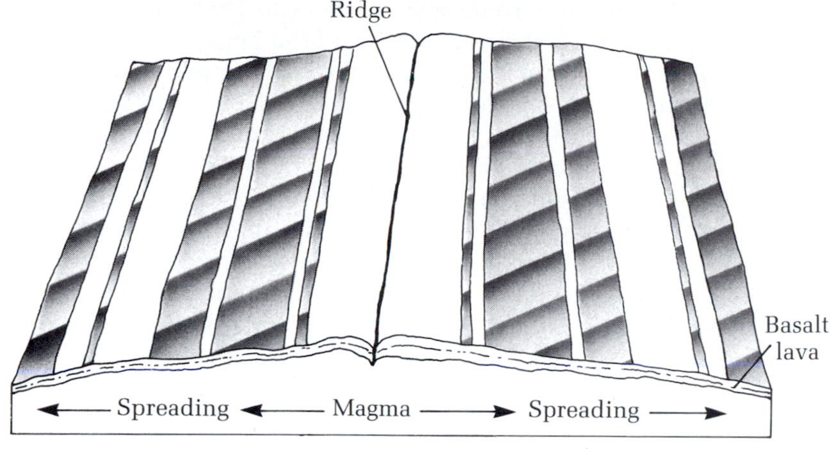

Figure 10.8 Magnetic polarity of regions adjacent to the Mid-Atlantic Ridge.

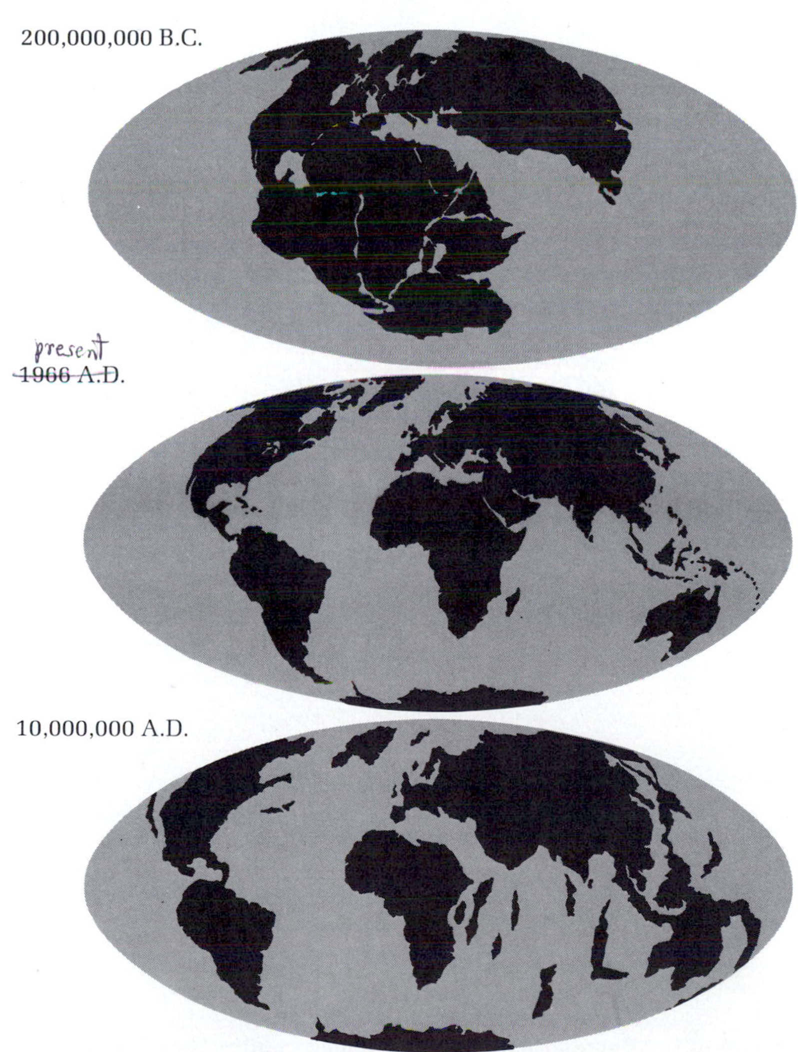

Figure 10.9 The continents of the world 200 million years ago (top), today (middle), and as they will be in 10 million years.

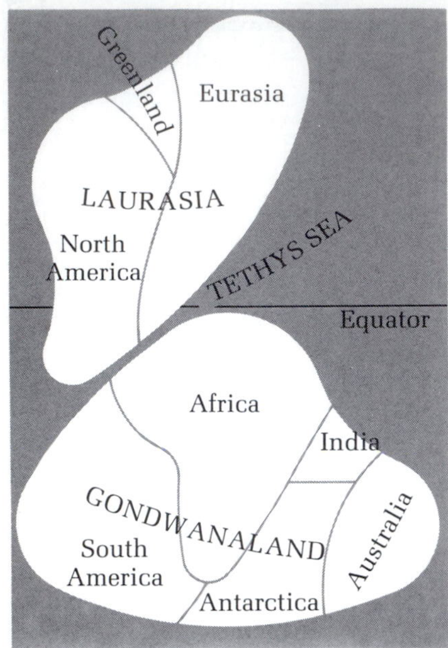

Figure 10.10 The two supercontinents, Laurasia and Gondwanaland, product of the breakup of the gigantic protocontinent Pangaea.

tion. About 200 million years ago it began to break up along the equator. Eventually a large sea, now called the Tethys Sea, separated the two parts, which, supercontinents themselves, are now referred to as Gondwanaland and Laurasia (Fig. 10.10). The present continents of South America, Africa, India, Australia, and Antarctica were contained within Gondwanaland, and the continents of North America, Greenland, and Eurasia were contained within Laurasia. The splitting of Gondwanaland and of Laurasia into these continents, which began about 150 million years ago, was caused by convection currents in the asthenosphere (created by hot material moving upward in some regions and cooler material moving downward in others). Magma, or liquid rock, ascending through the rifts created new ocean floor, which spread out away from the rift. Closer to the continents are regions called subduction zones, where the ocean bed disappears down into the mantle again (Fig. 10.7). Mountain ranges are usually seen in the regions just beyond subduction zones (e.g., the Andes in South America). Occasionally, continents collide, and exceedingly high mountain ranges are generated. India, a good example, was at one time completely surrounded by water; when it joined the mainland, the resulting pressure created the Himalayas.

We now have a fairly good understanding of the process of continental drift. As Figure 10.11 shows, the surface of the earth is made up of several plates; there are seven major ones and a few minor ones.

Figure 10.11 Boundaries of the earth's major plates; the dots represent the occurrence of earthquakes.

The continents ride on these plates as they move slowly in various directions. At one edge new material is generated; at the opposite edge it is disappearing back into the earth (Fig. 10.7). The surface of the earth is, in effect, being continually recycled. Figure 10.11 also shows that earthquake activity is concentrated along the plate edges; it was this activity, in fact, that allowed us to establish the plate boundaries.

ATMOSPHERE AND OCEANS

Planetary atmospheres, which will be discussed in later chapters, differ considerably. The earth's atmosphere is the only one that consists mostly of nitrogen (78.08%) and oxygen (20.95%). Of the four major regions into which we divide the atmosphere (Fig. 10.12), the

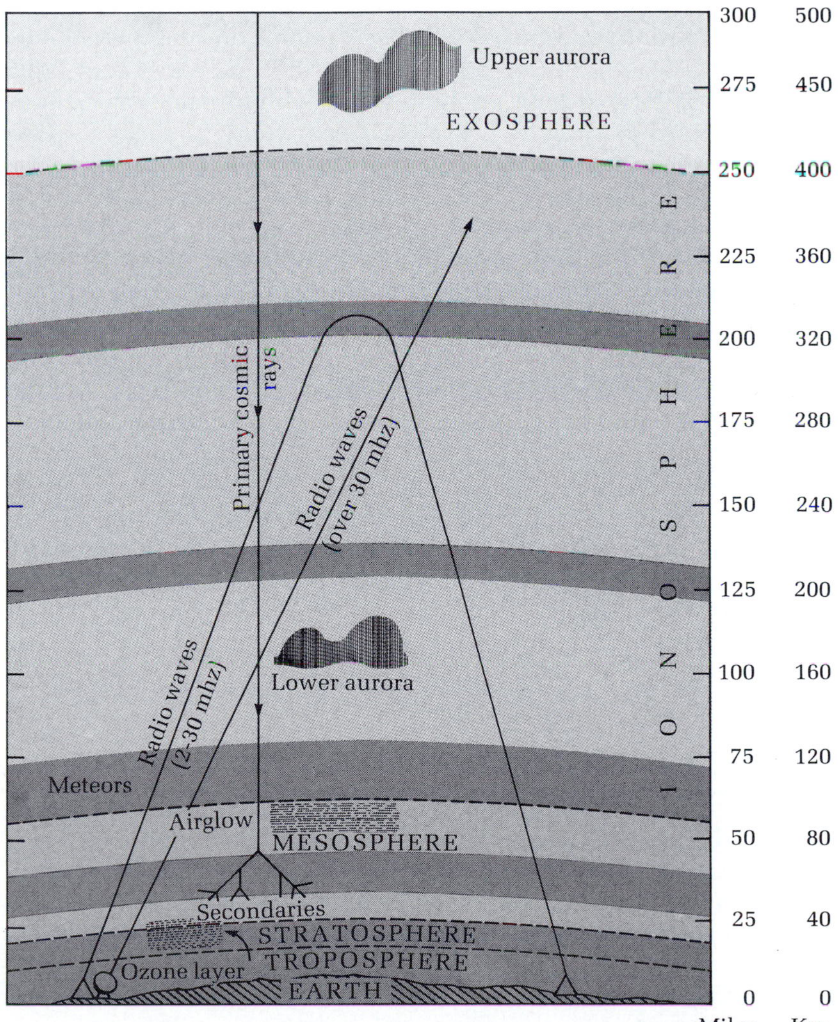

Figure 10.12 Layers of the earth's atmosphere.

lowest layer, which reaches to about 8–10 km above the surface, is called the troposphere. It is of particular importance to meteorologists, as it is where all our weather occurs. Extending about 30 km above the troposphere is the stratosphere, within which is a layer that protects us from the ultraviolet rays of the sun: the ozone (O_3) layer. (This is the layer thought to be endangered by exhausts from high-flying airplanes like the SST (supersonic transport plane) and by various chloro-fluoro-carbons used as propellant in some aerosol spray products. Destruction or diminution of the layer would permit more ultraviolet light to reach Earth, with a consequent increase in the incidence of skin cancer.)

Directly above the stratosphere is the mesosphere, and within it are the various layers of the ionosphere. The ionosphere is particularly important in radio communications (Chapter 8) because broadcast waves can be reflected from it. (Since radio waves move in a straight line, this bounce effect is what makes long-distance transmission—around the curve of the earth—possible.) Finally, above the ionosphere, we have the exosphere, the outermost layer, which tapers off to interplanetary space.

An important question in relation to our atmosphere is: Where did it come from? We saw in Chapter 7 that the earth (and other planets) was formed with an atmosphere of hydrogen and helium, which was swept away by the "solar gale." Many years later an era of vulcanism and outgassing released carbon dioxide, water vapor, and other gases into the region around the earth. As the earth cooled, the water vapor condensed to the surface, and the oceans began to form. Some of the carbon dioxide was absorbed by the water of the oceans, and the rest went into the creation of carbonate rocks (limestone); water acts as a catalyst in the reactions that create these rocks. Other gases such as methane and ammonia continued to accumulate, and eventually the earth had a second atmosphere, one we now refer to as the "primitive" atmosphere; it consisted of methane, ammonia, nitrogen, and water vapor, and scientists believe that the basic molecules of life formed in it (Chapter 27).

Our present atmosphere evolved from the primitive atmosphere. The oxygen that we now have is generally believed to have been created by marine vegetation, or algae, in the oceans. (As with plants in general, they absorb carbon dioxide, metabolize the carbon, then expel the oxygen.) It is likely that the ammonia present was absorbed into the oceans, leaving mostly nitrogen and oxygen—the major components of our present atmosphere.

MAGNETIC FIELD AND THE VAN ALLEN BELTS

As the name suggests, the magnetosphere is a region of plasma (charged particles) under the influence of the earth's magnetic field. We saw earlier how magnetic fields are represented as bundles of lines that emanate from one pole of a magnet and enter the other

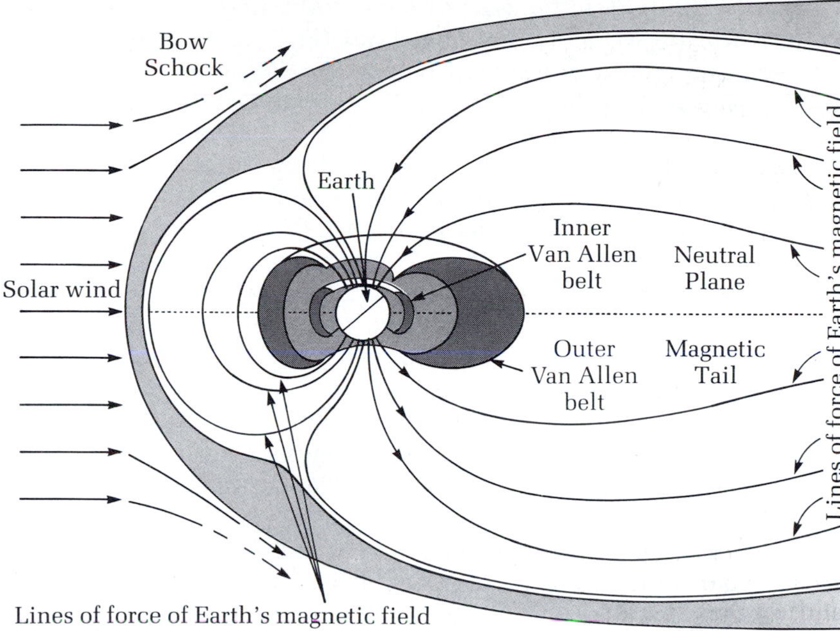

Figure 10.13 Cross section of magnetosphere (Van Allen belts) and Earth's magnetic field.

pole. The density, or number of lines crossing a unit area, determines the strength of the field. The earth's magnetic field is .6 gauss (G) near the surface at the poles. (Small toy magnets have field strengths of a few hundred gauss; a standard laboratory bar magnet, a field strength of several thousand gauss.)

The magnetic field of the earth is tilted about 11° with respect to the spin axis. It will not remain in this direction indefinitely, however; over long periods of time it wanders in a random but generally westerly direction. And, as we have noted (p. 176), every few hundred thousand years or so it completely reverses direction. We do not understand why this reversal occurs, but we do know what causes the magnetic field: circulating currents in the molten interior of the earth. Reversal of the field would seem to require reversal of these currents.

As Fig. 10.13 shows, the earth's magnetic field extends outward for thousands of miles. Any charged particles that enter this region will immediately come under its influence. A source of such particles, as we have noted (Chapter 8), is the solar wind.

In 1958 instruments aboard the Explorer I rocket indicated that there was a region of considerable radiation above the earth. Further details obtained with Explorer III showed that there were two radiation regions, now referred to as the Van Allen belts (Fig. 10.13).

When a charged particle enters a magnetic field perpendicular to the field lines, it experiences a side thrust that forces it into a circular orbit (Fig. 10.14). The radius of the orbit depends on the strength of the field and on the velocity, mass, and charge of the particle. On the other hand, a particle that enters at an angle to the field (as it is likely to do in the case of the earth) will go into a spiral or helical orbit down along the field lines (Fig. 10.14). If these lines converge, the radius of the orbit will get smaller and smaller as the particle approaches the region of convergence. Gradually the pitch, or distance between turns, of the helix will become smaller and smaller, until finally the pitch is zero. At this point the orbit will be perpendicular to the field lines and the particle will begin to spiral back toward weaker fields—that is, toward the region of divergence (hence, lower density) of lines. (This is called, appropriately, the mirror effect.) Once in a region of divergence, the particle will continue to the region of convergence at the other pole, where it reverses direction again and begins its spiral path back down the field lines.

Because the magnetic field of the earth converges at the poles, many of the charged particles that enter it will experience the mirror effect and, will spiral back and forth along the field lines, being reflected each time they approach a pole. The particles in this region are, in fact, trapped in the two Van Allen belts.

The inner Van Allen belt begins about 2000 km above the earth's surface and extends to about 3000 km. It is made up of both electrons and protons, with the protons much the more energetic. The outer belt begins at 6000 km and extends out to approximately 10,000 km; it too is made up of both electrons and protons, but in this case it is the electrons that are the more energetic.

As the particles spiral back and forth along the field lines, some are occasionally "dumped" into the upper atmosphere as a result of gusts in the solar wind. When they strike molecules in this region, they excite them, and the excitation energy is released in the form of a soft glow; we see it as the northern (aurora borealis; Fig. 10.15) and southern (aurora australis) lights.

SUMMARY: STAGES OF FORMATION OF THE EARTH

The earliest parts of the history of the earth are quite similar to those of the moon, but there are significant differences later on. It is convenient to divide the history of the earth into four stages, as follows.

Stage 1 The age of the earth according to radioactive dating (p. 163) and several other methods is about 4.6 billion years, the same as that of the moon. Shortly after it formed, while its composition was still homogeneous, an era of meteorite bombardment (discussed earlier in relation to the moon) began in the solar system. This bombardment soon turned the surface of the earth into a sea of molten lava. Eventually, the bombardment stopped, the surface cooled, and a crust formed. The crust had barely appeared when a second era of bombardment began, this time with much larger projectiles. Some of them created huge craters, in which, it is likely, the first oceans formed. The original atmosphere—hydrogen and helium—still surrounded the earth at this time.

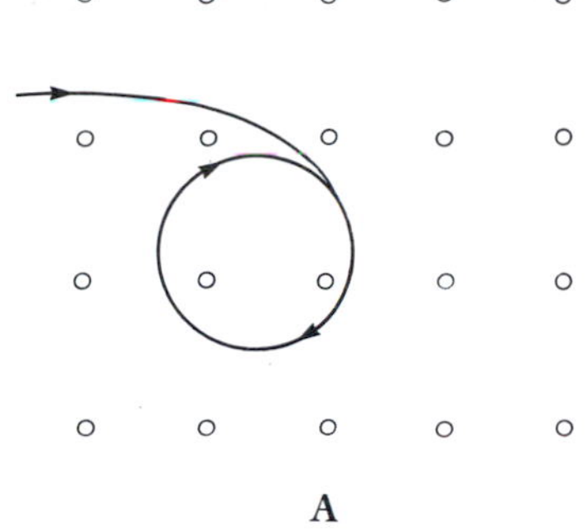

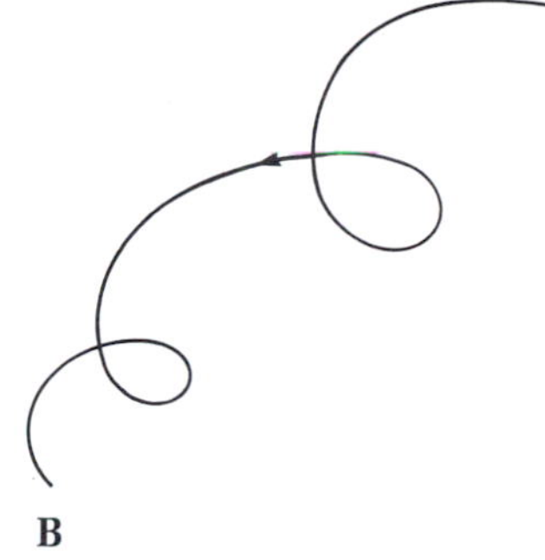

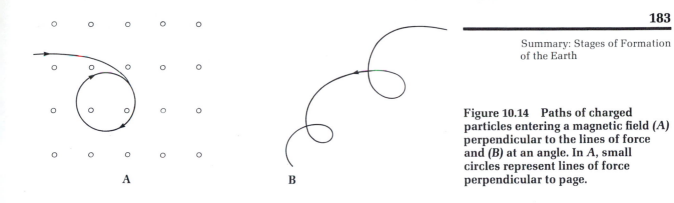

Figure 10.14 Paths of charged particles entering a magnetic field *(A)* perpendicular to the lines of force and *(B)* at an angle. In *A*, small circles represent lines of force perpendicular to page.

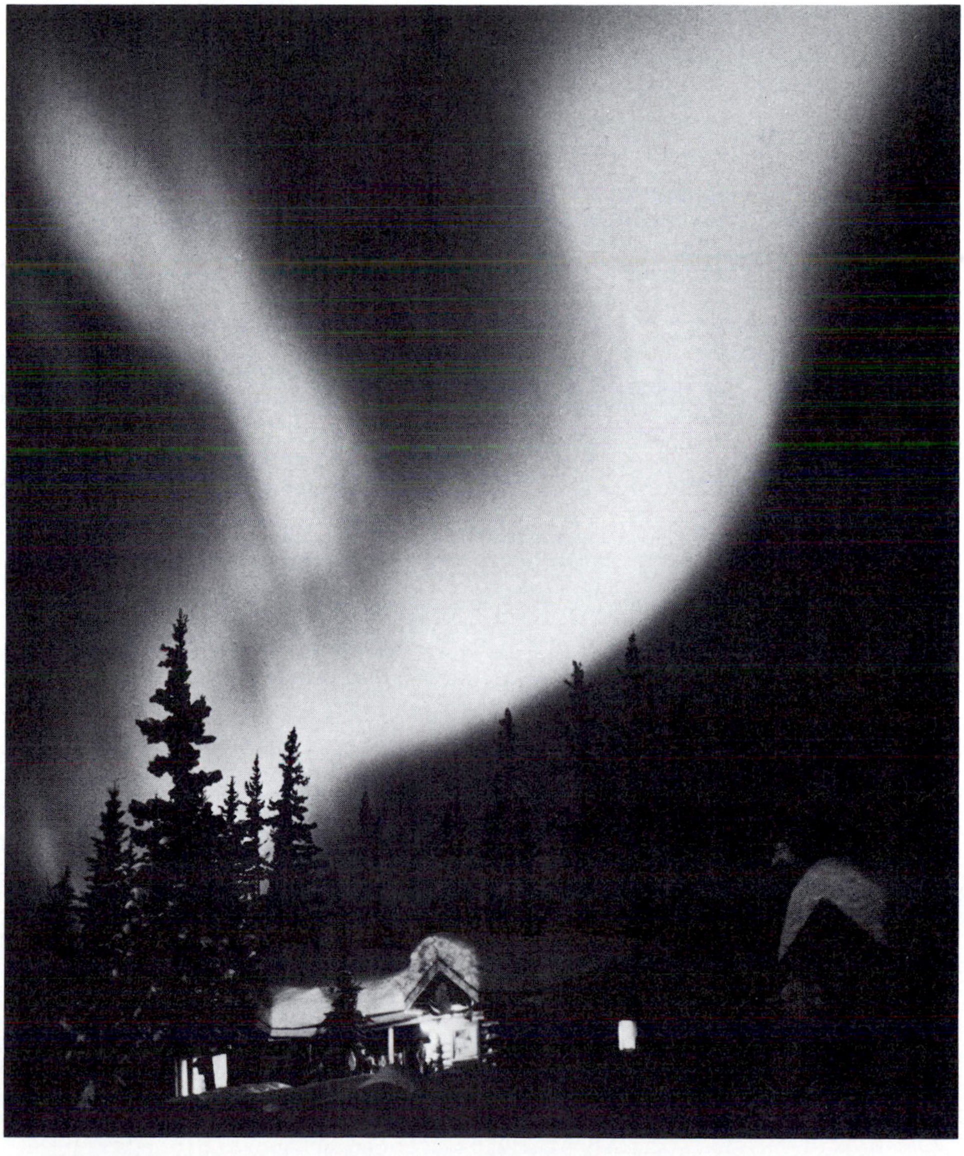

Figure 10.15 The aurora borealis, or northern lights.

Stage 2: Cooling and Differentiation Tens of millions of years later the surface had cooled considerably, but the interior was now heating, mainly from radioactive decay. (When a radioactive element emits a particle, considerable energy is transferred to the surrounding atoms, and over millions of years the amount of heat thus generated is significant.)

As the interior of the earth melted, heavier elements such as iron and nickel began to sink toward the center. However, some heavy elements, such as uranium and thorium, were "squeezed" out of the core and ended in the outer regions of the earth. The lightest material—the molten rock—was like a slag and floated to the top. Distinct layers were soon formed: a core of iron and nickel, a mantle, and an outer crust.

Nuclear reactions began in the sun during this stage, creating the solar gale that blew away the earth's atmosphere of hydrogen and helium.

Stage 3: Vulcanism About 3.7 billion years ago an era of volcanic activity began. Many different gases were given off by the volcanoes, and soon the earth had its second (the primitive) atmosphere. Water vapor from the volcanoes then condensed and accumulated into oceans; finally, much of the earth's surface was covered with water. This stage lasted about 2 billion years.

Stage 4: Formation of Plates The crust continued to thicken. Inside the earth convective currents broke the crust into plates, which began to move as the convective currents continued to stir them. The continents on these plates joined and split several times as the plates drifted with respect to one another. About 250 million years ago the present continents were part of one gigantic land mass; then the current separation began.

SUMMARY

1. The structure of the earth's interior has been determined through study of seismic waves. It consists of an inner and outer core (the outer one liquid) and a mantle. The outermost layer is the crust.
2. Wegener suggested in 1910 that the major continents of the earth were, at one time, part of a large supercontinent. Discovery of the Mid-Atlantic Ridge in the 1950s helped establish his ideas.
3. We now know that ocean floor spreading is pushing the continents away from one another at the rate of a few centimeters per year.
4. About 250 million years ago there was only one continent on Earth, now called Pangaea. Pangaea eventually broke into Laurasia and Gondwanaland. Our present continents arose from the breakup of Laurasia and Gondwanaland.
5. The earth's surface is made up of plates on which the continents ride.
6. Our atmosphere is made up mostly of nitrogen (78%) and oxygen (21%), with small amounts of argon and carbon dioxide. The various layers of the atmosphere are the troposphere (innermost), the stratosphere, the mesosphere, the ionosphere, and the exosphere (outermost).

7. The earth has a magnetic field of approximately .6 gauss at the surface near the North and South Poles. Particles from the solar wind interact with it and become trapped in two belts known as the Van Allen belts.

REVIEW QUESTIONS

1. Where are the focus and epicenter of an earthquake?
2. What is the difference between a secondary and a primary seismic wave? How is this difference of value to us?
3. Sketch the interior structure of the earth. Briefly describe the makeup of each section.
4. What evidence did Wegener have for his suggestion that the continents of the earth were once together?
5. Describe the Mid-Atlantic Ridge. What evidence do we have that magma is flowing out of it?
6. Briefly describe the evolution of the surface of the earth, beginning with its stage of 250 million years ago.
7. What is a plate? Why are earthquakes clustered along the edges of plates? What is a subduction zone?
8. What is the approximate makeup of the earth's atmosphere?
9. Name and briefly describe the layers of our atmosphere.
10. Describe the earth's magnetosphere.
11. What causes the northern and southern lights?
12. Briefly summarize the history of the earth.

THOUGHT AND DISCUSSION QUESTIONS

1. Discuss the various ways that we can determine the age of the earth. (Outside reading is required.)
2. What evidence do we have that the earth's surface is covered by moving plates? Using a map of the earth and Fig. 10.11, explain the origin of some of the major mountain ranges such as the Himalayas and Rocky Mountains.
3. The Atlantic seafloor is spreading at the rate of approximately 3 cm/yr. The African coast and the coast of South America are approximately 6400 km apart. When were they together?
4. Assume that the earth formed with a much higher spin rate (i.e., with a day that was shorter than it now is). How would this have affected its evolution (i.e., its internal structure, plate drift, atmosphere, oceans, and magnetic field)? Answer the same question assuming a much lower rate of spin. Answer the question assuming the earth formed closer and farther from the sun.

FURTHER READING

Calder, N. *The Restless Earth*. Viking, New York (1972).
Gamow, G. A. *A Planet Called Earth*. Viking, New York (1970).
Glen, W. *Continental Drift and Plate Tectonics*. Merrill, Columbus (1975).
Press, F. and Siever, R. (eds.). *Planet Earth*. W. H. Freeman, San Francisco (1974).
Wilson, J. T., et al. *Introduction to Continents Adrift*. W. H. Freeman, San Francisco (1972).

Aim of the Chapter

To describe how difficulties of observing and studying the inner planets were overcome and what has been learned from recent spacecraft observations.

chapter 11

The Inner Planets

The inner planets—Mercury, Venus, Earth, and Mars—are sometimes also called the terrestrial planets, because of their rough physical resemblance (they are generally small and solid) to Earth. They are distinguished from the low-density gas giants—also called the Jovian planets—which are discussed in Chapter 12.

MERCURY

The planet closest to the sun is Mercury, the second smallest and in many ways the planet whose conditions are most hostile to life as we know it. At high noon, surface temperatures soar to 700°K (800°F), but by midnight, they have become a chilling 100°K (−280°F). Until recently, little was known about Mercury's surface; only a few good photographs had been taken, and little surface detail is visible even in the best of them. It seemed likely that Mercury was a barren, moonlike world. On March 29, 1974, Mariner 10 passed within 6000 km (3730 mi) of the planet and the pictures it transmitted confirmed this. The resemblance was striking: Mercury looked like our moon (Fig. 11.1).

During much of the year it is difficult if not impossible to see Mercury. Though at times as bright as Sirius (the brightest star in Canis Major), it is so close to the sun that it is never seen with the naked eye except during twilight hours. The best time to observe it is when it appears farthest from the sun in the sky (the greatest elongation of its orbit). Favorable elongations occur twice a year—in early spring and early autumn. In spring Mercury is at its greatest eastern elongation and is seen as an evening star. The angle it subtends with respect to the sun (or horizon at sunset) at this time ranges from 18° to 28°, depending on its position in orbit and on the earth's position. The large difference is the result of the high ellipticity of Mercury's orbit; among the planets, only Pluto's is more elliptical.

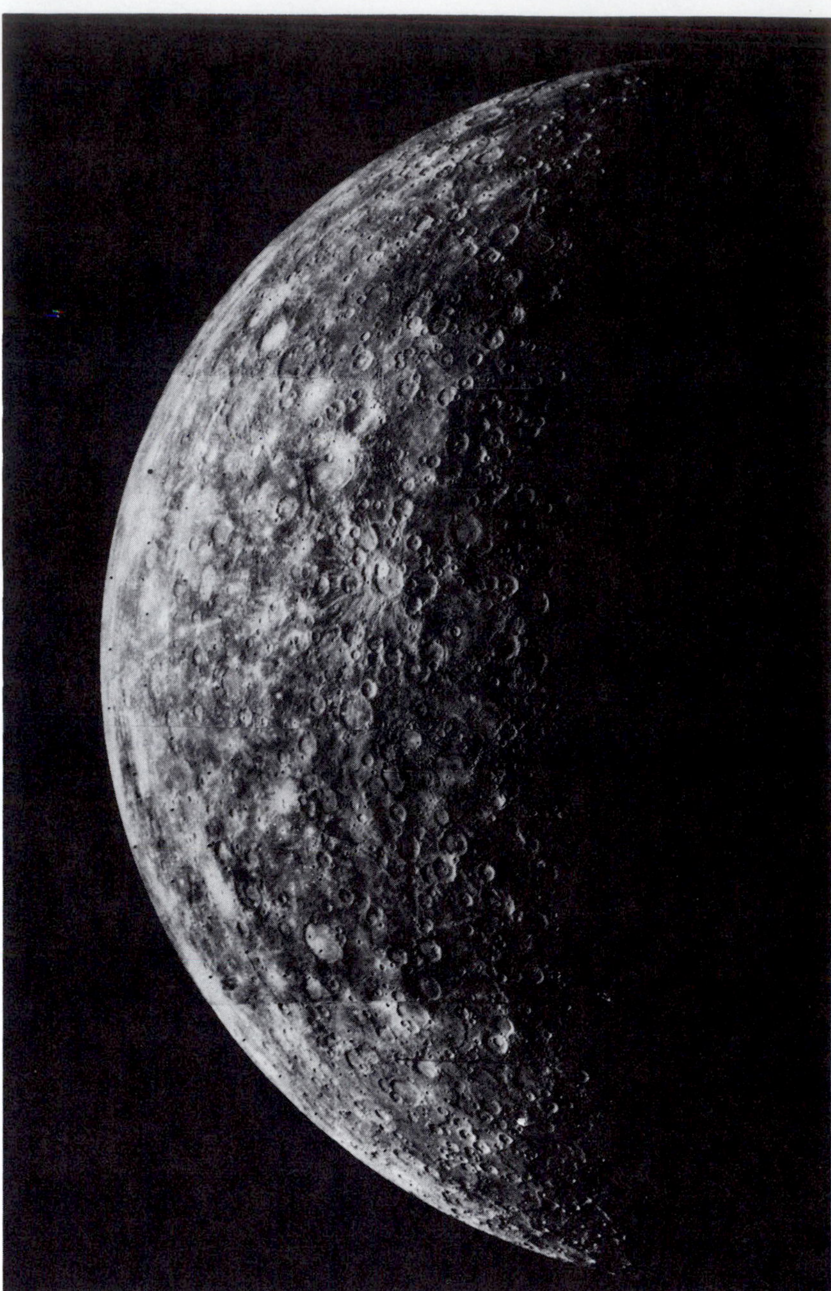

Figure 11.1 Mercury, seen from Mariner 10, shows a striking resemblance to the moon.

In early autumn a similar situation occurs when Mercury is at its greatest western elongation. At this time it is a morning star.

Astronomers prefer to observe Mercury during the daylight hours, when it is well away from the horizon and so can be viewed through less of the earth's atmosphere.

Rotation

Mercury is the fastest-moving planet. It has an average orbital speed of 48 km/sec (30 mi/sec) and takes a mere 88 days to complete an orbit. For years astronomers thought that it also completed a rotation

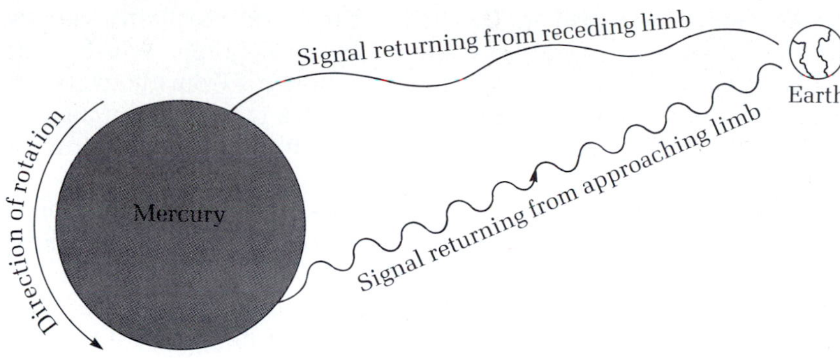

Figure 11.2 Determining Mercury's rotation with radar signals. Signals returning from the approaching limb have a shorter wavelength.

in the same time. This is not unreasonable considering its proximity to the sun. As in the case of the earth–moon system (p. 157), a small body sufficiently close to a larger one can become "locked" to it by the strong gravitational pull of the larger. Thus astronomers generally believed that Mercury was locked to the sun. Further verification seemed to come from the drawings and photographs that had been made of its surface features; fuzzy and indistinct as they were, they appeared quite similar even though they were made at different times. If Mercury was, indeed, tide-locked, it would be an incredibly hostile place; it would have both the highest and the lowest temperatures of any planet in the solar system. Being so close to the sun, any atmosphere Mercury might have retained after the solar gale (p. 127) would long ago have boiled off into the vacuum of space. With no atmosphere to speak of, its sun-facing side would be scorched, and the dark side would be near absolute zero.

However, in 1965 Roy B. Dyce and Gordon H. Pettengill, using the huge Arecibo radio telescope to measure the Doppler shift of a radar signal, proved Mercury's rotational period to be approximately 59 days. The techniques used (similar to those in the phenomenon "Doppler broadening," discussed in Chapter 6) are shown in Fig. 11.2: As Mercury turns on its axis, one limb is approaching us and one is receding. A radar pulse reflected from the approaching edge will decrease in wavelength; when reflected from the receding edge, it will increase. The difference between these signals and the incident signal tells us how fast the planet is rotating.

Physicist Guiseppe Colombo noted that Mercury's rotation (59) stood in relation to its revolution (88) approximately in the ratio of 2:3 and that, if the spin:orbit ratio was exactly 2:3, the period of rotation should be 58.65 days. This is now generally believed to be the case; thus Mercury rotates 3 times on its axis as it revolves twice around the sun.

Surface Features

Little was known about the surface of Mercury until the spacecraft Mariner 10 passed the planet in 1974. It reflected sunlight in much the same way that our moon does, and therefore some resemblance to our moon was expected. Lunarlike craters, ridges, plains, valleys, and basins were all clearly visible in the photographs sent back to Earth by Mariner 10 (Fig. 11.1). The craters ranged from about 200 km (120 mi) across to barely resolvable potholes (about 100 m across). As in the case of our moon they appeared to be formed mainly by meteoric impact.

Although the surface appeared moonlike, there were differences. The rims of the craters are generally thinner and not as high as those on the moon, and there are relatively smooth plains between the craters that are generally free of impacts. The lower rims are likely to be due to Mercury's higher gravity: Impact debris would not be thrown as far.

Another interesting feature not seen on the moon is the long, shallow cliff, or scarp, some of which are hundreds of kilometers long. R. G. Strom has suggested that the scarps might be "wrinkles" formed as Mercury cooled and shrank.

Figure 11.3 shows that large basins also occur on Mercury. The one shown is called Caloris Basin; it is curiously similar to one on the moon called Orientale Basin. Like Orientale Basin, it also was probably caused by the impact of a large meteorite, the force of which, some believe, was so great that a series of waves passed completely through the planet to produce a strangely grooved region (called the "Weird Terrain") on the other side of Mercury directly across from Caloris.

As in the case of the moon, Mercury's surface features were formed, for the most part, over 4 billion years ago and there is little erosion as it has no appreciable atmosphere. Mariner 10 detected an atmosphere about one billionth as dense as that of earth and composed largely of helium, nitrogen, oxygen, argon, and xenon.

Mercury has a weak magnetic field; at the surface it has about $\frac{1}{100}$ the strength of the earth's. Nevertheless, it is significant—and it is a puzzle. We believe that the earth's field arises from the motion of charged matter in its interior, but Mercury does not spin fast enough

Figure 11.3 Mercury's surface, showing, at left, a portion of Caloris Basin, ringed by craters. Mountains in this view are as high as 2 km.

to generate a field in this way. The only reasonable alternative is that it did so in the past and that the present field is a vestige of that. However, this would require a molten core, and many believe that Mercury is not large enough ever to have had one.

Internal Structure

To obtain a model of the internal structure of a planet, the astronomer must couple observational data with theory. Required are data about mass, density, temperature, shape (degree of oblateness), rotational velocity, and the motion of satellites. Knowledge of the probable makeup of the original solar nebula is also helpful. Particularly in the case of Mercury, useful information was obtained by Mariner 10. Since the surface of the planet was seen to be moonlike, and since we know the material on the surface of the moon to be of relatively low density, it seemed reasonable that the surface material on Mercury would also have a low average density. But the average density of Mercury is a high 5.45 g/cm^3, which would be explained by a core of particularly dense material—most likely, iron and nickel. According to calculations this core is large—about 80% of the mass of the planet. Thus, while the surface of Mercury is moonlike, the interior is strangely Earthlike (Fig. 11.4).

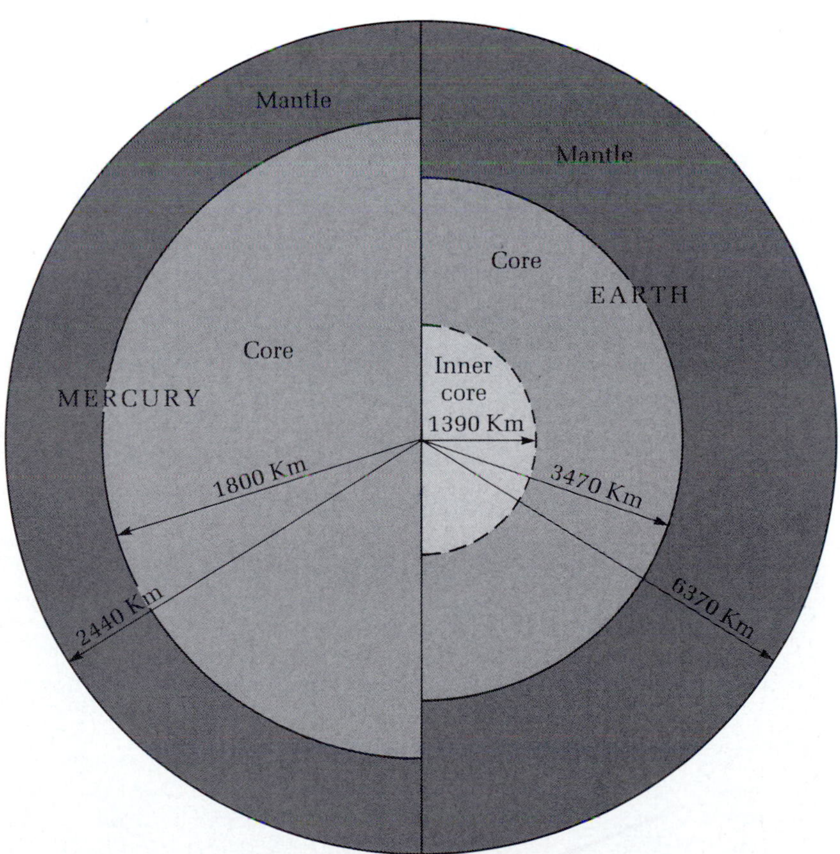

Figure 11.4 Structures of Mercury and Earth. Earth, much larger than Mercury, is shown reduced relative to Mercury.

VENUS

For centuries it was thought that Venus was potentially the most Earthlike of the planets. Some referred to it as the earth's twin—its diameter is 12,100 km (7520 mi), Earth's, 12,756 km (7950 mi). Many astronomers thought that there might be lush vegetation and perhaps even intelligent life beneath its blanket of dense, white clouds (Fig. 11.5; color). We now know that beneath these clouds there rages an inferno unmatched in the solar system except by the sun itself.

Venus's revolutionary period is 225 days. When visible it is one of the most brilliant objects in the sky; only the sun, moon, and an occasional comet are brighter. Like Mercury, it can be seen as both an evening and a morning star, but because its orbit is larger than Mercury's, it is seen at a greater height above the horizon than Mercury. At its most favorable maximum elongation it makes an angle of 48° with the sun, while at its least favorable one, it makes an angle of 47°. This small difference means that Venus's orbit is almost circular, and in fact it has the most nearly circular orbit in the solar system.

Venus seen through a telescope presents phases (Fig. 11.6), appearing like a moon. Figure 11.7 shows why this is so. When Venus is at Position 3 (near inferior conjunction), it appears as a large, narrow crescent, since we are, for the most part, seeing its night side. As it approaches superior conjunction, more and more of its surface becomes visible, but because it is now much farther away it appears fainter. Finally, when it is in a position in its orbit almost directly

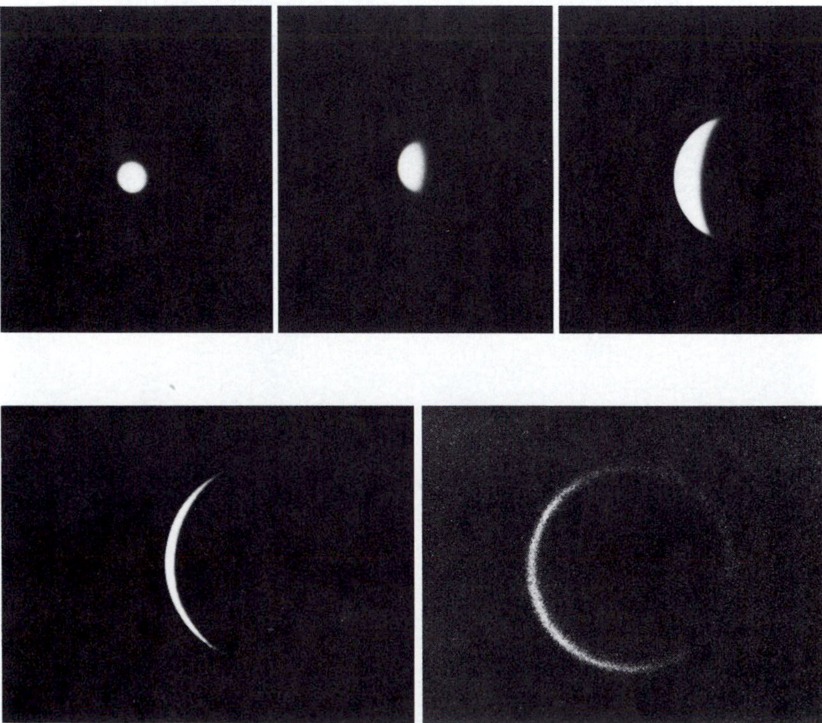

Figure 11.6 Phases of Venus.

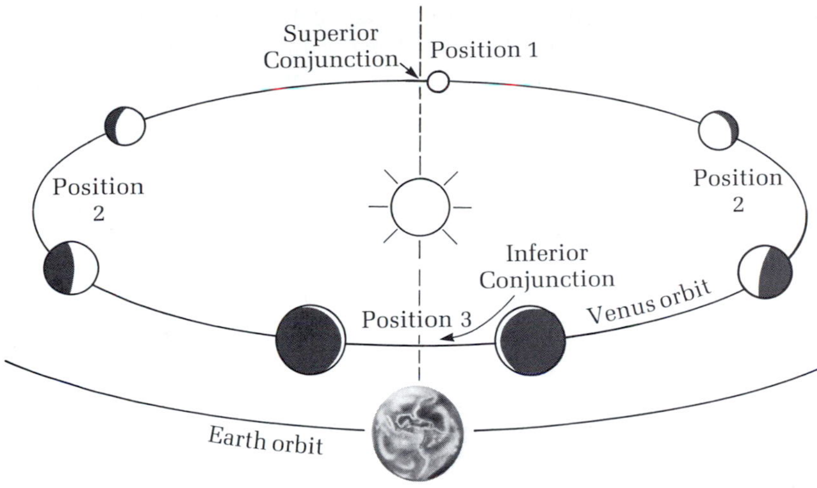

Figure 11.7 The mechanism of Venus' phases. When it is close to inferior conjunction we see it as a large crescent; at full phase it is much smaller because farther from Earth.

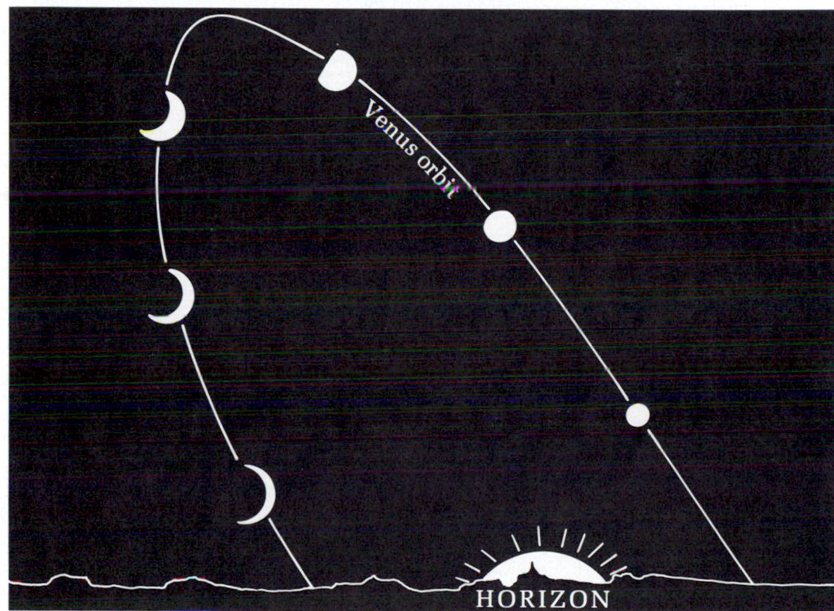

Figure 11.8 Phases of Venus relative to the horizon.

opposite the earth (Position 1, very close to superior conjunction), we see its entire disk, but at this point it is dimmest. Figure 11.8 shows how these phases relate to Venus's position above the horizon. When it is at its greatest western (or eastern) elongation, it will be seen in half phase, and not at its brightest; it is brightest when it makes an angle of approximately 39° with the sun.

Rotation

Because its surface could not be seen, little was known about Venus's rotation. In 1956, R. Richardson published data indicating that it had retrograde motion—that is, its spin was opposite that of all the planets in the solar system (except Uranus). Bernard Guinot confirmed Richardson's discovery in the early 1960s but found a much higher rotational velocity; the period, according to his measurements, was 4 days. However, Richardson's and Guinot's data pertained only to the clouds above Venus. In the early 1960s astronomers, using the Doppler shift of a radar signal reflected from its surface, found that Venus's period—much slower than that of its clouds—was a long 243 days. But, like the clouds, its direction of motion was retrograde.

What caused it to become retrograde? Venus is influenced by both the sun and the earth. It is close enough to the sun to have been tide-locked to it at one time, but it is the earth that has apparently determined its present period of rotation. It has recently been shown that Venus presents the same face to earth at each inferior conjunction, which means that Venus is tide-locked to the earth.

With such a strange rotation what might a day on Venus be like? The sun, of course, would rise not in the east but in the west, and it would do so not every 243 Earth days of its rotation but every 117 Earth days. (Because of the clouds one would not actually see the sun.) The difference between the rotational period and the length of the Venusian "day" has the same basis as sidereal time (Chapter 3, Fig. 3.24) and the corresponding relationship between Earth and the moon (Chapter 9, Fig. 9.1). Revolution also plays an important role.

Atmosphere

Even a casual look at Venus through a telescope shows that it is blanketed by dense clouds. Its albedo (reflectivity) is high, and in the visible region of the spectrum, no markings whatever can be seen. In 1932 Walter Adams and Theodore Dunham, Jr., showed that carbon dioxide is a major component of the Venusian atmosphere. Until the early 1960s, when W. M. Sinton and V. I. Moraz identified carbon monoxide, carbon dioxide was the only component known for certain. With this much oxygen available (even though bound to carbon) it seemed reasonable that there must be water vapor in the atmosphere, but almost a decade passed before it was identified spectroscopically—by Ron Schorn and several colleagues, who showed that about 0.01% of the Venusian atmosphere is water vapor.

From the mid-1960s on, several space vehicles—of the Venera (U.S.S.R.) and Mariner (U.S.A.) series—were successful in obtaining information about Venus. Thousands of photographs, taken in the ultraviolet, revealed bands of cirruslike clouds swirling around the planet. Then in December 1978, a probe from the U.S. spacecraft Pioneer Venus 2 landed on the surface and sent messages back to Earth for over an hour.

As a result of these explorations we are now relatively certain that the Venusian atmosphere is about 95% carbon dioxide, less than 5% nitrogen, less than 1% oxygen, and about 0.01% water vapor. Pioneer also indicated that small amounts of neon and argon are present, as well as sulphuric acid and solid sulphur. More recently Venera 12 and 14 detected chlorine, and, in addition, both hydrogen chloride and hydrogen fluoride have been detected spectroscopically

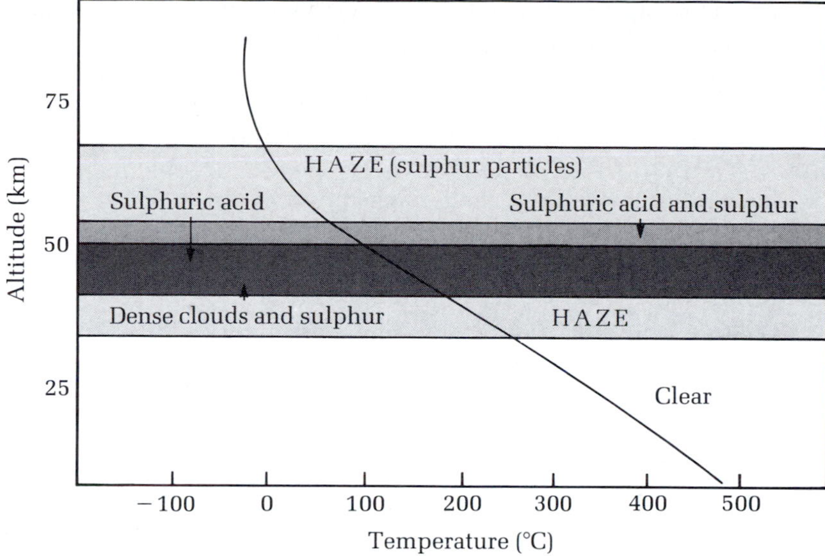

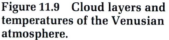

Figure 11.9 Cloud layers and temperatures of the Venusian atmosphere.

from Earth. The presence of sulphuric acid and hydrogen fluoride has, in fact, led to an interesting conjecture: The atmosphere of Venus may contain one of the most corrosive acids known—fluorosulphuric acid.

The sulphuric acid, sulphur, and chlorine are primarily in the Venusian clouds, which are quite different from those of Earth, not only in their makeup, but also in their distance above the surface (Fig. 11.9). There are three main layers. The uppermost layer, which consists mostly of haze, occurs at an altitude of about 68 km (42 mi). With decline in altitude the haze gradually thickens into a relatively dense smog, which is due to sulphur particles floating in the carbon dioxide atmosphere.

The second, next-lower cloud layer is at an altitude of about 56 km (35 mi) and consists of a mixture of sulphur particles, chlorine, and sulphuric acid droplets. Below it, at an altitude of about 50 km (31 mi), is the third layer of clouds, the most opaque of the three and the main reason we cannot see the surface of the planet. Sulphur particles, chlorine, and sulphuric acid droplets are present in this layer. If, as is probable, sulphuric acid drizzle occurs here, it must descend at an agonizingly slow rate and vaporize in the hot layers beneath it.

Below the third layer is another hazy region that gradually clears with descent, so that at about 31 km (19 mi) above the surface the atmosphere is relatively clear. In order to see the surface, however, one would have to descend to an altitude of about 7 km ($4\frac{1}{2}$ mi). Only about 1% of the sunlight that strikes Venus reaches this region, but this still gives light comparable to a dull, rainy day here on Earth.

Another layer of clouds, somewhat higher than the uppermost layer described above, is found near the polar regions. It forms a ring around the poles at a latitude of about 70° (Fig. 11.5). The region inside the ring is sometimes called the "polar hole."

Winds of up to 300 km/hr exist near the top of these clouds. Ultraviolet pictures taken by Mariner 10 show strong convective currents in this region, possibly similar to those that exist in Earth's atmosphere. The winds decline with altitude until, at the surface, they are negligible (about 6 km/hr). The pressure at the surface is almost 100 times what it is on Earth at sea level.

Because the pressure (and consequently the density) of the Venusian atmosphere is so high, it is highly refracting; light rays entering the atmosphere are severely bent—by 180°. For an observer on the surface this would give rise to several strange optical effects. For example, could he see the sun (and, of course, he could not), it would appear to flatten and elongate as it approached the horizon; finally, it would become a thin band stretched out entirely around him. Furthermore, regardless of where he stood on the surface, he would appear to be in a gigantic bowllike valley (again, in the unlikely case that he could see for any appreciable distance through the atmosphere).

Looking upward during the day, he would see an orange sky, which, as darkness fell, would glow. (We do not know what causes this.) Added to this strange view would be almost continuous bolts of lightning, most of which occurs in a zone that extends from an altitude of 5 km (3 mi) to about 10 km (6 mi). Venera reported up to 25 bolts per second as it passed through this region.

Temperature

Venus is an inferno; when the sun is overhead, the surface temperature is about 750°K, high enough to melt lead. Yet Venus is much like the earth in size and mass, and it is in the ecosphere, or life-zone,[1] of the sun. What accounts for the difference? Part of the answer lies in a phenomenon called the greenhouse effect. Although Venus receives twice as much sunlight as the earth (because it is closer to the sun), most of this light is reflected back into space, so that direct heating by the sun is not the cause of the extremely high temperatures. However, the small fraction of sunlight that penetrates the clouds strikes the surface and warms it. Once heated, the surface reradiates infrared waves, most of which, unlike the case for Earth, cannot penetrate the clouds. Thus the infrared radiation is reflected back and forth between the surface and the cloud deck and accumulates; this is the greenhouse effect (Fig. 11.10). Of course, if none of the infrared radiation escaped, the surface would heat up indefinitely. Obviously, some escapes, and a balance is reached that stabilizes the temperature at its very high level. On Earth, the surface temperature is about 50°F higher than it would be were there no greenhouse effect.[2]

[1]The region around the sun within which the range of temperatures is suitable for life as we know it. See Chapter 27.

[2]Actually, the greenhouse effect is misnamed. Though the glass walls of a greenhouse do stop infrared radiation (*out* of the greenhouse), the important heating effect comes from the fact that the warmed air cannot escape from the greenhouse. The glass helps by stopping infrared, but a greenhouse made of rock-salt panes—which let infrared pass through—would work nearly as well. In fact, such a greenhouse was built to test this point, and it functioned admirably.

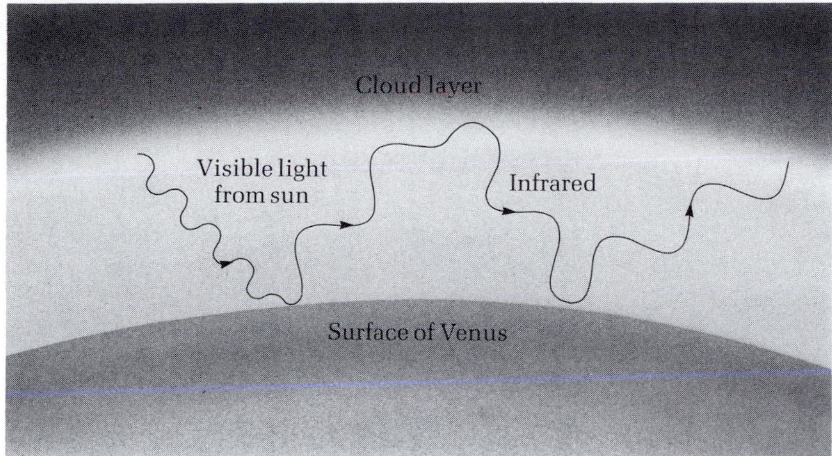

Figure 11.10 Mechanism of the greenhouse effect.

Until recently it was generally believed that the greenhouse effect could account for all the heat that was observed near the surface. However, Pioneer Venus found that more energy was being radiated upward from the lower layers of Venus's atmosphere than they were receiving as sunlight. It seems almost as if heat is being transferred from cool regions to hot regions—a phenomenon not possible under normal conditions. The explanation is not known.

Surface Features

Because we cannot see the surface of Venus, we must rely on information obtained from spacecraft and on radar probing of the surface (Fig. 11.11). Radar probing is based on the principle that a radar beam from Earth striking an elevation—in this case on the surface of Venus—will return in a slightly shorter time than one reflected from, say, a valley floor. Measurement of the frequency and resolution of the returning beam are also useful in determining structures. In addition to the radar mapping that has been done from Earth, Pioneer Orbiter has been mapping the surface since it went into orbit in December 1978. As a result, we now have a fairly good indication of what Venus looks like.

Venus is remarkably round; this is undoubtedly due to the fact that it rotates too slowly to have an equatorial bulge. Moreover, its features vary little in altitude from its mean radius. The Venusian crust is old, and it may be extensively covered with boulders. Photographs obtained by Venera 9 and 10 showed boulders, some quite sharp (Fig. 11.12), others flatter and rounded, and confirmed that they were made up mostly of granite. Venera 13 and 14 indicated that the rocks were similar in composition to terrestrial oceanic basalt and are yellow-orange. Some volcanic ejecta were also found.

Like the earth, Venus also has lowlands and highlands, but only 10 to 20% of Venus is lowland compared to 75% of the earth. The Venusian lowlands are generally smooth, but large circular regions believed to be impact craters have been detected on them and else-

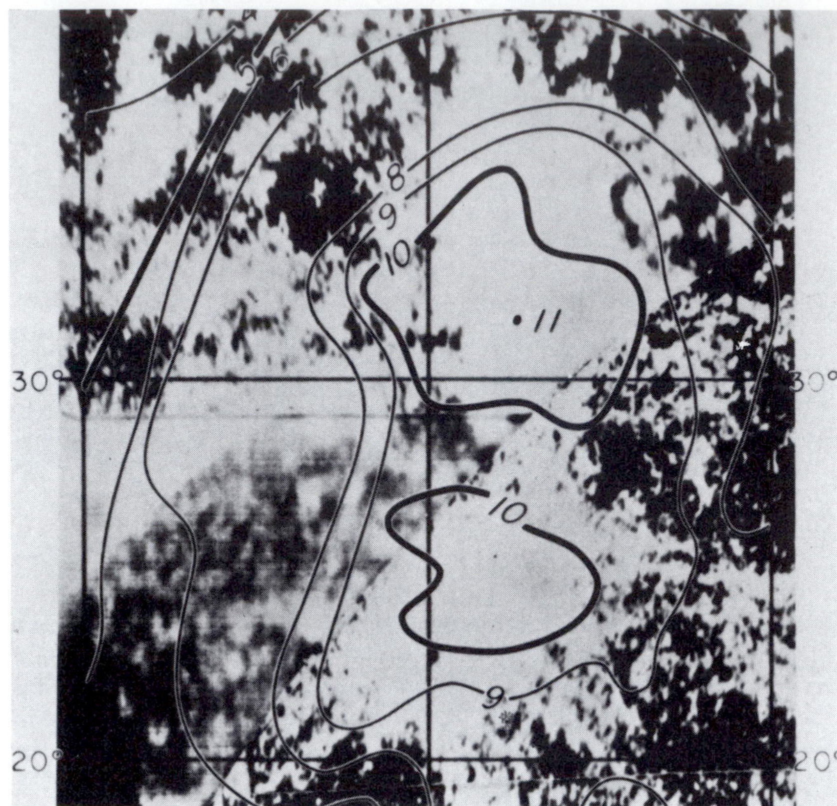

Figure 11.11 Radar map of part of Venus.

where. They vary from 480 to 800 km (300–500 mi) in diameter and are generally much shallower than lunar craters. When one considers the thick Venusian atmosphere, this is perhaps to be expected.

There are two large continents (raised land masses surrounded by flat areas) on Venus: Ishtar Terra and Aphrodite Terra (Fig. 11.13). They are similar to continents on earth, except that large mountain ranges are conspicuously absent, although there are individual mountains. The highest, Mt. Maxwell, is in fact higher (above the mean surface level) than Mt. Everest is above sea level. Mt. Maxwell has an altitude of about 10 km and is located near one end of Ishtar Terra. Although it is probably a volcano, there is no evidence of recent flow. To the east of Maxwell there is a large caldera (basin) almost 1000 km (about 600 mi) across.

Figure 11.12 The first photograph of the surface of Venus, taken from the Soviet craft Venera 9.

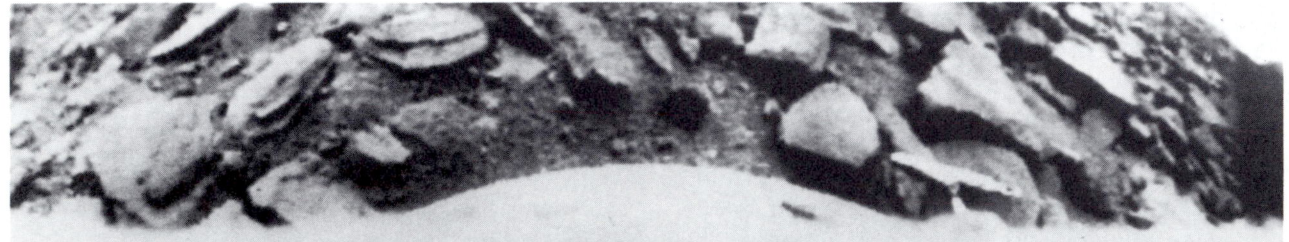

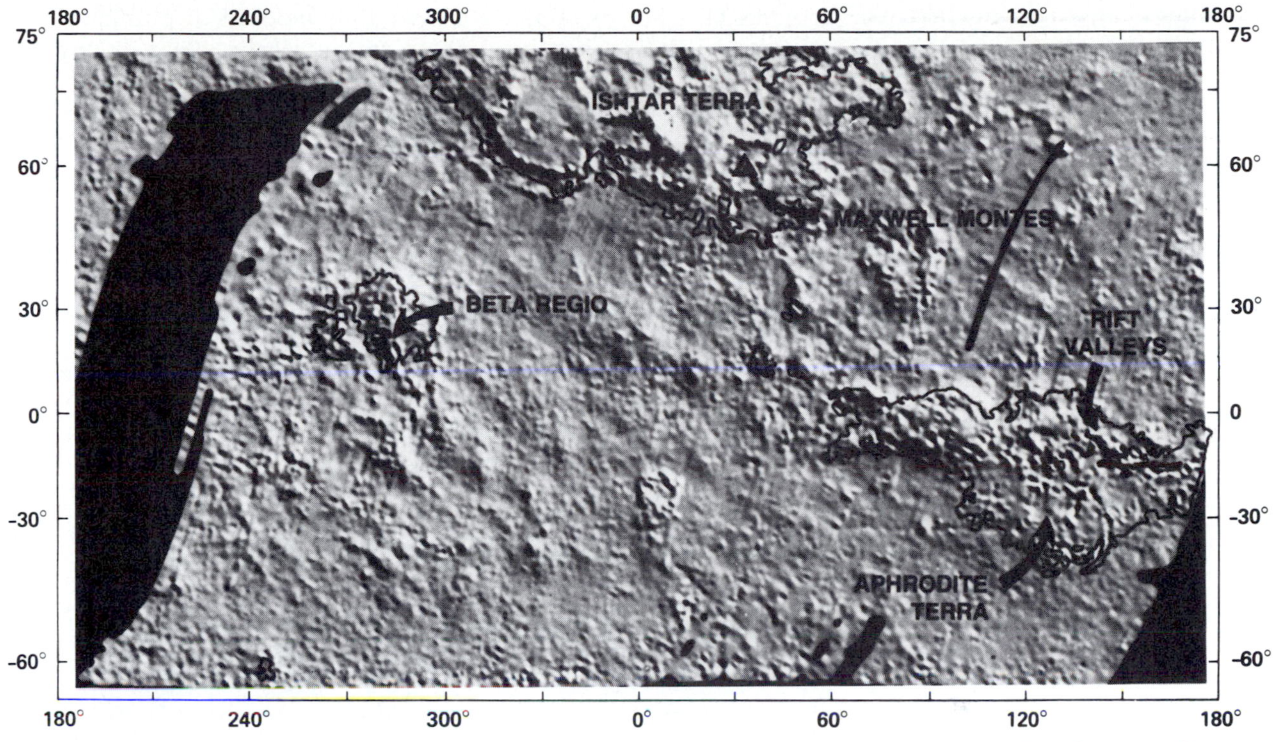

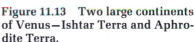

Figure 11.13 Two large continents of Venus—Ishtar Terra and Aphrodite Terra.

The other continent—Aphrodite Terra, somewhat smaller than Ishtar Terra—has no mountains that rival Mt. Maxwell but has some peaks about 3600 m (12,000 ft) high. There are no impact craters on either continent, an indication that they are much younger than the rest of the surface.

There are two other elevated regions, which could perhaps be called small continents, on the planet. One of them, Beta Regio, believed to be an extinct volcano, is about 4 km high and in many ways reminiscent of Olympus Mons on Mars (see next section). The second region, Alpha Regio, is one of the roughest and, consequently, brightest areas on the planet.

In addition to continents there are several great rift valleys. As in the case of the earth, these may have been caused by plate tectonics, though there is little, if any, evidence for plate movement at present. It is possible, of course, that there was a small amount in the past.

MARS

Mars has long been a favorite of science fiction writers. In perhaps the first and best-known such tale, H. G. Wells's "War of the Worlds," strange creatures from Mars, out to conquer Earth, were defeated by our bacteria. Mars's popularity in fiction may have had some of its roots in fact: For example, a tremendous amount of publicity has, at one time or another, been given to the Martian "canals" by the press. It is perhaps little wonder that so many people panicked (in the belief that we were under attack by Martians) when "War of the Worlds" was broadcast on radio in 1938.

It soon became clear, however, that the fiction could be no more than that; Mars could not support a higher form of life, for the evidence seemed to be that it was a vast desert, and any life on it could be only of primitive form. The first flight of a space vehicle showed that the landscape was even more harsh and barren than expected and, moreover, covered with craters (Fig. 11.14).

Even through a telescope, Mars is difficult to see clearly from Earth; it is considerably smaller (6795 km in diameter compared to Earth's 12,756 km), and both our atmosphere and the haze and dust in the Martian atmosphere hinder our view. Considerable patience and a night of "good seeing" are needed before distinct markings finally

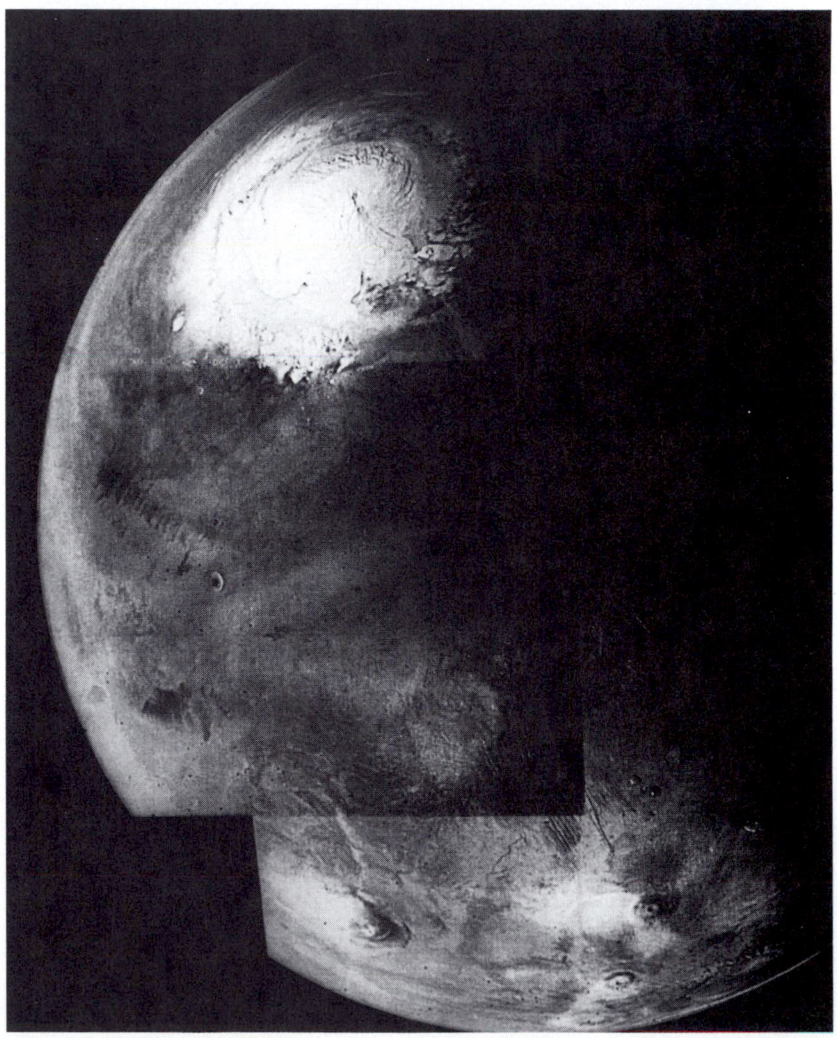

Figure 11.14 Mars in a Mariner 9 photograph.

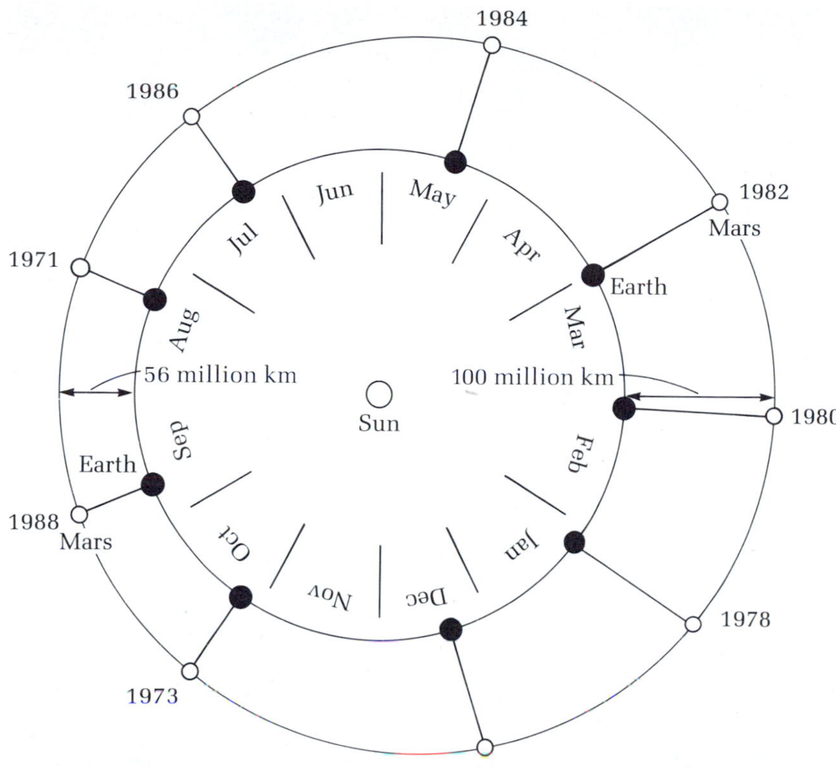

Figure 11.15 Oppositions of Mars from 1971 to 1988.

become visible. In addition, it is only at certain times that Mars is ideally situated for viewing. These occur when Mars is directly opposite Earth from the sun in its orbit; at this position Mars is said to be in opposition. Because of the ellipticity of the orbits (particularly that of Mars), some oppositions are more favorable than others (Fig. 11.15). The most favorable occur near Mars's perihelion; on these occasions the two planets may be as close as 56 million km (35 million mi). Oppositions of this type occur only once or twice every 15 to 17 years, usually in the month of August or September. Oppositions in general, on the other hand, occur every 780 days. During an unfavorable one, the two planets can be separated by as much as 100 million km (62 million mi).

With the development of the telescope, Mars took the spotlight; it was the only planet whose surface could be clearly seen. In 1659 Christian Huygens sketched dark markings resembling an area we now call Syrtis Major. A few years later G. D. Cassini determined its period of rotation to be approximately 24 hours by watching the movement of the dark markings. We now know that the Martian day is indeed very close to ours—24 hours, 37 minutes, 22 seconds; its orbital period is two of our years.

When William Herschel turned his giant reflector toward the red planet, he was amazed by its similarity to Earth and was sure it was inhabited. It even appeared to have seasons like Earth's (Fig. 11.16).

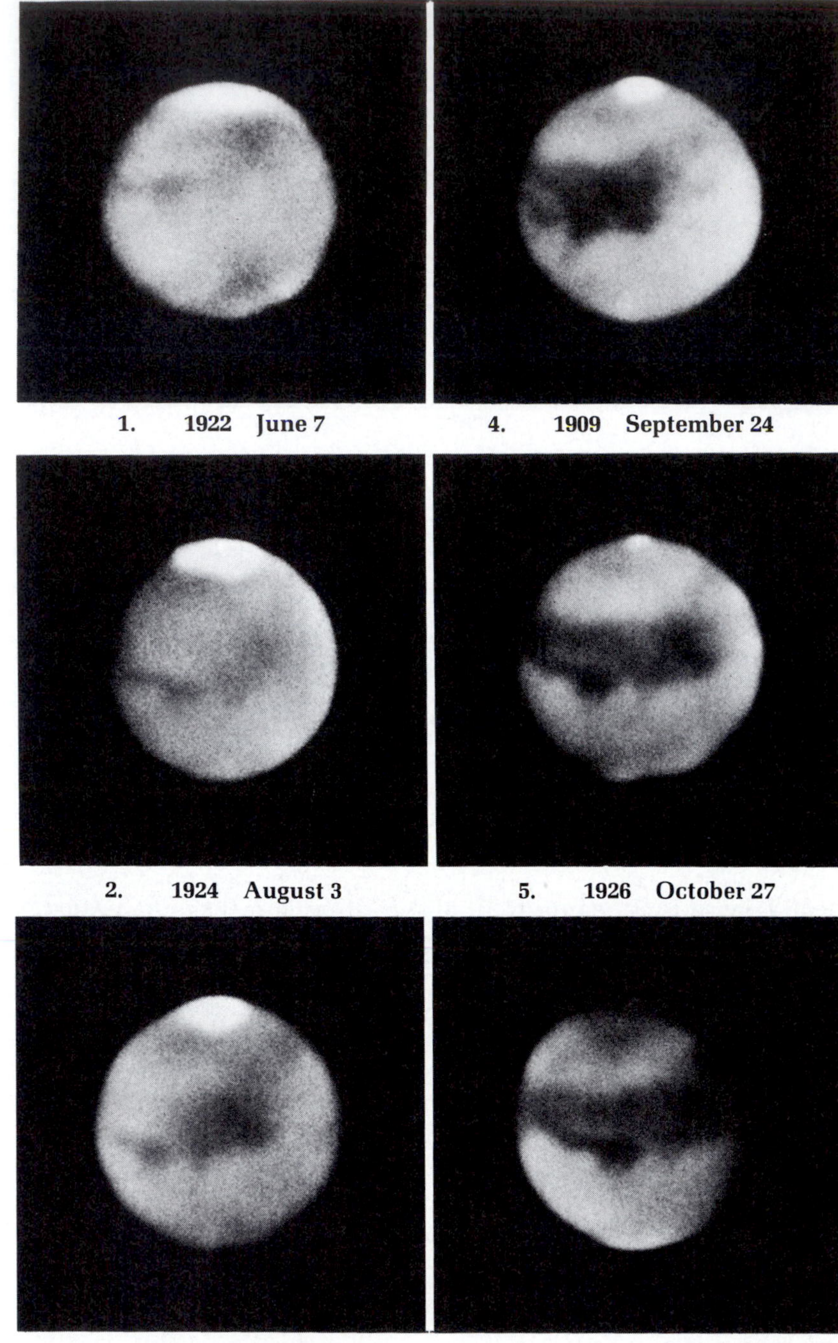

Figure 11.16 Mars as seen from Earth, showing seasonal changes.

Watching over a period of months, he saw the polar caps recede and a wave of darkening spread over the equatorial regions of the planet. Herschel thought this darkening had to be vegetation and, perhaps, cultivated fields. He also noted that, since there seemed to be a cycle of seasons, Mars must have its axis tilted at some angle other than 90° to the plane of its orbit. We now know that its angle of tilt is 25°, quite close to the earth's 23.5°.

In 1877, an opposition year, astronomers around the world were scrutinizing Mars's surface. Asaph Hall of the U.S. Naval Observatory was more interested in whether Mars had a moon. (Several people, including Kepler and Jonathan Swift, had made predictions—really little more than guesses—that there would be two moons close to the surface.) Hall finally did locate two tiny moons, which will be discussed on page 205.

At the same time that Hall was searching for moons, Giovanni Schiaparelli was looking for something quite different. A few years earlier a fellow Italian, Father Secchi, had made drawings showing faint lines on the planet. Schiaparelli was scanning the planet looking for these *canali* (channels) and, indeed, found them; in fact, he found many more than Secchi had, and he also drew them as being much more pronounced and considerably thicker. These "canali" (changed in the American press to "canals") fascinated the public. The implication was established: Mars must be inhabited!

In America in 1894 wealthy Bostonian Percival Lowell, convinced that there were beings on Mars and fascinated at the prospect, set up an observatory outside Flagstaff, Arizona, to study Mars. Soon recognized as the world expert on the Red Planet, he made thousands of drawings, many as imaginative as they were skillful. According to his drawings (Fig. 11.17) the canals were straight or smoothly curved and joined at what he called "oases." They varied in length from a few hundred to thousands of kilometers; some were even double, but none was ever photographed.

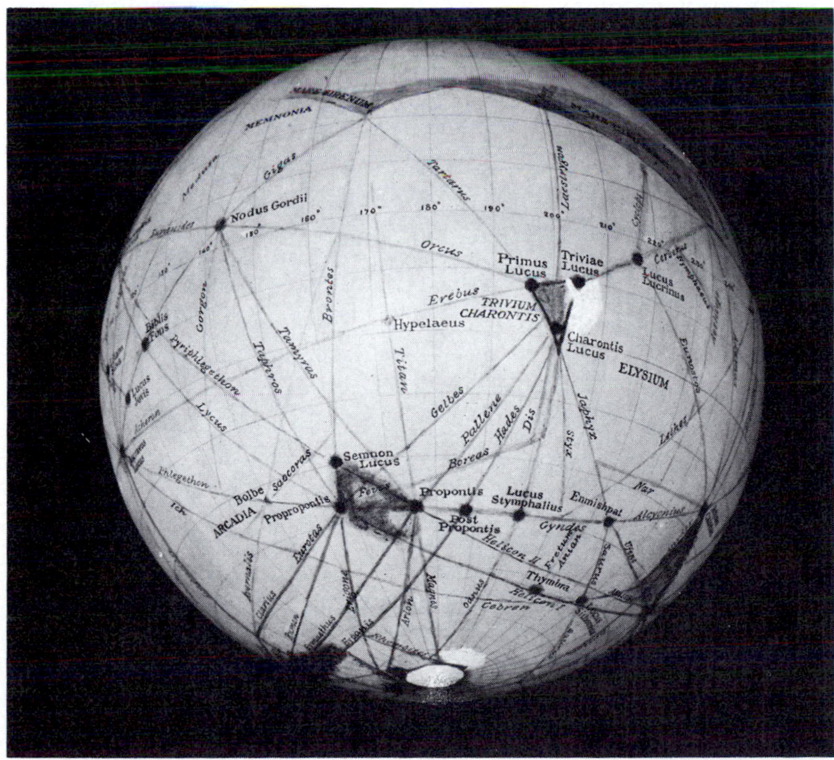

Figure 11.17 Martian channels drawn by Percival Lowell.

Lowell was convinced that Mars was a dying planet inhabited by a race of beings who built gigantic canals to conserve the last of their water. Each year as the polar caps thawed (he assumed they were water ice), the water was brought via the canals to the equatorial regions, where it was used to irrigate crops.

Some astronomers saw Lowell's canals; a number were convinced that Lowell was seeing things and had an overactive imagination. The debate raged for years, even when larger telescopes went into operation. When the 100-in reflector at Mt. Wilson went into operation, it showed an image very similar to the one obtained at the Lowell Observatory.

Although the image of Mars on the Mt. Wilson plates was still fuzzy, the new reflector provided enough light to study it spectroscopically. Nevertheless, this analysis did not end the debate; spectra showed that the Martian atmosphere was quite different from that of Earth—mostly carbon dioxide. Low temperatures were much lower than those on Earth: A balmy temperature of 293°K (68°F) at midday, with the sun directly overhead, could plunge at night to 187°K (−123°F). Life capable of building canals could not survive for long here, but there was a chance that lower forms could.

The Mariner Flights

The controversy over life was far from over when the first space vehicle to Mars (fourth in a series called "Mariner") was launched in November 1964. Astronomers, not sure what to expect when Mariner 4 arrived at Mars, were shocked to see craters everywhere. Were the whole planet covered to the same extent (only a small section was viewed), Mars would be cratered as much as the moon. But since Mars, unlike the moon, had an atmosphere, most of its

Figure 11.18 Mariner 9 photograph of Mars.

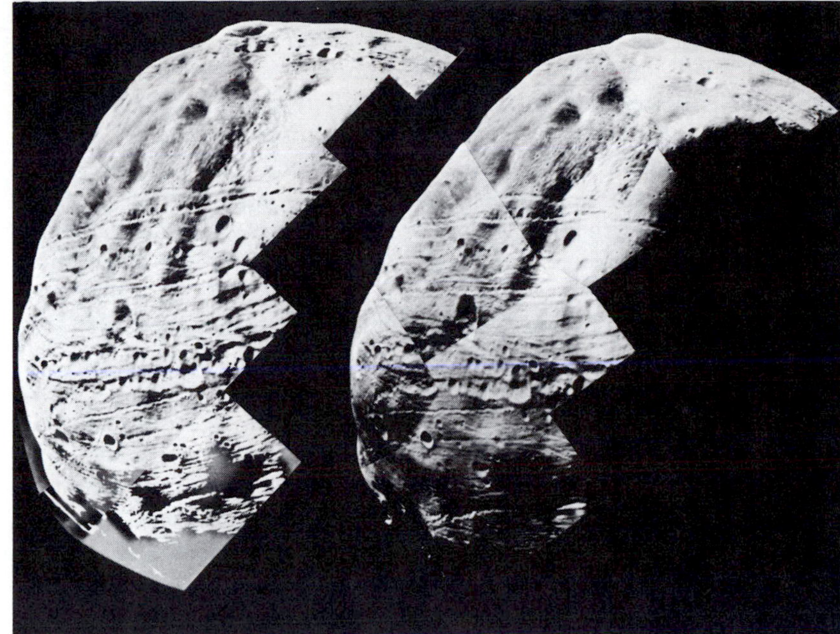

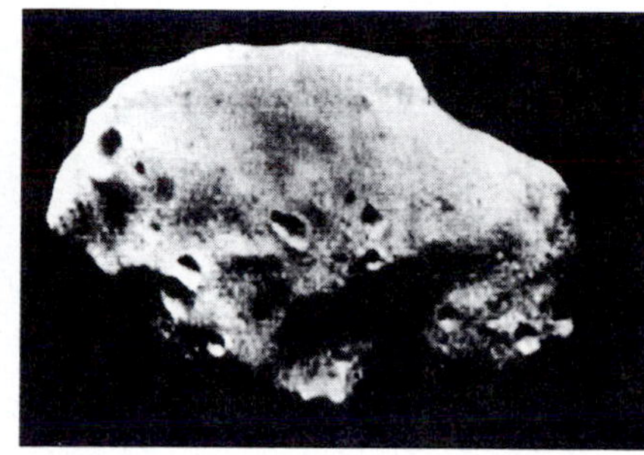

Figure 11.19 Phobos, the smaller of Mars' two moons, as photographed by Mariner 9 (right), and in a closer view (left) from Viking 1. The grooves are 100–200 m wide and about 10 m deep.

craters should have been eroded, as they are on Earth. However, because the atmosphere of Mars is much thinner than that of Earth (pressure near the surface less than $\frac{1}{100}$ that on Earth at sea level), erosion occurs at a much slower rate. There are dust storms, but apparently they do little to hasten erosion.

To determine the extent of cratering, Mariner 6 and 7 were launched in 1969. The areas viewed again showed many craters. In 1971, Mariner 9 was put into orbit around Mars with the much more ambitious mission of photographing the surface for months (Fig. 11.18). Because the face of Mars was completely obscured by a gigantic planet-wide dust storm, the cameras were aimed instead at Mars's two moons, Deimos and Phobos. Deimos and Phobos—about 16 km (10 mi) and 25 km (15 mi) across, respectively, appeared to be heavily pocked with craters, and Phobos showed long grooves (Fig. 11.19). Phobos, only 6000 km above the surface of Mars, has a period of 7 hours, 40 minutes, which means it crosses the sky three times a day. Deimos, at a distance of 20,000 km, orbits every 30 hours.

By late January 1972 the dust storm began to subside, and the surface of Mars became visible. Many volcanoes were seen, four particularly large. The largest, now called Olympus Mons (Fig. 11.20), is 500 km across at the base and is capped with a crater 65 km in diameter. Its height, 25,000 m, is twice that of Mt. Everest (measured from sea level).

The northern hemisphere of Mars, its volcanic region, is quite different from the heavily cratered southern hemisphere, photographed in earlier Mariner flights. It is apparently much younger, and there is considerable evidence of lava flows. The major reason for the great differences in size between Martian volcanoes and those of Earth is the absence of plate tectonics on Mars. As we saw in Chapter

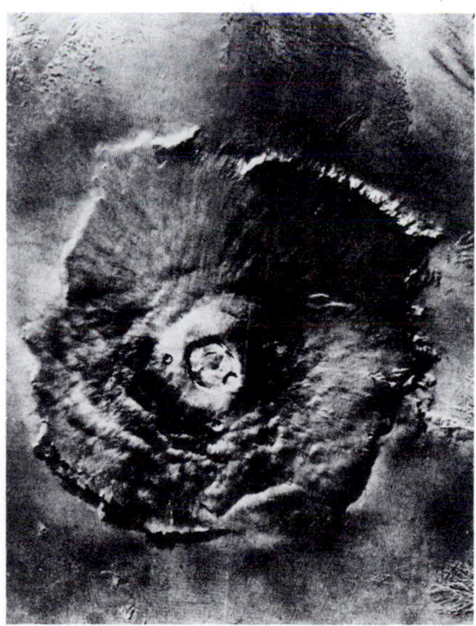

Figure 11.20 Olympus Mons, largest volcano on Mars. The caldera, or basin, at the center, is about 25 km above the surface.

10, the earth's geology is dominated by plate tectonics. The motion of the continents and the formation of mountains and of the ocean depths are all controlled by the motion of plates. Volcanoes on earth can grow only for a few hundred thousand years before the plates on which they rest move them away from their sources of magma (molten rock) and they become extinct.

On Mars, with no plate movement, volcanoes remain indefinitely over the source of magma continue to grow as long as magma is available. Their great height is also attributable to Mars's crust, believed to be twice as thick as Earth's and so able to support a greater mass. Mars's lower gravitational pull is also a factor. But even if the source of magma is available indefinitely, the height to which a volcano can rise is still limited by hydrostatic pressure (from the interior of the planet) by which the magma is forced upward. We may, in fact, be seeing this limit in the largest volcanoes: all have approximately the same height (about 27 km). Also of interest, but unexplained, is the fact that the three largest volcanoes rest on an elevation in Mars's crust called the Tharsis bulge. It has a maximum height of approximately 10 km above the surface and is about 6000 km across.

Mariner 9 also disclosed a gigantic canyon (Fig. 11.21), now called Valles Marineris, running along Mars's equator. If placed in the United States, the canyon would extend from coast to coast and beyond. It is 5000 km (3100 mi) long, 120 km (75 mi) wide, and 6 km (3.7 mi) at its deepest. (The Grand Canyon is 150 km long.) The canyon is the one place on the planet where there is some evidence of an early stage of plate tectonics; it may have been caused by the separation of two plates.

Perhaps the most startling discovery of Mariner 9 was what appear to be old riverbeds. These channels have all the characteristics of riverbeds: shape, sandbars, an apparent downhill flow, and, in many, the dendritic appearance (like the veins of a leaf) typical of riverbeds (Fig. 11.22).

Even before the Mariner flights it was known that wind velocities were high on Mars. Annual dust storms were a clear indication. Mariner 9 provided further evidence of high winds; sand dunes and wind-blown landscapes in several areas (Fig. 11.23). Before Mariner it was assumed by many that the wave of darkening throughout the equatorial regions each spring might be a lower form of vegetation. Astronomers now believe, however, that it is due to the fierce Martian winds, which annually produce a gigantic dust storm that covers a large fraction of the face of Mars. Winds up to 150 km/hr (90 mi/hr) blow the fine sand from the surface up to heights of 30 km (19 mi). In short, it seems that sand is blown away from the equatorial regions in the Martian spring, exposing the darker sand beneath it; later it is blown back into these regions.

But why are the storms so gigantic compared to earth storms? This is caused by a kind of feedback mechanism. As more dust is picked up by the wind, the intensity of the wind increases, and this, in turn, picks up even more dust. Eventually, though, the atmosphere

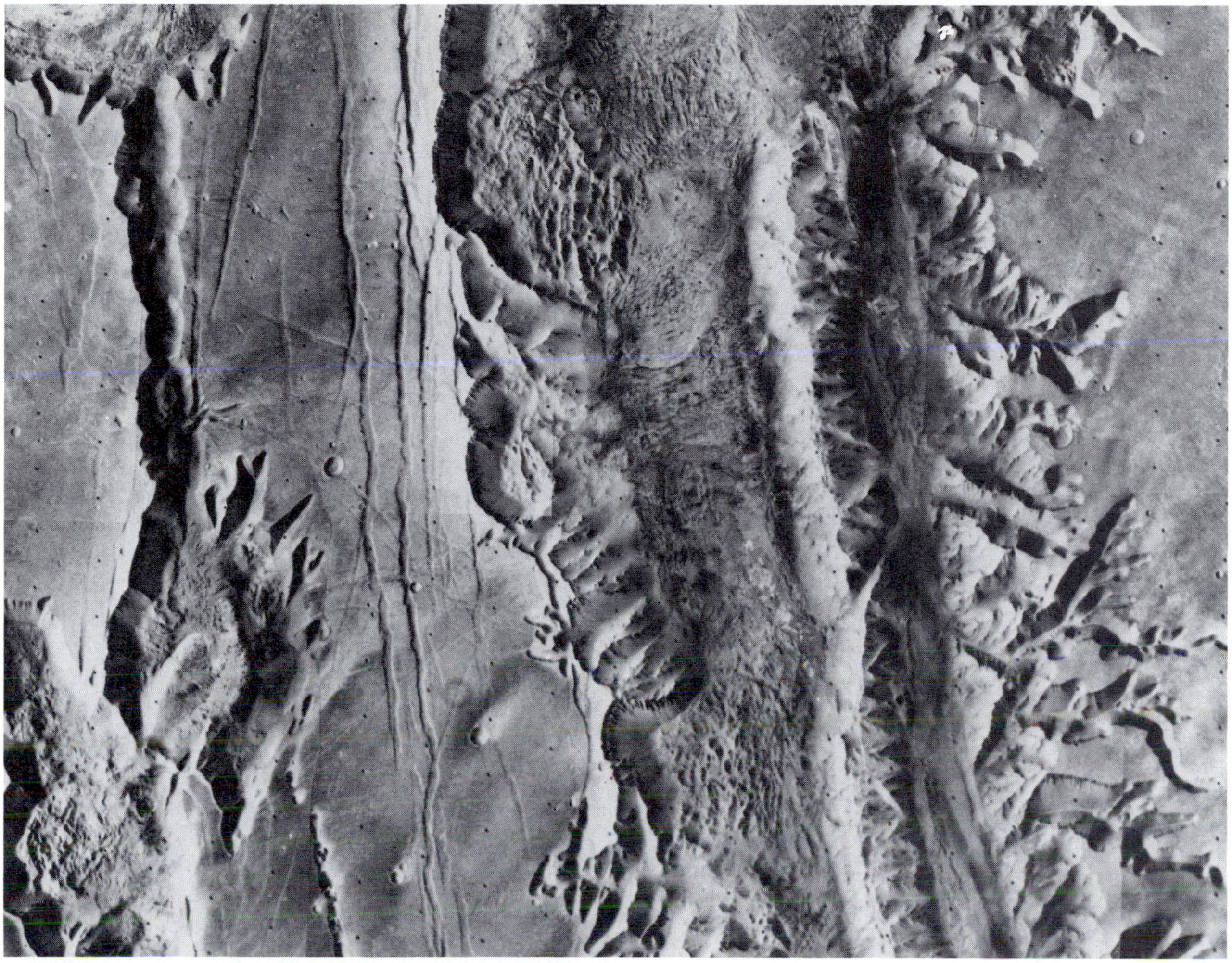

Figure 11.21 A portion of 5000-km-long Valles Marineris.

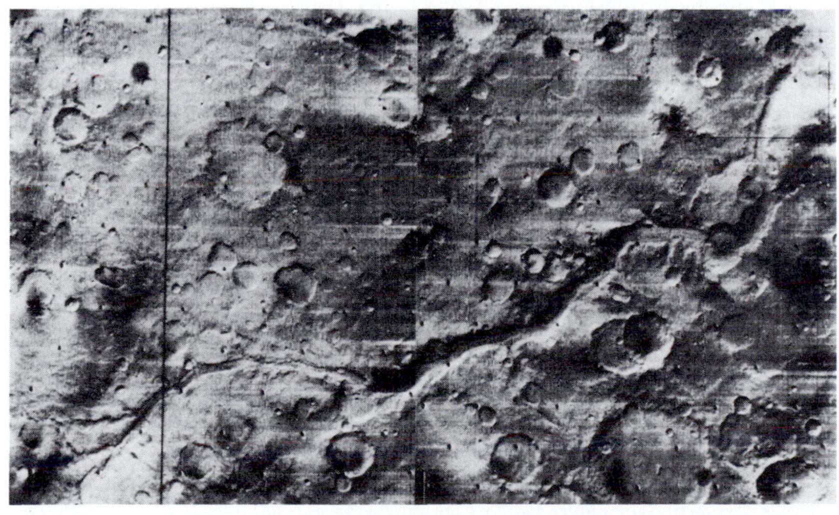

Figure 11.22 (left) A meandering channel on Mars that has the characteristics of a river bed.

Figure 11.23 Martian sand dunes, each about 1–2 km across.

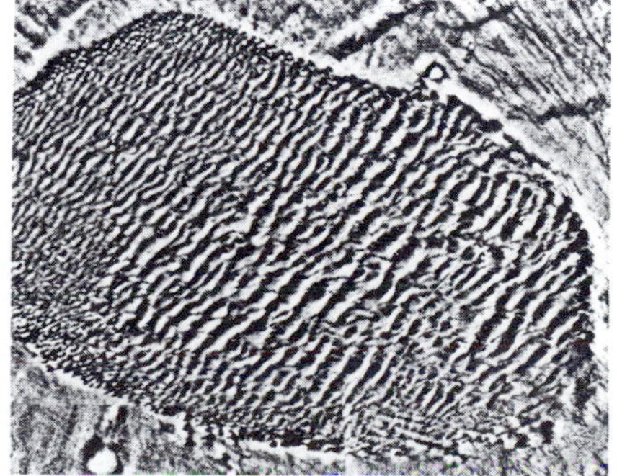

can keep no more suspended, and the dust begins to settle. In two to three months the storm is over; most, but not all, of the dust settles.

Viking

Though scientists were sure there were no higher forms of life on Mars, a landing was still needed to check for other forms. On July 20, exactly 7 years after the first manned landing on the moon, the Viking 1 craft landed on Mars.

The area around Viking was quite flat (Fig. 11.24; color). No mountains were visible, but a few small dunes could be seen in the distance. A field of boulders extended in all directions. Beneath the boulders was fine, flourlike sand (Fig. 11.25), and both boulders and sand were red. Because of the fine red sand that was continually circulating in the atmosphere, the sky was pink, and sunsets were purple. The atmosphere, as expected, consisted mostly of carbon dioxide (95%), but small amounts of nitrogen, argon, carbon monoxide, oxygen, and water vapor were also found. The pressure, as mentioned earlier, was less than $\frac{1}{100}$ that on Earth (at sea level). Temperatures ranged from a high of 244°K (−20°F) to 187°K (−123°F), and the soil contained silicon, iron, calcium, and, as expected, considerable ferric oxide (iron rust).

While Viking 1 was checking conditions on the surface, Orbiter 1 (still in orbit) was photographing other regions of the planet, in particular, the polar caps. For years astronomers had believed that they were frozen carbon dioxide. As "summer" approaches in the northern hemisphere of Mars, the north polar cap recedes. Orbiter observed the cap when it had been reduced to its minimum size and found that the temperature was well above that of solid carbon dioxide. Since there was still a relatively large residual cap, it seemed that it could only be composed of water. The upper layer of the polar cap is apparently carbon dioxide snow that, when it has melted, reveals a layer of water ice that lay beneath it. Much of the water of the planet may, in fact, be held here.

There is other evidence for water (in solid form) on the planet. The chaotic terrain at the end of Valles Marineris appears to be a region that has sunk or collapsed when ice beneath the surface melted and the resulting water flowed away. And, of course, the numerous dry riverbeds also indicate that there was considerable water on the surface at one time. The abovementioned dendritic appearance of many of these river systems (Fig. 11.26) indicates that rain may have fallen in the past.

Despite the evidence of water in Mars's past and present, there were none of the canals that Lowell and Schiaparelli had seen. Even the large canyon and numerous old riverbeds are too small to have been seen from Earth. Furthermore, the positions of the canals in most of the early drawings generally do not coincide with the positions of the riverbeds and canyon. The canals led from the poles to the equatorial regions; most of the riverbeds and the canyon, on the other hand, lie in the equatorial regions. It is likely, then, that what Lowell saw were chance alignments of various markings on the surface.

Figure 11.25 Martian rocks and sand. Considerable drifting of the sand can be seen.

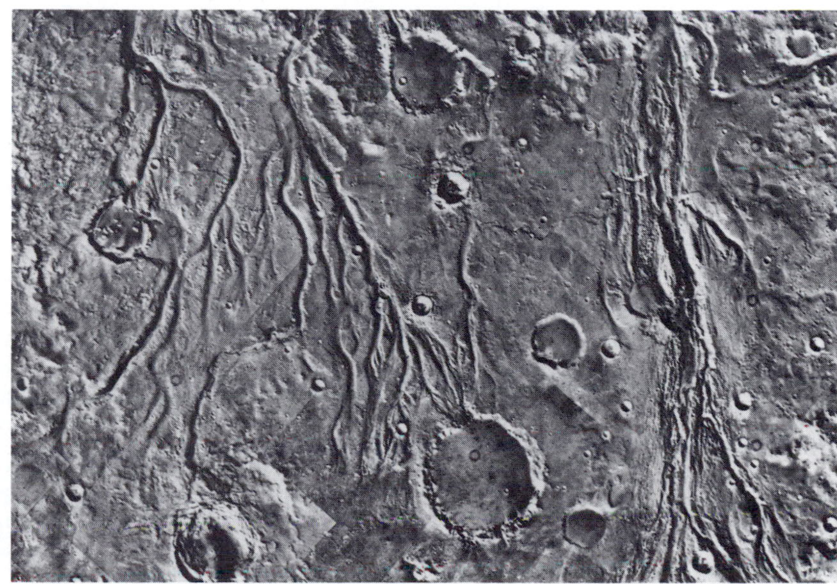

Figure 11.26 Evidence of water flow on the surface of Mars. The area shown is about 180 km on a side.

Early History

Mars is, in many ways, even more intriguing than we had anticipated. Although there appears to be no life on it, or only the possibility of lower forms (see below), it is still a geologist's dream, and we are likely to learn much about the earth and the solar system from studying its surface. Its early history seems to be different in many ways from that of Earth, and this is reflected in its soil and rocks, which are generally quite different. The only place on Earth where somewhat similar rocks and soil are found is in dry volcanic deserts.

There is no doubt that there was once liquid water on the surface of Mars in considerable amounts. Some of the riverbeds, for example, are gigantic by earth standards—over 50 km (30 mi) across. Today water cannot flow on the surface of Mars; in general, temperatures are too low, but, even when otherwise, water resulting from melting ice must evaporate immediately into the atmosphere because of the low atmospheric pressure. For water to have flowed at one time, then, Mars would have to have had a much denser atmosphere. We now believe this to be the case.

The atmosphere of Mars, like that of Earth, was expelled from volcanoes early in its history. (Volcanism is believed to have commenced on Mars when it was about 1 billion years old; the period did not last as long on Mars as it did on Earth because Mars is much smaller.) Light gases such as hydrogen escape rapidly into space. On Mars, with its relatively weaker gravitational pull, heavier gases such as nitrogen also escape, but slowly; a gas such as argon, which is particularly heavy, does not escape to any degree. Almost all the argon that has ever been expelled on Mars is still in its atmosphere. This gives us a measure of the total amount that was degassed in the past. Since volcanoes release a known fixed ratio of gases, we can determine (from the amount of argon present now) approximately how much nitrogen, carbon dioxide, and so on were expelled in the past. The amount of argon on Mars (about $\frac{1}{25}$ Earth's) seems to indicate that there was much more nitrogen in Mars's atmosphere at one time than there is now. If this was the case, water would have flowed on the surface.

Looking at the planet today, we see considerable evidence that water is still there, much of it in the polar caps, but a considerable amount beneath the surface, possibly in the form of a layer of permafrost. Other evidence of the presence of water may be seen in a difference between the Martian craters and the lunar craters: The ejection patterns around lunar craters appear to have been caused by airborne debris, which is what one would expect if a high-speed projectile struck dry sand, for example. On Mars, on the other hand, most of the large craters (larger than 1 km) are surrounded by flow patterns (Fig. 11.26), which indicates that a "melting" occurred when the projectile struck. Astronomers now believe that the heat generated by the explosion melted the permafrost, generating "mud" that flowed outward from the blast.

Soon after Mars began to heat internally (as a result of radioactivity), numerous volcanoes pushed upward. Although in many cases volcanic activity did not reach the surface, the ground was heated, and, as a result, trapped ice melted and water rushed out onto the surface and began to flow. Much of it eventually evaporated, giving rise to rain over large areas. Later, as most of Mars's nitrogen was lost to space (because of the weakness of its gravitational field) and as the atmospheric pressure decreased, the water went back into the ground and froze. Oddly enough, there is evidence for several such eras as this. It is easy to see, for example, that some riverbeds are much older than others: Some are extensively pitted with craters, some have only a few pits, and others have almost none. According to a recent study, which substantiates this conclusion, Mars passes through periodic ice ages such as the present one because of changes in its orbit and its axial tilt. It is likely that there was liquid water on its surface between these ice ages.

Most of the craters that we now see on Mars were probably made about a billion years or so after it was formed. There is evidence for an era of extensive bombardment throughout the solar system at this time. But, like the moon, Mars also has regions that obviously were hit by much larger objects—possibly asteroids. As in the case of the moon, we call these regions maria; the Hellas basin is a good example. The more powerful impacts are believed to have occurred after the extensive bombardment era, perhaps 3 billion years ago. The force of the blow, in many cases, was so great that extensive melting occurred, as attested to by associated lava flows.

Life

Because of the considerable evidence that Mars once had surface water, it seems reasonable that there might be a lower form of life there. The major objective of Viking was to establish whether this was the case. To this end five procedures were undertaken: a visual search; an analysis of the soil for biological molecules; and three experiments in which soil samples (some mixed with nutrients) were heated and analyzed for changes that life processes would have produced under the circumstances.

The first project, of course, gave a negative result; the second, the search for biological molecules, was also disappointing. The last three tests gave initial results that were encouraging. The changes that occurred when the soil was processed were similar to those that occur in terrestrial soil under the same conditions. However, further checks showed that the changes were not a result of biological reactions but rather of chemical ones with which we were not familiar.

The last hope was for life beneath the surface. Because Mars has a very thin atmosphere, UV radiation from the sun—destructive to most life—reaches its surface. The best protection from this radiation would be afforded by rocks, and so a rock within reach of Viking's mechanical arm was moved and the soil beneath it tested—with the same results: no life.

Although the results of all these tests do not necessarily rule out life on Mars, they make the case for it very weak. The areas around both landing craft were checked, and both gave the same negative result. It is possible that there is life at a lower level in the soil, or perhaps at some other region of the planet. We still cannot be sure.

SUMMARY

1. Mercury, the innermost and second smallest planet of the solar system, has a rocky, cratered, moonlike surface where temperatures are extreme (ranging from 100°K to 700°K). It completes an orbit in 88 days and rotates in 59 days.
2. The first closeup photographs of Mercury (made by Mariner 10) showed numerous craters, long, shallow cliffs or scarps, a large basin now called Caloris Basin, and many other features. An exceedingly thin atmosphere—about one billionth that of Earth—and a very weak magnetic field were detected. Mariner 10 also determined that Mercury has a large iron-nickel core.
3. Venus is similar in size to the earth. With the exception of the sun and moon it is usually the brightest object in the sky (when visible). Because its orbit lies within that of the earth, it presents a complete cycle of phases.
4. Venus revolves around the sun in 225 days and rotates in a retrograde sense in 243 days.
5. Because Venus is blanketed by dense clouds, we see its face as a featureless disk. There are three distinct cloud layers, beneath which is a clear region of intense heat (about 750°K). The high temperatures are mainly a result of the greenhouse effect.
6. The atmosphere of Venus consists mostly of carbon dioxide plus nitrogen, oxygen, and water vapor. The clouds consist of sulphur particles, chlorine, and sulphuric acid droplets. The atmospheric pressure at the surface is approximately 100 times that of the earth (at sea level).
7. The surface of Venus is relatively flat, but there are some mountains and numerous craters. Rocks on the surface have been photographed.
8. Mars has a diameter of 6800 km. It revolves around the sun in approximately 2 earth years and rotates in approximately 24 hours.
9. Temperatures on Mars range from about 293°K at midday on the equator to about 187°K at midnight. The Martian atmosphere is thin—less than $\frac{1}{100}$ that of the earth at sea level.
10. Mars's two moons, discovered by A. Hall in 1877, are Deimos (about 16 km across) and Phobos (about 28 km across).
11. The Martian surface is heavily cratered and consists mostly of sand and rock. Many surface features are visible, two of the most prominent being Olympus Mons, a volcanic crater about 25 km high, and Valles Marineris, a huge canyon about 5000 km long.
12. Flights have been made to Mars by Mariner 4 (first), Mariner 6 and 7, Mariner 9 (which went into orbit and extensively mapped the surface), and Viking 1 and 2.
13. There is strong evidence that there was once water on Mars. Old riverbeds are visible, and in summer the polar caps consist mostly of frozen water.
14. Tests for life were made by Viking 1 and 2; none was found.

REVIEW QUESTIONS

1. Why is Mercury usually so difficult to observe?
2. Is Mercury tide-locked to its moon? Discuss.
3. Describe how astronomers determined the rotational period of Mercury.
4. What is the 2:3 coupling? How is it likely to have come about?
5. Compare the internal structure of Mercury to that of Earth.

6. Explain why Venus exhibits phases. Where is it in relation to Earth when we see it in full phase?
7. Compare the Venusian atmosphere to Earth's atmosphere. What are the differences?
8. Describe the makeup of the Venusian cloud layers. What do these clouds look like through a UV filter? Compare the patterns to those of Earth. Explain.
9. Explain the greenhouse effect. Is this effect important only on Venus?
10. Describe how radar can be used to probe the surface of Venus. What have we learned as a result of such probes?
11. Why are the Martian craters generally flatter than those on the moon?
12. Describe Olympus Mons from the viewpoint of an observer on the surface of Mars. Do the same for Valles Marineris.
13. How do we now explain the wave of darkening that appears on Mars each spring? What did astronomers once think it was?
14. What are some of the important results of Mariner 9?
15. Describe the sky as seen from the surface of Mars. What would a sunset look like?
16. At one time astronomers thought the polar caps of Mars were composed entirely of solid carbon dioxide. What is the present view?
17. What do we now believe the "canali" seen by Schiaparelli were? Discuss some of the ideas that Lowell put forward to explain the "canals."
18. Did Viking establish beyond a doubt whether there is life on Mars? What are some of the problems that life forms would encounter?
19. Describe the evidence we now have for early eras of rainfall and flowing water on the surface of Mars.

THOUGHT AND DISCUSSION QUESTIONS AND PROJECTS

1. If traveling at 10,000 km/hr in a straight line through space, how long would it take you to get to Venus at inferior conjunction, and at superior conjunction (assuming you avoid the sun)? Is this speed practical? Why?
2. Assume you have been set down on the surface of Venus for one Venusian day. Describe what you would be likely to see and experience during this time (e.g., sunrise, sunset, clouds, objects visible, winds, temperature).
3. Compare and contrast Mercury, Venus, and Mars. What properties do they have in common? How do they differ? List the properties in each case that are favorable to life, and those that are unfavorable.
4. *Project:* Observe Venus once a week for several weeks with a small telescope. Sketch the phase you see each time.
5. *Project:* Observe Mars with a small telescope. Sketch what you see.

FURTHER READING

Arvidson, R. E.; Binder, A. B.; and Jones, K. L. "The Surface of Mars," *Scientific American,* p. 76 (March 1978).
Carr, M. "The Volcanoes of Mars," *Scientific American,* p. 32 (January 1976).
Hartmann, W. K. "Viking on Mars: Exciting Results," *Astronomy,* p. 6, (January 1977).
Leovy, C. "The Atmosphere of Mars," *Scientific American,* p. 34 (July 1977).
Murray, B. (ed.). *Mars and the Mind of Man.* Harper & Row, New York (1973).
Murray, B. "Mercury," *Scientific American,* p. 58 (September 1975).
Oberg, J. E. "Venus," *Astronomy,* p. 6 (August 1976).
Scientific American. *The Solar System.* W. H. Freeman, San Francisco 1975).

Aim of the Chapter

To introduce the gas giants (and Pluto) and to describe how they differ from the inner, terrestrial planets discussed in Chapter 11.

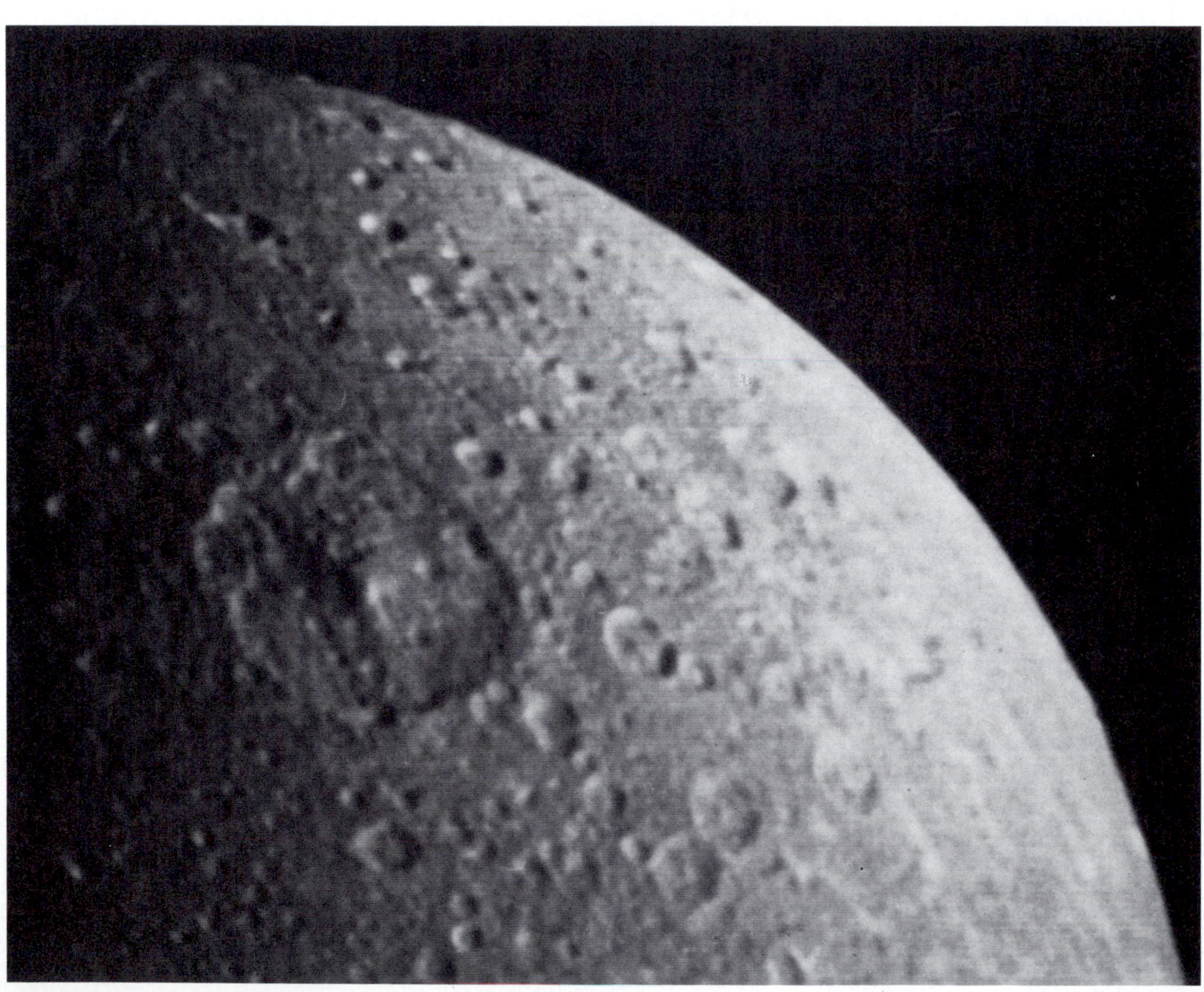

chapter 12
The Outer Planets

The outer planets are quite different from the inner, terrestrial planets (Chapter 11). With the exception of Pluto they are gas giants of low density composed mostly of hydrogen and helium and characterized by low temperatures and high spin rates. The gas giants are called the Jovian planets, after Jove, monarch of the Roman gods.

JUPITER

With a mass $2\frac{1}{2}$ times that of all the other planets combined, Jupiter is the giant of the sun's family. It is over five times as far from the sun as the earth, yet we still see it as one of the brightest objects in the sky. Seen through binoculars it is accompanied by as many as four tiny points of light nearby; they are its moons. Galileo, delighted at his first sight of this, said, "It looks like a solar system in miniature." Indeed, it was this sight that convinced him of the validity of the Copernican model of the solar system. Here was proof that everything did not revolve around the earth after all.

The colorful bands that lie parallel to Jupiter's equator can be seen even through a small telescope (Fig. 12.1). The lighter ones are now referred to as zones, the darker ones as belts. The zones are usually white or light yellow, the belts orange, brown, or red. Many interesting features are superimposed on the belts—a number of small spots, and a particularly interesting red one considerably larger than the rest. In 1665 Cassini used this red spot to determine the rotational period of the planet; he found it to be approximately 10 hours.

With a rotational period of only 10 hours it is obvious that Jupiter is spinning at a tremendous rate compared to Earth. A point on its equator has a linear velocity 30 times that of a corresponding point on Earth. This high velocity gives Jupiter a markedly oblate shape.

Figure 12.1 Jupiter's Great Red Spot and one of the planet's satellites (dark circle) in a Pioneer 10 photograph.

But Jupiter does not rotate as a solid. Near its equator it rotates in 9 hours, 50 minutes, while near the poles it rotates in the slightly slower time of 9 hours, 55 minutes. What we are seeing, of course, is only the outer atmosphere of the planet. Beneath the clouds of that atmosphere, as determined from the rotation of its radio sources, which are linked to its magnetic field, its period is the same as at the poles.

Jupiter has 318 times the mass of Earth and 11 times its diameter, giving a density of 1.3 gm/cm^3—hardly more than that of water (1.0 gm/cm^3), and certainly much less than Earth's. Jupiter, then, is composed of much lighter materials than Earth, and cannot have a large iron core. The elements most consistent with this density are hydrogen and helium, and astronomers are now convinced that Jupiter is made up mainly of these. Other substances, such as methane, ammonia, and water vapor are present, but in small quantities.

Jupiter, unlike a small planet, such as Mars, has sufficient gravitational pull to retain its hydrogen. In the case of Jupiter, moreover, the effect of the much higher gravitational field is supplemented by the temperatures of its outer regions, which, being relatively low—on the order of 130°K (−220°F)—result in an average velocity of the molecules composing its atmosphere that is lower than on Earth.

The abovementioned large red spot used by Cassini to determine Jupiter's rotational period is still visible. Now called the Great Red Spot (Fig. 12.2), it has changed over the years from brick red to pale pink, and it has even vanished at times, but its location—the "Red Spot Hollow"—is still visible even when the spot itself is not. It is oval, about 14,000 km (8700 mi) wide by about 40,000 km (25,000 mi) long. At one time it was thought that the Red Spot might be associated with subsurface phenomena, possibly a volcano. Some thought it might be caused by a large depression on the surface that created a whirlpool of gas above it. We now know that these ideas cannot be correct (see below), and recent spacecraft observations indicate that the spot is a gigantic and very long-lasting cyclonic storm.

Figure 12.2 The gigantic cyclonic storm called the Great Red Spot.

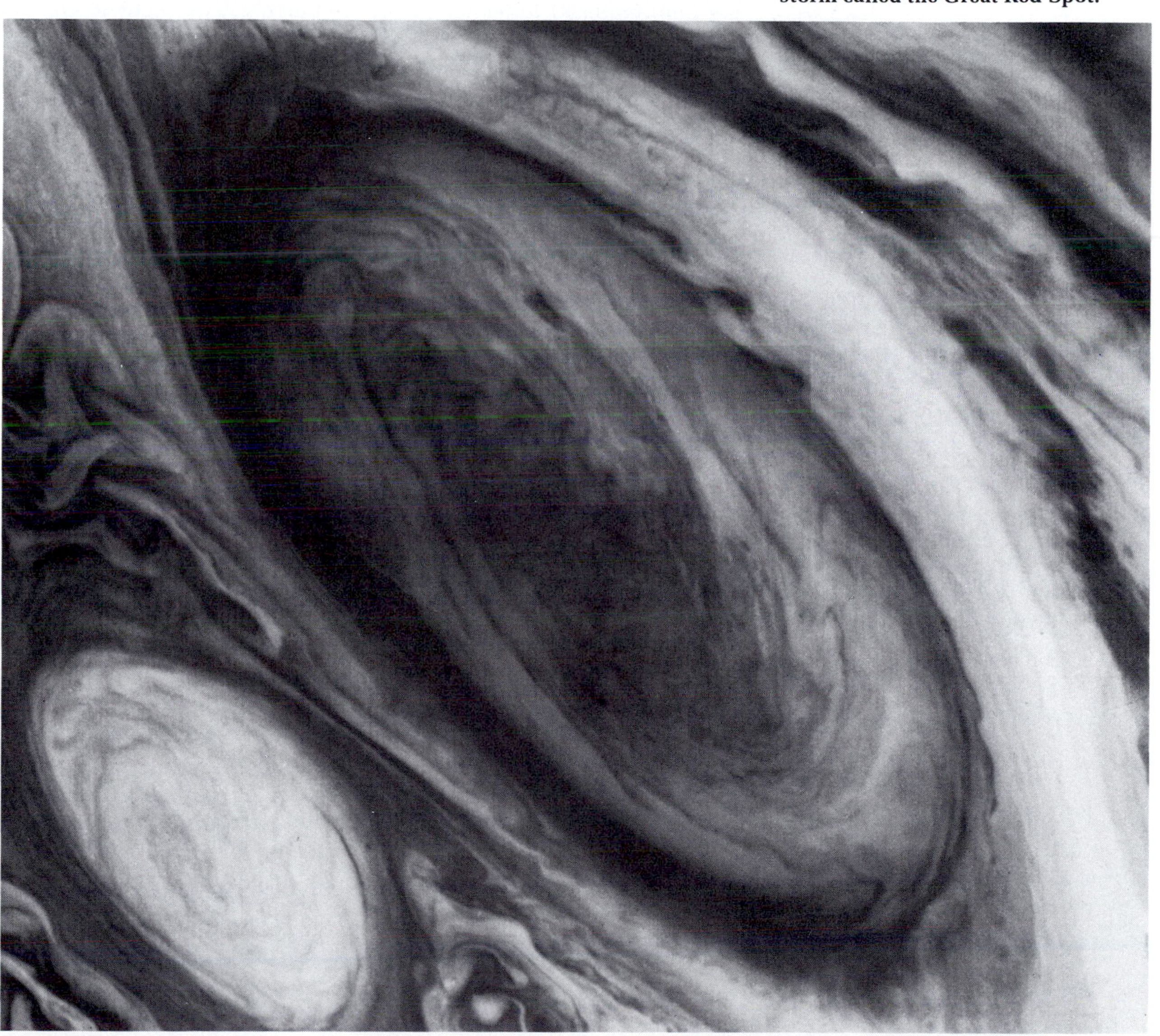

Jupiter, first examined with radio telescopes in 1955, is a strong source of three kinds of radiation—infrared thermal radiation, decimetric radiation, and decametric radiation. Infrared thermal radiation is due to the heat generated in the interior of the planet. Jupiter emits about two and one half times as much radiation as it receives from the sun. The source of this energy, most astronomers believe, is gravity, which is still pulling the mass of the planet slowly inward and, in the process, converting gravitational energy to heat energy.

Decimetric radiation, called so because it has a wavelength of approximately .1 m, comes from the region surrounding the planet; it has characteristics of synchrotron radiation, which occurs when high-speed electrons, encountering a magnetic field, spiral around the field lines, emitting photons (i.e., radiating). The presence of synchrotron radiation implies a magnetic field, and, in fact, Jupiter has one about 10 times as strong as Earth's. This field is similar to Earth's in that it is not oriented along the spin axis of the planet but is tilted from it by about 10°; it differs from Earth's in that it has opposite polarity with respect to the ecliptic: Its north magnetic pole is where our south magnetic pole is. Furthermore, charged particles are trapped in the magnetic field, and so Jupiter, like Earth, also has Van Allen belts. The third kind of radiation, decametric (because it has a wavelength of approximately 10 m), is intermittent and appears to come from a region closer to the planet, where streams of electrons are probably responsible for it. This radiation also correlates with the position of Jupiter's moon Io: When Earth, Jupiter, and Io are aligned, decametric radiation is low; when Io, Jupiter, and Earth make an angle of 90° on one side of Jupiter (or 60° on the other side), its intensity is high. The phenomenon is believed to be due to Io's motion through Jupiter's magnetic field and to the consequent disturbance of the electrons trapped there.

In 1979, with the discovery of two moons by Voyager space craft, the total of Jupiter's moons reached sixteen; it seems likely, though, that several more will be discovered. The fifth, after Galileo's original four, was discovered in 1892 by E. E. Barnard and is only about 240 km (150 mi) in diameter. It is called Amalthea. Of the subsequently discovered moons, most are only a few tens of kilometers in diameter.

Jupiter's moons seem to fall into three general groups. The five closest to the planet—Amalthea, Io, Europa, Ganymede, and Callisto—constitute the inner group; with the exception of the abovementioned Amalthea, they are much larger than the others. The outer four have retrograde motion and are probably asteroids captured by Jupiter. The third group is composed of all those that lie between the inner and outer groups. All of these move in the correct way (prograde) with respect to Jupiter, are small, and also may be captured asteroids.

In 1973 and 1974 Pioneer 10 and 11 swept by Jupiter and accumulated an enormous amount of information. One of the first things discovered was that the radiation around Jupiter was much stronger than anticipated—a thousand times as strong as the known lethal dose for humans.

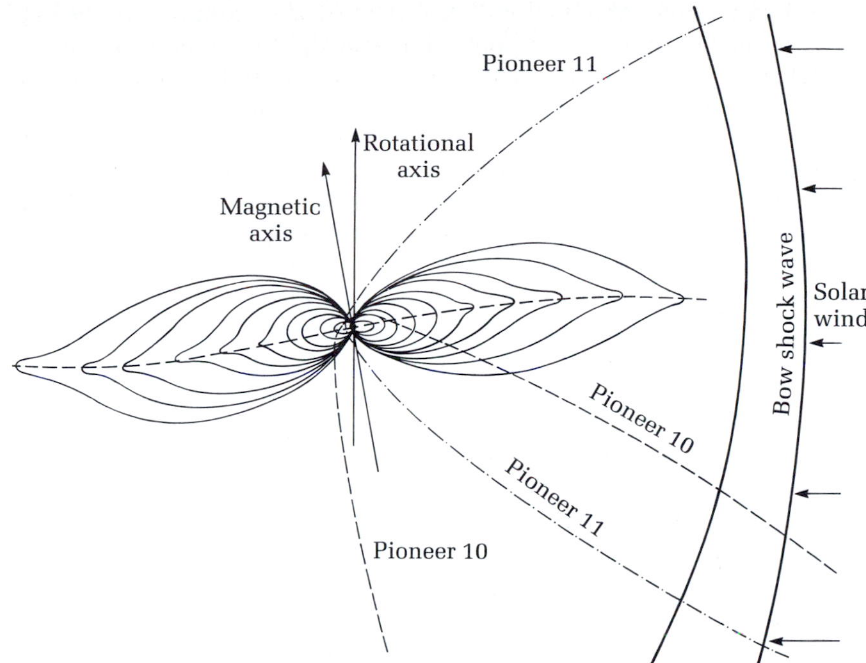

Figure 12.3 Jupiter's magnetosphere, which expands and contracts continuously in response to the varying pressure of the solar wind.

Figure 12.3 shows the overall structure of Jupiter's magnetosphere; its shape is similar to that of the earth's but flatter in the outer regions. The moons Io and Amalthea are in the doughnut shaped inner region of the field. This inner field is quite "stiff" and rotates rigidly with Jupiter itself, but the outer field is flexible and tends to fluctuate considerably, moving inward and outward very rapidly. Like Earth's magnetosphere, Jupiter's interacts with the solar wind, forming a bow wave—a kind of shock front—that diverts the solar wind around the field. As is also the case with Earth's magnetosphere, Jupiter's draws along with it highly energetic charged particles. Most of this plasma is protons, but nuclei of oxygen, carbon, nitrogen, sulphur, neon, sodium, and iron are also present. With the exception of sodium these are generally the same elements found in the solar wind.

Pioneer 10 disclosed Jupiter's zones and belts as they had never been seen. The zones are about 20 km (12 mi) higher (above Jupiter's surface) than the belts; they are also cooler, and the clouds associated with them appear to rise. The belts are descending regions of gas. Jupiter apparently has large convection currents that are somehow generated in the lower parts of the atmosphere and move upward at the zones and downward at the belts.

Pioneer also reported on Jupiter's atmosphere, which it confirmed to be mostly hydrogen (85%) and helium (about 14%). Methane and ammonia were also detected, as they had been from Earth. Earth-based measurements also tell us that water vapor, ethane, and acetylene are present along with a few other minor components.

With the data from Pioneer we now know that Jupiter's atmosphere consists of three distinct layers of cloud, with clear areas separating them. The upper cloud layer consists of ammonia crystals at a temperature of about 130°K (−220°F). Approximately 25 km below this is a layer of ammonium hydrosulphide crystals at a temperature of about 250°K (−9°F). The third layer, another 25 km lower, consists of ice crystals and perhaps some liquid ammonia.

Though we have not seen below this region, we can construct a model (Fig. 12.4) based on the data we have. First, we know that Jupiter is made up mostly of hydrogen; we also know that pressures and, therefore, temperatures increase with penetration into its interior. Earthlike temperatures prevail about 60 km below the upper cloud layer. Primitive life forms may exist here. With descent toward the center, pressures and temperatures continue to increase, and there is a gradual transition from gaseous to liquid hydrogen.

The region of liquid hydrogen extends for perhaps 20,000 km but gradually, with further increases in pressures and temperatures, changes again—this time to a region of liquid metallic hydrogen, which conducts electricity. It is currents in this metallic hydrogen that generate Jupiter's magnetic field, as the earth's is generated in its iron core.

There may be a small rocky core at Jupiter's center. This is expected because Jupiter was formed of the same solar nebula as the rest of the planets, and any heavy elements would collect in this region. Some or all of this core may be molten; the temperature is expected to be at least 30,000°K.

In 1979 Voyager 1 and 2 came within 300,000 km of Jupiter's surface and produced phenomenally high-quality images of the planet. Eddies and currents in the clouds stood out clearly. Numerous small storms were visible along the edges of the bands, and in the largest storm of all—the Great Red Spot—granulation and clouds could be seen near the center. It appeared that only the outer regions rotated—with a period of about 12 hours.

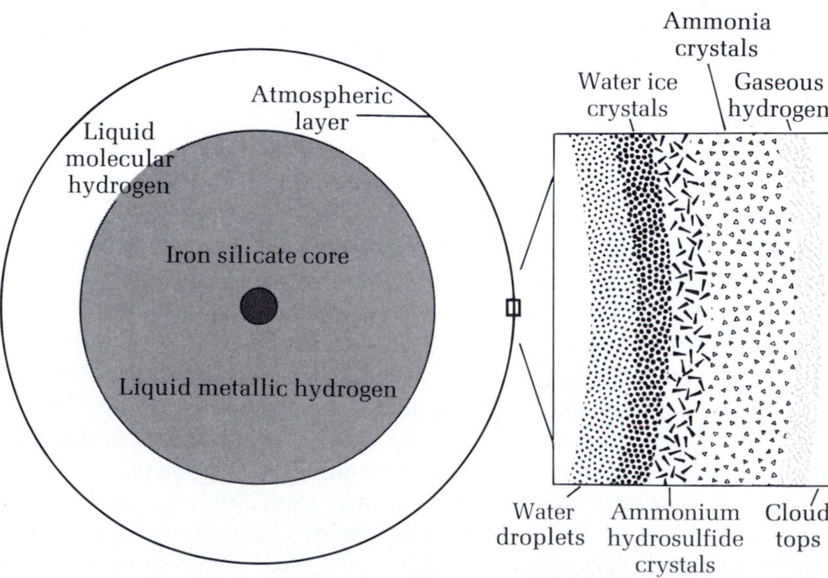

Figure 12.4 Diagram of Jupiter's interior. Right-hand portion is enlargement of boxed segment (small rectangle) of atmospheric layer.

A ring was also detected that appears to be about 10,000 km (6200 mi) wide and perhaps 30 km (19 mi) thick. Its width is actually quite uncertain; we may be seeing only the bright section of a ring that extends all the way to the planet's surface.

Each of the moons (Fig. 12.5; color) was seen to have a distinctly characteristic surface. To many Io looked like a particularly colorful pizza. Its surface, obviously quite young, showed no craters, although there were volcanic eruptions that dwarfed anything seen before. At least two were larger than the most violent volcanic eruptions ever recorded on Earth. Eight giant eruptions, with plumes and fountains rising 70–280 km, were recorded; seven of these were still visible four months later.

Io still has active volcanoes mainly because it is so close to Jupiter. Tidal forces frictionally heat its surface by moving it in and out with each orbital revolution. Astronomers believe that beneath its outer solid surface there is a layer of liquid sulphur on which floats a mixture of solid sulphur and liquid sulphur dioxide.

Io's abovementioned interaction with Jupiter's magnetosphere is believed to be related to this volcanic activity by the following mechanism. We know that Io has a gaseous cloud of sulphur and sodium associated with it; in fact, this torus (doughnut shape) of gas seems to extend entirely around Jupiter, and astronomers believe its origin lies in the volcanic eruptions that occur on Io. There is no water on Io, and these eruptions are composed of rocks, sulphur, and a fine spray of sulphur dioxide. This mixture is shot hundreds of kilometers above the surface; the rocks and sulphur fall back to the surface of Io, but the sulphur dioxide remains aloft. The plasma associated with Jupiter's magnetic field, traveling at a tremendous velocity relative to Io (Jupiter spins much faster than Io revolves), collides with the sulphur dioxide in Io's atmosphere and ionizes[1] it. Once ionized, the sulphur dioxide quickly comes under the influence of Jupiter's magnetic field and is drawn along by it, producing the torus of gas that encircles Jupiter. Strangely, Io is not always in the center (in cross section) of it; this is partly because Io orbits in Jupiter's equatorial plane while the torus is oriented in the equatorial plane of Jupiter's magnetic field, which, as we saw earlier, is tilted about 10° to the spin axis. Thus, as Jupiter and Io spin, the plasma torus wobbles and Io weaves a complex up-and-down pattern within it as it moves through it.

In addition, a strong concentration of magnetic field lines from both the north and south poles of Jupiter also passes through Io. Since electrons move most easily along field lines, it was expected that there would be an enormous current along these lines. However, Voyager observed no such current; it is not known why.

From Earth-based measurements we know that Io and Europa are roughly twice as dense as Ganymede and Callisto and, therefore, probably consist of rock, with perhaps some ice. Io has an average density of 3.5 gm/cm^3, and Europa has an average density of 3.0 gm/

[1] An ion is an atom or molecule that has lost or gained an electron or electrons and so is left with a positive (lost electron) or negative (gained electron) electrical charge.

cm^3. (The density of our moon is 3.3 gm/cm^3.) Europa is quite similar to Io in that it has a relatively young surface, with few craters. Its major feature, covering much of its surface, is long, flat, dark stripes. These are 15–60 km wide and many kilometers long. Occasionally there is a central bright line.

Because Europa reflects sunlight as water ice does, astronomers believe that it is covered with a layer of ice. The stripes appear to be fractures in the ice, and the bright center lines may be regions where the ice has been pushed upward from beneath the surface by tidal forces similar to those that heat Io's surface. Europa also has some mottled terrain, in which the ice crust seems particularly thin.

Ganymede, the largest of the moons (with diameter 5300 km and density 2 gm/cm^3), also has a particularly strange feature: side-by-side grooves, 10–20 km (6–12 mi) apart, that look almost like huge dune-buggy tracks. Ganymede is also different from the inner two moons in that it has craters, though they are much shallower than those on Earth's moon. This is due to the fact that Ganymede's surface is mostly water ice, so that, over a long period, crater rims would sink as a result of "flow." The icy crust is believed to be about 100–150 km thick; the core is probably rock. Between the rock core and icy crust there may be a region of water. There is also evidence that a thin layer of a dark material covers the ice around many of the craters. The impacts that caused the craters appear to have blown this "soil" away, revealing a lighter material (ice) beneath.

Callisto (with density 1.6 gm/cm^3), the outermost of the Galilean moons, is covered with overlapping craters; it may have the most craters per unit area in the solar system. Callisto also has several gigantic "bull's-eyes" on its surface, where large asteroids obviously struck it. The waves that rippled out from the center when the asteroids struck can still be seen, apparently frozen into the icy surface. These no longer have any height and are visible mainly because they have a slightly contrasting color. At the center of each of the bull's-eyes is a giant ringed basin, the only areas on the planet not saturated with craters. These basins are quite shallow, again a result of the icy surface, as Callisto, like Ganymede, has an icy crust surrounding a water core.

Two other phenomena of interest were also observed by Voyager. In the region close to the north and south magnetic poles, auroral emissions were noticed; Jupiter obviously has northern and southern lights. Powerful lightning bolts were also seen in the upper cloud layers.

SATURN

With a ring system (Fig. 12.6) that extends almost 77,000 km from its surface, the most inspiring and beautiful of the planets is Saturn. Although it is almost twice as far from the sun as Jupiter, it is brighter than most stars and can easily be seen on any clear evening for several months of the year.

Figure 12.6 The rings of Saturn as they appear at various points in the planet's orbit; seen edge-on (upper right) they virtually disappear.

Saturn, smaller than Jupiter, has a diameter of 120,000 km (75,000 mi). But, like Jupiter, it spins rapidly on its axis and therefore is also oblate. Also like Jupiter, it has complex differential rotation, the equatorial clouds rotating with a period of 10 hours, 2 minutes, while those near the poles have a period of approximately 11 hours. Radio bursts detected by Voyager show that the planet itself rotates once every 10 hours, 39 minutes. A diameter nine times that of Earth (and, so, a volume 729 times) coupled with a mass only 95 times that of Earth give Saturn an amazingly low density—.7 gm/cm^3—making it probably the least dense planet in the solar system. (A possible exception is Pluto.)

When seen through a telescope, Saturn, like Jupiter, has colorful bands, differentiated as "zones" and "belts." Generally yellow and tan, their average temperature is about 100°K, with the belts approximately 5° warmer (and darker) than the zones.

A considerable amount of methane has been detected on Saturn, much more than on Jupiter. Although ammonia is also present, because of the low temperatures in the outer region it may exist only in solid form, as a kind of snow.

In 1980 Voyager 1 found that Saturn's atmosphere was in some ways even more violent than Jupiter's. Detected in a number of the equatorial belts were winds up to 400 km/sec, which tapered off to nearly zero at the belt-zone boundaries. Like Jupiter, Saturn appears to have cyclonic storms.

As would be expected, the most prominent element in Saturn is hydrogen (about 85%), which with helium makes up most of the planet. One gas (besides methane) that has been detected spectroscopically is ethane.

Again like Jupiter, Saturn probably has an outer region of gaseous hydrogen that surrounds a sea of liquid hydrogen. And, again, with increasing depth and consequent increasing pressure the liquid hydrogen becomes liquid metallic hydrogen. Finally, at the center of the planet there may be a small rocky core.

In 1979 Pioneer 11 verified a weak magnetic field—about $\frac{1}{20}$ as strong as Jupiter's. Unexpected was the fact that Saturn's magnetic field axis is aligned exactly with the rotational axis, which is the case for no other planet. Also verified by both Pioneer and Voyager is the fact that Saturn radiates into space over $2\frac{1}{2}$ times more energy than it receives from the sun.

Saturn's most striking feature is, without doubt, its ring system. Galileo, first to see it, thought it was two moons, one on either side of the planet. He was puzzled, however, when they did not move, and even more puzzled when they suddenly disappeared. In 1655, when better telescopes had become available, Cassini recognized it as a ring system. We now know that this system is approximately 274,000 km (170,000 mi) across and so thin (about 10 km) that stars can be seen through it.

In 1675 Cassini discovered a gap that appeared to separate a bright ring, which we now refer to as Ring B, from a dimmer outer ring, called Ring A (Fig. 12.7). Later a third ring, Ring C (sometimes called the Crepe ring), was discovered inside Ring B. These are the only rings readily seen from Earth, but we now know that there are four others, both inside and outside these; the final order (outward from the planet) is: D, C, B, A, F, G, and E.

The rings are in the plane of the equator, which is tilted approximately 27° with respect to the plane of the orbit. This means that, as Saturn moves around its orbit, we see the rings at various angles (Fig. 12.6). Twice during the $29\frac{1}{2}$-year orbital period we see them edge-on, and at these times they literally disappear from our view. (This is, of course, what happened when Galileo noted their disappearance.)

In 1859 Maxwell showed that, because of their closeness to Saturn and the resulting tidal forces, the rings could not be solid but must consist of billions of tiny particles. Spectral analysis shows, in fact, that they are small particles, probably varying from a few

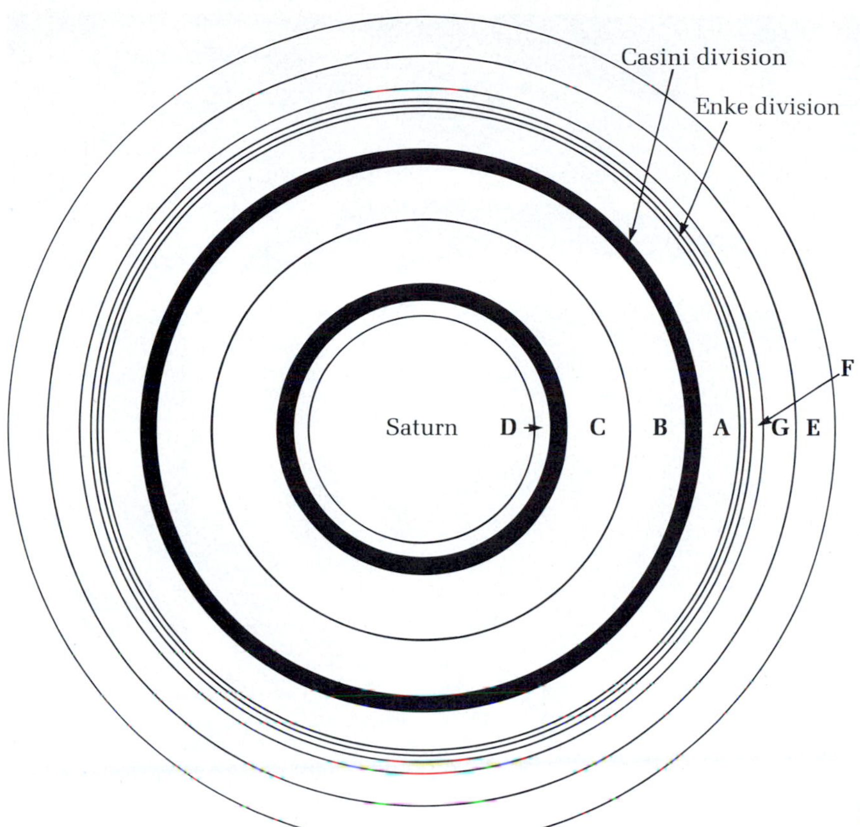

Figure 12.7 Plane view of Saturn's rings.

centimeters to perhaps a meter. They also appear to be composed of ice or to be ice-covered. Their formation is related to the Roche Limit, the distance from any planet within which the tidal forces are so great that a moon that moved into this region would soon explode into billions of pieces. Thus, the rings may once have been a moon that came within Saturn's Roche Limit; it may also be material, left over from the nebula, that was not able to combine into a moon. Its total mass is, interestingly, about that of our moon.

Our best view of the rings has come from Voyager (Fig. 12.8; also color), which showed, among other surprises, that the Cassini division (between Rings A and B) was not empty but had about 20 small rings within it. The entire ring system, in fact, is much more complex than was previously believed.

In Ring B there are also mysterious "spokes," whose origin is a puzzle, since the particles within the ring have varying velocities. According to Kepler's law (p. 30), particles closer to the planet have a higher velocity than those farther out; this would destroy any spokes that might arise. It may be that the spokes are composed of small particles lifted out of the ring by electrostatic forces created by Saturn's magnetic field. Also puzzling is the asymmetry of one

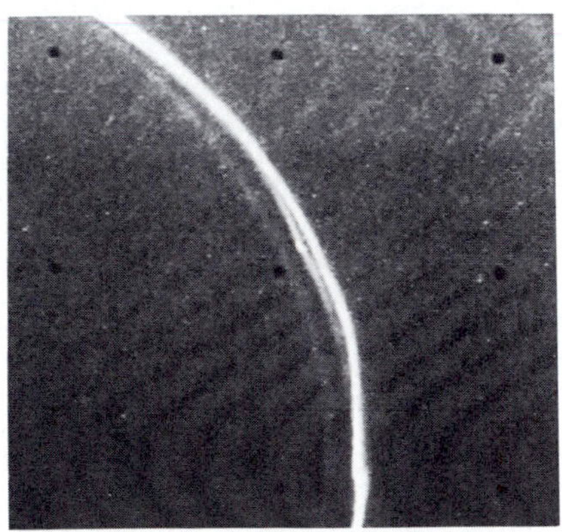

Figure 12.8 At top, Saturn's rings in Voyager photograph. Lower photograph shows braids in the F ring, believed caused by the moons within the ring.

of the rings, which is thicker on one side than the other. The shape of Ring F, clumped and strangely braided, provided another surprise. The braiding now appears to be related to the presence in the region of two moons only recently discovered.

For many years Saturn was thought to have ten moons (p. 119). Pioneer discovered two more but was unable to find Janus, one of the last moons discovered (in 1966) from Earth. However, Janus was located by Voyager 1, as were several new moons. Moons 13 and 14 were discovered orbiting just inside and just outside Ring F, and 15, a tiny moon about 100 km across, was discovered about 800 km outside Ring A. Two others, 16 and 17, have been found in the same orbit as Tethys, another of Saturn's moons.

The moons were seen clearly for the first time by Voyager. Of most interest was Saturn's largest moon, Titan. With a diameter of approximately 5118 km, it is one of the largest moons in the solar system, and—an important distinction—it is the only moon known to have an atmosphere. Photographs of Titan taken by Voyager were disappointing, showing a vast, featureless orange ball (Fig. 12.9), the surface completely hidden by clouds which differed only in that the north polar region was darker than the rest.

Methane, the only gas that had been identified from Earth, was detected as composing only about 1% of the atmosphere; most of the rest is nitrogen.

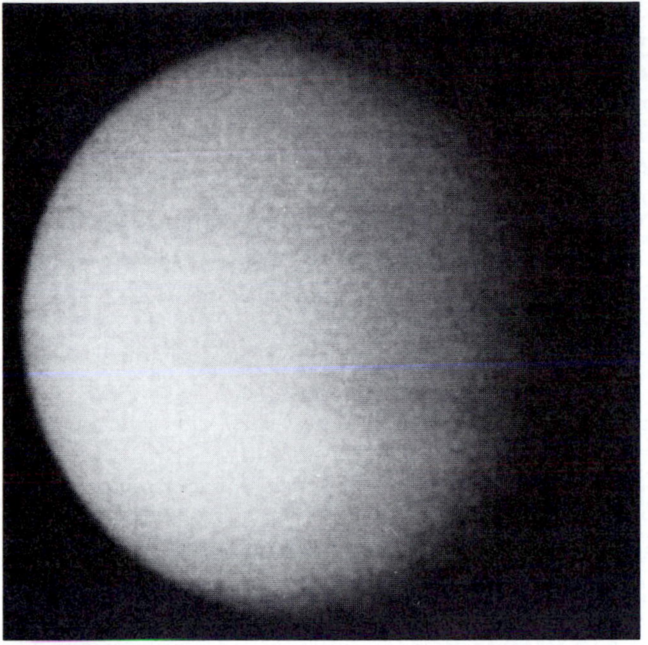

Figure 12.9 Titan, at left, and Dione at right in a closer view than Titan.

Attempts at radio detection of the surface were unsuccessful; no surface was found. The atmospheric pressure at the surface is probably about $1\frac{1}{2}$ times that of Earth at sea level, and the temperature is about 92°K (−180°C) according to estimates based on the measurements that were taken. There was also some evidence of the greenhouse effect; the temperatures were slightly higher in the lower layers of the atmosphere.

It is quite possible, if the temperature is sufficiently low at Titan's surface, that there is an ocean of liquid nitrogen beneath the clouds. It is also possible that there is liquid nitrogen rain, as the clouds are likely to be made up of droplets of liquid nitrogen.

All of Saturn's other moons were also photographed by Voyager. They appeared to be composed of mixtures of ice and rock and to have bright, icy surfaces. Of all the moons, the one that looked most like ours was Dione (Fig. 12.9). Many had characteristic features: Mimas, a huge crater about 130 km across; Tethys, a gigantic, 750-km-wide trench; and Iapetus, one hemisphere much darker than the other.

URANUS

The planets studied so far have been known from antiquity. As far as we know, the first recorded "discovery" was of Uranus (Fig. 12.10). This occurred the night of March 13, 1781, when William Herschel, busy mapping the stars, came upon an object significantly different from the others; it looked like a small, diffuse disk. Night after night Herschel watched it, plotting its movement. Calculations soon left no doubt that it was a planet. As the first discoverer of a planet since

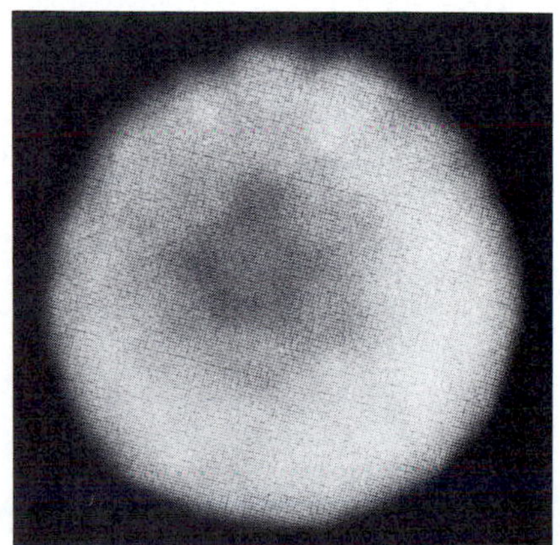

Figure 12.10 Uranus in a photograph taken 80,000 feet above the earth.

the beginning of recorded history, Herschel's fame spread overnight; many wanted to name the planet after him, but tradition prevailed and it was named Uranus. (In mythology, Uranus is the father of Saturn, as Saturn is the father of Jupiter.)

Uranus has a mass about 15 times that of Earth and a diameter about four times Earth's, making its density 1.2 gm/cm³. Like Jupiter and Saturn, it is made up mostly of light gases; methane and molecular hydrogen have been detected spectroscopically, and it seems reasonable that water vapor is also present. At the low temperature of the outer layer (56°K, or −360°F), any ammonia would likely have "snowed out" to lower levels.

The outermost region of Uranus's atmosphere is assumed to consist of a transparent layer of molecular hydrogen; a thick cloud layer of methane is assumed to lie directly below this. The green cast of the planet is, no doubt, caused by the methane. Most of the atmosphere below this is molecular hydrogen, which with increasing depth eventually becomes liquid.

The average density of Uranus is higher than that of Saturn, so it cannot consist entirely (or almost entirely) of hydrogen and helium, as Saturn does. Astronomers feel that a large fraction of the planet—perhaps half—is an icy material. At the center there is probably a rocky core, and there is some evidence that the core may be molten.

Perhaps the strangest feature of Uranus is its axis of rotation, which is oriented so that it is almost in the plane of its orbit (Fig. 12.11). This means that for 21 years (one quarter of its total year, which is 84 earth years) one pole, say the north pole, is in total darkness as the other pole points toward the sun. Twenty-one years later, the axis is perpendicular to the direction of the sun, and the whole planet (or at least most of it) has a normal day and night (approximately 23 hours). After another 21 years, the south pole is in total darkness.

Like Venus, Uranus has retrograde spin—but barely. This is because its spin angle with respect to the orbital plane is greater than 90° (98° to be exact).

In 1977 Uranus occulted (eclipsed) a dim star. Such things as size and oblateness can be calculated precisely during occultations

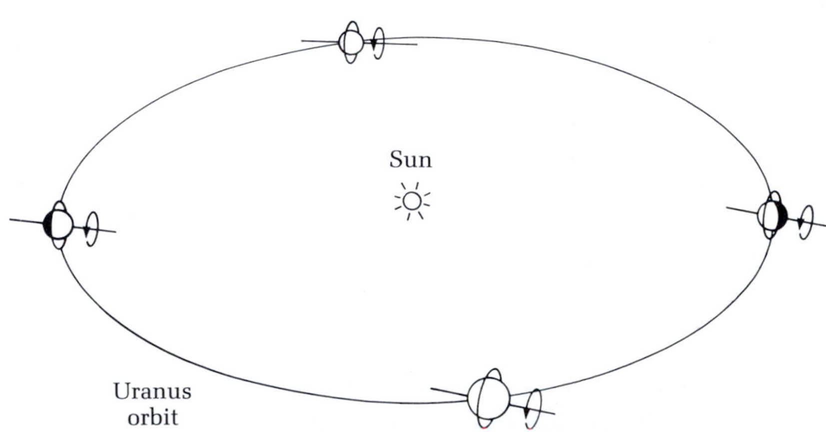

Figure 12.11 Positions of Uranus' axis as the planet orbits the sun.

of this type. In this case astronomers noticed several slight fluctuations in the light of the star as it passed on each side of Uranus. The fluctuations were symmetric, which could mean only that Uranus has a ring. Detailed study of the fluctuations showed, in fact, that Uranus has several narrow rings separated by relatively large gaps.

Uranus has five moons (Fig. 12.12); all have been named (Table 7.3). Their orbits are aligned with Uranus's equator, so that, because of the unusual orientation of Uranus's axis of rotation, they appear to be revolving in a strange way. All have regular orbits; none is retrograde and none exceedingly eccentric.

Figure 12.12 Uranus and three of its satellites.

NEPTUNE

Neptune is similar to Uranus in size and appearance, but its density, 1.54 gm/cm^3, is slightly greater. An occultation in 1968 allowed calculation of Neptune's diameter as 47,200 km and showed that it is quite oblate. Its rotational period is approximately 22 hours. Both methane and molecular hydrogen have been detected spectroscopically, but it is assumed that frozen ammonia and frozen water vapor

Finding Neptune

For a few years after its discovery Uranus appeared to follow its predicted orbit. But gradually it began to deviate until, in about 30 or 40 years, it was considerably off course. This convinced astronomers that there was another planet somewhere beyond Uranus that was measurably affecting its orbit. The calculation of its position was taken up separately by two men—J. C. Adams in England and U. J. Leverrier in France.

Adams started his work shortly after graduating from Cambridge. Less experienced than Leverrier, he took a little longer to perform the complicated calculations. When they were completed, he took them to the Royal Astronomer, Sir George Airy; Airy was too busy to see him, so Adams left his calculations. Eventually Airy examined them but was unimpressed: How could a young graduate like Adams possibly predict the position of a planet? Airy sent Adams a test question to determine his competence in celestial mechanics (mathematics of orbits). The question was a relatively easy one and Adams, insulted, ignored it.

A few months later Leverrier's prediction arrived from France. Airy compared it with that given him by Adams. Although they were less than 1° apart, Airy still acted cautiously and sent the predictions to a nearby observatory for verification. There, a slow, footdragging search was finally undertaken.

Leverrier, too impatient to wait for verification, sent his prediction to the director of the Berlin Observatory, J. Galle. Galle received the letter on September 23, 1846; that night he looked for the planet and within half an hour found it.

In England, Airy proceeded to botch things further for Adams by saying nothing about the existence of his prediction. Eventually news got out, and the French immediately cried "fraud," as there was considerable rivalry between England and the continent. Proof of Adams's work was presented, however, and both men are now credited with the discovery.

Did Galileo See Neptune?

Although Neptune was not recognized as a planet until 1846, it was seen earlier by French astronomer J. Lalande, who, thinking it was a dim star, plotted it on a star map in 1795. There is now evidence, however, that it may have been seen even earlier than this. In the late 1970s Charles Kowal, testing the accuracy of calculations of Neptune's orbit, searched early records that might give Neptune's position. A listing of all the planetary occultations since 1557 (published in 1977 by S. C. Albers) showed an occultation of Neptune by Jupiter in January of 1613. Kowal realized that Galileo had been studying Jupiter's satellites at that time and should have seen Neptune in his field of view. Searching through Galileo's records for December 1612 and January 1613, Kowal and a colleague found that Galileo had, indeed, recorded an apparent "star" in his field of view at that time.

Of particular interest to Kowal was the fact that the predicted position did not coincide exactly with the position Galileo recorded. The object was plotted about one minute of arc to the west of where Neptune would have been according to its present theoretical orbit. Although it might seem reasonable to attribute this to the crudeness of Galileo's instruments, it is known that Galileo used a micrometer grid at this time, which would make an error of this magnitude unlikely. In fact, Kowal believes that the discrepancy is not in Galileo's figures but rather in our present calculations of Neptune's orbit.

Figure 12.13 Neptune and its satellites: Triton (just north of Neptune) and Nereid (indicated by arrow).

are present. Neptune's internal structure is, no doubt, quite similar to that of Uranus, and like Uranus, Neptune has a slight green cast. Also, like both Jupiter and Saturn, it shows a thermal excess, radiating more energy than it receives from the sun.

Neptune has one of the strangest satellite systems known. One of its three moons, Triton (Fig. 12.13), is huge (about 4500 km). Its orbit around Neptune is highly inclined, and its motion is retrograde. Calculations show that its orbit is changing faster than any other in the solar system; an observer on Neptune would probably be able to see a significant change in only a few decades. Another of Neptune's moons, Nereid, is tiny compared to Triton, and it too has a peculiar orbit: Its eccentricity is the greatest in the solar system.

PLUTO

With the discovery of Neptune, astronomers thought that the deviations from Uranus's predicted orbit would be completely accounted for. That they were not led to speculation that there was another planet beyond Neptune. Shortly after the turn of the century Percival Lowell (p. 203) calculated its most likely position and began searching for it.

After a few years, Lowell gave up. Later, on the basis of new calculations a second search was conducted from 1914 until Lowell's death in 1915, again unsuccessfully.

In 1929 the search began again, this time with the then new instrument, the blink-comparator. In earlier searches astronomers had looked for an object that presented a disk—as Uranus and Neptune (to some degree) do. With the blink-comparator considerably more attention could be paid to smaller, starlike objects. In this device two photographs, taken several days apart, are alternately illuminated. When this is done, any object that moved—even slightly—during the interval between exposures will appear to jump. Unfortunately, because asteroids, as well as planets, appear to jump, numerous checks had to be made to eliminate asteroids.

Finally, in February 1930, after thousands of plates had been scanned in this way, a jumping object was found that did not appear to be an asteroid. Its verification as a planet was announced on March 13, 1930, Percival Lowell's birthday. The planet was called Pluto—in Greek mythology, the god of the underworld.

It was later discovered that the calculation of the predicted position was in error, so that the discovery was, in a sense, an accident. Furthermore, once astronomers began studying the planet, it was determined that it was far too small to cause any significant perturbation of Uranus's orbit: Pluto has a mass only about $\frac{1}{600}$ that of the earth; an object seven times the earth's mass was needed. This seems to indicate that there might be another planet beyond Pluto, but several searches have shown none; the mystery remains.

Pluto is so far from the sun that the sun would look like a bright star in the sky. Its surface is in perpetual darkness, and, with temperatures hovering near 40°K (−460°F), it must be a very forbidding place. (Its average surface temperature is 37°K; its daytime temperature is approximately 43°K.) Pluto's period of revolution is about 248 years, and, it has the most eccentric orbit of any of the planets, exceeding even that of Mercury. In addition, its orbit is the most inclined to the plane of the other planets (17°). Because of its high orbital eccentricity it sometimes passes inside the orbit of Neptune. (They will never collide because the orbits do not actually intersect; the two planets, in fact, are never closer than 18 A.U. to one another.) This is the only place in the solar system where such an orbital relationship occurs. Pluto is, in fact, in this section of its orbit now.

Figure 12.14 Two photographs of Pluto (arrows) taken 24 hours apart. The slight shift in position would appear as a "jump" in a blink-comparator.

In photographs (Fig. 12.14) Pluto looks like a dim star. It presents no discernible disk, and therefore its size is difficult to determine accurately. In 1965 Pluto came extremely close to occulting a dim star—so close, in fact, that astronomers could determine that its diameter was definitely less than 6800 km (4225 mi). In 1978 J. Christy noted what appeared to be a bump on one edge of Pluto's photographic image. Three days later the bump was on the other side; then, six days later, it was back in its original position. Careful study soon showed that the object was a moon; it was named Charon (in Greek mythology, the ferryman who carried the souls of the dead to Hades, the kingdom of Pluto). Charon's period of revolution has been shown to be the same as Pluto's period of rotation—6.38 days, and its average distance from Pluto is approximately 17,000 km.

The discovery of Charon led to a better estimate of the mass and (indirectly) the diameter of Pluto. If we know the orbital period and mean distance of a satellite from its parent planet, we can determine the sum of their masses; if we can then determine the ratio of the two masses, we can determine them individually (p. 267). The result of this calculation showed that Pluto was much lighter than astronomers had expected, its mass, as noted above, being only about $\frac{1}{600}$ that of Earth. (Prior to this time it was believed that Pluto has a mass about $\frac{1}{10}$ Earth's.) The calculation also showed that Charon had a mass approximately $\frac{1}{10}$ that of Pluto.

The next problem was to determine Pluto's diameter. A planet's brightness, reflectivity, and diameter are interrelated: We know that Pluto's visual magnitude or brightness is 15.2 (Chapter 14). Infrared spectral measurements of Pluto have shown that methane is present, and it is now generally believed that the surface is coated with methane frost, the reflectivity of which is estimated at approximately 70%. Taken together with the brightness of 15.2, this gives a diameter of 2600 km. Another technique, called speckle interferometry (p. 266),[2] has given a similar result (about 3000 km). If Charon has the same reflectivity as Pluto, its diameter is about 1500 km.

On the basis of the volume (from the diameter calculated above) and the known mass of the planet we can make a reasonable guess at its internal structure. Pluto is probably composed mostly of water ice, with perhaps a silicate core.

Although the outer planets are all gas giants, Pluto has a solid surface more like a moon. On the basis of this and other evidence astronomers have suggested that Pluto may be an escaped moon of Neptune.

The other evidence: The fact that Pluto passes inside Neptune's orbit, and the relation between the orbital periods of Neptune and Pluto, which is 2:3 (i.e., Neptune orbits three times to Pluto's twice).

Kuiper and Lyttleton have suggested that Pluto, while a moon of Neptune, may have interacted with Triton, forcing it into a retrograde orbit. As a result of the interaction, Pluto may have left Neptune and taken up an orbit around the sun.

SUMMARY

1. Jupiter is the largest planet in the solar system. It has a mass (and volume) greater than that of all other planets combined.
2. One of Jupiter's most striking features is the colorful bands that encircle it. The dark bands are called belts, the light ones, zones.
3. Jupiter is quite oblate, a consequence of its high rotational speed. It rotates on its axis in approximately 10 hours.
4. Most of Jupiter is made up of hydrogen and helium, but methane, ammonia, and water vapor are present in its outer regions.

[2] In speckle interferometry a computer reassembles images from many short exposures.

5. The Great Red Spot has been visible ever since Jupiter was first observed through a telescope. It is about 14,000 km wide by 40,000 km long and is now generally believed to be a long-lasting storm.
6. Jupiter emits several different types of radiation—heat radiation, decimetric waves from the region surrounding the planet, and decametric waves from the region closer to the surface.
7. Jupiter has a strong magnetic field. Because of this field it is surrounded by radiation belts similar to the Van Allen belts that surround Earth.
8. There are three distinct cloud layers above Jupiter. Ammonia is common within them. Below these clouds is a region of liquid molecular hydrogen about 25,000 km thick. Interior to it is a region of liquid metallic hydrogen. There is probably a small rocky core at the center.
9. Jupiter has 16 known moons. The four largest—Io, Europa, Ganymede, and Callisto—were recently photographed by the Voyager spacecraft. Io is of particular interest in that it has active sulphur volcanoes.
10. Saturn is the second largest planet in the solar system. It has the lowest average density of all the planets (a possible exception is Pluto), so low that it could float in water.
11. Saturn has colorful atmospheric bands. It is made up mostly of hydrogen and helium, with some methane in the outer regions. Its internal structure is probably similar to that of Jupiter.
12. Saturn's most striking feature is its ring system. It is 274,000 km across and so thin that stars can be seen through it. It was probably formed when one or more moons came within Saturn's Roche Limit and broke up.
13. Saturn has 17 known moons. Its largest, Titan, is of particular interest in that it appears to be the only moon in the solar system with an atmosphere.
14. Uranus was discovered by W. Herschel in 1781. It has a mass 15 times that of Earth. Its atmosphere is mostly hydrogen, with methane clouds near the outer edge. A large fraction of its interior may be ice.
15. Uranus's strangest feature is the tilt of its axis, which lies near the plane of the planet's orbit. It has 21-year seasons.
16. The position of Neptune was predicted by both Adams and Leverrier. The planet was observed by Galle of the Berlin Observatory.
17. Neptune is similar to Uranus in size, appearance, and internal structure. It rotates in approximately 22 hours. It has three moons; one of them, Triton, is the third largest in the solar system.
18. Pluto's position was predicted by P. Lowell. It was found in 1930 by C. Tombaugh. Pluto is composed mostly of water ice; it is the smallest planet in the solar system and has one moon, Charon.

REVIEW QUESTIONS

1. What is the difference between a zone and a belt? Describe each.
2. Why is Jupiter oblate?
3. Describe the Great Red Spot in detail. What is it?
4. What is the difference between decametric and decimetric waves?
5. Describe the major groups of Jovian moons. Are any captured asteroids?
6. Describe Jupiter's magnetosphere. What is the bow shock wave?

7. Describe the three cloud layers in Jupiter's atmosphere.
8. Briefly describe Jupiter's interior structure.
9. List some of the discoveries of Voyager 1 and 2.
10. What is the internal makeup of each of the Galilean satellites (Io, Europa, Ganymede, and Callisto)?
11. Explain how Io interacts with Jupiter's magnetosphere.
12. Describe Saturn's internal structure.
13. How thick is Saturn's ring system? What evidence do we have for this?
14. Why do Saturn's rings sometimes disappear?
15. What is the Roche Limit? Is there a Roche Limit around the earth? What will eventually happen as a result of it?
16. Describe Saturn's largest moon. What is particularly noteworthy about it?
17. Who discovered Uranus?
18. Describe the seasons that occur on Uranus and explain why they occur.
19. How do Neptune and Uranus differ? In what ways are they similar?
20. Describe the events that led to the discovery of Pluto. Who predicted its position? Was the prediction correct? Who discovered it?
21. What evidence do we have that Pluto has a moon?

THOUGHT AND DISCUSSION QUESTIONS AND PROJECTS

1. Assume that you are in a space vehicle visiting Jupiter. Describe what you (or your instruments) would see (or detect) as you passed through the magnetosphere, the outer atmosphere, and into the lower atmosphere of the planet.
2. Compare and contrast the surfaces of the four Galilean satellites (Io, Europa, Ganymede, and Callisto). Explain how the markings may have come about. Can you think of any other than those given in this chapter?
3. Compare and contrast the ring systems of Jupiter, Saturn, and Uranus. Discuss two ways they may have come about.
4. Describe a year on Uranus in detail (i.e., length of daylight and darkness, length of season, direction of apparent motion of the stars). Describe what you think would happen on Earth if its axis were suddenly tilted at the same angle as Uranus's.
5. *Project:* Using a pair of binoculars or a small telescope, plot the position of Jupiter's four largest moons over several evenings. Using the plots, explain the motions of the moons.
6. *Project:* Using a small telescope, sketch Saturn's rings. Can you see divisions in the rings? At what angle are the rings tilted?

FURTHER READING

Belton, M. J. S. "Uranus and Neptune," *Astronomy,* p. 6 (February 1977).
Berry, R. "Mysterious Pluto," *Astronomy,* p. 14 (July 1980).
Gore, R. "Saturn: Riddle of the Rings," *National Geographic,* p. 3 (July 1981).
Hunten, D. M. "The Outer Planets," *Scientific American,* p. 130 (September 1975).
Morrison, D. "Io," *Astronomy,* p. 6 (February 1976).
Scientific American. *The Solar System.* W. H. Freeman, San Francisco (1975).
Soderblom, L. A. "The Galilean Moons of Jupiter," *Scientific American,* p. 88 (January 1980).

Aim of the Chapter

To describe objects other than the planets and their moons that are part of the solar system. To show how study of these objects helps in understanding the origin of the solar system.

chapter 13

Minor Components of the Solar System

Bode's law (p. 117) may, at first, have seemed to be little more than a number game. But the more Bode and other astronomers thought about the strange sequence, the more they became convinced that it had some validity, and if it was valid, there should be a planet between Mars and Jupiter.

On New Year's Eve of 1800 Sicilian astronomer Guiseppe Piazzi, while making a map of the sky, came upon an object that did not appear on earlier maps. Although he thought, at first, that it was a comet, it looked more like a dim but clearly defined star.

Piazzi reported its position to Bode, who became convinced that it was the planet that had been predicted to lie between Mars and Jupiter.

The announcement of the discovery was published during the summer of 1801. Karl Gauss, one of the world's great mathematicians, calculated the orbit, and several months later the object was found at the position he predicted. Piazzi named the object Ceres, after the patron goddess of Sicily.

ASTEROIDS

Astronomers studying the object soon saw that it was exceedingly small—much smaller than any of the other planets in the solar system. The following year, another object about the same size and also between Mars and Jupiter was observed by Olbers. In 1804 another was discovered, and in 1807 a fourth. Bode's law had not predicted this, and astronomers began to speculate that these objects were the larger fragments of a planet that had been destroyed.

The objects discovered after Ceres were named Pallas, Juno, and Vesta. Ceres, with a diameter of 1000 km (620 mi), is the largest of the four. Pallas is slightly less than 600 km (373 mi) across, Vesta about 550 km (342 mi), and Juno about 300 km (186 mi). Because they

Figure 13.1 A multiple exposure of Icarus. The streaks of light indicate its position at four-minute intervals.

are so much smaller than the other planets, they are now referred to as minor planets or asteroids.

Astronomers continued to find more and more asteroids. By 1850, thirteen had been discovered. Asteroid hunting then became the fashion, and numerous astronomers joined in the chase. By 1890 the total was 300, and when photography was introduced as an astronomical technique in 1891, many more were found. (Because they move in relation to the "fixed" stars, they appear as lines on a long photographic exposure, as may be seen in Fig. 13.1.) Although about 2000 have had their orbits plotted, there are probably many more, perhaps as many as 100,000. Today much of the novelty of discovering an asteroid is gone, and astronomers do not often trouble themselves to calculate the orbit of every streak they see on a photographic plate.

Aside from the four largest, almost all asteroids are little more than large rocks in space; most are not spherical. About 250 are known to have diameters larger than 100 km; the rest are smaller. Most asteroids orbit in the region between Mars and Jupiter, now referred to as the asteroid belt. However, some venture inside the orbit of Mars, and a few even pass the earth's orbit.

Astronomers no longer hold the earlier view that these objects are part of an exploded planet. The main reason for their view is the exceedingly small total mass of the asteroids—no more than about $\frac{1}{50}$ the mass of Earth. In addition, asteroids in different regions of the asteroid belt are made up of different material, and it is unlikely that this would be the case if the asteroids had resulted from a shattered planet. Astronomers now believe that asteroids are leftover material from the original solar nebula.

Chapter 7 described how the planets formed when fragments of the solar nebula struck one another at relatively low speeds and coalesced. In the region between Jupiter and Mars, however, the forces acting on the fragments were different, for, in addition to the strong solar gravitational pull, there was also a gravitational pull from Jupiter. (We are assuming, of course, that Jupiter had already formed at this point.) This tended to increase the speed of fragments to the point where, when they collided, they shattered instead of coalescing. Many of the remnants of the shattering were no doubt deflected out of the asteroid belt into other regions, where they caused the extensive cratering seen throughout the solar system. The pieces not deflected—many of the smaller ones and a few of the larger—are still orbiting in the asteroid belt.

It is not the case, as this might seem to imply, that the asteroid belt is strewn with boulders of all sizes. When the Pioneer spacecraft (and also Voyager) passed through the asteroid belt on their way to Jupiter, they found only that grain-sized particles were more common—about three times as abundant as they are in the region near Earth, and not nearly as hazardous to pass through as we had believed.

Composition

Spectral analysis indicates that the asteroids are composed mainly of rock, with embedded metals. However, there is a large range of reflectivities (albedos) among them, indicating that there are basic differences in composition. Ceres, for example, is almost as dark as coal, while Vesta has a light surface with a high reflectivity. It has been shown, in fact, that most of the asteroids fall into two classes: dark and light. Astronomers suspect the presence of carbon or carbon compounds in the dark ones.

A special class of meteorite (discussed later)—carbonaceous chondrite—also has a high carbon content, and these meteorites may be pieces from dark asteroids. The light asteroids are believed to be composed mostly of light-colored stone.

Almost all of the inner asteroids (those near Mars) are of the light variety, whereas most of the outer ones are of the dark variety. This tells us that the original solar nebula was not uniform in structure. In fact, the material we see in this region today is probably still quite similar to the way it was 4 billion years ago.

Trojan and Apollo Groups

Not all asteroids orbit in the asteroid belt. In the late 1700s French mathematician J. L. Lagrange discovered that there were two points along Jupiter's orbit where objects such as asteroids should be focused as a result of the gravitational forces exerted by Jupiter and the sun. One point was 60° degrees ahead of Jupiter (Fig. 13.2), and the other was 60° behind. We now refer to these points as Lagrangian points. A search of these two regions disclosed, at first, several aster-

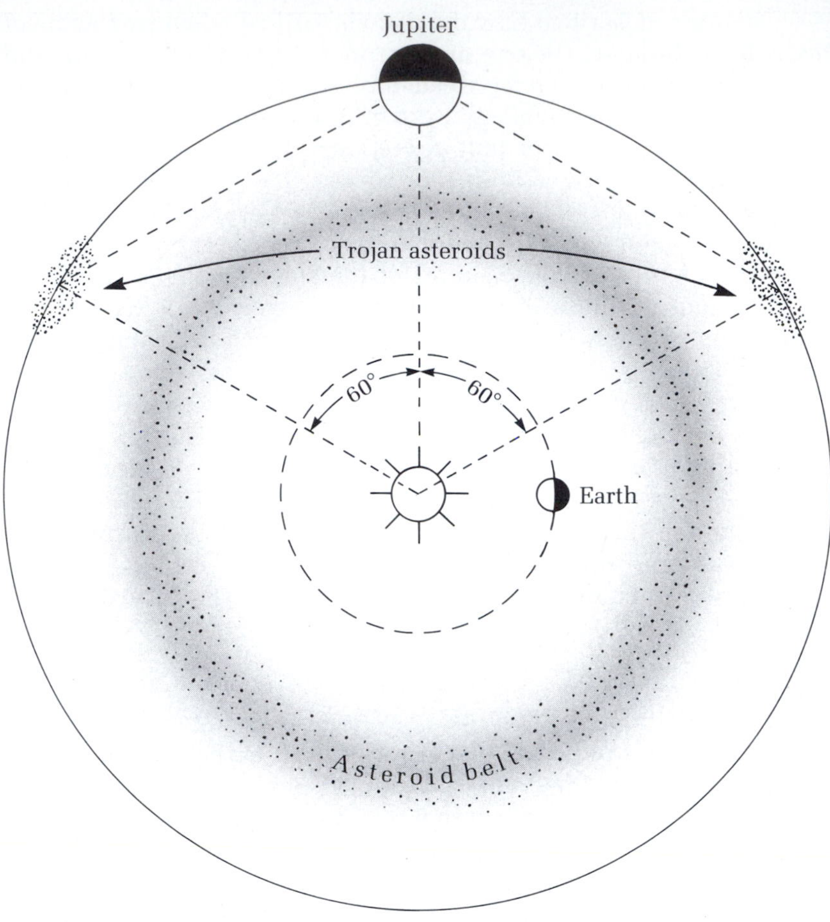

Figure 13.2 Positions of the Trojan asteroids.

oids and, over some time, a relatively large number; these are now called the Trojan asteroids.

Other asteroids, called the Apollos, actually pass, at some point in their orbit, within the earth's. One, a small asteroid called Icarus, even passes within Mercury's orbit. With the number of asteroids passing Earth a collision is, of course, quite possible; fortunately, serious collisions do not occur often. Astronomers estimate that we are hit by an asteroid having a half-mile or better diameter every million years or so. There have been a number of near misses recently. In June 1968 Icarus passed at a distance of about 6 million km (3.7 million mi); Hermes passed Earth in 1937 at a distance of 650,000 km (400,000 mi). The most recent near-miss was an object that skimmed our atmosphere in the northern United States in 1972; Astronomers believe it may have been about 20 m across.

The surface of the moon shows what would happen were an asteroid to strike Earth; the remnants of many such collisions are clearly visible there. Debris would be thrown out for hundreds of kilometers; the accompanying explosion would be a gigantic and awesome event, at least equivalent to the explosion of a hydrogen

Chiron

In early November 1977, Charles Kowal, using a blink comparator, was examining plates that he had taken in late October. Examining one of the points that had moved—possible evidence of a comet or asteroid—he found that it did not look like a comet yet was small—200–800 km across. It seemed that it might be an undiscovered planet, but so small ($\frac{1}{10}$ the size of Mercury) that it could only be classified as a "miniplanet."

Brian Marsden found that its image had been recorded on plates made as far back as 1895. Using the plates, he calculated its orbit, which was much more circular than the orbits of most asteroids or comets. Except for a small section just inside Saturn's orbit, most of the object's orbit was between Saturn and Uranus. Its period was 50.7 years.

Officially, the object could not be classified as an asteroid; all asteroids were inside Jupiter's orbit (most in the asteroid belt). R. C. Smith argued that it might be the largest of a new group of asteroids, a group that orbits beyond the asteroid belt, and that they may have been captured from the regular asteroid belt.

But Chiron, as the object is now called, may not be with us forever, and it is likely that it was not here from the beginning of the solar system. S. Oikawa and E. Everhart of the University of Denver have shown that its orbit is unstable. It will probably be thrown completely out of the solar system in the next few million years by the gravitational influence of either Saturn or Jupiter.

bomb. (The exact magnitude of the explosion would of course depend on the size of the asteroid and its speed relative to Earth.) Were it to land in the ocean, it would generate a tidal wave hundreds of meters high that would destroy numerous coastal cities.

For years there has been speculation that a catastrophe such as this may have occurred about 65 million years ago, presumably causing the extinction of dinosaurs and various other species that the fossil record indicates occurred about that time. Luis and Walter Alvarez and their associates have recently sparked a resurgence of interest in this hypothesis by showing that layers in certain outcroppings of rock have an exceedingly high concentration of iridium. Iridium is rare on Earth but relatively common in chondrite meteorites, which contain approximately a thousand times more iridium than terrestrial rocks. Alvarez believes that a meteorite of this type (estimated at about 10 km across) disintegrated in our atmosphere and that the resulting dust stayed in the stratosphere for several years, cutting off sunlight and consequently killing vegetation needed by various animal species (including the dinosaurs).

Because of their relative closeness to Earth it has been suggested that we mine asteroids for their iron and nickel. Such an expedition would be likely to find asteroids quite irregular bodies. (Gravitational forces are not strong enough to shape such small objects into spheres.) Most large asteroids are literally jagged, rough, flying mountains. A person on one 8 km (5 mi) or less across could—with a reasonable run and jump—actually launch himself into space.

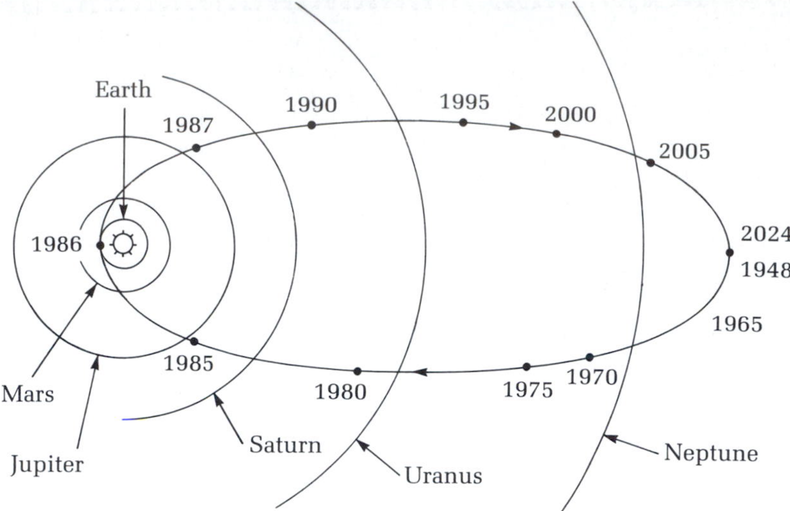

Figure 13.3 Orbit of Halley's comet.

COMETS

Every few years strange apparitions appear in the sky. Their long, glowing tails and ghostlike appearance terrified many of the ancients, who looked upon them as omens of disaster—frequently the end of the world; at best they were taken to herald famine or war. We now refer to these objects as comets.

At one time it was believed that comets were an atmospheric phenomenon. However, Tycho Brahe showed, after studying the comet of 1577, that it was well beyond the moon. Still, they were feared. A better understanding of comets came in the early 1700s, when Edmund Halley studied several, using Newton's law of gravity to calculate their orbits (Fig. 13.3). A series of rather bright ones, which had appeared in 1531, 1607, and 1682, was of particular interest to him. He calculated the orbit of the 1682 comet and showed it had a period of approximately 76 years; this meant that it and the comets of 1531 and 1607 were one and the same. Continuing back at intervals of 76 years, he discovered that it had, in fact, been seen numerous times. He went on to predict that it would return in 1758, and, on Christmas Eve of that year—16 years after Halley's death—it did.

Composition

Our present notions concerning the makeup of comets are based on a model (sometimes called the "dirty snowball" or "dirty iceberg" model) presented in 1950 by Fred Whipple. According to this model the only solid part of a comet is a small nucleus, about 10–15 km (6–9 mi) across, that consists of frozen gases such as methane, ammonia, and carbon dioxide, along with frozen water and grains of dust.

As this nucleus approaches the sun, the frozen gases in the outer

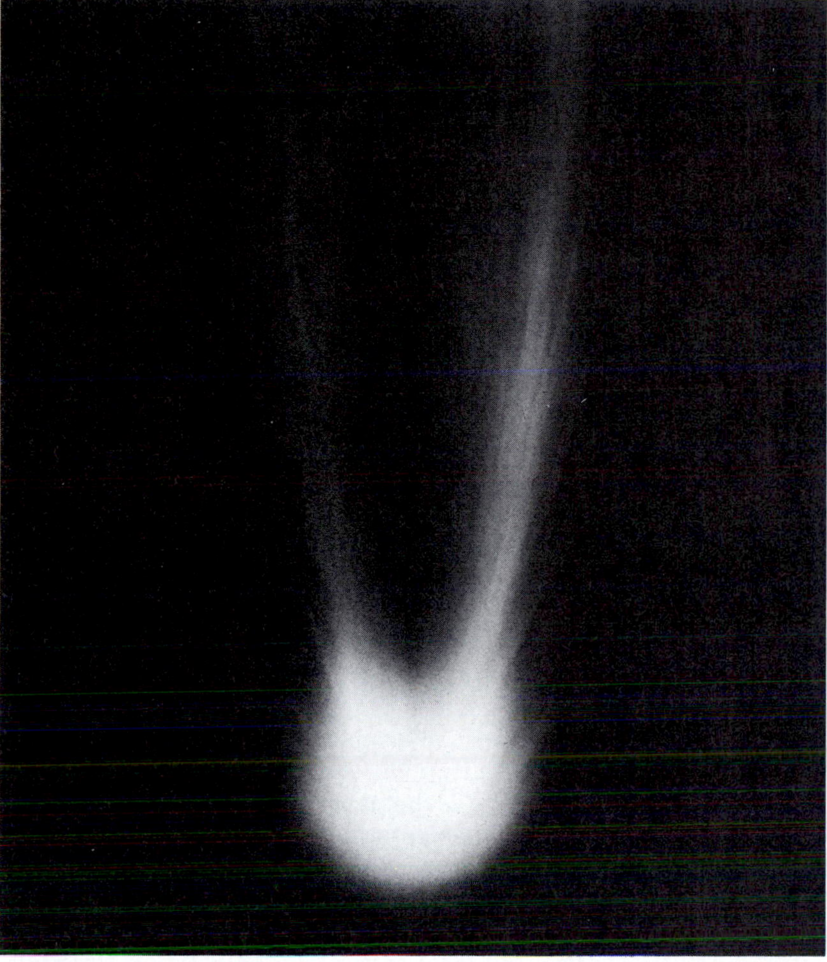

Figure 13.4 The head, or coma, of Halley's comet.

layer begin to evaporate. Over a period of days, or perhaps weeks, a large coma forms around the nucleus; it can grow to as large as 100,000 km or more across (Fig. 13.4). The evaporating gases form a long, flowing tail, sometimes many millions of kilometers in length. The comet gets brighter and brighter as it approaches the sun, then begins to dim as it swings around and begins to recede. Finally, after a few weeks, it is too dim to see with the naked eye.

Part of the (apparent) glow of the comet is reflected sunlight. Just as we see the planets by reflected sunlight, so, to a large extent, do we see comets as a result of it. But some of the light is emitted by electrons in the evaporated gas, which are excited by solar ultraviolet radiation as the comet approaches the sun.

In many cases the long tail that trails behind the head of a comet can be seen to consist of two distinct parts, the gas tail and the dust tail. The gas tail consists of ionized gases, which interact with the ions of the solar wind and are forced into a position pointing directly away from the sun (Fig. 13.5). This tail may be slightly lumpy in appearance. As the gas evaporates from the outer shell of the nucleus,

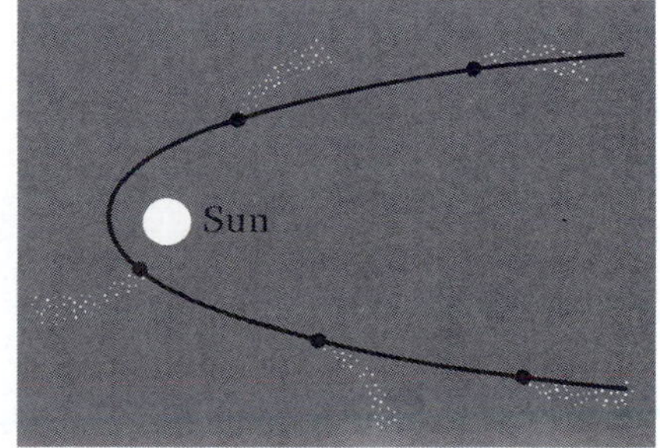

Figure 13.5 Relation of comet tail to sun.

dust particles and small grains are released; these also form a tail, but because it is composed of uncharged particles it is not seriously affected by the ions of the solar wind; however, it is affected by a pressure caused by the photons from the sun, which pushes them into a long, curving tail behind the comet. This tail is usually much smoother and more straight-edged than the gas tail.

In 1970 astronomers were able, for the first time, to observe a comet from an orbiting satellite. A study of its spectra showed that the head was surrounded by a large hydrogen cloud. Astronomers believe that the water that evaporates is acted on by ultraviolet light from the sun to separate it into its component hydrogen and oxygen. The released hydrogen then forms a huge halo—millions of kilometers in diameter—around the head of the comet.

The Oort Cloud

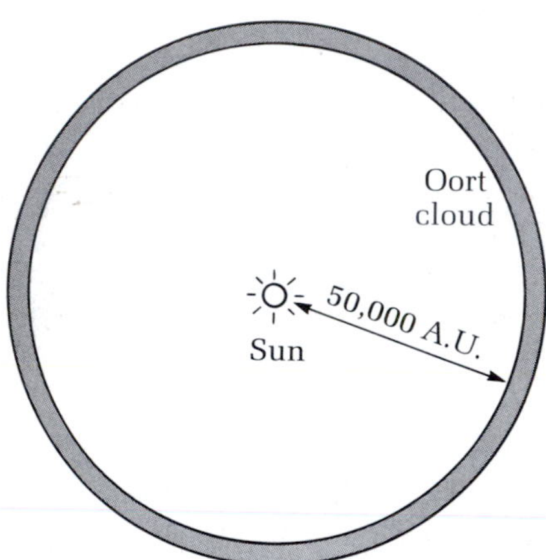

Figure 13.6 The Oort cloud.

Comets that are visible to the naked eye are relatively rare and are seen only every few years. Astronomers may observe half a dozen or more per year that are too dim to be seen with the naked eye. There are two basic types of comets—long-period and short-period. Halley's Comet is typical of the short-period ones, orbiting the sun every 76 years. Long-period comets may take thousands—in some cases, even millions—of years to make a single orbit; the comet Kohoutek, which appeared in 1974, is a good example of this type.

In 1950 Dutch astronomer J. Oort proposed an explanation of such large differences in periods that is now accepted by most astronomers. He suggested that the solar system is surrounded by a cloud of comet nuclei lying about 50,000 A.U. (about 1 light year) from the sun (Fig. 13.6). (This distance is a large part of that to the nearest star.) There are millions, perhaps billions of potential comets in this cloud (called the Oort cloud), and most orbit in it undisturbed. Occasionally one is perturbed (perhaps by a nearby star) into a new orbit that takes it very close to the sun. It may take thousands of years to traverse its strange, new, elongated (cigar-shaped) orbit, but eventually it will enter the inner solar system (where the planets orbit) and, as it approaches the sun, will develop its characteristic tail.

If it is not perturbed again, it will leave the inner solar system and will not return for thousands of years–if ever. However, it may be perturbed in the inner region by either Jupiter or Saturn and may then take up a much smaller orbit that lies entirely within this region; in effect, it will have become a short-period comet (Fig. 13.7).

Unlike most objects in the solar system, comets are not confined to the plane of the planetary orbits. The Oort cloud is a spherical shell, and comets may come into the inner solar system from any direction. About half have retrograde motion.

Sun Grazers

There is strong evidence that the nuclei of comets are quite fragile. In fact, many have broken into several pieces. This breakup occurs

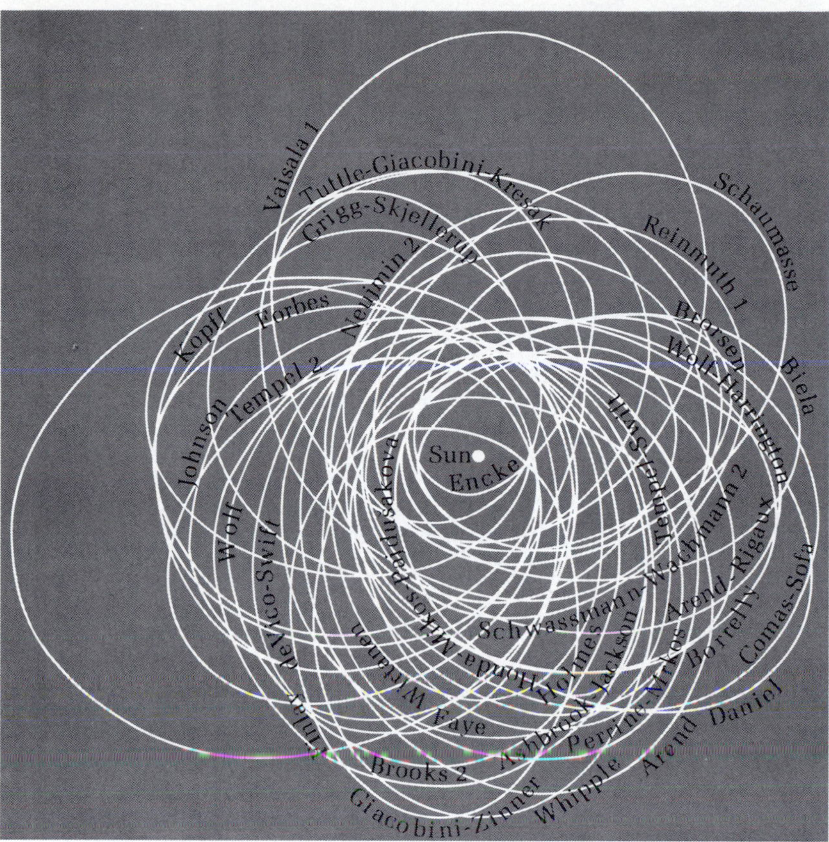

Figure 13.7 Orbits of short-period comets.

most frequently in a group of comets called "sun grazers," which approach the sun very closely—within one solar diameter. Several of the visible comets of recent years—the comet Ikeya-Seki, for example—have been of this type.

As the sun grazer approaches the sun, tidal forces across its surface increase significantly and may become strong enough to rip it apart. One of the strangest breakups of this type occurred in the nineteenth century, when the comet Biela broke into two pieces during its 1846 perihelion passage; on the next passage two side-by-side comets were seen, and on the next they were seen again, but much farther apart. The comets did not return on their next scheduled passage, but a few years later, the earth was bathed in a brilliant shower of "falling stars." Astronomers showed that this shower was the remains of Biela, whose orbit had been changed slightly and that somehow had been shattered.

Other comets that have recently broken up are the abovementioned Ikeya-Seki and the comet West; Ikeya-Seki broke into two pieces after its 1965 passage, and West broke into four pieces after its passage of 1976.

METEORS

As mentioned above, comets release a considerable amount of dust and debris as they approach the sun and occasionally break up entirely. This and the material left over from the solar nebula (most of which is in the asteroid belt) make the solar system a dusty place. Indeed, there is direct evidence that this is the case: the several "falling stars" or meteors (Fig. 13.8) that can be seen on any clear night, are actually tiny grains of sand or debris (called meteoroids), most with a mass less than 1 gm. Before midnight about three per hour can be seen, but after midnight this number increases quite significantly to as many as 15 per hour, and, on certain nights of the year, as many as one per minute. Meteors strike the earth's atmosphere at extremely high speeds (12–72 km/sec, or 27,000–160,000 mi/hr), and, as they plunge toward the surface, they are heated to white-hot temperatures by air friction. Most begin to glow about 100 km above the earth's surface but burn out by the time they are 50 km above it. Worldwide, about 25 million strike the earth's atmosphere each day.

Some of these grains are probably particles left over from the solar nebula, but most are the debris left by the partial evaporation of comets. We see more after midnight than before as a result of Earth's orbital motion, which is bringing it into encounters with meteors moving in the opposite direction (or at an angle) to its own—somewhat as the windshield of a moving car encounters far more rain than does the rear window (Fig. 13.9).

Every so often the earth passes directly through the orbital path of a present or former comet, and when this section of the orbit is strewn with evaporated debris, we see a brilliant display—a meteor shower, in which large numbers of "shooting stars" brighten the night sky. Observed carefully, all will appear to come from a single point in the sky called the radiant, which is actually the path along which we are passing through the meteor debris (Fig. 13.10). Showers are usually named after the constellation that contains the radiant. Several showers and their dates of occurrence follow.

Shower	Date of maximum display
Quadrantids	January 4
Lyrids	April 22
Perseids	August 12
Draconids	October 10
Orionids	October 21
Taurids	November 8
Leonids	November 17
Gemenids	December 12

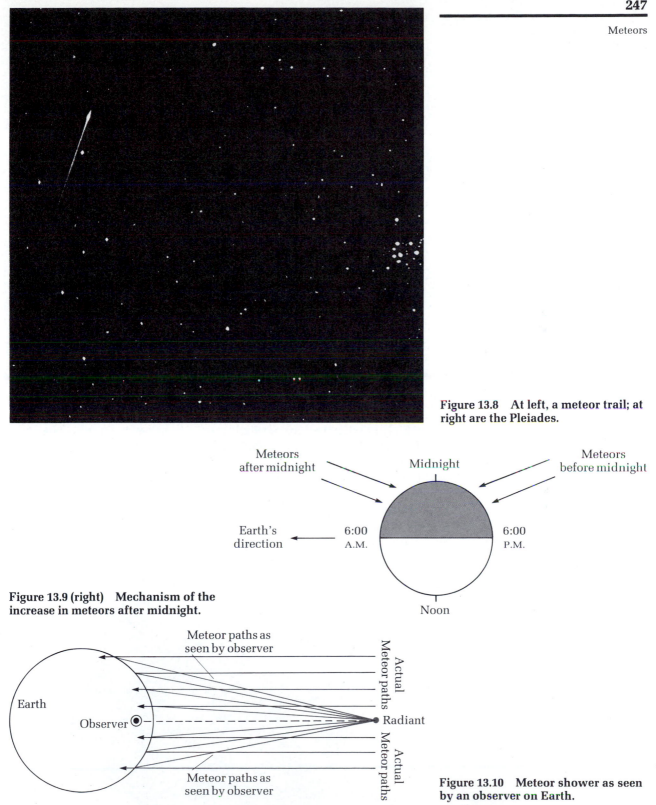

Figure 13.8 At left, a meteor trail; at right are the Pleiades.

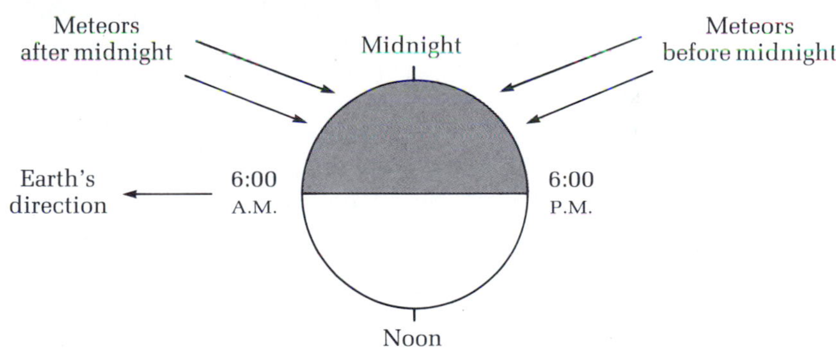

Figure 13.9 (right) Mechanism of the increase in meteors after midnight.

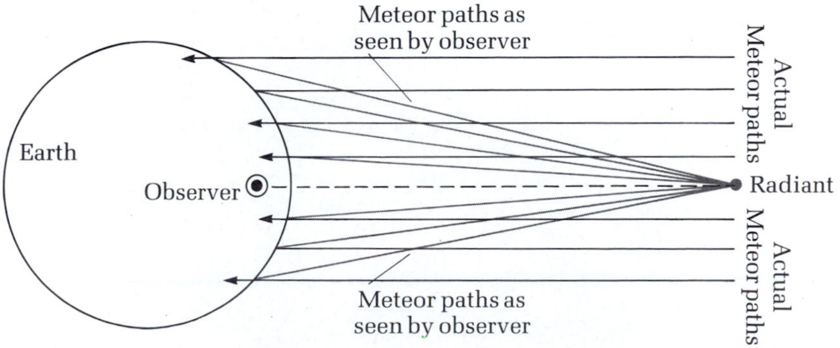

Figure 13.10 Meteor shower as seen by an observer on Earth.

METEORITES

The particle that causes the meteor phenomenon (called a meteoroid) is usually quite small; occasionally, though, a large piece of material that has perhaps broken off an asteroid strikes our atmosphere, and part of it manages to survive all the way to the surface of the earth. When this occurs, we see a fiery blaze of light in the sky—a fireball. If the object is large enough (Fig. 13.11) and has adequate velocity, it can cause a tremendous explosion and may leave a large crater. The remnant of the object left on the surface of the earth is called a meteorite.

Meteorites have been found in literally every country on Earth; there are three main types: stones, irons, and stony-irons. As the names suggest, stones consist mostly of a stony material, irons consist mostly of iron, and stony-irons contain a mixture of both.

Most—perhaps as much as 90%—of the meteorites that strike the earth are stones. But these look so much like ordinary earth rocks that they are usually difficult to identify. About 80% of those found are irons. Most of the stones are chondrites, which contain shot-sized spherules of silicate called chondrules. A particularly important subclass of this group is made up of chondrites (called carbonaceous chondrites) that have a high carbon content.

Because many asteroids are dark and probably contain a considerable amount of carbon, it is believed that carbonaceous chondrites are chips or pieces from these dark asteroids. They may, in fact, be similar to the material that was in the original solar nebula, which is why they are so highly prized by astronomers. The most famous carbonaceous chondrite of recent years is the Murchison meteorite, which fell near Murchison, Australia, in 1969 and contained organic molecules—amino acids, compounds basic to living matter.

Most recent meteorite collisions have occurred in sparsely populated areas and produced little damage. Although a near-miss occurred in Iowa in 1923, no one had been reported to have been

Figure 13.11 A meteorite that fell near Allende, Mexico.

hit by a meteorite until 1954, when a woman in Alabama was injured slightly by a meteorite that crashed through the roof of her home. The meteorite was about the size of a canteloupe. A meteorite of similar size came through the roof of a house in Connecticut in 1982.

The largest known meteoritelike[1] explosion of recent times occurred in Siberia in 1908. (There is evidence of much larger ones in the distant past: one in Antarctica and one in Hudson Bay.) This Tunguska explosion (see Box) caused widespread damage over hundreds of square miles of sparsely populated forest. If it had fallen in a large city, it would certainly have caused a major catastrophe.

The Tunguska Explosion

At 7:17 A.M. on July 30, 1908, a deafening explosion rocked a sparsely populated region along the Tunguska River in central Siberia. A gigantic pillar of fire, glowing so brightly that it could be seen for hundreds of miles, rose out of the explosion. Searing thermal currents rushed across the countryside, knocking over trees, scorching vegetation, and literally flattening everything in their path. Thousands of reindeer were killed. Hundreds of miles away roofs were blown from houses and people were swept off their feet. Shock waves from the blast were recorded around the world. This was one of the largest explosions ever to take place on the surface of the earth.

Strangely, an interval of almost 20 years elapsed before anyone visited the blast site. The first expedition to the region was headed by Russian scientist Leonid Kulik. Kulik and his group, expecting to find the remains of a large meteorite, were shocked and surprised by what they saw instead: trees blown down and sheared off over approximately 1000 square miles, but no sign of a crater; the whole area had simply been flattened.

Kulik established that the trees had not been burnt as in a forest fire but had apparently been scorched from above, as if there had been a sudden flash of intense heat from the sky. This led him to believe that the explosion had taken place a few miles above groundlevel. He also noted that the devastated area was in the shape of a butterfly—almost as if it had been caused by a string of explosions.

Over the years there has been considerable speculation about what caused the explosion. Initially it was thought to have been a gigantic meteorite explosion, though of course there was no significant depression or crater, and no characteristic meteorite remains. A comet colliding with Earth was suggested, but no tail had been seen. Explanations soon became quite exotic. Could it have been a small blob of antimatter floating through space? (When matter and antimatter meet, they annihilate one another in a large explosion.) Others have proposed that a spaceship collided with Earth. Still others have suggested a black hole as one cause. Of course it would have to have been a tiny one—a primordial black hole (Chapter 20).

Scientists now feel reasonably sure as to what caused the explosion. One of the major clues was traces of carbonaceous chondrite that were found at the site. On the basis of this scientists believe that the explosion was caused by a large carbonaceous chondrite object—perhaps a meteorite or burnt-out comet nucleus—that collided with and blew up in the earth's atmosphere. The nucleus would have exploded into a cloud of black soot—and, indeed, "black ash" was seen immediately after the explosion. There is also evidence for several other smaller explosions of this type around the world. One took place near Revelstoke, British Columbia, in 1965. A thin layer of black ash was spread over freshly fallen snow for many square miles.

[1] "Meteoritelike" because there is still considerable controversy concerning its cause; many believe it was caused not by a meteorite but by a large, carbonaceous chondrite object.

In 1947 another meteorite hit Siberia, laying waste an elliptical area about 5 × 1.6 km (3 × 1 mi).

The largest meteorite crater in the United States is near Flagstaff, Arizona. Called the Barringer Crater (Fig. 13.12), it is 1300 m (about 1 mi) across and about 180 m (600 ft) deep. It is believed to be the result of a collision that occurred some 25,000 years ago.

Figure 13.12 The Barringer crater, near Flagstaff, Arizona, is the largest meteorite crater in the United States.

ZODIACAL LIGHT

By terrestrial standards interplanetary space is essentially a vacuum, though as mentioned above, with considerable dust dispersed in it. Evidence of this dust, apart from the numerous meteors that can be seen each night, can be observed by looking toward the western horizon at twilight on a clear evening in the spring, or at the eastern horizon just before sunrise. The glow extending upward along the ecliptic is called zodiacal light. Broad at the base, it tapers to a diffuse point at some distance above the horizon. Astronomers have shown that it is light reflected from tiny dust particles that orbit the sun. It is broadest at the base because it is in the direction of the sun and the concentration of particles is greatest here.

SUMMARY

1. Bode's law gives the approximate position of the planets out to Uranus. It predicts a planet between Mars and Jupiter, but instead a large number of asteroids are found here.
2. The asteroids are believed to be leftover material from the original solar nebula.
3. The Trojan asteroids follow (and precede) Jupiter in its orbit. The Apollo asteroids pass inside the earth's orbit.
4. Comets have a nucleus (a few kilometers across) of various types of ice. When this nucleus approaches the sun, it evaporates partially, creating a coma and a tail. The tail always points away from the sun.
5. The Oort cloud is a cloud of comets that has the form of a spherical shell around the solar system. It lies about 1 light year from the sun. The comets we see come from this cloud.
6. Sun grazers are comets that approach very close to the sun. The tidal forces of the sun usually break them up.
7. The meteor phenomenon is caused mainly by tiny grains of sand striking our atmosphere at high speeds. The glow is a result of the high temperatures that are generated by friction as the grain passes through our atmosphere. Most meteoroids burn out before they reach the surface. When a large number of these grains strike our atmosphere, we see a meteor shower.
8. Occasionally a particularly large object strikes our atmosphere. If it manages to reach the surface of the earth, we call it a meteorite.
9. Meteorites are classified into three major groups according to composition: irons, stones, and stony-irons.

10. Giant meteorites have struck the earth in the past, leaving large craters. The largest in the continental United States is near Flagstaff, Arizona. A particularly large meteoritelike explosion occurred in Siberia in 1908.
11. A faint glow, called the zodiacal light, can be seen along the ecliptic. It is light reflected from dust particles that orbit the sun.

REVIEW QUESTIONS

1. Is the asteroid belt predicted by Bode's law? Discuss.
2. Why does the solar system have an asteroid belt? What is this material?
3. Describe the various types of asteroids. How do they differ?
4. What are the Trojan asteroids? What are the Apollos?
5. Describe the makeup of a comet. Where does most of the mass reside? What causes it to glow?
6. Why does a comet change in appearance as it approaches the sun? What happens to it?
7. What is the difference between the gas tail and the dust tail?
8. Describe the Oort cloud.
9. What frequently happens to sun grazers? Why?
10. What is the difference between a meteor, a meteoroid, and a meteorite?
11. Why are more meteors seen after midnight than before?
12. Why is there a radiant associated with a meteor shower?
13. What types of meteorites strike the earth most? Is this the type we generally find? Why?
14. What are carbonaceous chondrites?
15. Describe the Tunguska explosion. What is believed to have caused it?

THOUGHT AND DISCUSSION QUESTIONS AND PROJECTS

1. Describe how astronauts on an asteroid about 1 mile across would move, what they would be likely to see, what elements they might find, and any other aspects of their experience. What could they see and measure were they to follow a comet in space?
2. What would happen if:
   (a) the tail of a comet passed across the surface of the earth?
   (b) the coma of a comet passed across the surface of the earth?
   (c) the nucleus of a comet struck the earth?
   (d) an asteroid (approximately 1 km across) struck the earth?
   (e) a typical meteoroid struck the earth?
   (f) a carbonaceous chondrite (meteorite) struck the earth?
3. Give the relative positions of the earth, the sun, and a comet when:
   (a) the tail of the comet is approximately opposite the comet's direction of travel.
   (b) The tail of the comet is in the direction of travel.
   (c) The tail of the comet is at right angles to the direction of travel.
4. *Project:* Using a magazine such as *Astronomy* or *Sky and Telescope,* find out what meteor showers occur during the month and observe them. From what constellation do they appear to emanate? Explain.

FURTHER READING

Grossman, L. "The Most Primitive Objects in the Solar System," *Scientific American*, p. 30 (February 1975).

Hartmann, W. K. "The Smaller Bodies of the Solar System," *Scientific American*, p. 142 (September 1975).

McCall, G. J. *Meteorites and Their Origin*. Halstead, New York (1973).

Morrison, D. "Asteroids," *Astronomy*, p. 6 (June 1976).

Richardson, R. "The Discovery of Icarus," *Scientific American*, p. 106 (April 1965).

Whipple, F. L. "The Nature of Comets," *Scientific American*, p. 48 (February 1974).

part 4

Stars and Stellar Evolution

In Chapter 8 we studied a star of particular importance to us, our sun; in this part we turn to stars in general. After many years of investigation, astronomers now feel they understand stars quite well. Although much is still to be learned, many important tools have been developed in the last few decades. Computers in particular have been of considerable value in analyzing data and allowing the astronomer to build theoretical models of stars.

In the first chapter of this part—Chapter 14—we talk about the various properties of stars, beginning with a listing and discussion of some of the most important of these. It is important to note in this what astronomers can, and cannot, measure.

The main topic of discussion in Chapter 15—in a sense an introduction to Chapter 16—is energy. We have talked about energy in earlier chapters and have seen that it can be defined as the ability to do work. It takes many different forms: The energy of a speeding car, for example, can do considerable work (damage) to a guard rail should the car hit it, and a large weight suspended from a pulley can lift an elevator when it is released. In this chapter, however, we will be concerned mostly with nuclear energy, energy released by the nucleus in certain reactions. We will discuss two methods of releasing it—fission, basis of the atomic bomb, and fusion, basis of the hydrogen bomb and, more importantly, the process by which energy is released in the stars. The latter part of the chapter gives a brief introduction to the "strange" world of elementary particles, whose importance in astronomy will be treated later.

Chapter 16, the central chapter of this section, describes the life cycles—in most cases, hundreds of millions of years—of stars (most of the details on how they die are given in Chapters 18 and 20).

Chapter 17 completes and supplements the picture of the life cycle up to the death of a star.

Aim of the Chapter

To show the special properties of stars and how these properties are measured.

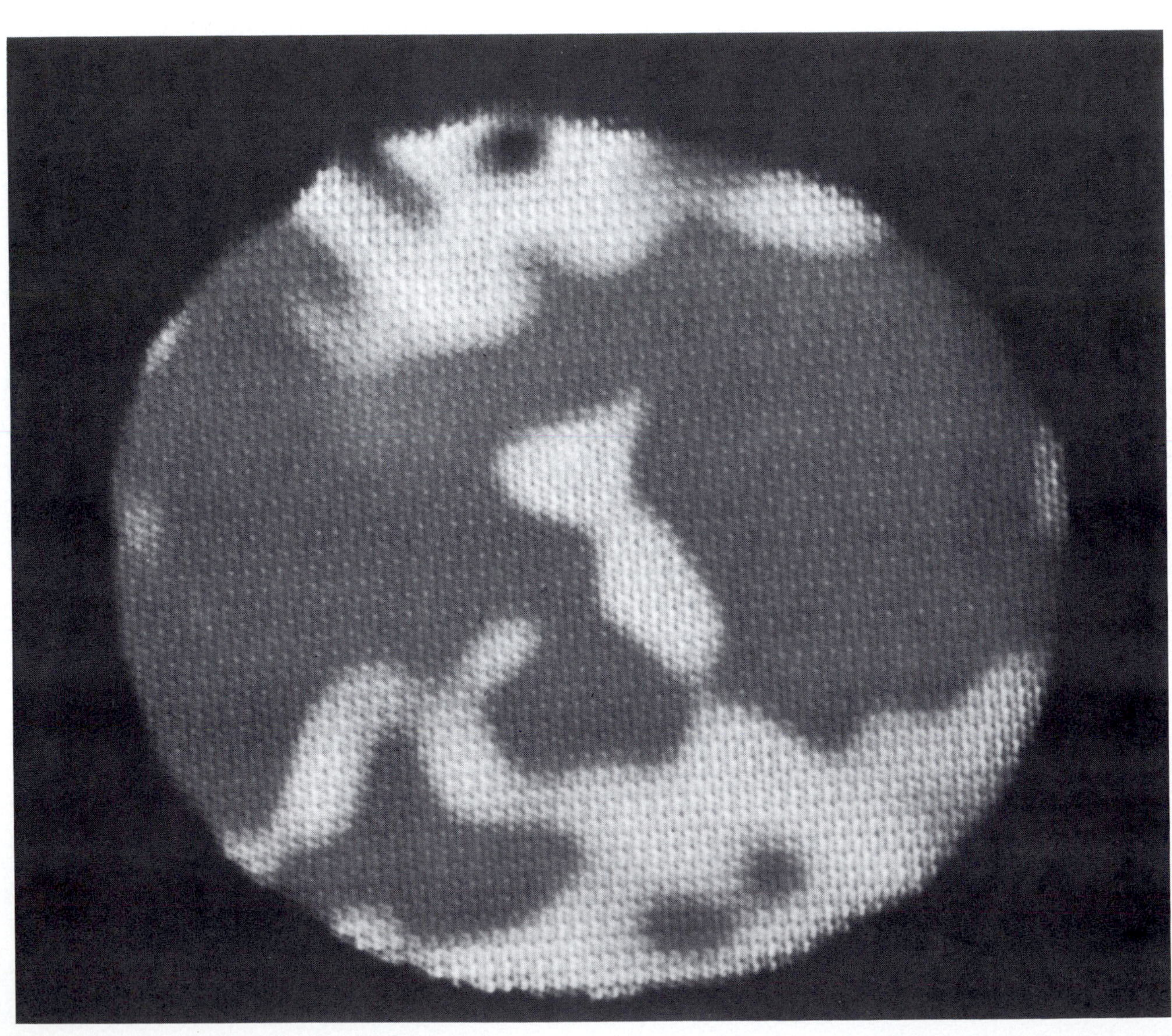

chapter 14
The Stars

Stars show a great range in size. Some are gigantic (if you placed a star such as Betelgeuse where the sun is, it would engulf all the inner planets out to perhaps Mars); some special types of stars are smaller than the earth. Stars are also of different colors and brightnesses: red, blue, dim, bright—but all are luminous balls of gas held together by gravity.

Astronomers are interested in gathering information about stars in order to answer such questions as how stars change over millions of years, what finally happens to them, and why some explode while others do not. However, stars are not easy to study. They are so distant that no telescope can show any details of their structure; there is only a point of light to examine. But, as we noted earlier, this point of light, studied with many different techniques and devices, provides an astonishing amount of information.

The chief properties of a star in which we are interested are its magnitude or brightness, distance, motion, color and temperature, size or diameter, mass, rotational speed, and composition and atmosphere makeup.

MAGNITUDES

The Greek astronomer Hipparchus was the first to devise a scale of brightness. He divided the stars he could see into six classifications, calling the brightest "1st-magnitude," and the dimmest, "6th-magnitude." This is roughly the way we do things today; the major difference is that we now have photometers that allow us to measure a star's intensity (brightness) accurately; Hipparchus could only estimate and compare.

We call this intensity (the brightness we see) the apparent magnitude of the star. Because Hipparchus lacked an effective measuring device, not all his 1st-magnitude stars were of the same actual brightness; some were significantly brighter than 1st-magnitude (as we know it today). To account for these stars, we have extended the scale through zero to negative numbers. Thus, a star of magnitude −1 is brighter than one of magnitude +1; Sirius, our brightest star, has a magnitude of −1.4. When the telescope was invented, it brought into view many stars dimmer than the 6th magnitude, so that the other end of the scale also had to be extended. With the 5-m Palomar reflector we can now photograph stars slightly dimmer than the 23rd magnitude. Representative examples are our north star, Polaris, which has an average magnitude of 2.5 (it undergoes a slight periodic change); Vega, the bright blue star that is overhead in North America during the summer and whose magnitude is 0.0; and Deneb, in the constellation Cygnus, which has magnitude 1.3. Others are as follows:

Object	Apparent magnitude
Sun	−26.7
Full moon	−12.5
Venus (at brightest)	−4.4
Naked-eye limit (rural areas)	6
Naked-eye limit (city)	3 – 4
Limit with binoculars	9 – 10
Limit with 6-inch telescope	12 – 13
Visual limit with 5-m reflector	20
Photographic limit with 5-m reflector	23.5

Note that we can even apply the scale to our sun and moon.

The apparent-magnitude scale was standardized in 1856 by Norman Pogson. Noting that 6th-magnitude stars were almost exactly one hundred times as dim as 1st-magnitude ones, he showed that there was a ratio of approximately 2.51 ($2.51 \times 2.51 \times 2.51 \times 2.51 \times 2.51 \approx 100$) in light intensity between successive numbers of the scale.[1] In other words, a 1st-magnitude star is 2.51 times as bright as a 2nd-magnitude one, a 2nd-magnitude one is 2.51 times as bright as a 3rd-magnitude one, and so on.

The apparent-magnitude scale is of considerable value to the astronomer, although, obviously, we do not see true relative brightness. Stars are at different distances from us, so that looking at them is like looking at a group of illuminated light bulbs of different wattage, each at a different distance from us. To see the stars' actual relative brightness, all would have be at the same distance from us. We could, of course, adjust for differences in distance—were we able to determine them (see below). Furthermore, in looking at a star we see only its visible radiation, but stars radiate at many different (and *non*visible) wavelengths. A scale that sums the radiation at all wavelengths has been devised: it is called the bolometric scale.

[1]The exact ratio is $\sqrt[5]{100} = 2.51188$.

STELLAR DISTANCES

Although stars are so far away that it is extremely difficult to determine their distances, several methods have been developed to do so. One of the simplest uses the concept of parallax, which has been described on p. 21 and in Fig. 2.3.

As Fig. 14.1 shows, the nearer a star, the larger the parallax (and accuracy). Since increasing the distance between observation points also increases parallax, astronomers use the greatest separation obtainable, making observations from opposite points in the earth's orbit (i.e., six months apart), which constitutes a separation of 2 A.U. (Fig. 14.1). A star at a distance at which 1 A.U. subtends an angle of parallax of 1 second of arc (1 arcsec) is 1 parsec (or about 3.26 light years) away.[2] An obvious limitation of this method is that the star whose parallax we are measuring must be relatively close, in astronomical terms. In addition, there must be some very distant background stars. (Unfortunately, the number of stars close enough for this method to work is only a few hundred; about 700 have now been measured to an accuracy of 10% or better.)

We referred earlier to the problem of determining actual relative brightness of stars—of how they would look if we could view them

[2]Thus the distance to a star is the reciprocal of its parallax in seconds of arc: A star with an angle of parallax of $\frac{1}{2}$ arcsec is 2 parsecs away, one with an angle of $\frac{1}{5}$ arcsec is 5 parsecs away, and so on.

Figure 14.1 Determining the distances to relatively close stars by means of parallax.

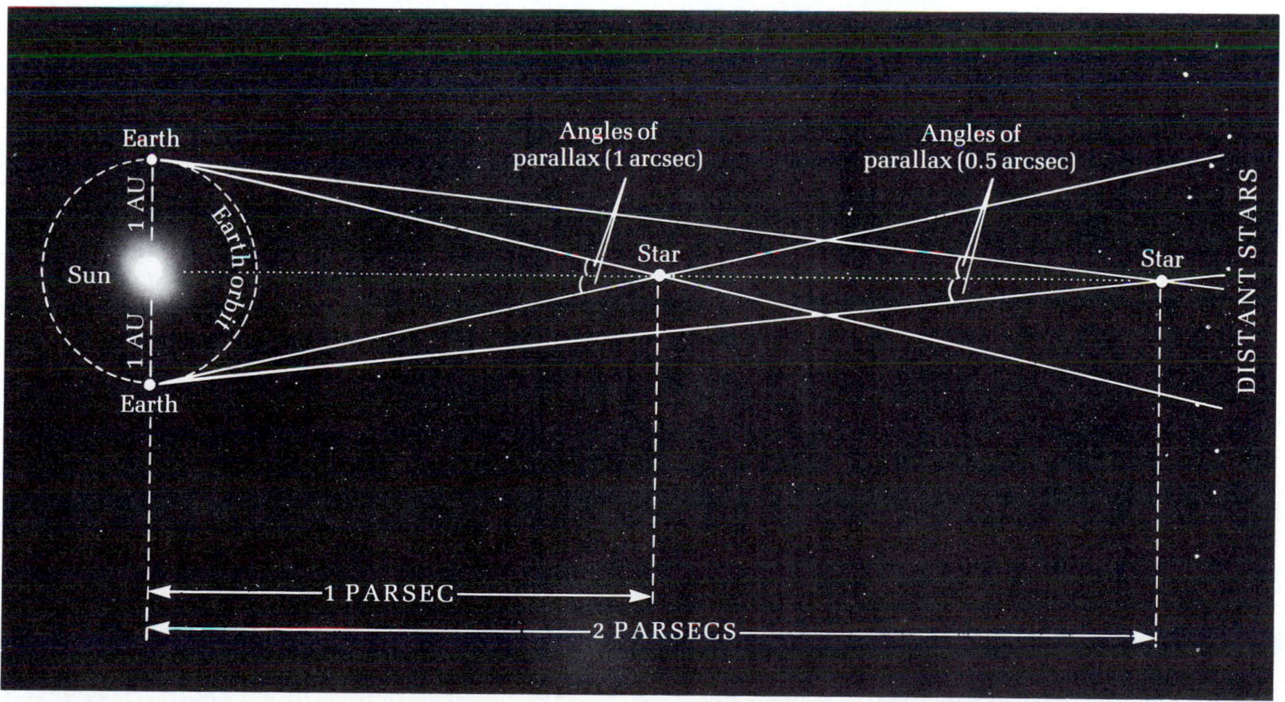

all at the same distance. Astronomers have resolved this by selecting a distance of 10 parsecs, at which the magnitude that stars have would be called their absolute magnitude. It is easy to see that stars closer than 10 parsecs have an absolute magnitude less than their apparent magnitude, while stars farther away than 10 parsecs have an absolute magnitude greater than their apparent magnitude.

The relationship between the apparent and absolute scales can be calculated as follows:

$$M = m + 5 - 5 \log D$$

where M is absolute magnitude; m, apparent magnitude; and D, distance in parsecs. Using this equation, we can easily determine the distance to a star if we know its apparent and absolute magnitudes.

PROPER MOTION

Unfortunately, parallax is not as straightforward and simple to measure as it might seem. Superimposed on it are several other motions. One of these motions was noticed in 1718 by Halley when he was studying early Greek star charts. He discovered that a number of the brighter stars had apparently moved from the position recorded by the Greeks; in particular, Arcturus, Procyon, and Sirius had all shifted considerably. To Halley it seemed incredible; we now know, of course, that stars do move, despite their tremendous distances, in amounts that can, over a long period, be detected. Astronomers refer to this as the proper motion of the star. A few hundred stars show proper motions of more than 1 second of arc per year. Most proper motions, though, are exceedingly small—so small that we usually measure them in seconds of arc per *century*.

It is convenient to split the actual motion of a star into two components (Fig. 14.2)—radial and tangential. Radial motion—approach or recession—is in the line of sight and, as we saw in Chapter 6, can be determined from the Doppler shift of the spectral lines. Tangential motion can be determined by carefully measuring the change in position of the star over long periods of time (e.g., 20 years). Taken together, these give true motion in space.

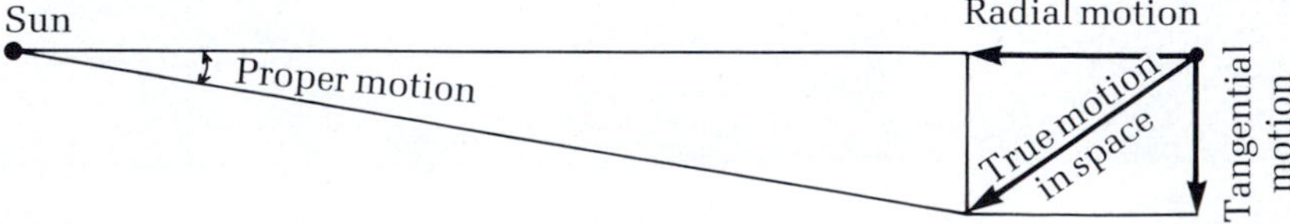

Figure 14.2 Proper motion and its relationship to true motion in space. Radial motion and tangential motion are components of true motion.

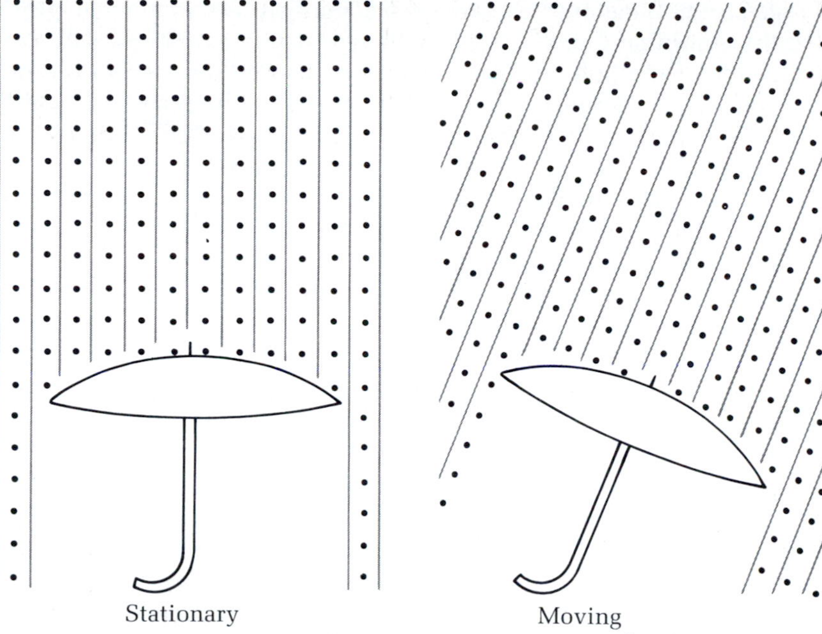

Figure 14.3 Effect of aberration. When moving, the umbrella must be tilted if it is to intercept the raindrops, which are falling vertically.

Proper motion and parallax can be distinguished easily: Since it is caused by the earth moving around the sun, parallax is cyclic. (It has a period of one year.) Proper motion, on the other hand, is always in the same direction and is therefore cumulative. However, there are other motions to disentangle.

In 1728 English astronomer James Bradley discovered another phenomenon affecting motion that must also be taken into consideration; it is called aberration, and a simple analogy can be made between it and a common experience: When it is raining, but there is no wind and you are standing still, you must hold an umbrella directly over your head to shield yourself (Fig. 14.3). If you are walking, however, you must tilt the umbrella slightly in your direction of travel to keep the rain off; in effect, you are walking into the rain as it falls. (The apparent slanting of vertical rain seen from a moving vehicle is another example.) Starlight is, in this respect, like rain; since the earth is moving, we must point our telescopes slightly in its direction of travel if we are to meet the starlight head-on. Aberration is, in fact, a considerably larger factor in the motion of stars than either parallax or proper motion.

COLOR AND TEMPERATURE

We saw in Chapter 6 that stars vary considerably in surface temperature. A blue star, for example, is much hotter than a red one, and naturally it emits most of its visible radiation in the blue region of the spectrum. But stars emit radiation at many different wavelengths, and the dominant ("peak") frequency of a star is not necessarily the frequency of the color we see when we observe the star. Our sun, for example, peaks in the blue-green region yet appears yellowish white. This means, then, that the wavelengths on either side of the peak affect what we see, and consequently we must be careful in using color as a temperature indicator.

In the case of our sun our view is also affected by the considerable scattering of radiation that occurs in our atmosphere. The photons corresponding to short wavelengths (toward blue) are scattered far more than those corresponding to long wavelengths (toward red); this reduces the blueness we might otherwise expect to see.

A straightforward observation of color is a useful guide to the surface temperature of a star, but there are better ways to make approximations. One of the simplest involves what is called the color index of the star. Our eyes are most sensitive to the yellow-green region of the spectrum; standard photographic plates, on the other hand, are most sensitive to blue light. This means that a photograph of a section of the sky that contains stars of different colors will appear quite different from what we see. The blue stars will stand out in the photograph; the red ones will appear dimmer. If we subtract the magnitude we see for a given star from the magnitude given by the blue-sensitive photographs, we get the color index of the star. In practice, of course, we do not rely on our eyes; we use a yellow-green filter (of wavelength approximately 5400 Å) that gives an equivalent magnitude called the photovisual magnitude. The color index (C.I.), then, is expressed as $m_p - m_{pv}$ where m_p is the magnitude as measured on the blue-sensitive photographic plate and m_{pv} the photovisual magnitude. The approximate surface temperature of the star can be calculated from the color index. (Surface temperature can also be determined—generally with more accuracy—from a star's spectral emission curve and its spectral lines.)

In recent years the single filter for determining m_{pv} has generally been replaced by a set of three filters called the UVB set. (U stands for ultraviolet, B for blue, and V for visible). They can be used in combination with a photomultiplier tube to determine two different color indices of stars: B − V and U − B, each of which are related to the temperature of the star. And with infrared devices now readily available, color indices such as R − I (red minus infrared) are occasionally determined.

Spectral Classification

In the late nineteenth century Edward Pickering developed the objective prism (a large prism or objective grating placed in front of the objective lens of a telescope), which made it possible to take a spectrogram of many stars at once. Pickering began a spectral classification project that employed approximately a dozen women, many of whose tasks were so tedious that we might think of them as his "computer," although in their time they were jokingly referred to as Pickering's harem.

In his first attempt Pickering classified spectra according to the strength of their hydrogen absorption lines. Groups of spectra were labeled A, B, C, and so on, with A the strongest. One of the workers, Annie Jump Cannon, pointed out that with this scheme, there were distinct jumps or discontinuities in the nonhydrogen lines. Eventually Cannon found a scheme that would give continuous transitions, but by then the letters of the old classification scheme had become quite well established. The new sequence (retaining the old letters, but now based not on the strength of the absorption lines but on the temperatures reflected by the spectra) became O B A F G K M. (Henry Russell of Princeton University invented the following mnemonic for this sequence: *Oh be a fine girl; kiss me.*)In this new sequence, then, stars of Type O are the hottest and those of Type M, the coolest. Each type is further subdivided ten times, so that designations become B5, A9, and so on. Details of the various spectral types are given in Table 14–1, and a number of representative spectra are shown in Fig. 14.4. Our sun is spectral Type G2.

Table 14–1. Stellar Spectral Types

Type	Spectral Characteristics	Typical Temperature[3] (degrees Kelvin)
O	Ionized helium lines	50,000
B	Hydrogen, helium lines	25,000
A	Hydrogen lines strong, calcium lines and some weak ionized lines	12,000
F	Hydrogen lines growing weak, calcium lines growing strong	8,000
G	Iron lines now strong; calcium lines also strong	6,000
K	Band spectra due to the formation of molecules appear	4,500
M	Molecular bands strong	3,500

The first Draper Catalog[4] came out in 1890; it contained 10,000 entries. Antonia Maury, a niece of Henry Draper's and the only trained astronomer among Pickering's corps of assistants, noting that the lines within a given spectral type sometimes varied considerably in width, divided them into three additional groups (*a*, *b*, and *c*) according to width, the *a*'s being the widest.

[3]Near the lower limit of the respective class.

[4]Named after physician-astronomer Henry Draper, whose widow funded Pickering's classification project.

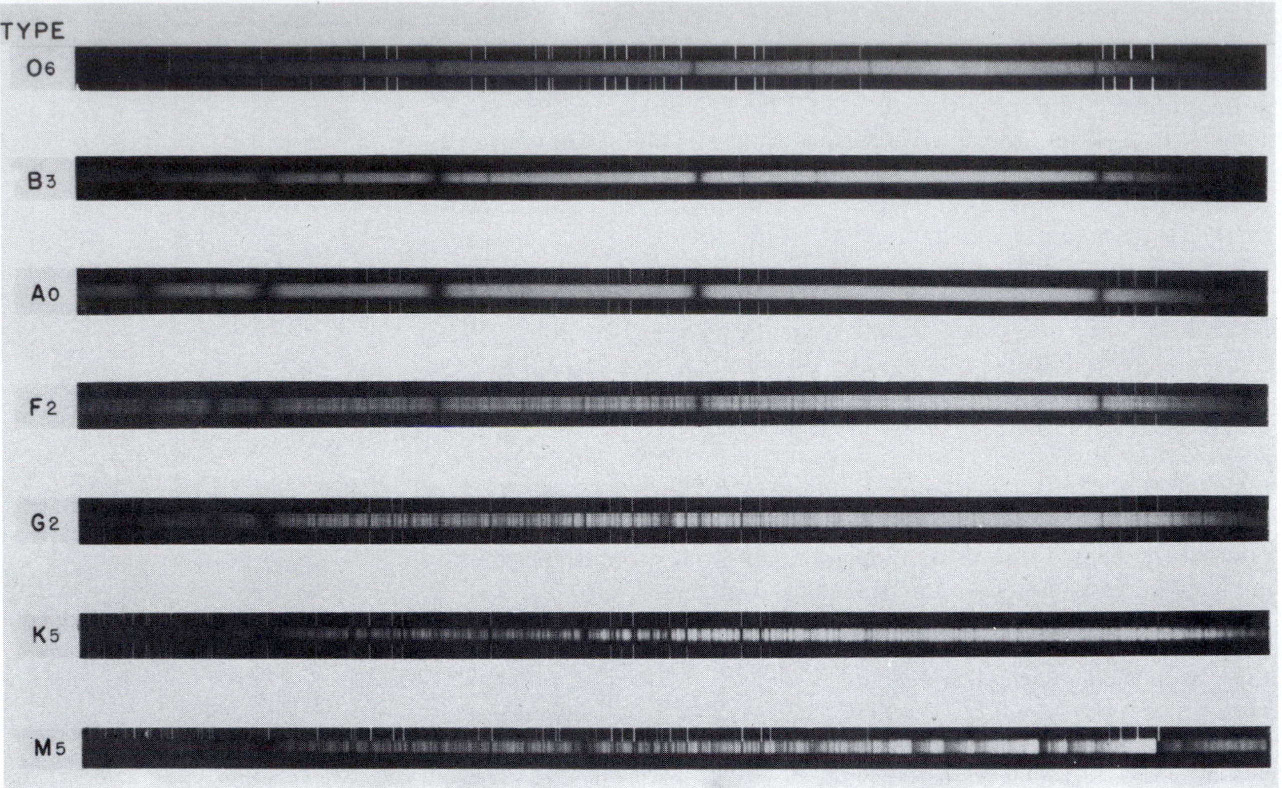

Figure 14.4 Representative spectra of stars from Type O through Type M.

The H-R Diagram

For several years there was little interest in Maury's discovery. In 1905, however, Danish photochemist Ejnar Hertzsprung found that he could distinguish red stars of large absolute magnitude from tiny, faint ones by using a filter and that among the red stars all the *c* types were intrinsically bright, whereas the *a* types were dim. Henry Norris Russell, examining the same spectral records used by Maury and Hertzsprung, also noticed that there were two classes of red stars—dwarfs and giants. In 1913 he published a plot of absolute magnitude versus spectral type for several hundred stars. In it the red giants were in the upper right-hand region and the red dwarfs in the lower right-hand region. A year earlier an assistant of Hertzsprung had published a similar diagram, and so the two, Hertzsprung and Russell, are now jointly credited with the discovery of the diagram, now called the Hertzsprung-Russell—or H-R—diagram (Fig. 14.5). (Hertzsprung is reported to have suggested that it be called simply the color-magnitude diagram.)

Hertzsprung and Russell probably did not realize at the time how important their diagram would eventually become; most astronomers now agree that it is the key diagram of astronomy. In it most stars fall on a diagonal band called the *main sequence*. Starting at the lower right-hand corner, we find small red stars. Further up along the main sequence are yellow stars, like our sun. Still further up are larger, hot, blue stars, and to their right (off the main sequence) are even larger stars—the giants and supergiants. Finally, in the lower left-hand corner are the white dwarfs. (Figure 14.5 shows a simplified version of the H-R diagram in which well-known stars, including our sun, are plotted.) Although in our figure surface temperature is plotted against luminosity, the horizontal axis could be spectral

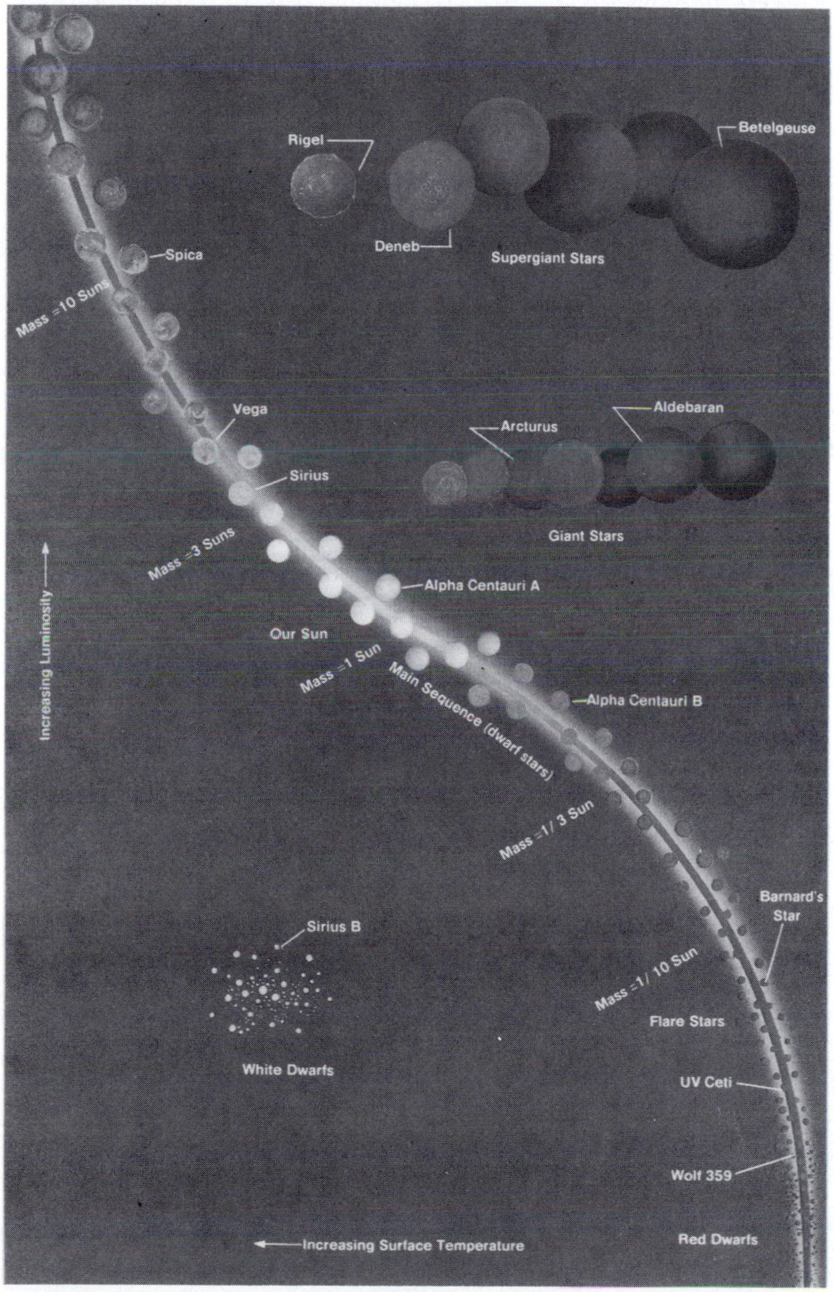

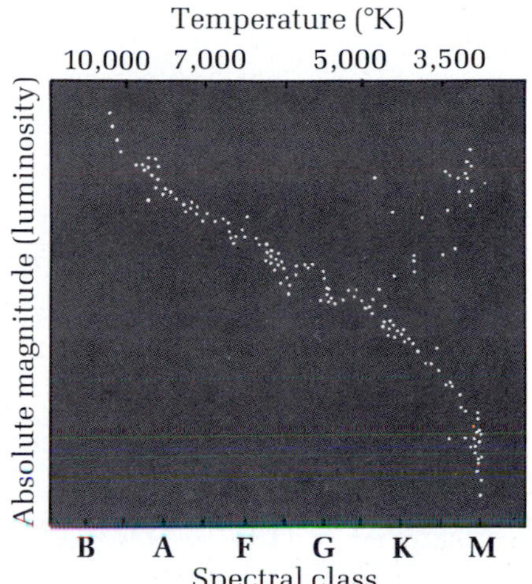

Figure 14.5 At top right the Hertz-sprung-Russell (H-R) diagram, and at left is a simple representation of the H-R diagram showing many well-known stars, including our sun.

type or color and luminosity can be replaced with absolute magnitude; the resulting diagrams are equivalent.

The significance of the H-R diagram can be seen by making an analogy with the distribution of people in which height and weight (horizontal axis) are plotted as the two fundamental variables that correspond to surface temperature and brightness in the H-R diagram. There will be a "main sequence" of people of average weight for their height: short people at the bottom, those of medium height in the middle, the tallest at the top. But, of course, there are people who are tall and severely overweight and a few who are short and extremely thin. They will not lie in this "main sequence," just as red giants and white dwarfs do not lie in the main sequence of the H-R diagram. It should be noted that this analogy is not perfect in that weight (the horizontal axis) in the people diagram increases to the right, while temperature (the horizontal axis in the H-R diagram) increases to the left.

SIZES OF STARS

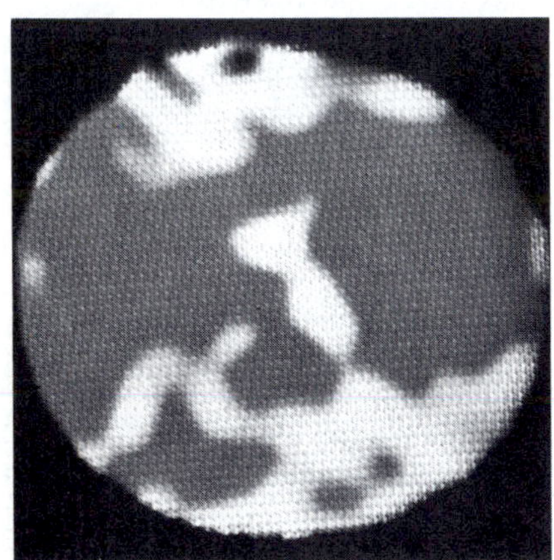

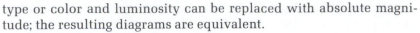

Figure 14.6 Image of the surface of the giant star Betelgeuse obtained by speckle interferometry. The nonuniformity is probably caused by convection currents.

The sizes of stars—so far away that we cannot see their disks (with one exception, noted below)—are determined, for the larger stars, by interferometry (Chapter 5). Using it, astronomers have determined that some stars are gigantic; ϵ Auriga B, for example, has a diameter over 2000 times that of our sun, and Betelgeuse, in Orion, has a diameter about 500 times as large. We saw in Chapter 5 that interferometry has also been used in a technique called speckle photography to obtain an image of the surface of Betelgeuse (Fig. 14.6); in this method a computer is used to reassemble a number of short exposures into a single composite.

But neither of these techniques can be used to measure small stars. For such a star, if its luminosity and temperature are known, its diameter can be estimated from the Stefan-Boltzmann law, according to which, the hotter a star is the more it radiates per unit area. Thus measurement of a star's radiation output can give the area of the disk emitting it.

STELLAR MASSES

To determine stellar masses, the astronomer turns to binary systems—two stars revolving around their center of mass. Most of the bright, apparently single stars visible to the eye are actually binaries. Polaris has a faint companion that can easily be seen with a 20-cm telescope; Sirius also has a faint companion. In many cases, although we cannot see the companion even with a large telescope, there is a shifting back and forth of the spectral lines of the star (Fig. 14.7) that betrays its presence. The shifting is a result of the Doppler effect as the star alternately approaches and recedes from us. Such systems are called spectroscopic binaries.

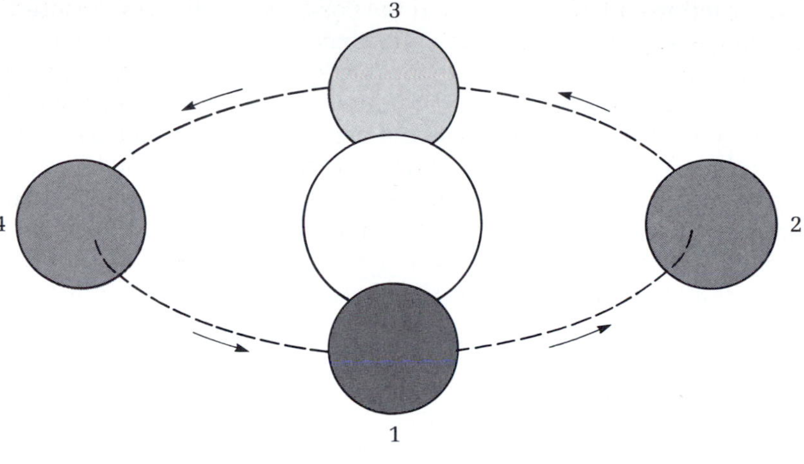

Figure 14.7 Spectral lines of a binary system. Shift is to the right as the star recedes from us, to the left as it approaches. Positions 1 and 3 indicate no movement.

In the first step in determining the masses in these binary systems we measure the distance between the two stars (having selected a system in which this is possible) and observe their period of revolution; this gives the sum of their masses ($m_1 + m_2$). Next, we determine the distance of each of the stars from their center of mass; this gives the ratio m_1/m_2. With these two quantities in hand a simple algebraic calculation gives the masses separately.

Plotting the masses of these stars against their absolute magnitudes (Fig. 14.8) shows an approximate relation between mass and luminosity. This relationship is important, for it can be used to estimate the mass of any main-sequence star as long as its absolute magnitude is known—and, of particular importance, the star need not be part of a binary system.

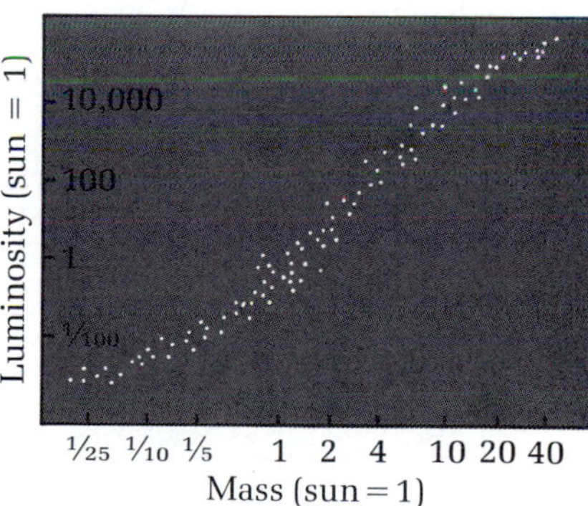

Figure 14.8 The mass-luminosity diagram, by means of which the approximate mass of a star can be determined from its luminosity.

COMPOSITION AND OTHER PROPERTIES OF STARS

A star's spectrum gives us information about its atmosphere, though uncertainly, since the intensities of the lines change significantly (Chapter 6) with temperature, pressure, and other factors, and the lines of many elements present in the star may not be visible at a given temperature.

There are only indirect ways of assessing the interior of a star; as described in Chapter 15, it is possible to determine what nuclear reactions are taking place in the core, and, with these reactions known, the conditions and elements present can be determined.

Such reactions tell us that stars are composed mostly of hydrogen with some helium, but many other elements are also present in small quantities. Each of these elements (including the hydrogen and helium) is a kind of fuel; when it burns, another element is left as "ash." Thus, as the star ages, its composition changes, and consequently its luminosity and surface temperature change. This means, of course, that its position in the H-R diagram will also change; in other words, *as a star ages, its position in the H-R diagram moves*. This is an important point and is taken up in Chapter 16.

There are many other properties of a star that can be determined—most from spectra. In Chapter 6, for example, we saw that the Zeeman effect tells us whether a star has a magnetic field and what the intensity of this field is. The Doppler effect tells us how fast a star is spinning. Several atmospheric properties can also be determined from spectra: for example, pressure and ascending and descending currents.

SUMMARY

1. The apparent magnitude of a star is its apparent brightness as expressed in a magnitude system. The absolute magnitude of a star is the apparent magnitude the star would have were it at a distance of 10 parsecs.
2. The simplest method of determining the distance to a star uses the concept of parallax. Nearby stars appear to move slightly relative to distant ones as the earth moves in its orbit.
3. Over a long period the nearer stars appear to move relative to the background sky. Astronomers refer to this motion as the proper motion of the star. The space motion of a star is composed of its radial and tangential motions.
4. The color index (C.I.) is the difference between the photovisual magnitude and the photographic magnitude. (Other color indices can also be defined.) It can be used to determine the temperature of a star. A star's spectral emission curve and its spectra also give its temperature.
5. Stellar spectra are classified according to the lines that are present. In order of descending temperature the types are O, B, A, F, G, K, M.
6. The most important diagram in astronomy is the H-R diagram. It is a plot of luminosity (absolute magnitude) vs surface temperature (spectral type).
7. The approximate size of a star can be determined using either interferometry or the Stefan-Boltzmann law.
8. The mass of a star can be determined from the mass-luminosity diagram. Other properties of a star such as rotational speed, composition, and structure of atmosphere can also be determined from spectra.

REVIEW QUESTIONS

1. What is the difference between the apparent and absolute magnitudes of a star?
2. What are the present limits of the apparent-magnitude scale? Give some magnitudes of representative objects.

3. How much brighter is a 2nd-magnitude star than a 4th-magnitude one?
4. Hold your index finger at arm's length and look at it with one eye, then the other. Estimate the angle between your eye and the two apparent positions of your finger. How does this angle change as you bring your finger closer to your eye?
5. Describe proper motion, parallax, and aberration.
6. What is the difference between radial and tangential motion?
7. Who was first to determine the distance to a star? By what technique?
8. A star that appears bright to the unaided eye may appear quite dim in a photograph. Explain. How is this useful?
9. The photographic magnitude of a star is 4.26; its photovisual magnitude is 3.84. What is its C.I.?
10. What are some of the main spectral lines in an A-type star? An F star? A B star? An M star? Our sun?
11. If the spectral lines corresponding to a particular element are not present in a star's spectrum, does this necessarily mean that the element is not present? Explain.
12. Discuss the significance of Antonia Maury's discovery.
13. Describe characteristics of the stars (temperature, size, appearance) moving from bottom to top of the main sequence in a representative H-R diagram. Are there stars off the main sequence? What do they look like?
14. How do astronomers determine the mass of a star?
15. What is a spectroscopic binary? Can we see its components as separate stars in a telescope? How do we know they exist?
16. Besides the above, what properties of a star can be determined?

THOUGHT AND DISCUSSION QUESTIONS AND PROJECTS

1. Determine the distance (in parsecs and in light years) of a star whose absolute magnitude is 3 and whose apparent magnitude is 10. If a star's absolute magnitude is twice its apparent magnitude, will it be brighter or dimmer if it is moved to a distance of 8 parsecs?
2. How far away is a star with a parallax of .035? How far away would a star with half this parallax be?
3. If you were measuring parallaxes from the moon, what corrections would have to be made? What changes would occur were you measuring parallaxes from Mars?
4. *Project:* Given that Polaris has an average apparent magnitude of 2.5, Sirius has a visual magnitude of -1.40, and Vega has a magnitude of 0.0, select several bright stars and estimate their magnitude. Compare your results to the known values.

FURTHER READING

Aller, L. H. *Atoms, Stars and Nebulae,* Revised Ed. Harvard University Press, Cambridge, Mass. (1971).

DeVorkin, D. H. "Steps Toward the H-R Diagram," *Physics Today,* p. 32 (March 1978).

Page, T. and Page L. W. *Starlight.* Macmillan, New York (1967).

Philip, A.G.D. and Green, L. C. "Henry Norris Russell and the H-R Diagram," *Sky and Telescope,* p. 306 (April 1978).

Aim of the Chapter

To describe how stars generate energy via nuclear reactions and to introduce some of the elementary particles that will be important in later chapters.

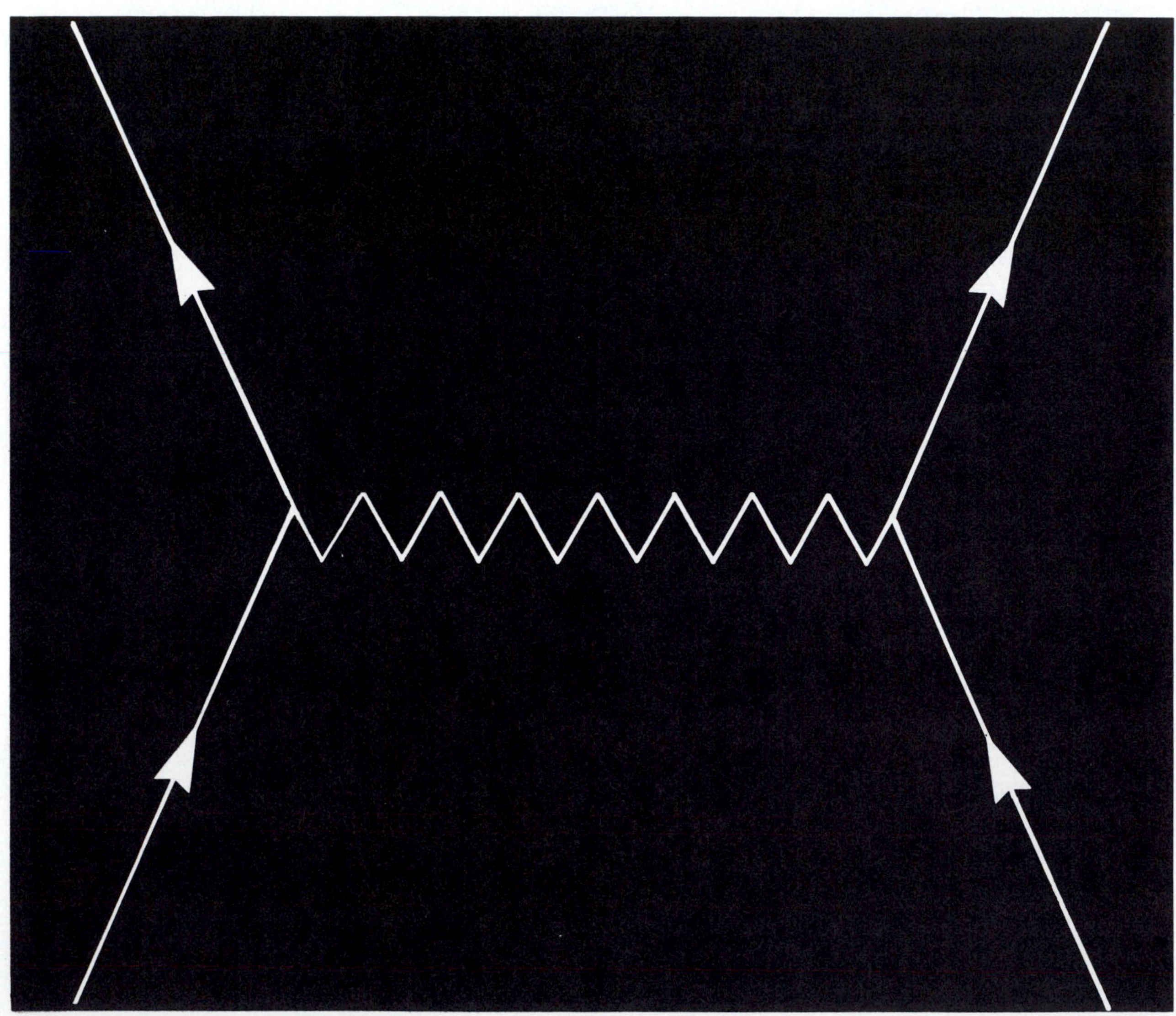

chapter 15

Nuclear Energy

By 1930, astronomers had learned a considerable amount about the stars: they knew approximately how distant many were; their approximate size, surface temperature, and mass; and, in general, what they were made of. But a vital key was still missing: What was the source of their energy? According to geological evidence our sun had been shining for millions of years (we now know that it has been shining for *billions* of years) and there was no sign that it was running down. How was this possible? Indeed, when we estimate its total energy output, the feat is even more amazing: Although it is 93 million miles away, and we catch only an insignificant portion of its total output, at about midday every square meter of the earth (perpendicular to incoming light) receives enough energy to turn a liter of water into steam every 25 minutes. If we could somehow convert all this energy efficiently into, say, electrical energy, our energy problem on Earth would be over.

Certainly this energy is not being generated by ordinary combustion. Were the sun made entirely of coal, it would burn itself out in about 1500 years. Since nuclear energy seems a possible source, it is useful to examine forces involved in the makeup of the atom. It is easily shown that the "electrical" force that holds the electron in its orbit is not sufficiently strong to account for the energy of the sun.

However, the force that holds the nucleus together cannot be electrical, for the protons that compose it are of like (positive) charge, and like charges repel. At a point where the distance between protons becomes extremely small (about 10^{-13} cm) the repulsive electrical force is overcome, and they are pulled toward one another with incredible force. Separating them would be almost impossible. The force holding the protons together, called the nuclear force, is much stronger than the electrical force but acts over a much shorter range. Actually the foregoing description is simplified: Were two protons forced together as described above, a nuclear reaction would occur, converting one of the protons into a neutron and releasing a couple of other particles, but there still would be a nuclear force holding the proton and the newly created neutron together.

The nucleus, then, is held together by a particularly strong force (note that there are nuclear forces acting between protons, between protons and neutrons, and between neutrons). We distinguish it from another nuclear force that is important in radioactivity by calling it the strong nuclear force; the latter force is called the weak nuclear force.

FISSION AND FUSION

Is this strong nuclear force, then, the source of the energy of the stars? And if so, how is this energy released? The first clue that this might be the case came in the mid-1930s, when Italian physicist Enrico Fermi (Fig. 15.1) was attempting to produce new elements by bombarding the heaviest known nuclei, those of uranium, with various particles. Fermi succeeded in producing new heavy nuclei but, with-

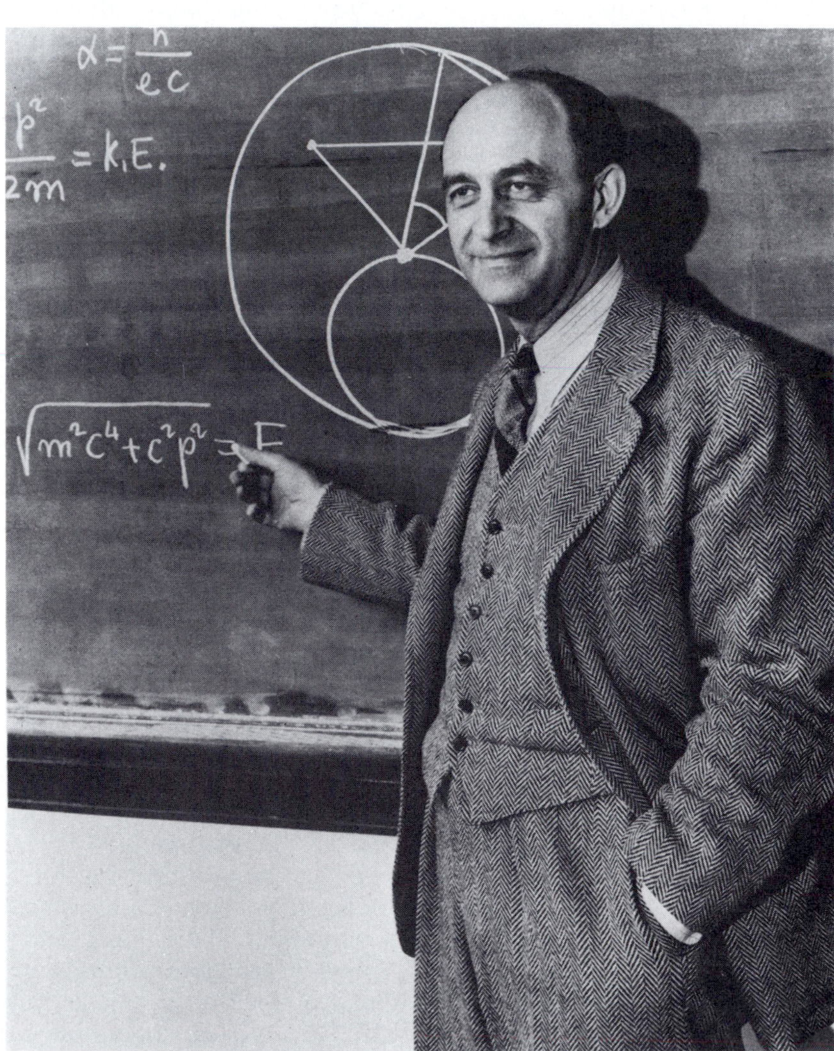

Figure 15.1 **Enrico Fermi, whose discoveries in nuclear physics were crucial to understanding the energy systems of the stars.**

out realizing it, also stumbled on the most significant discovery of the decade—nuclear fission.

In Germany Otto Hahn and Fritz Strassmann took up where Fermi left off. They bombarded uranium with neutrons and discovered that not only new heavy nuclei were being produced, but much lighter ones—such as barium—as well. In Sweden, exiled German scientists Lise Meitner and Otto Frisch quickly repeated the experiment and concluded that, when the uranium nucleus absorbed the neutron, it somehow became unstable and broke into two nuclei of approximately the same mass. (Energy was released in the reaction.) Meitner called the process "nuclear fission," and the resulting nuclei, fission products.

Soon after this, Fermi discovered that neutrons were among the fission products. This meant that, under the proper conditions, the reaction could be sustained, for these neutrons would cause other nuclei to fission. The result would be a chain reaction, and if this chain reaction continued long enough, a considerable amount of energy would be released—an amount that, in effect, would constitute an atomic bomb.

Astronomers felt that nuclear fission might be the clue to the energy of the stars. However, uranium was the necessary element in producing experimental fission, and there was virtually no uranium in stars. Was there a similar process that involved light nuclei and also produced energy? As mentioned above, when two protons are brought extremely close together, a nuclear reaction occurs in which one of the protons changes into a neutron, a proton-neutron combination (called a deuteron) is formed, and two other particles are released. In other words, there is a release of energy when light nuclei are brought together just as there is when a heavy nucleus breaks into two smaller nuclei. This process is called fusion, and it is the process that fuels stars.

ENERGY CYCLES

Fusion is, indeed, taking place in the sun, but it is not simply the fusion of two protons into a deuteron; it is the conversion—via fusion—of hydrogen into helium. In effect, four hydrogen nuclei come together to produce a helium nucleus. However, the process is considerably more complicated than simple collision; a cycle of reactions is needed. The cycle for the sun—known as the proton-proton cycle, and discovered in the 1930s by Hans Bethe—occurs in the core of a star if the temperature is between about 10 and 15 million °K. The high temperature provides the energy (the higher the temperature the greater the velocity of the particle) required to overcome the large mutually repulsive force of the protons.

The first step of the proton-proton cycle (Fig. 15.2) is the creation of a deuteron. This deuteron, hit by a proton, produces helium three—He^3 (an isotope of helium). Finally, two He^3 nuclei fuse to form He^4. The time between steps here is of the order of billions of years.

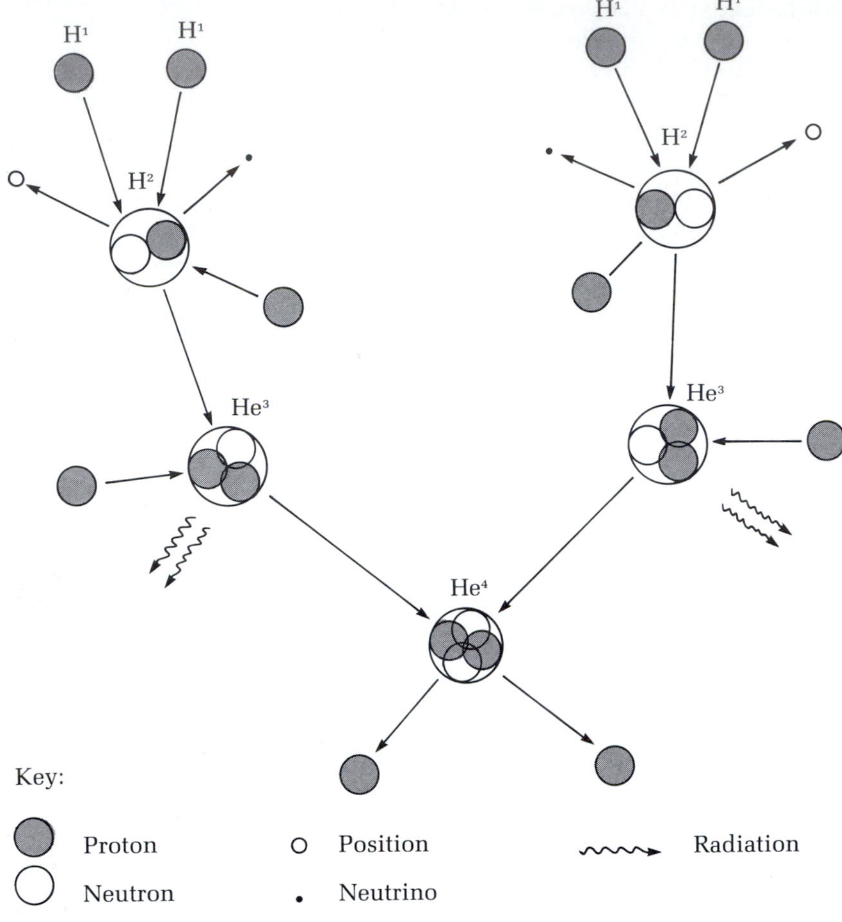

Figure 15.2 The proton-proton cycle, in which four hydrogen nuclei (H^1) fuse to form one helium nucleus (He^4). In the first step, two hydrogen nuclei form a deuteron (H^2); in the second, the deuteron and another hydrogen nucleus form He^3; in the last step, two He^3 form He^4.

This is no hindrance to the process, however, since there are billions upon billions of particles going through the cycle at any time; hence billions are reaching each stage every second. In fact, 512 million metric tons of hydrogen are converted to 508 million metric tons of helium every second in the sun. This difference in mass is the source of the energy produced. Einstein showed that mass and energy are equivalent forms, and, if nuclear energy is created, there must be a corresponding loss in mass (Chapter 19). The difference of 4 million metric tons here appears as energy.

In stars with a core temperature greater than 15 million °K a different cycle is operating. Hydrogen is still being converted into helium, but a kind of catalyst—carbon—is now needed to make the reaction go. The cycle is therefore called the *carbon cycle.* (Since nitrogen and oxygen can also trigger this cycle, it is sometimes called the CNO cycle.)

In extremely hot stars (core temperature greater than 10^8 °K) we encounter another cycle, called the triple α process. In this cycle three α particles (helium nuclei) come together to form a carbon nucleus, and, as before, they do not just collide but require an intermediate step.

OTHER PARTICLES, OTHER FORCES

So far in our discussion of atomic reactions it has been necessary to treat four different types of particles: electrons, protons, neutrons, and photons. All of these are involved in one way or another with the reactions that fuel the energy cycles of the sun and are therefore important in astronomy. In 1928 Paul Dirac predicted a particle that would be similar to the electron in all respects except charge; it would have a positive charge. The particle—called the positron—was found in 1933 by physicist Carl Anderson. It was found that, brought together, the electron and the positron annihilated one another—with the release of considerable energy. Because of this, the positron is now referred to as the antiparticle of the electron. Indeed, it is now known that all particles have antiparticles, and when any particle meets its antiparticle they annihilate one another.

In the same year that the positron was discovered, German physicist Wolfgang Pauli suggested that another particle—needed to explain certain nuclear reactions—might exist; Fermi called it the "little neutron" or neutrino. But the neutrino was as elusive as a rainbow, and detection was particularly difficult because it had no charge and was believed to travel at the speed of light. It was, however, finally found in 1953; later its antiparticle was also detected.

All particles discovered to this point could be classified as light (like the electron) or heavy (like the proton). In 1935, however, Japanese physicist Hideki Yukawa postulated existence of a medium-weight particle—called the meson—that would have a mass between that of the electron and the proton. Its discovery, 13 years later[1], began a flood of discovery of particle after particle, all short-lived. They were grouped into various classes, but, when the number approached 300, some physicists became convinced that there must be a simpler underlying scheme.

The particles are now divided into hadrons (heavy) and leptons (light). Leptons compose a small and relatively simple class that includes the electron and its antiparticle, the muon and its antiparticle, and a particle called tau (τ) and its antiparticle; each of these is paired with its particular neutrino (e.g., electron neutrino, muon neutrino).

The hadrons can be subdivided into the baryons (heaviest) and the mesons (medium heavy) particles. Of these two groups the largest by far is the baryons; the proton and neutron belong to it, as do many other particles (which would take pages to describe) generally designated by Greek symbols.

In the early 1960s physicists Murray Gell-Mann and George Zweig suggested that the hadrons were made up of three other particles, called "quarks"—the up quark (u), the down quark (d), and the strange quark (s), each, like all particles, having antiparticles—respectively, $\bar{u}$, $\bar{d}$, and $\bar{s}$. According to the theory all baryons are combinations of three quarks: For example, the proton consists of two u

[1]The μ-meson or muon, discovered 3 years after Yukawa's prediction, was at first thought to be the particle he had predicted. It turned out to have the wrong properties; the π-meson discovered in 1948, however, had the correct properties.

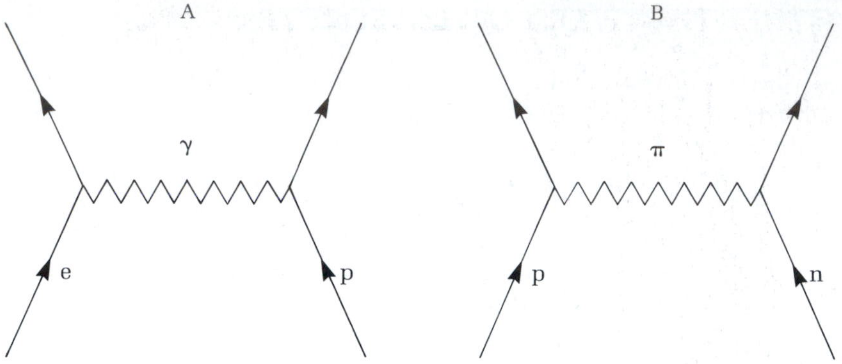

A B

γ π

e p p n

Figure 15.3 *A* **represents a collision between an electron and a proton;** *B*, **a proton-neutron collision.**

quarks and a d. All mesons are combinations of a quark and an antiquark; the π-meson, for example, has the recipe $u\bar{d}$.

As we have noted, protons and neutrons are held together in the nucleus by a force—the strong nuclear force. What holds the quarks together? In all there are four basic forces of nature: the electromagnetic force, the strong nuclear force, the weak nuclear force, and the gravitational force. We know that an electromagnetic force operates between, say, an electron and a proton. Physicists picture this force as resulting from the back-and-forth transfer of photons between the particles. (A simple analogy might be made with a boatload of girls on a lake passing beachballs back and forth to a boat full of boys; in effect, the attraction is a result of the transfer.) The photon transfer between a proton and an electron is represented in a standard diagram of elementary particle theory in Fig. 15.3a.

In the case of the strong nuclear force the entire mechanism is the same, though the particle transferred is a π-meson. This is represented in Fig. 15.3b. Similarly the particle for the weak nuclear field is known as the W particle, and that in the case of the gravitational force is called the graviton. (Gravitons have not yet been found, but W particles may have been detected in 1983.)

In the case of quarks, then, if they exist in groups, as has been postulated, then there must be another force in nature. Physicists call it the color force, and with it they also had to introduce "colored" quarks. As a result of the introduction of the concept of color the family of quarks was increased to nine: there was now a red, a green, and a blue one for each of u, d, and s. The particle of the color force is called the gluon (the *glu*e that holds the quarks together)[2], and each has two colors associated with it. A proton, then, according to this picture, consists of two u quarks and one d quark, each colored differently. (Of particular importance is the fact that the three colors must add up to white.) Also, of course, there are gluons moving back and forth between the quarks, holding them together.

Though quarks have not been detected, the elegant simplicity

[2]The gluon is now considered to be the particle actually being transferred in strong interactions. When two particles (i.e., a proton and a neutron) come sufficiently close, their quarks interact with an exchange of gluons. The overall interaction is equivalent to an exchange of mesons. (Remember: mesons are made up of a quark and an antiquark that exchange gluons.)

of the theory has endeared it to many; others are skeptical of the existence of quarks, and some have been from the beginning.

Ken Wilson has suggested that quarks are held together by "strings" that are assumed to be made up of gluons. David Politzer added the idea that if the string is pulled (i.e., if we try to pull a quark out of a particle) the force increases as tension on the string increases. This means that the force between two quarks increases as their separation increases, exactly the contrary of the situation with other forces. Further, if the string breaks an antiquark appears at its end (and an antiquark plus a quark is a meson).

In 1974 two groups simultaneously discovered a new, unpredicted particle, now called the ψ/J particle, which could exist only if there was a fourth quark. Several scientists had argued the existence of a fourth quark for years, and it was added and labeled charm (c). Then two more were predicted (called bottom and top), so that, with them we now have:

Ⓤ Ⓓ Ⓢ Ⓒ Ⓑ Ⓣ

$$\text{(u)} \quad \text{(d)} \quad \text{(s)} \quad \text{(c)} \quad \text{(b)} \quad \text{(t)}$$

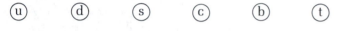

and their antiparticles:

$$(\bar{u}) \quad (\bar{d}) \quad (\bar{s}) \quad (\bar{c}) \quad (\bar{b}) \quad (\bar{t})$$

each of these in three colors, for a total of 36 quarks!

But do all these particles really play an important role in astronomy? Though so far the quarks do not, there are indications that they may; even now they are seriously discussed in relation to the early universe. They may play a role in such unexplained astronomical objects as quasars (Chapter 12). The other particles—neutrinos, various antiparticles, and so on—do play an important role in astronomy today, as we shall also see in Chapter 25.

SUMMARY

1. The nucleus of an atom is held together by a force called the strong nuclear force. It is much stronger than the electromagnetic force, but of much shorter range.
2. Nuclear fission occurs when a heavy nucleus, such as that of a uranium atom, becomes unstable and breaks into two nuclei of approximately the same mass.

3. Nuclear fusion occurs when two light nuclei come together to form a heavier one. Energy in stars is generated by fusion.
4. The energy cycle for generating helium from hydrogen in stars with a core temperature between about 10 and 15 million °K is called the proton-proton cycle. Other cycles (in hotter-burning stars) are the CNO and triple α cycles.
5. There are two main classes of particles: hadrons and leptons.The hadrons are further subdivided into the baryons (heavy particles) and mesons (medium-weight particles). The leptons are light particles.
6. Gell-Mann and Zweig theorize that all hadrons are made up of quarks. There are six types of quarks—up, down, strange, charmed, bottom, and top. Each has an antiparticle.

REVIEW QUESTIONS

1. How is the nucleus held together if the protons in it repel one another via the electromagnetic force?
2. What is the difference between fission and fusion? Which is most important in relation to stars?
3. Discuss the proton-proton cycle. Why is temperature important in relation to it? If there are exceedingly long times between steps, how can the cycle generate so much helium?
4. What is the relationship between the electron and the positron? What is a neutrino? Does a neutrino have an antiparticle?
5. Describe the three classes into which particles are divided.
6. What are the four basic forces of nature? Which is the weakest? The strongest?
7. Explain how these forces operate (according to recent theories).
8. What is the color force? What is its relation to the "strings" in the nucleus?
9. Describe the various types of quarks.
10. Can a quark be pulled from the nucleus? Why (according to a recent theory)?

THOUGHT AND DISCUSSION QUESTIONS AND PROJECTS

1. The reaction shown in Fig. 15.2 can be represented in a much simpler way. Show how. (Consider input and output only.)
2. The relative magnitudes of the four basic forces of nature are as follows:

Strong nuclear	10^{39}
Electromagnetic	10^{37}
Weak nuclear	10^{24}
Gravitational	1

Speculate on what the universe would be like if, other forces remaining the same:
a. The magnitude of the gravitational force were increased by 10^{12}.
b. The magnitude of the electromagnetic force were decreased by 10^{12}
c. The nuclear force were zero.

FURTHER READING

Glashow, S. L. "Quarks with Color and Flavor," *Scientific American*, p. 38 (October 1975).

Nambu, Y. "The Confinement of Quarks," *Scientific American*, p. 48 (November 1976).

Parker, B. "The First Second of Time," *Astronomy*, p. 6 (August 1979).

Aim of the Chapter
To explain how stars are born and how they live out their lives.

chapter 16

The Birth and Evolution of Stars

Plotted on the H-R diagram (introduced in Chapter 14) are points corresponding to stars of many different sizes and types: supergiant stars such as Antares, stars of medium diameter such as our sun, and tiny dwarf stars, such as the companion of Sirius. Are all these stars related in any way? Do they, like humans, go through a life cycle, so that the stars we see are at many different stages of their life? Indeed, astronomers now have considerable evidence that this is the case.

Compared to humans, however, stars live an exceedingly long time. Even short-lived stars live for millions of years, and average stars, like our sun, live for billions. This being the case, we cannot, of course, follow their life cycles by watching them; the human life span, even the entire time that man has been on earth, is but the blink of an eye on the cosmic scale of time. However, although we cannot see stars change (with a few exceptions), we can learn from the billions of stars we can see that are at various stages of their lives; we have "still" pictures of them, and we are like a person looking at such stills outside a theater and trying to determine the plot of the movie from them. With only a few pictures this would, no doubt, be quite difficult, but as you looked at more and more, the likelihood of being correct would increase.

Spread out on a table, a number of our movie photos can be arranged into what seems the correct order. With this sequence established we can then fill in what we think happened in the gaps between the photographs. Of course, if we then find some photographs that do not fit into our idea of the story, we have to start over. Eventually, though, with enough pictures and perseverance we are likely to succeed in piecing together the sequence of events.

Astronomers face a parallel problem: They have thousands of photographs (and other data) of stars at various stages of their lives and aim to construct from them the proper sequence—the life story of a star.

Since the H-R diagram is a plot of points corresponding to stars at various ages, it seems reasonable to start with it as the best place

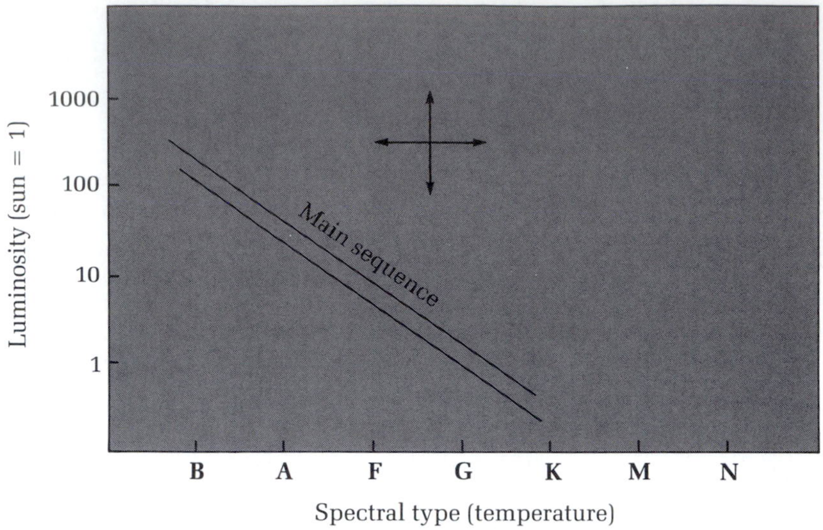

Figure 16.1 Movement of a point on the H-R diagram. Note that luminosity increases in a direction away from the origin, while temperature increases toward the origin.

to represent this life story. As a star ages, it changes physically, and, therefore, its position in the diagram changes. Over a long period it will trace out a line or track that astronomers refer to as the evolutionary track of the star.

Suppose we are considering a particular point on the H-R diagram (Fig. 16.1). What happens if the star brightens, but keeps its surface temperature constant (which will occur only if it increases in size)? The point representing the star will move directly upward in the diagram. Similarly, if the star dims while keeping constant surface temperature (shrinks in size), its point will move directly downward. On the other hand, if the star's surface temperature increases but it gets no brighter, its point moves to the left, and if its surface temperature decreases with no change in brightness, its point moves to the right.[1]

We now know, then, that the points on the H-R diagram correspond to stars of different ages and that each of these points moves slowly over a very long period. How and where they move is, of course, the basic problem of stellar evolution. The answer, arrived at over many years, is now fairly clear, though still not complete. To find this answer, astronomers have had to rely on both observation and theory. Stellar models were built, and predictions were made from them with the help of computers. These predictions were then analyzed in light of the observations. When they did not agree, new models were tried until finally there was agreement.

STELLAR MODELS

Early Static Models

To understand how a star forms and later evolves, the astronomer must build a model, not a model in the ordinary sense of the word, but a theoretical one, built from mathematical formulas.

[1]The H-R diagram tells nothing about the core temperature of the star. In fact, when surface temperature decreases, there is usually an increase in core temperature.

The foundations of stellar model building were laid by Arthur Eddington, an English astronomer-physicist. Around 1917 Eddington began to look at the problem of stellar structure. He soon realized that radiation pressure is an important mechanism inside a star; coupled with gas pressure, it could supply the outward force needed to balance the inward gravitational pull.

Eddington discovered that radiation is transferred from the core outward via the process of emission and absorption. On the basis of this and the gas laws[2] (composition also plays a minor role) he then calculated the luminosity of a star of given mass. The result is the now well-known mass-luminosity relation, which tells us that all main sequence stars of the same mass (and composition) have the same luminosity.

Eddington made many other contributions: He calculated an approximate temperature profile for the interior of a star; he was the first to realize that a star is gaseous throughout, and he realized that there must be an exceedingly long-lived source of energy at the core. He did not, of course, know the details but felt that this source must be subatomic.

Eddington's book *The Internal Constitution of the Stars* was published in 1926. It summarized most of his work in the area and is now considered a classic. In the same year, H. N. Russell and H. Vogt independently arrived at a result now known as the Russell-Vogt Theorem, which can be stated as follows: *If a star is in hydrostatic and thermal equilibrium,*[3] *its structure is uniquely determined by its mass and composition.* In other words, if we know the mass of the star and how much hydrogen, helium, and other elements are in it, we can calculate a model of the star. Everything about the star is then known.

Computer Models

To compute the model of a star, we must begin with two basic assumptions: first, that the fundamental gas laws are satisfied, and second, that the star is in thermal and hydrostatic equilibrium. (This means that there are no abrupt changes in temperature or pressure throughout the star, it does not pulse or change erratically in any way, and all energy generated in the core is transported outward and eventually radiated at the surface.)

Then we divide a sphere (theoretically) into about 100 to 200 concentric shells of equal thickness. Our problem is to determine all possible properties—pressure (P), temperature (T), total mass (M), and luminosity (L)—of each of the shells. (The luminosity is a measure of how much energy is being transported through the shell.) These characteristics are presumably all known for the surface of the star (they can be either calculated or measured), and it is assumed that those for the center of the star are also known. (Actually, we are not certain of the latter, but reasonable guesses can be made.)

The problem now is to calculate these quantities for each shell using a set of equations that tells us how they change from shell to

[2]The gas laws give the relation between the pressure, temperature, and density of the star.

[3]These terms are discussed below.

shell. The calculations are tedious (which is why they are usually done by computer). Starting with the outer solution, we move inward, calculating new values as we go; at the same time we start at the center and move outward. Somewhere in between, the two solutions overlap. This overlap is carefully examined. If it does not match, we adjust the starting values (usually the ones for the center) slightly and try again until they do match. Actually, there are several variations on this procedure, but the idea is the same in each case, and the end product is also the same: a stellar model.

Using models of this type, astronomers have determined that stars can exist only within a certain range of mass. If the mass is too small (the critical value is about $\frac{1}{10}$ $M_{\odot}$)[4], nuclear reactions are not triggered in the collapse and we end with a planet. Jupiter is a good example of an object of this type; with about 70 times its present mass it would be a star. A collapsing mass can also be too large. If it is greater than about 100 $M_{\odot}$, the outward pressure eventually becomes so great (as the mass collapses) that the star literally blows itself apart. Observational evidence seems to be in agreement with this; no stars larger than about 70–80 $M_{\odot}$ have been found.

Evolutionary Models

How do static models (the kind described above) change in time? How does our model evolve? Assume that a star is initially at some given point in the H-R diagram and we have a static model of it. One of the things the model tells us is how fast the star is using up its fuel; with this information we can calculate a new static model for some later time; then, with this new model, we can calculate still another one for an even later time, and so on. The time interval we choose between models will vary considerably depending on where the star is in the H-R diagram. (As we will see later, stars spend much more time in certain regions than in others.)

If we calculate a large number of successive static models and position them properly in the H-R diagram they will fall along a line, the evolutionary track of the star. Japanese astrophysicist C. Hayashi was one of the first to calculate evolutionary tracks. He concentrated on stars moving into the main sequence (Fig. 16.2).

Starting at the top, we can see that large, massive stars move almost directly across the H-R diagram toward the main sequence. This means, of course, that their brightness changes little during this stage of their life. Less massive stars, such as our sun, move downward at first, then hook upward into the main sequence (Fig. 16.3). Dwarf stars, much smaller than our sun, move directly downward throughout this phase.

In computing these tracks Hayashi noticed that there was a vertical strip, to the right of the main sequence, beyond which no stars could form. Now called the Hayashi strip (Fig. 16.4), this region corresponds to small masses, and objects to the right of it cannot become stars because they are not massive enough to trigger nuclear reactions.

[4]A reminder: $M_{\odot}$ = mass of our sun.

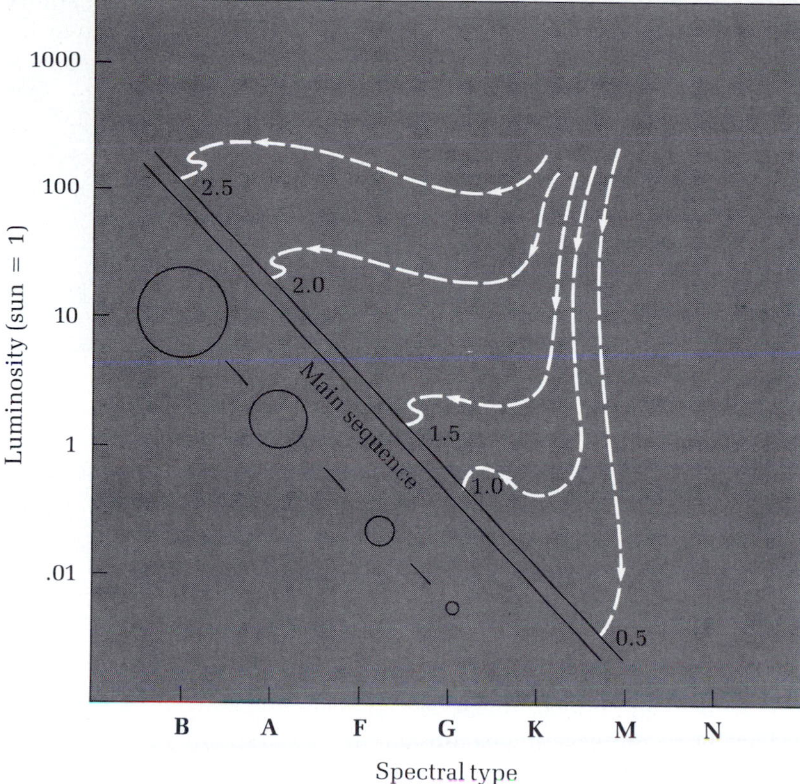

Figure 16.2 Evolutionary tracks of several stars of various masses up to the main sequence. Masses range from $0.5M_{\odot}$ to $2.5M_{\odot}$. Circles indicate increase of size with luminosity.

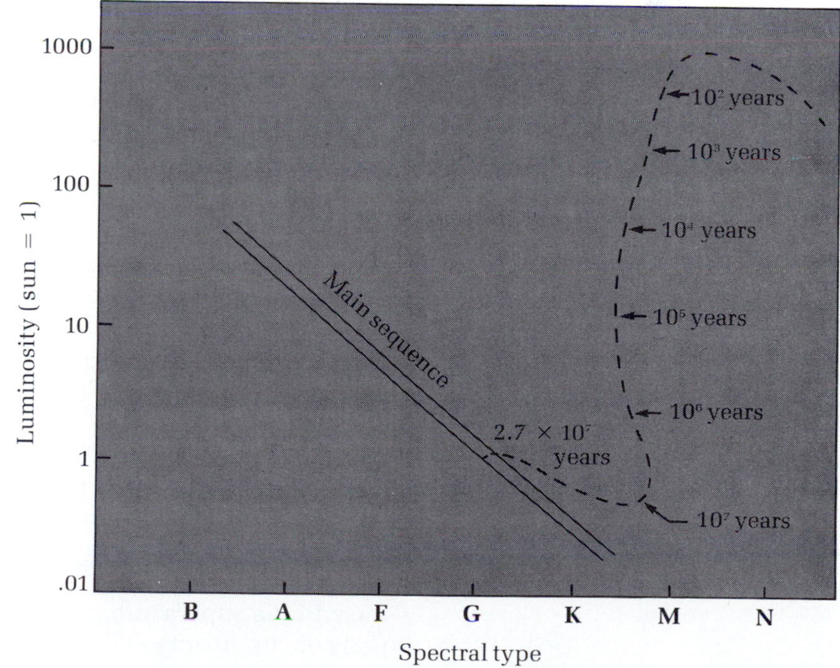

Figure 16.3 The evolutionary track of the sun up to its present position in the main sequence. Time required to reach various points is indicated at right.

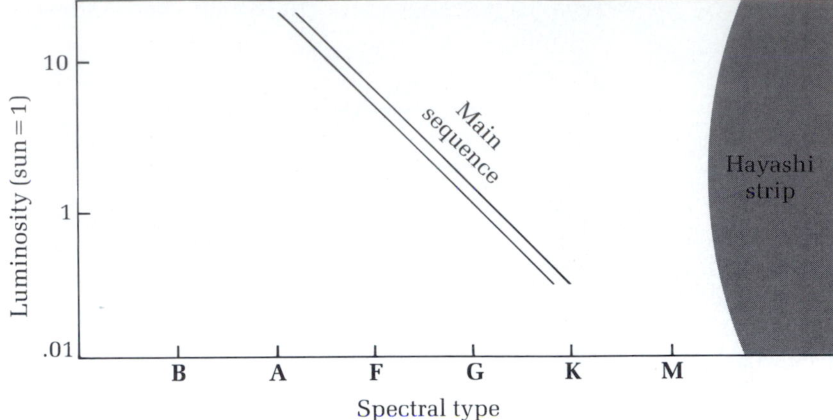

Figure 16.4 The Hayashi strip, beyond which no stars can form.

With this information on how astronomers create evolutionary models, we turn now to how they use it to explain the life cycle of a star.

THE INTERSTELLAR MEDIUM

Stars are born in gigantic gas clouds—the interstellar medium. This interstellar medium is made up mostly of hydrogen, but other elements such as helium, nitrogen, carbon, and oxygen usually are also present, along with particles of dust. (The exact makeup of this material has a considerable effect on the makeup of the star; this will be discussed later.)

In 1919 James Jeans, theorizing about the interstellar medium, pondered a serious problem: If the temperature was near 0°K (as it is in interstellar space) and the gas composing the medium was homogeneous, the forces would be the same on all sides of a given particle, so that inhomogeneities (which would be required if changes were to take place) would not develop. However, he realized that even at exceedingly low temperatures atoms are not stationary; they move around slowly and occasionally collide, the number of collisions dependent on the density of the gas. On the average the atoms and particles of the gas cloud are uniformly distributed, but occasionally excesses develop in certain areas, and once these excesses occur they tend to attract other particles gravitationally. Jeans showed that this would continue until the gas cloud was broken into a large number of globules. It is these globules that eventually give us the stars.

Our present theory of star formation derives from Jeans's. Although more general, its basic idea is the same. We will begin a discussion of it by considering our galaxy. As we saw in Chapter 1, it is a spiral with long, trailing arms. According to one theory these arms developed as a result of shock waves (like the shock waves produced by a supersonic airplane) that spread throughout our galaxy early in its history. (These waves still maintain the arms.) These same waves are also credited with the imbalance that was needed to break the cloud into globules. There are still many difficulties with this theory, but it is gaining acceptance.

Whether the breakup was caused by shock waves or was simply a result of the building up of small density fluctuations (as in Jeans's theory) is not certain. In any case, once the globules formed, there would be no way of stopping their collapse, for gravity would pull them inward toward their center of mass in a gradually accelerating infall. Further breakup might occur along the way, but eventually there would be numerous collapsing gas clouds well on their way to becoming stars.

GLOBULES AND DARK CLOUDS

At this stage because the temperature of these large, collapsing clouds is still low, they are not radiating light; but they are visible in contrast against background stars and gas. These clouds are grouped into two general classifications—dark nebulae (or clouds) and Bok globules, named after Bart J. Bok. Figure 16.5 shows dark nebulae,

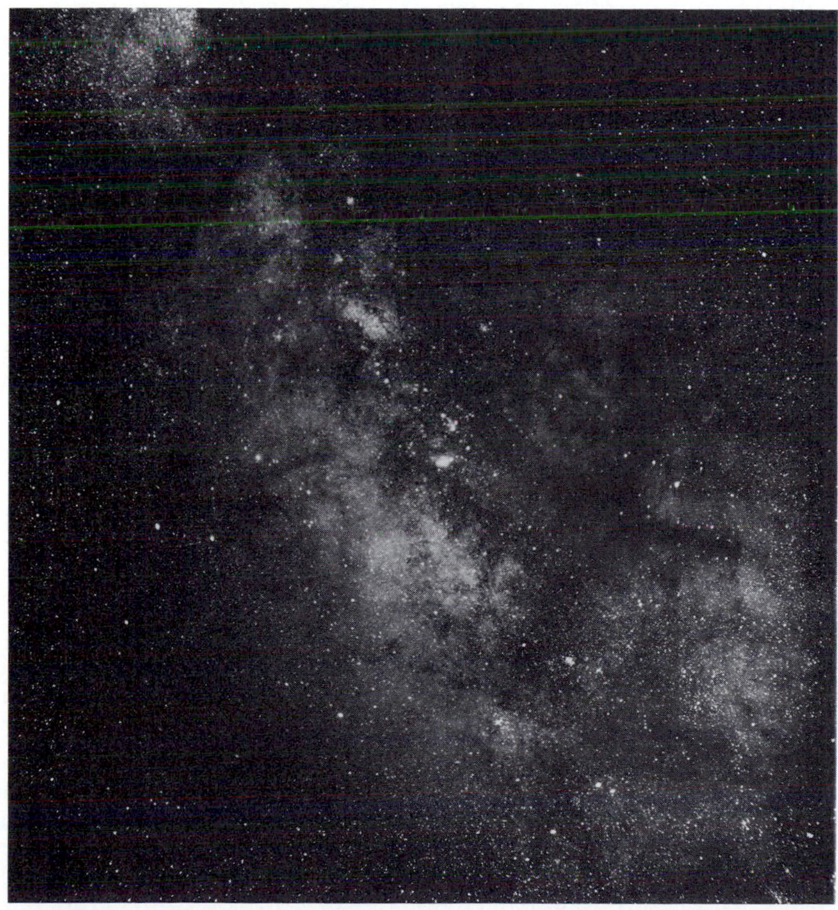

Figure 16.5 A portion of the Milky Way in which several dark clouds can be seen against the star clouds.

which appear almost like holes in the star cloud but are actually foreground clouds of gas and dust. (Light from the stars behind them is being absorbed by dust in the nebulae or dark clouds.) Clouds of this type are usually about 10 or 12 light years across. Within 1000 light years of the sun there are at least a dozen of them.

Bok globules are even more numerous than dark clouds. According to Bok there are at least 100 within 1000 light years of us. They are considerably smaller than dark clouds; a typical one might be 1 light year across. In general they are best seen against bright nebulae; both the Rosette Nebula (Fig. 16.6; color) and the Orion Nebula contain several.

At this stage the cloud could not be plotted in the usual H-R diagram unless the horizontal axis were extended to the right beyond the usual limit of low temperatures.

PROTOSTARS

The gas cloud continues to collapse in on itself. Before long individual particles and atoms falling toward the center have reached incredible velocities, and collisions have become much more frequent. The cloud shrinks in size a thousand times; its core temperature is perhaps as high as 50,000°K at this stage. As infall continues, its surface temperature reaches 2000°K, then 3000°K, and it is now luminous; soon, though its surface temperature may be only 3500°K, it will be 100 times as bright as our sun (surface temperature, 6000°K). This is because, being so much larger (and consequently having a much larger surface), it radiates off much more energy.

Although the object now makes its appearance in the H-R diagram (upper right-hand corner), it is still not a star but a protostar. Soon the protostar begins to decrease in luminosity for, as gravity continues to pull it inward, the increasing density of the gas increases its opacity (ability to absorb radiation), until finally much of the radiation is trapped—absorbed, reemitted, then absorbed again as it makes its way slowly through the gas. With, consequently, much less energy being lost (per unit time), the protostar heats up quickly.

It is now moving downward in the H-R diagram. For the first few thousand years this downward motion is quite rapid, but gradually it slows as the core increases in pressure and temperature. Finally, when the core temperature reaches approximately 10 million °K, the object is almost stationary in the diagram. Then the proton-proton cycle (or possibly the CNO cycle, depending on the exact position of the star in the diagram) is triggered, and the star—for that is what it has become—begins to burn[5] hydrogen.

It has taken our protostar 10 million years to reach this point in the H-R diagram; it will take it (now a star) another 15 million or so to enter the main sequence. Once it is in the main sequence, however, it will stay there (assuming it is about the size of our sun) for several billion years. As we will see later, the actual time spent here depends on the mass of the star.

[5] A reminder: "Burning" is used here in a very restricted sense: changing one element (hydrogen) into another (helium). As we saw in Chapter 15, this occurs via nuclear reactions.

OBSERVATIONAL EVIDENCE

So far we have been describing what happens according to theory, but is there observational evidence to back this theory up? Have we, in fact, actually seen protostars? From the above discussion it seems reasonable to assume that a protostar would look like a luminous sphere surrounded by a halo or cocoon-like envelope of gas and dust. They would be most likely to be found in dense clouds of interstellar matter and, since their surface temperatures would be about 3000–4000°K, would be associated with strong infrared radiation.

Stars of this type have been seen. In 1936 a star appeared in the constellation Orion; it was embedded in gas and dust and had all the properties of a very young star. (It is now called FU Orionis.) In the early 1940s Alfred Joy noticed several strangely erratic stars that had many of the properties of young stars. Joy named them after the brightest of the group—T Tauri (Fig. 16.7).

Figure 16.7 A young cluster in Monoceros. Many of the stars are T Tauris.

T Tauris have dense, highly active outer atmospheres and seem to be ejecting matter; indeed, they may be in the process of throwing off their cocoons. Their spectral lines are broader than those of most other stars, indicating that they may be spinning rapidly, and there are many emission lines present, whereas most stars only show absorption lines. These emission lines may be a result of hot gases rising rapidly to the surface. T Tauris are erratic and unpredictable, sometimes changing brightness significantly in only a few hours.

There is little doubt that they are very young stars. They are found to the right of the main sequence where young stars should be, and they are always found near or in large gaseous clouds. In fact, they usually occur in groups—frequently called T associations—in these clouds.

When infrared astronomy came into its own, astronomers began to search the skies for infrared sources—in particular those associated with nebulae. An important source found by E. Becklin and G. Neugebauer in Orion, and now called the BN object in their honor, was similar to a T Tauri but a much stronger source of infrared radiation. Detailed study showed that it was a large gas cloud (1500 times as large as our sun) in collapse. Some astronomers now believe that it may become a star within 20 years!

More recently (1977) a research team headed by Rodger Thompson discovered an infrared source, known as MWC 349, which may be the strongest evidence we now have for the condensation theory. It is a disk of collapsing gaseous matter that is believed to be less than 1000 years old.

Most objects of this type are now called "cocoon nebulae," since they appeared to be immersed in a gaseous halo. Several smaller but distinctively different systems were found in 1946 by George Herbig and Guillermo Haro and are now known as Herbig-Haro objects; about 40 have been discovered. No star is seen in them, yet they appear to be an early stage of a star. Many astronomers believe that they will eventually evolve into T Tauris.

Recently (1939, 1970, 1975) several T Tauris suddenly brightened by many magnitudes. Though they were first thought to be novae, it was noted that their rise in brightness was slower than that of a nova and that they also faded much more slowly. In addition, shortly after the sudden brightening a hook-shaped nebulosity appeared near the outer region of the star. This is now called the Fuoro phenomenon (after FU Orionis, one of the stars that underwent it).

INTO THE MAIN SEQUENCE

Because our sun is an average, run-of-the-mill star we will consider its evolutionary track into the main sequence in some detail.

As shown in Fig. 16.3, it enters the H-R diagram shortly after it is formed (less than 100 years). At this stage it is several hundred times as bright as it now is, and perhaps 200 times as large. At this point it is still a protostar but is collapsing rapidly. Within 10,000 years it is down to 100 times its present brightness; after 100,000 years it is only ten times as bright. Its surface temperature—approximately 4000°K—changes little during this time.

It continues to shrink and grow dimmer until it is about 10^7 years old. During this time its core temperature steadily increases until, finally, it is about 10 million °K. This is a critical temperature: nuclear reactions are suddenly triggered, and hydrogen begins to burn. An outward pressure quickly builds up as radiation begins to pour out. Soon it is equal to the inward gravitational pull, and the star reaches equilibrium. It now moves slowly upward and to the left, entering the main sequence in about 27 million years. (Astronomers refer to the locus of points where stars first enter the H-R diagram as the zero-age main sequence or ZAMS.) It will remain here for about 9 billion years; this is the quiescent stage of its life as hydrogen burns peacefully and few changes occur. It will not, however, remain in exactly the same position in the main sequence during this 9 billion years, but its motion will be slight.

The internal structure of the sun is now stabilized, and it will remain largely unchanged for millions of years. We can easily calculate and examine this structure using the techniques described earlier. Figure 16.8 shows the result of a calculation of this type. The temperature, density, and mass are given for various positions within the sun. Note, in particular, the tremendous difference in the values with movement inward; the sun is obviously much denser near the core than it is in the outer envelope. Nuclear reactions take place only in this region (the core extends to one quarter of the sun's radius).

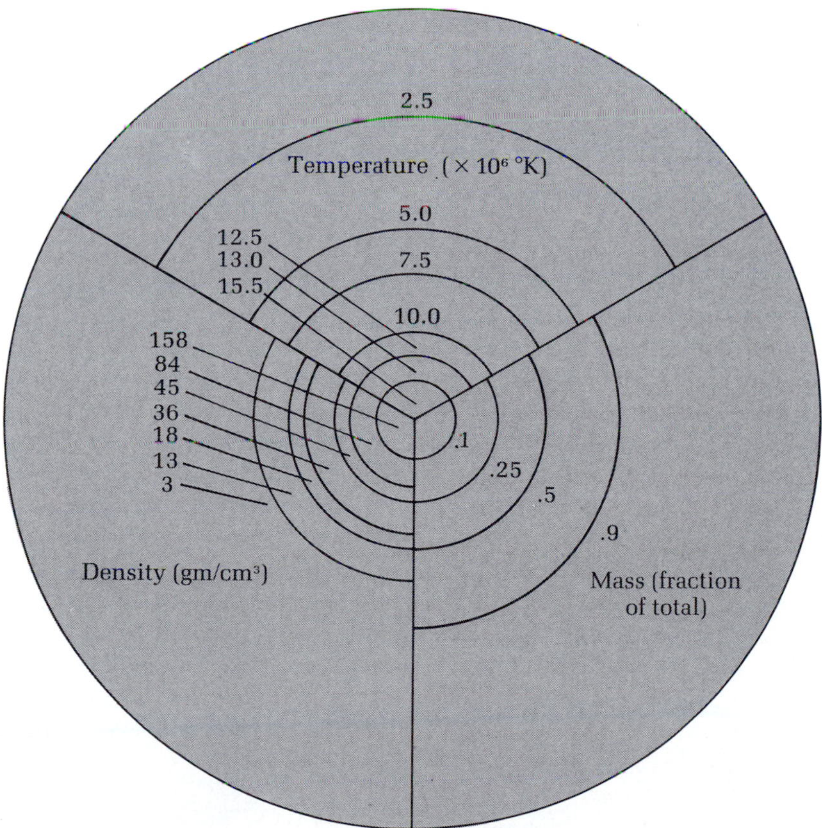

Figure 16.8 The internal structure of our sun. For the next 4 to 5 billion years few changes will take place in this structure. The upper section shows temperatures for various regions in millions of degrees Kelvin; the innermost value is the central temperature.

FROM THE MAIN SEQUENCE TO INSTABILITY

As we look down the main sequence, we see an array of stars ranging from bright, exceedingly massive ones to tiny red dwarfs. The large stars near the top obviously have considerably more fuel to burn than their smaller counterparts further down, but surprisingly, they burn it faster and, consequently, stay in the main sequence for a shorter period. This is because the core temperature of a massive star is much higher than that of a small star, so that it burns its fuel much faster. Comparing a star like this to one like our sun would be like comparing a huge truck to a Volkswagen.

The stars near the top of the main sequence stay there only a few million years, then move off to the right. Stars further down, such

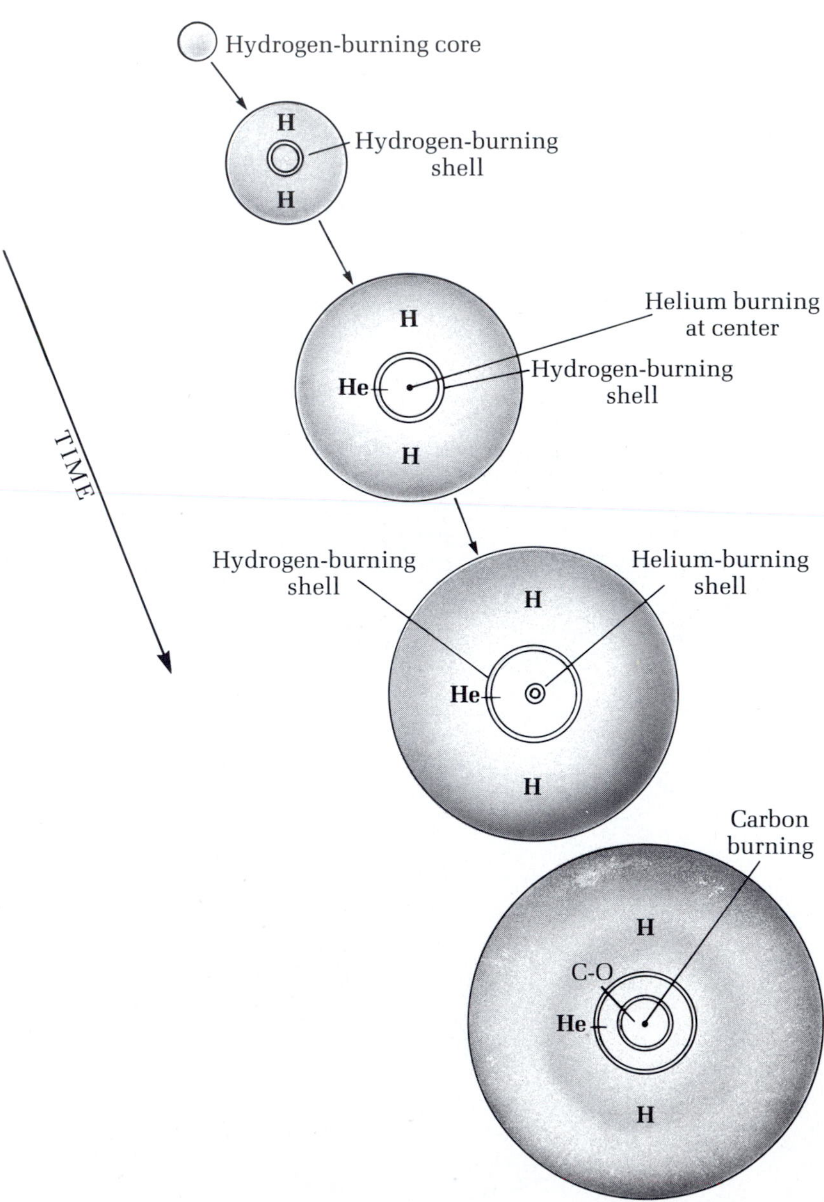

Figure 16.9 Cross-section of an evolving star (not to scale). Note the growth in number of burning shells with increasing age of the star.

as our sun, stay there several billion years, and some still further down (e.g., red dwarfs) will remain so for 40 or 50 billion years (and more). Some of these stars have been there almost since the universe began (since the first stars formed).

In a later chapter we will see that it is important that large stars evolve fast, for this is where heavy elements are being generated, and if solar systems like ours are to form, these stars must evolve fast and somehow distribute their elements into space.

There are actually two different burning cycles along the main sequence. Our sun burns hydrogen via the proton-proton cycle (p. 273), but stars slightly more massive (1.2 $M_\odot$) than our sun use the hotter CNO cycle.

As hydrogen continues to burn, helium is generated. Heavier than hydrogen, it accumulates in the core, and, as burning continues, the core grows, until finally the remaining hydrogen is burning in a shell around it (Fig. 16.9).

Helium continues to build up until it can no longer withstand the tremendous pull of gravity and begins to shrink. We could almost say that it is being overcome by its own weight. As it contracts, its temperature increases, causing the hydrogen to burn even more furiously. Finally, when about 12% of the hydrogen has been burned, the star begins to move slowly out of the main sequence to the upper right (Fig. 16.10).

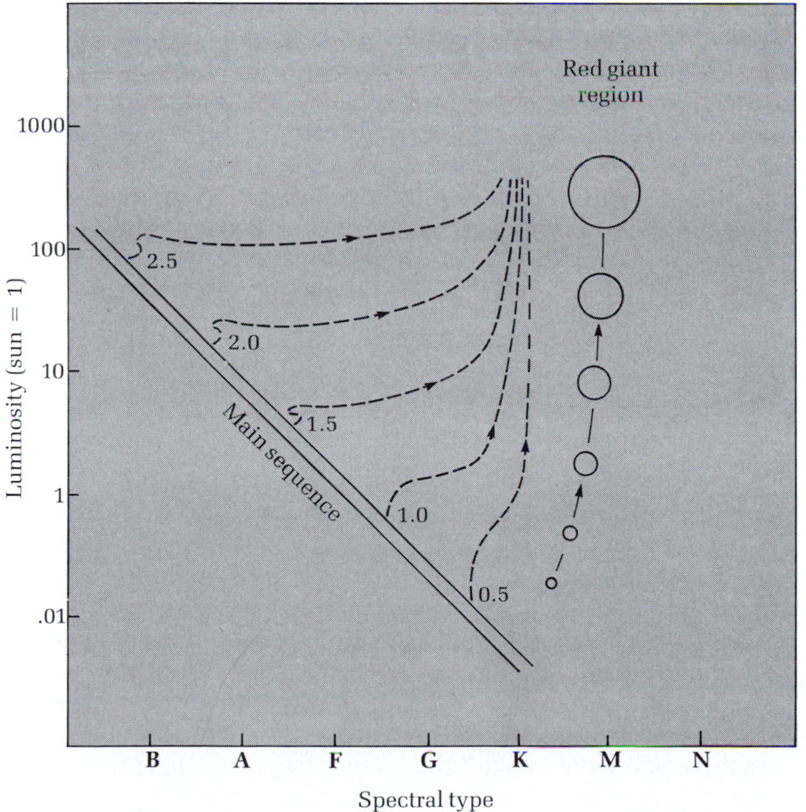

Figure 16.10 Evolutionary tracks of several stars of various masses away from the main sequence. Note the gradually increasing size of the stars.

LEAVING THE MAIN SEQUENCE

The star's evolution from this point on depends, of course, on its mass, but also on its constitution. W. Baade has shown that stars can be divided roughly (on the basis of composition) into two classes: Population I and Population II. (For now we will only point out the important differences in these two classes; further details will be given in a later chapter.)

Population I stars consists mostly of hydrogen with some helium and about 2–3% heavier elements. Population II stars also consist mostly of hydrogen, but they contain less helium than population I stars and only about $\frac{1}{50}$ the amount of heavier elements; in general, they lie in the bottom half of the main sequence.

When Population I stars leave the main sequence, if they are not already burning hydrogen via the CNO cycle, they soon begin to. Population II stars, however, never enter this cycle; they are deficient in carbon, and, as we saw earlier, carbon is needed in the CNO cycle.

As the star leaves the main sequence, radiation continues to pour out of the core at an ever-increasing rate. The outer layers of the star are heated and begin to expand outward, but as they swell, they cool. Thus, rather than brightening, the star's surface grows cooler and redder, its temperature falling quickly. The star continues to move to the right and upward in the H-R diagram, its exact path depending on its starting position in the main sequence. Stars near the top move mostly to the right, whereas stars near the bottom move almost directly upward. Several tracks (calculated by astrophysicist I. Iben) are shown in Fig. 16.10.

Consider our sun as a specific example again. Its evolutionary track, also shown in Fig. 16.10, shows that when it leaves the main sequence it will move almost directly upward at first; then its luminosity will remain constant for a while, and finally, it will brighten rapidly. The constant-luminosity phase will begin in about 4.5 billion years and will last for about a billion years. After that the sun will expand rapidly, becoming a red giant in about 300 million years.

THE RED GIANT STAGE

In this phase the star is in the upper right-hand region of the H-R diagram. Although this giant is perhaps 160 million km (100 million mi) across, all its energy is being produced in a tiny shell of burning hydrogen, barely 3000 km (about 2000 mi) thick, that surrounds a core of helium about 30,000 km (about 20,000 mi) across. The outer envelope is now so tenuous and extended that gravity barely holds it to the star; by earth standards, most of it is a good vacuum.

The bright red star Antares, in the constellation Scorpius, is an example of a star of this type. It is so large that if we placed it where the sun now is, it would engulf the planets as far out as Mars.

THE HELIUM FLASH

As burning continues in red giants, core temperatures approach another critical value—100 million °K. The events that occur now depend again on the mass of the star. Consider one with approximately the mass of the sun. The helium in the core is so compressed at this stage that it is extremely rigid; though still gaseous, it acts like a steel ball, and nuclei are far closer together than they usually are. Astronomers refer to matter of this type as degenerate matter. It has many strange properties, one being the ease with which it conducts heat. (Degenerate matter is discussed further in Chapter 18.)

At the critical temperature (100 million °K) a new burning cycle—the triple α process—is triggered. Helium begins to burn in the center, but the core being degenerate, heat spreads rapidly from the center throughout it. Soon the entire core is hot enough for helium burning, and, like a wildfire or gigantic fuse, the burning area begins to spread.

In ordinary stars the core would expand and cool, thereby controlling the burning. Because the core is so rigid it cannot expand, and so there are no controls. Soon the core explodes, blasting material out past the hydrogen-burning shell, putting out its fires. The explosion—called the helium flash—cools the interior and, for a time, the nuclear furnaces are shut down. Strangely enough, the surface of the star remains unchanged, and there is no direct evidence of the tremendous turmoil within. There is soon evidence of it, however, when the star begins to dim; in the H-R diagram it begins to move downward and to the left (Fig. 16.11). For thousands of years it continues moving in this direction, but gradually order is restored. The exploded helium falls back into the core; the hydrogen begins to burn in a shell again, and finally the helium begins to burn relatively gently at the center. On the H-R diagram the star begins to move upward again toward its original position.

We have been talking about a star with approximately the mass

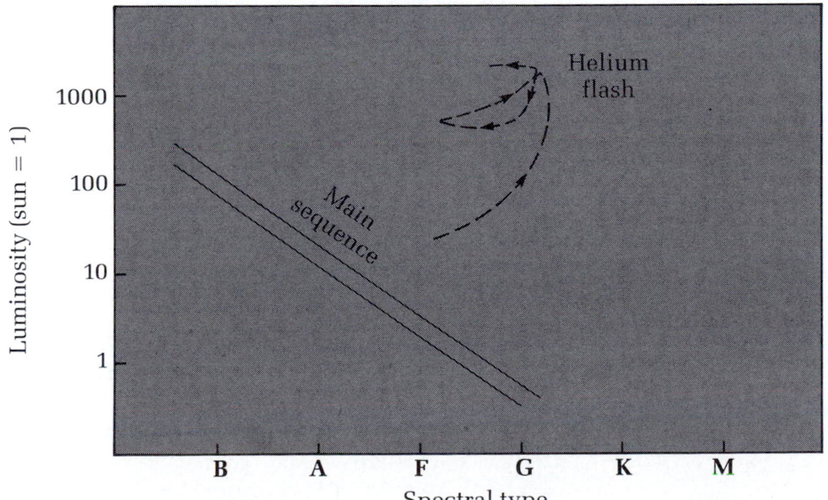

Figure 16.11 A section of the evolutionary track of a star with a mass about that of our sun. Note that it moves downward after the helium flash, then upward again.

of our sun. Stars that are sufficiently massive (greater than about 5 $M_\odot$) do not undergo a helium flash; in these helium burning begins gently. For stars much smaller than the sun there is also no helium flash, for helium burning is not triggered in them.

AFTER THE HELIUM FLASH

Most of the details of what happens after the helium flash will be treated in later sections, but a general overview will be given here. On the H-R diagram the star moves to the position it had before the helium flash (assuming a star of about 1 $M_\odot$) and then begins to move to the left. What happens after this depends on its mass: If it is about the mass of the sun, it will trigger only helium burning and nothing further; however, if it is exceedingly massive (greater than 10 $M_\odot$), it will continue to trigger cycle after cycle.

As helium burns, it also leaves an "ash" of carbon-oxygen (C-O), which accumulates in the center. With increasing pressure, temperatures in the C-O core finally become high enough so that the carbon is ignited. There can even be a carbon flash (with a star in a certain mass range), or the carbon may burn gently, leaving oxygen as its "ash." Finally the star is burning on many levels (Fig. 16.12) and the core is iron. Strangely, things can now go no further; iron cannot burn. What follows will be discussed in later chapters.

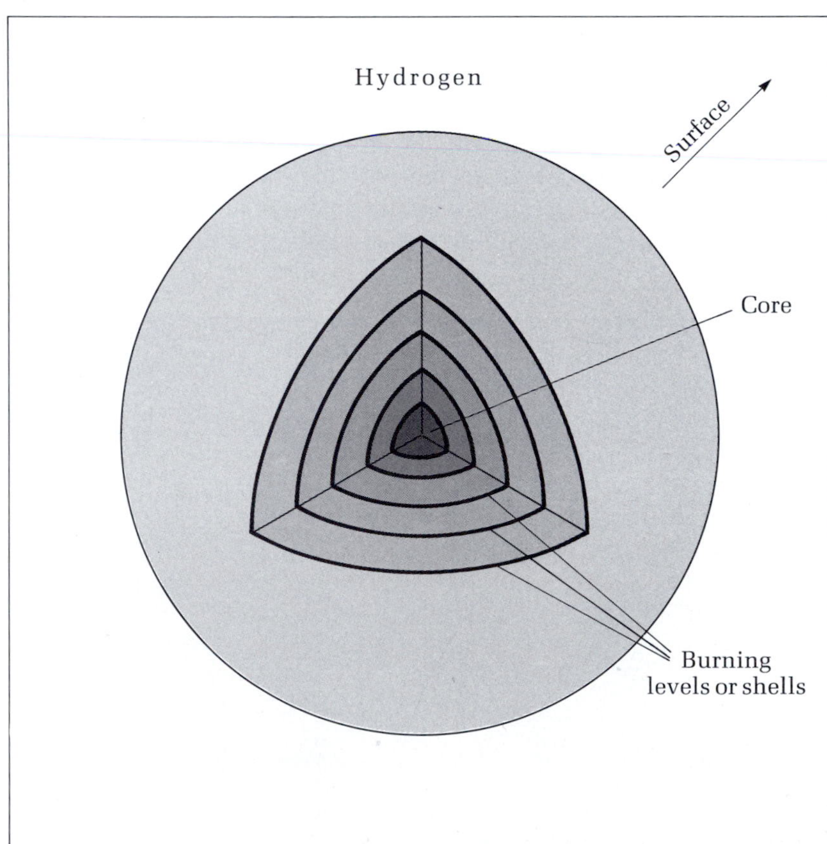

Figure 16.12 Schematic diagram of the burning levels in the core of a massive star in which, eventually, the core is iron, which cannot burn. The surface of the star is assumed to be beyond the edge of the page.

The Missing Neutrinos

Until about ten years ago astronomers were reasonably confident that they understood the reactions that go on in the cores of stars. In 1968, however, Raymond Davis of Brookhaven National Laboratory began an experiment that would allow him to test these reactions for the case of the sun. According to the accepted theory at that time, large numbers of neutrinos are released. Most of these neutrinos pass directly out of the sun, and most of those that come in our direction also pass directly through the earth. But occasionally one interacts with an atom of chlorine, changing it into an isotope of argon. To use this particular reaction to monitor the neutrinos, Davis installed a large tank, filled with C_2Cl_4, in a deep mine in South Dakota.

According to Davis's original calculation, the earth above the detector would shield it from unwanted particles but several neutrinos would interact with the chlorine atoms in the fluid each day, changing them to argon. Surprisingly, only about one fifth the number of neutrinos predicted were observed.

For many years astronomers had no idea why this was the case. However, in 1980, F. Reines, H. Sobel, and E. Pasierb noticed that the neutrinos produced in nuclear reactors appeared to exist in three different "states"—that is, there appeared to be three different kinds of neutrinos. It was then determined that only one of the three types would be detectable in Davis's experiment. Thus, if neutrinos from the sun act the same as those produced in nuclear reactors (and we have no reason to suspect that they do not), we have a reasonable explanation of Davis's result.

Another recent discovery also helps explain Davis's findings: It has been found that there are several different nuclear reactions creating neutrinos in the sun. The most common reaction produces neutrinos with energies lower than those Davis is able to detect. Only the rare reactions produce neutrinos with detectable energies. This means that most neutrinos are passing undetected through Davis's apparatus.

Which of these two explanations is most important? Astronomers are uncertain, but an experiment is now under way in which gallium rather than chlorine will be used. With it, neutrinos of much lower energy can be detected; we should have an answer soon.

THE SUN AND EARTH: OUR FUTURE

There will be no major changes in our sun in the near future. It will remain stable for a few billion years, and there will be no significant changes in the average temperature of the earth. (This does not rule out a few ice ages, however.) Our sun has now been in the main sequence between 4 and 5 billion years; in another 4 billion years it will begin to leave. Initially, its expansion will be quite rapid; then it will slow and remain relatively constant for about 5 billion more years. After that it will again expand rapidly. Its outer envelope will expand until it engulfs Mercury, then Venus. As it approaches Earth, it will appear initially as a gigantic, red, oversized sun, growing larger and larger in the sky. Eventually it may enclose Earth, but well before this happens Earth will have become uninhabitable; the oceans will have boiled off into the atmosphere and most (and eventually all) of the atmosphere will have disappeared into space. The surface of Earth will look quite different; where there once were lakes and oceans there will now only be seas of molten lava.

Figure 16.13 The fate of the earth—a frozen wasteland, with the sun just a bright star in the sky.

Soon after this the sun will begin to shrink. Temperatures will moderate and Earth may even harbor life again, but if so it will be short-lived.

In the end, as with all things, our sun must die. We will see in a later chapter that it will slowly collapse until it is little more than a bright star in the sky. Earth will by then have become a frozen wasteland. (Fig. 16.13).

SUMMARY

1. Stars go through a life cycle just as humans do. It is convenient to use the H-R diagram when discussing this cycle.
2. To construct a computer model of a star, it is assumed to be divided into a number of shells. A set of equations is then solved telling how changes in pressure, temperature, mass, and luminosity take place from shell to shell. Using these models, astronomers can calculate evolutionary tracks.
3. Stars are born out of the interstellar medium. Dark clouds and globules form from this material as a result of gravity.
4. Dark clouds are thought to condense into small units called Bok globules, which condense into protostars.
5. A protostar is much larger and more luminous than the star it eventually collapses to. The collapse time is of the order of several million years.
6. The protostar becomes a star when it triggers nuclear reactions, which it does just before it reaches the main sequence.
7. A star spends most of its life in the main sequence. Large, hot stars near the top spend the least time here; small stars near the bottom spend the most.
8. When the star's fuel (hydrogen) starts to run out, the star moves to the upper right-hand region of the H-R diagram and becomes a red giant.
9. If the star is about the mass of our sun, an explosive event called the helium flash will occur while the star is in this region of the H-R diagram.
10. Eventually the star burns helium peacefully, leaving C-O ash. If it is massive enough, it may burn C, then O, etc., up to iron. It burns these elements in shells around the core.

REVIEW QUESTIONS

1. Select some point in the H-R diagram. Mark several directions away from this point, then tell what change would be taking place in a star as it moved in each of these directions.
2. Does the H-R diagram tell anything about the core of the star?
3. What are the main components of the interstellar medium?
4. How do globules form from the interstellar medium?
5. What is the difference between a Bok globule and a dark cloud?
6. How can a protostar be much brighter than a star when it has a much cooler surface?
7. Where do protostars lie in the H-R diagram? Discuss.
8. What evidence do we have for the existence of protostars?
9. What are the characteristic features of a T Tauri star?
10. Explain how astronomers build stellar models.
11. State the Russell-Vogt Theorem in your own words.
12. Two stars are spectral types M and A. Which leaves the main sequence first? Why?
13. When does a protostar become a star?
14. Has the sun ever been dimmer than it now is? Explain.
15. What is the Hayashi zone?
16. What cycle is the sun using at present to burn hydrogen? What other cycles will it trigger?
17. What is the difference between a Population I and a Population II star?
18. Describe a red giant. What are some of its properties?
19. Is the helium flash visible at the surface of the star? Why?
20. Using the H-R diagram, describe the events that occur after the helium flash for a very massive star.

THOUGHT AND DISCUSSION QUESTIONS

1. At one time it was believed that stars evolved into the main sequence, then moved or "slid" down it. The theory was called the slide theory. Give several arguments showing that this theory cannot be correct.
2. As a star evolves, changes occur both at the surface and internally. Describe in detail the changes that are occurring both internally and externally as a star of mass 1 $M_\odot$ leaves the main sequence and moves toward the red giant stage. Do the same for a $5M_\odot$ star and a $10M_\odot$ star.
3. A star is in equilibrium when it radiates at its surface all the energy that it generates at its core. What happens if it begins to generate more energy than it radiates? Less energy? Explain.
4. Assume that the Greeks had been able to photograph stars and nebulae. What changes would we now see in most stars were we to compare the photographs with the present state of the stars? What types of objects would change the most? What would these changes be?

FURTHER READING

Bok, B. J. "The Birth of Stars," *Scientific American,* p. 48 (August 1972).

Cohen, M. "Stellar Formation," *Astronomy,* p. 66 (September 1979).

Dickman, R. L. "Bok Globules," *Scientific American,* p. 66 (June 1977).

Seeds, M. "Stellar Evolution," *Astronomy,* p. 6 (February 1979).

Strom, S. and Strom, K. "The Early Evolution of Stars," *Sky and Telescope,* pp. 279, 359 (May/June 1973).

Zeilik, M. "The Birth of Massive Stars," *Scientific American,* p. 110 (April 1978).

Aim of the Chapter

To describe further the later stages in the life story of a star.

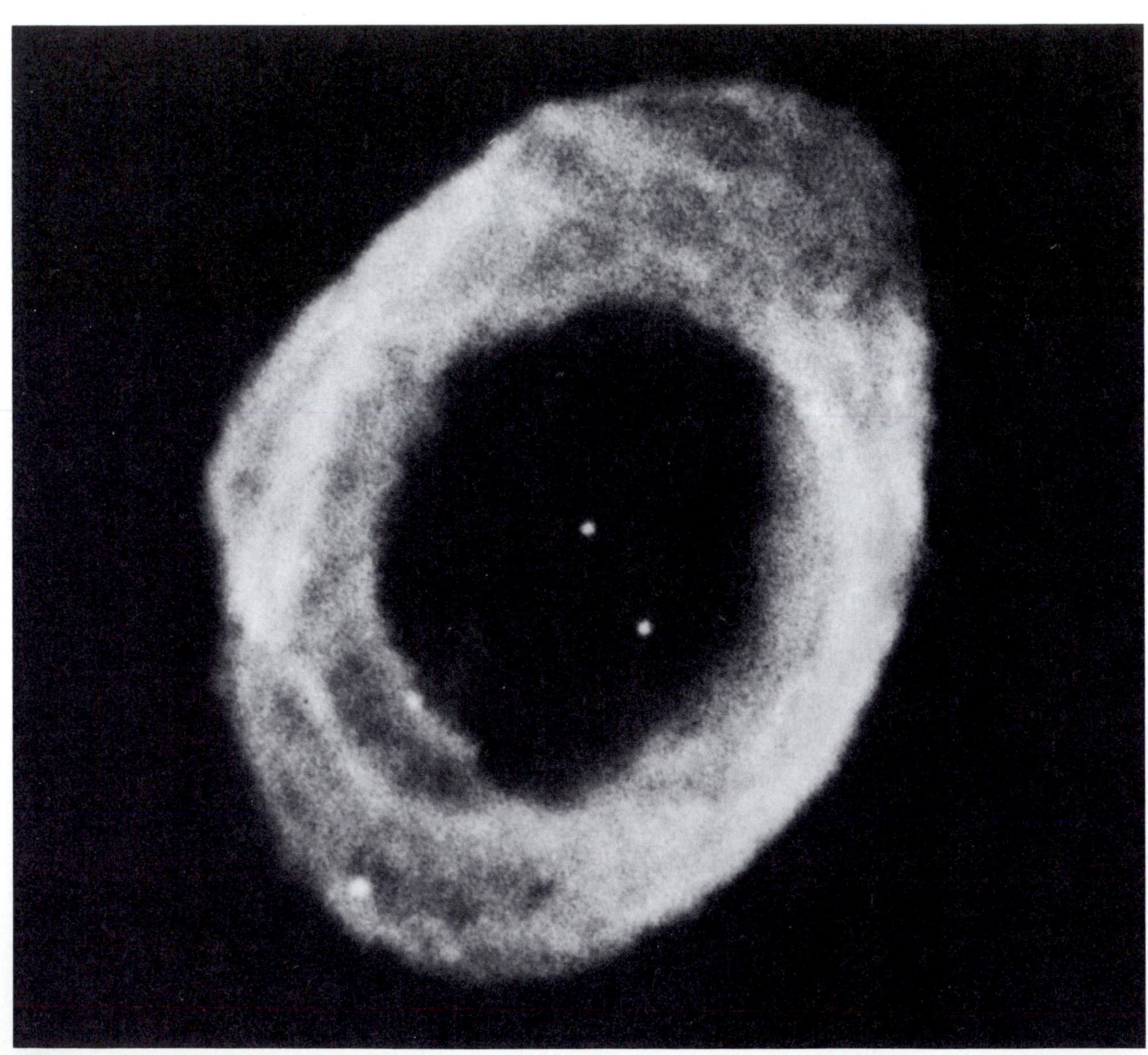

chapter 17

Later Stages of Evolution

We have seen how a star is born and how it moves into the main sequence in the H-R diagram, staying there for perhaps billions of years. We also know that after much of its fuel is used up it moves off to the right, increasing in size as it goes, until finally it becomes a red giant. As a star moves away from the main sequence, it may pass through what is called the instability strip (Fig. 17.1), a region of tremendous internal turmoil.

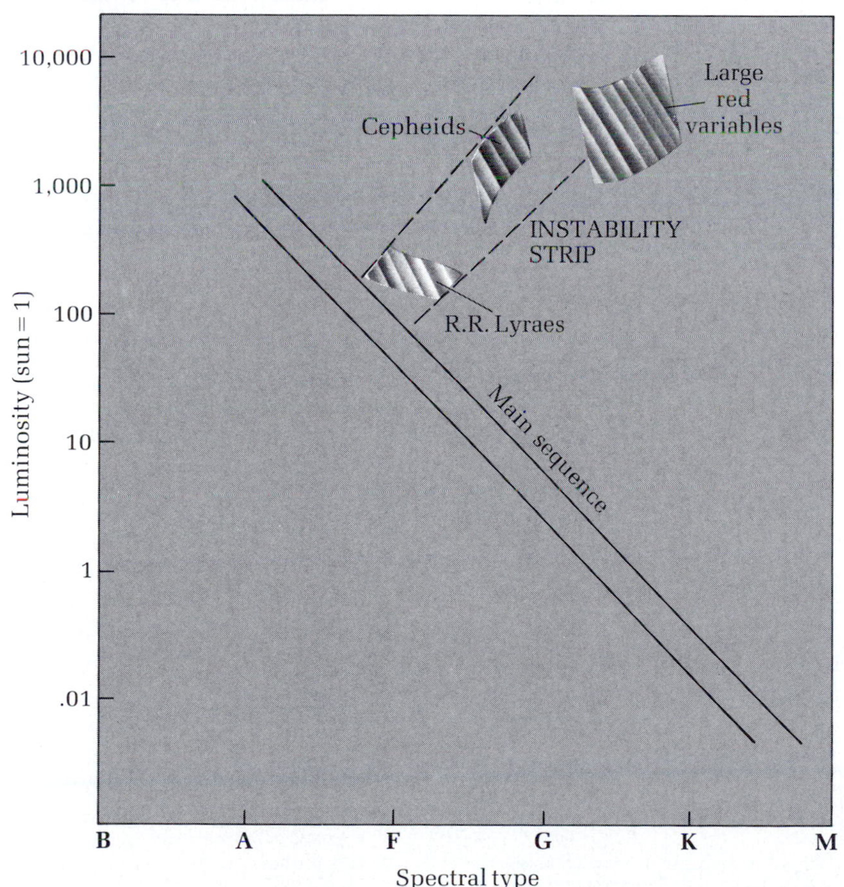

Figure 17.1 **The instability strip (area enclosed by dotted lines). Positions of RR Lyraes, cepheids, and large red variables are shown relative to the main sequence.**

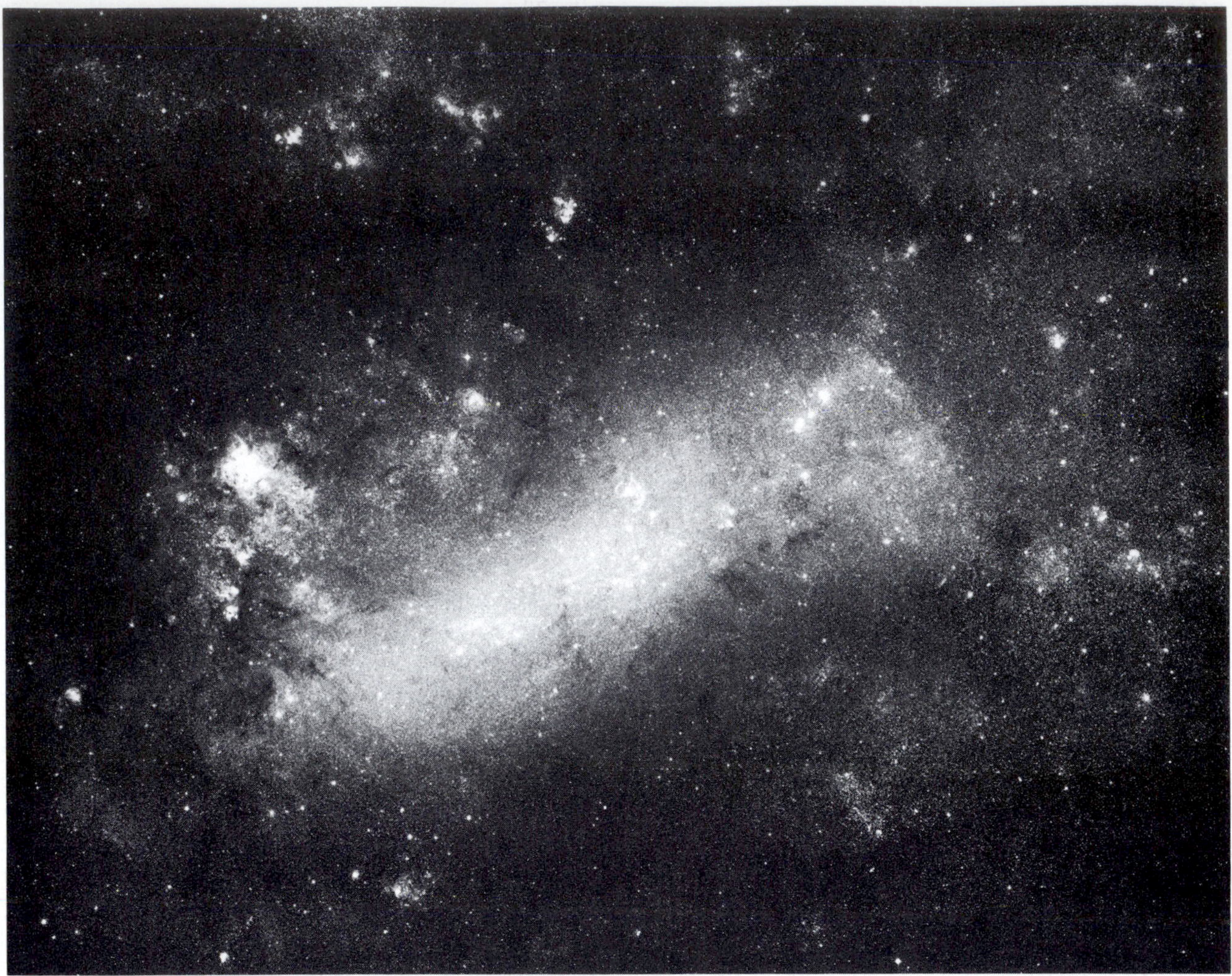

Figure 17.2 The Large Magellanic Cloud, the cloud of stars in which many of the more luminous stars were noticed by Henrietta Leavitt to vary in brightness. A study of these stars led to discovery of the period-luminosity relationship.

VARIABLES

As the star enters the instability strip region it becomes unstable, its equilibrium is temporarily upset, and it may begin to expand radially. This latter expansion, which increases its brightness, is checked within days, when the star returns to its original magnitude—only to begin the cycle again. Stars that vary periodically in this way are called variables.[1]

But what causes this variation? In most stars radiation makes its way slowly outward from the core to the surface, where it disperses into space. However, conditions in the instability strip create a layer of ionized helium near the surface of the star, and this layer temporarily traps the radiation; in effect, it absorbs it. As a result, pressures build, and the layer begins to expand. When the expansion reaches a certain point, conditions are arrived at that allow the radiation to

[1]Stars also vary in magnitude for reasons other than those described here. Also, not all variables are periodic; some flare up sporadically.

pass through, and the star then returns to its original state. This cycle (called the period of the variable) repeats itself over and over.

One of the first to observe stars of this type was Henrietta Leavitt. In 1908, while examining photographs of the Magellanic Clouds—two irregularly shaped galaxies in the southern hemisphere (Fig. 17.2)—she noticed that several of the brighter stars had apparently changed in magnitude. She soon found that they had periods ranging from a day or so to about 50 days, and that the brighter the variable, the longer the period. Since the stars were in a group, they were, to a good approximation, all at the same distance. This meant that there must be a definite relationship between the period and the luminosity (true brightness) of the star; we now refer to this as the period-luminosity relation.

Hertzsprung soon noticed that the variables that Leavitt was studying were similar to one called δ Cephei that was studied much earlier by John Goodricke. We now call this class of variables cepheids, after δ Cephei, which is the brightest of the group. Hertzsprung also realized that since they were, effectively, all at the same distance, we were, in a sense, seeing their absolute magnitudes in relation to each other. But absolute magnitude and apparent magnitude can give us distance (p. 260). Were we, then, able to obtain an independent measurement of the distance to a single cepheid, it would permit us to "calibrate" the distances to *all* cepheids. None of the cepheids, however, was close enough for the parallax method (Chapter 14) to be applied. It was, therefore, many years before an independent measurement was obtained, but when it was, an important new tool became available to the astronomer—a yardstick with which one could measure the universe. To summarize how this yardstick is obtained: 1) a cepheid is found; 2) its average apparent brightness and its period are noted; 3) using the calibration mentioned above and the cepheid's period the absolute magnitude of the cepheid is determined (Fig. 17.3); 4) with the absolute and apparent magnitudes known, the distance to the cepheid is determined using the formula on p. 260.

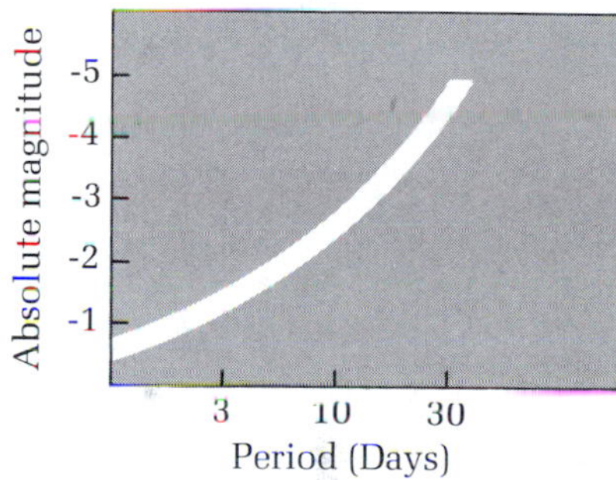

Figure 17.3 The period-luminosity relationship for cepheids.

Cepheids are intrinsically bright stars, yellow supergiants with absolute magnitudes between −1.5 and −5. Of the 700 or so known, all have periods between about 1 and 50 days. A typical light curve for a cepheid (a plot of the magnitude of a star versus time) is shown in Fig. 17.4. It is noteworthy that the change in brightness is small and

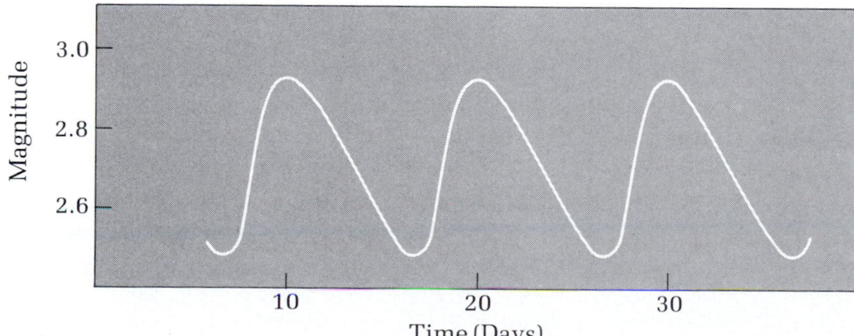

Figure 17.4 A representative light curve for a cepheid variable. The period of the variable is the distance (time) from any point to the next corresponding point, as from one maximum (peak) to the next.

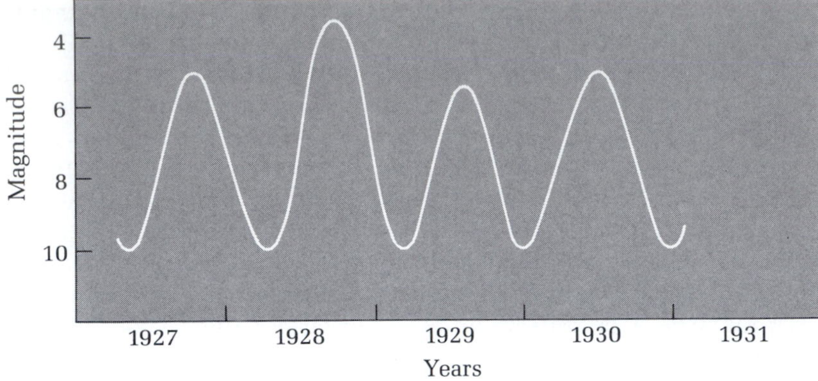

Figure 17.5 Light curve of the long-period variable Mira Ceti (Mira in the constellation Cetus).

that the curve has a distinct shape, the brightness increasing rapidly, then falling off relatively slowly. One of the better known cepheids is Polaris, our North Star; it has a period of 4 days and fluctuates in apparent magnitude between 2.5 and 2.6.

Spectroscopic studies show that the radii of variables increase about 5 to 10 percent with brightness. In other words, there is an actual pulsation in the volume of the star—hence the designation "pulsating variable." Surface temperature also fluctuates by 5 to 10 percent.

Pulsating variables with periods of less than a day (called RR Lyraes, after the brightest of their group) also exhibit a period-luminosity relation, but one not well defined. Nevertheless, because all have absolute magnitudes of approximately 0.0 they are also useful measuring rods: Like illuminated 100-watt light bulbs scattered over a field, since the absolute brightness of each is known, it is easy to determine the distance to each.

At the other end of the cepheid period scale are variables with periods longer than 50 days, called long-period variables. (Some have periods exceeding 700 days.) Unlike shorter-period variables that undergo luminosity changes of only a magnitude or so, these giants change by as much as 8 magnitudes. One of the first of this type to be discovered was Mira in the constellation Cetus. (It was called "Mira the Wonderful" by early Arabian astronomers because of its great variability—almost third magnitude at its maximum and not visible a few months later.) Its period is approximately 332 days (Fig. 17.5), and its spectrum also changes during this time, indicating surface temperature changes.

NOVA

In August 1975 a student called my attention to a bright "new star" in Cygnus. By the time I saw it—several days later—it was barely visible. The "new star" was actually not new, but a nova, most of which are stars that brighten rapidly—sometimes by 8 or 9 magnitudes—then dim slowly. This nova, however, had decayed relatively rapidly back to the limit of the eye (about sixth magnitude). Nova Cygni, as this nova was called, brightened from about magnitude 14 to approximately magnitude 2 in about two days. Fifteen or so days later it is back to magnitude 8 (Fig. 17.6).

Careful examination of a nova with a telescope shows an expanding envelope around it, indicating an explosion of some type. But, since the star eventually gets back to its original magnitude, very little material is actually blown off in the explosion; in fact, only about $\frac{1}{1000}$ of the total mass is lost.

Novae are relatively common in our galaxy, about 30 to 50 occurring each year; most are too dim to be seen with the naked eye. There are three general types: classic, recurrent, and dwarf. In classic novae there has been, as far as we know, only one explosion; magnitude increase in this case is generally quite large. Recurrent novae are repeaters, and dwarf novae are those that have much smaller explosions; they occur in many stars.

Until the early sixties astronomers could only guess at what causes these explosions. In at least one case a binary system was involved, but beyond that little was known. Subsequently, Robert Kraft discovered that up to 70 percent of novae were associated with binary systems and that the binary systems themselves were of a particular type, in each case consisting of a main sequence star, like our sun, and a white dwarf—a small hot star with no nuclear furnaces. (White dwarf stars are discussed in Chapter 18.) In each case the two stars were close together, their periods of revolution averaging about 4 hours.

Of several theories advanced to explain the explosion mechanism, interest eventually centered on the thermonuclear runaway theory, which postulates that hydrogen gas from the main-sequence star is pulled by gravitational forces onto the surface of the white dwarf, eventually diffusing down to the core, where thermonuclear reactions are triggered. Were this to happen in an ordinary star, its surface would expand and cool, thereby quenching the reaction. But a white dwarf does not expand when heated; it continues to increase in temperature. This, in turn, causes the reaction rate to in-

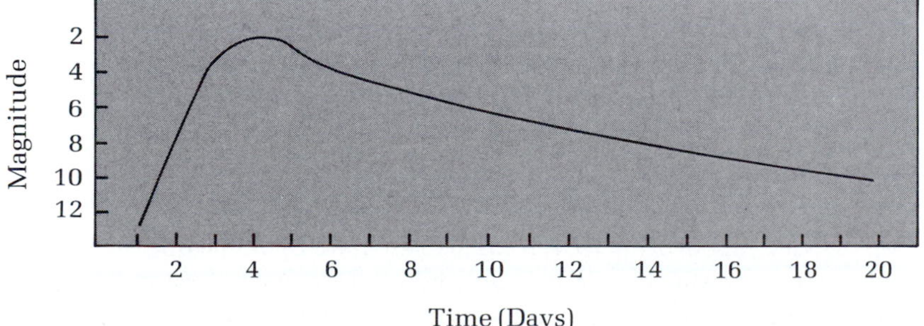

Figure 17.6 Light curve of Nova Cygni (1975).

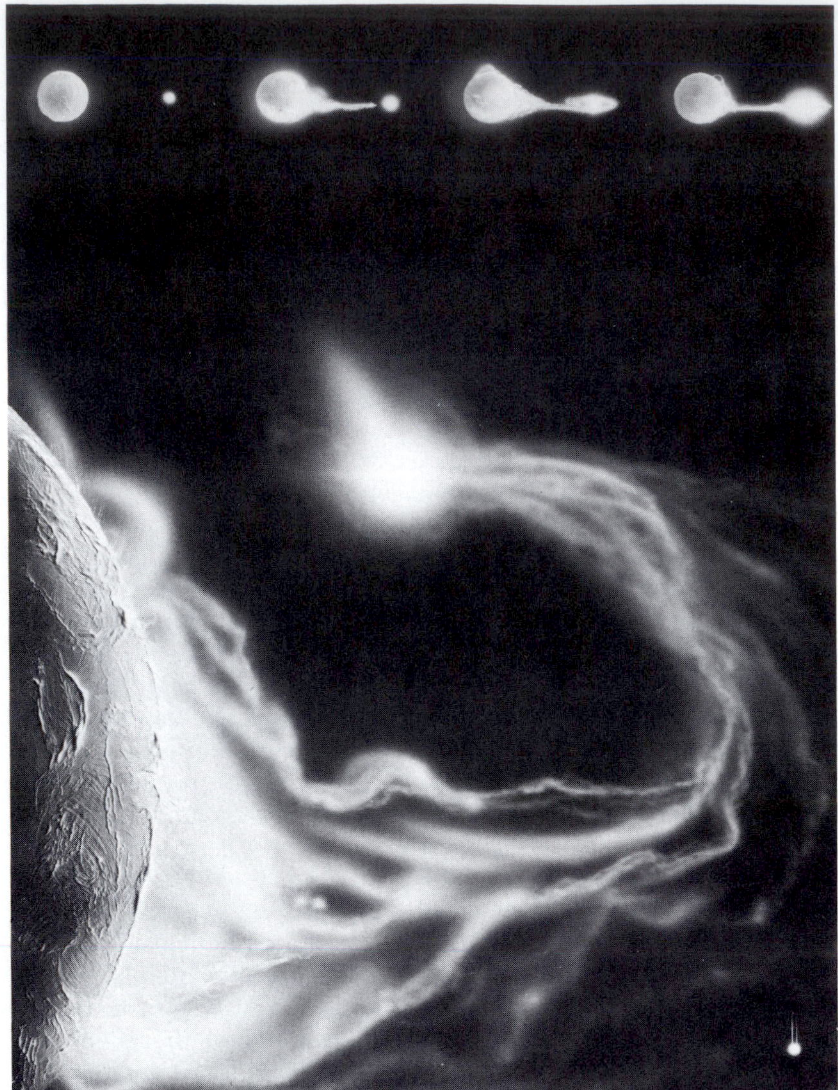

Figure 17.7 A nova explosion according to the accretion theory: As matter from the main sequence star spirals to the white dwarf and strikes the surface an explosion occurs.

crease, until the explosion occurs, blowing the outer layer of the star into space.

This mechanism was accepted until the late sixties, when a flickering was noticed in several binary systems. Further examination of novae then showed that the white dwarf was surrounded by a whirling ring of hydrogen (called an accretion disk), which was gas from the main-sequence star spiraling onto the dwarf. In 1973, in light of this new information, astronomers formulated the accretion theory, according to which, when hydrogen is pulled off the main-sequence star it spirals around the white dwarf, gradually moving inward toward it. As it gets closer to the white dwarf it heats rapidly; finally when it hits the surface of the dwarf the entire system suddenly flashes into a brilliant nova (Fig. 17.7).

Astronomers now believe that this phenomenon occurs only in the case of recurrent and dwarf novae; the thermonuclear-runaway process is still considered to be the one responsible for classic novae.

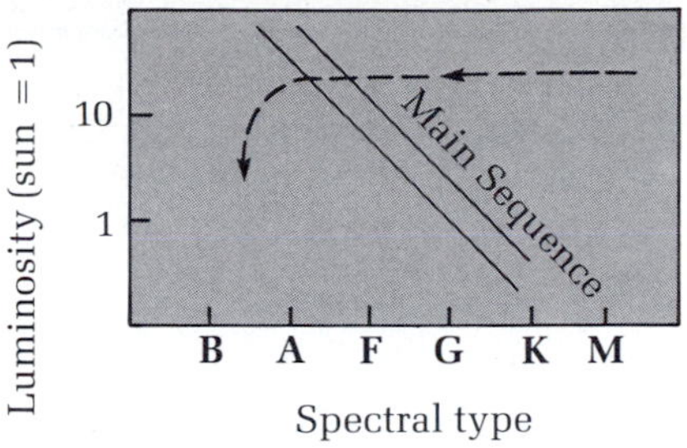

Figure 17.8 Path of a star after it leaves the upper-right region of the H-R diagram.

PLANETARY NEBULAE

What happens to a star after it leaves the instability strip, or for that matter, what finally happens to any star that has moved into the upper right-hand region of the H-R diagram? Eventually all these stars begin to move to the left. Sometimes they move back and forth a few times, but sooner or later they move all the way to the left—or at least most of the way (Fig. 17.8). Their exact path, as we have mentioned many times, depends on their mass. Consider, then, a star of a particular mass, say less than 4 $M_\odot$, with a hydrogen-burning ring and a helium-burning ring, which means its core is C—O that (because its mass is less than 4 $M_\odot$) will not burn.

This star is indeed a bloated giant: near its core are the two burning rings and, around them, a thin atmosphere of hydrogen that extends out for millions of kilometers—so far that gravity barely holds it to the core. The outer regions of this envelope are quite cool—perhaps 3000–4000°K.

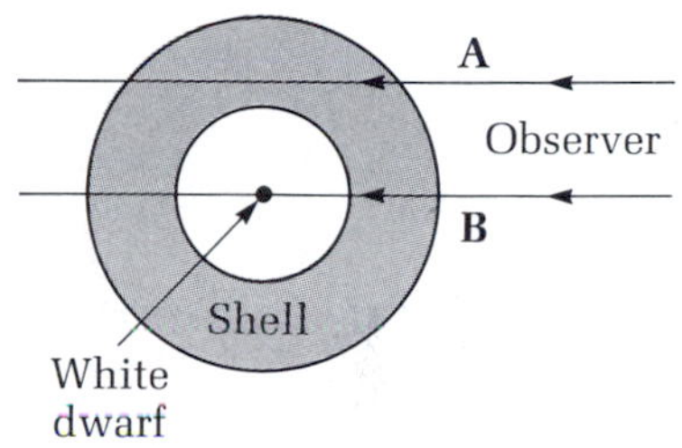

Figure 17.9 Two views through a planetary nebula. Although it is actually a three dimensional shell expanding in all directions, to us it looks like a smoke ring. The figure shows why: when you look through along line *A* you are seeing a longer pathlength of gas than when you look along *B*, hence it appears more dense in region *A*.

As the star continues to expand, its outer layers continue to cool, until finally nuclei and electrons begin to recombine. Photons are released into the envelope as a result of this recombination; some are absorbed but, eventually, a considerable amount of heat is developed. This expands the envelope even more, causing the recombination rate to increase. Soon a runaway process is in full swing, and the envelope is literally pushed off into space. Although it is actually a three-dimensional shell expanding in all directions, to us it looks like a smoke ring—for reasons shown in Fig. 17.9.

As the cool layers move off into space the hot core lies exposed, and the star's surface temperature increases from about 3000° to perhaps 50,000°K. However, because the star is now much smaller, its luminosity stays roughly the same. In the H-R diagram it moves rapidly to the left through the main sequence (Fig. 17.8); eventually it becomes a white dwarf. From a distance, systems of this type look like wispy, doughnut-shaped clouds centered on faint stars (Fig. 17.10). We call them planetary nebulae—a name left over from when Herschel and other early astronomers who first saw these objects noted their similarity to planets.

About one star in a million is a planetary nebula at any given time. One of the most popular astronomical objects in the northern hemisphere for the amateur is the planetary nebula called the "Ring Nebula" in Lyra (Fig. 17.10).

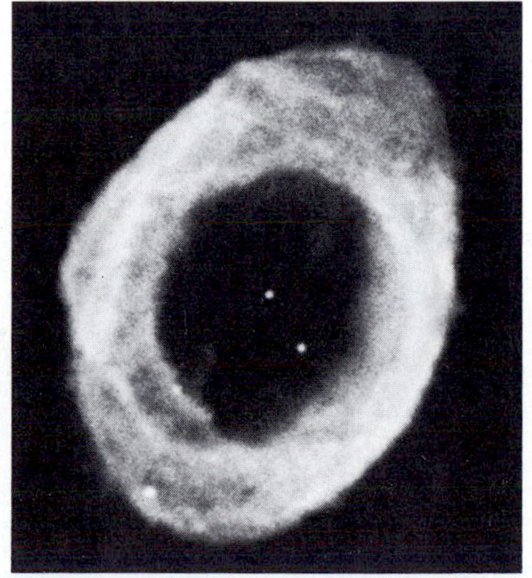

Figure 17.10 The "Ring Nebula" in Lyra.

CLUSTERS

We have discussed the life cycle of a star and have shown how astronomers go about determining it. There are groups, called clusters, that constitute evidence of the cycle as we have described it. They are groups of stars of different mass, all approximately the same age, that condense out of the same gas cloud.

Astronomers divide clusters into two main types: galactic, or open clusters, and globular clusters. Galactic clusters consist of from about 50 to a few hundred stars. Approximately 1000 of these clusters are known at present; it is likely that there are many in our galaxy we cannot see. You may be familiar with the best known of those we can see; clearly visible to the naked eye as a cluster of six stars in the winter sky, it is called the Pleiades (Fig. 17.11; color). Through a small telescope we see many more than six, though even then we are not seeing all of them; photographic exposures indicate that the cluster may contain as many as 150 stars. The brightest stars in this cluster are blue, which means they are extremely hot, and probably quite young. Long photographic exposures bear this out; they show filamentary nebulosity (characteristic of youth) surrounding several of the stars.

Nearby in the sky is the Hyades. In contrast to the youthful Pleiades, the Hyades is an old cluster, with much cooler stars; most are spectral Type K. In the Pleiades, on the other hand, the stars are mostly Type B.

How are galactic clusters useful in understanding the life cycle of a star? Suppose the stars in a cluster have just entered the main sequence, and that these stars vary considerably in mass. The main sequence will then have a good distribution of points along it, as shown in Fig. 17.12A. As the cluster ages, however, the uppermost stars will begin to move (Fig. 17.12B) to the right of the main sequence. (As has been pointed out, they evolve the fastest because they are the most massive main-sequence stars.) As the stars continue

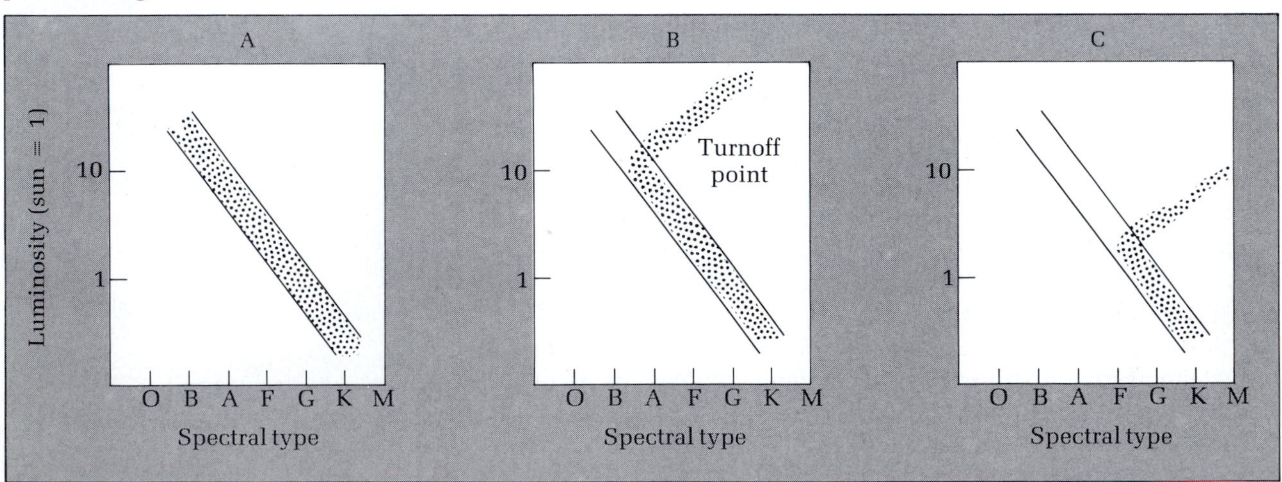

Figure 17.12 Changes with age in the main sequence of a galactic cluster. In *B*, uppermost stars begin to move to the right; *C* shows the turn-off point moving downward.

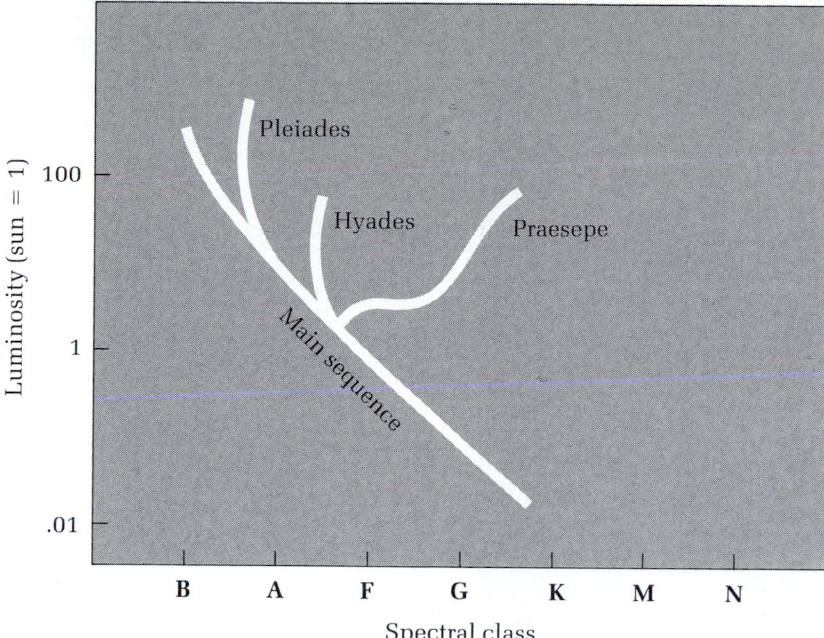

Figure 17.13 Simplified diagram of three galactic clusters. Luminosity and spectral-class values are approximate.

to age, the turnoff point moves downward (Fig. 17.12C). Since we know the masses of the various stars along the main sequence (i.e., at a given surface temperature and luminosity), and, since, from our theory, we also know their main-sequence lifetimes, the turnoff point tells us something very important—*the approximate age of the cluster:* If the turnoff point is near the top of the main sequence, the cluster is young; if near the bottom, the cluster is old. Figure 17.13 shows three open clusters plotted on the same H-R diagram. It is easy to see from this that the Pleiades is relatively young, and Praesepe relatively old.

H-R diagrams of open clusters also tell us the distance to the cluster. Suppose you have a plot of apparent magnitude versus spectral type for a particular cluster. Fit the main sequence of this plot over a standard H-R diagram such as that of Fig. 14.5, where absolute magnitude is plotted. When the two main sequences are aligned the absolute magnitudes of the stars in your plot can be read off. With both absolute and apparent magnitudes known you can then easily determine the distance to the cluster.

Globular clusters (Fig. 17.14) usually contain hundreds of thousands of stars—from about ten thousand to a few million. Approximately 100 have been observed in our galaxy, but it is believed that there are many that cannot be seen because of gas and dust; the total may exceed 1000. Their H-R diagrams (Fig. 17.15) indicate that they are exceedingly old. The turning point, close to the bottom of the main sequence, tells us that they are all approximately 13 billion years old, which means that these clusters were formed about the same time as our galaxy. Their age, in fact, gives us a good indication of the age of our galaxy.

Figure 17.14 Globular cluster in Hercules (M 13), about 35,000 light years from Earth.

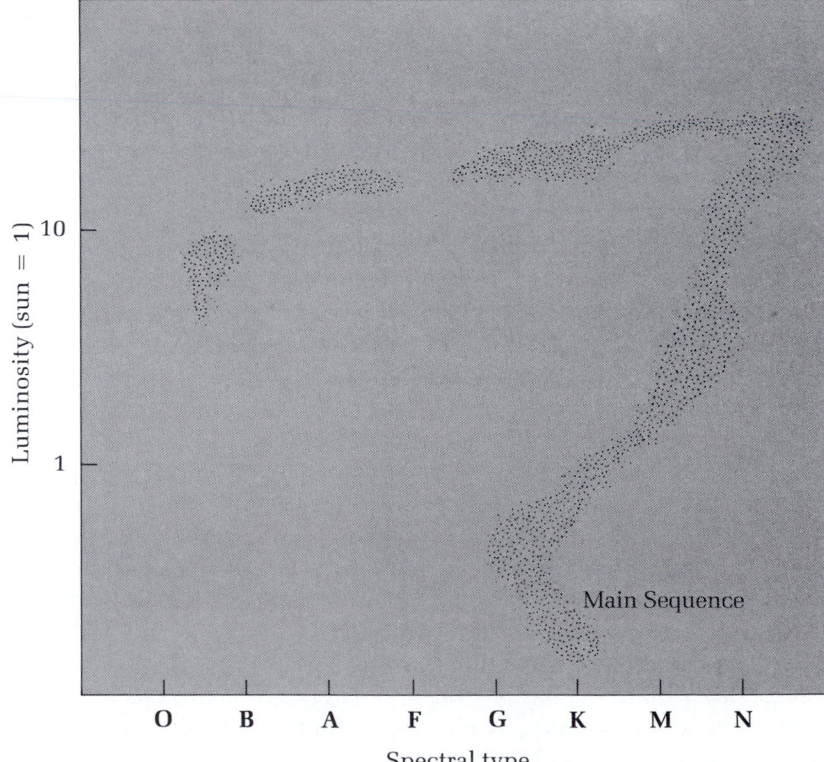

Figure 17.15 H-R diagram of a globular cluster.

Distances to globular clusters, determined from the cepheids in them, tell us that most globular clusters lie outside the galactic disk, either above or below it, forming a sort of halo around our galaxy. From some distance away, they might look like a swarm of bees. Halos of this type may be common to all galaxies; the Andromeda galaxy also has one. Clues to their origin may be found in the history of our galaxy, which, as we have seen, was presumably much larger in its early form than it is now, and consisted mostly of gas. Astronomers believe that the globular clusters may have formed from clouds in the outer regions of this original nebula. Then, as the Galaxy continued to evolve, most stars moved closer to the center, leaving the globular clusters in the outer regions.

Observations of individual stars indicate that the clusters are indeed old. There are no hot blue stars, only small red ones, and there is no gas or debris between the stars; the clusters have been swept clean. Furthermore, the stars are all of the Population II kind.

In general, most of the clusters are elongated rather than spherical. This seems to indicate that they rotate slowly. (If there was a large overall rotation, the cluster would gradually flatten out like our galaxy.) Studies have shown that the stars are, indeed, orbiting the center of mass of the cluster, but that all have different orbits; most, in fact, are quite complex because of internal perturbation, near collisions, and so on.

One of the most interesting features of globular clusters is the recent discovery that several emit strong, erratic X-ray pulses. By 1975 at least five had been shown to do so. These pulses are not periodic, but usually build up rapidly, sometimes increasing in intensity by a factor of 30 or more in a fraction of a second, then dying off more slowly.

What causes them? There has certainly been no shortage of theories. The best, at the present time, strangely enough, is that a black hole (black holes are briefly discussed in Chapter 1, and in greater detail in Chapter 20) is at the center of the globular cluster, and that when gas from stars, or perhaps entire stars, are pulled into it, bursts of X-rays are emitted.

SUMMARY

1. When a star enters the instability strip it becomes unstable and may begin to pulse. This causes its brightness to change.
2. A star that changes in magnitude is called a variable. There are two types of variables: irregular and regular. Cepheids are examples of regular variables; they have periods ranging from about 1 to 50 days.
3. The period-luminosity relation discovered by Henrietta Leavitt gives us a method for determining the distance to a cepheid if its period is known.
4. RR Lyraes are regular variables with periods of less than 1 day. All have absolute magnitude of approximately 0.0.

5. Large red variables have periods longer than 50 days. They change considerably in magnitude (in some cases by as much as 8 magnitudes).
6. A typical nova brightens in a few hours and decays back to its original magnitude in a few months.
7. There are three main classes of novae: classic, recurrent, and dwarf.
8. The thermonuclear runaway theory explains the explosion in the case of classic novae; the accretion theory explains the explosion in recurrent and dwarf novae.
9. A star with a C–O core may slowly push its outer envelope off into space. When this occurs it is called a planetary nebula.
10. Clusters are of two main types: open or galactic, and globular. Galactic clusters contain from 50 to a few hundred stars; globular clusters contain from about ten thousand to a few million.
11. The H-R diagram of a galactic cluster is useful in determining the cluster's age and distance.
12. The H-R diagram of a globular cluster also gives the cluster's age. Most galaxies appear to have halos of globular clusters around them.

REVIEW QUESTIONS

1. Explain one of the mechanisms of pulsation for a star in the instability strip.
2. What is the period-luminosity relation? How can it be used to determine the distance to a cepheid?
3. How can the distance to an RR Lyrae variable be determined?
4. What are the differences among a classic, a recurrent, and a dwarf nova?
5. What are the major differences between the thermonuclear runaway theory and the accretion theory?
6. Explain why some stars become planetary nebulae while others do not.
7. What happens to the remnant star of a planetary nebula after its outer envelope is blown off into space? Describe this star.
8. Why do the expanding envelopes in planetary nebulae appear as rings to us?
9. What is the major difference between a galactic and a globular cluster? Which is generally older?
10. Give some examples of galactic clusters. Discuss them with respect to age, appearance, etc.
11. Explain how the H-R diagrams of galactic clusters are useful.
12. Where do most globular clusters lie in relation to our galaxy?

THOUGHT AND DISCUSSION QUESTIONS AND PROJECTS

1. A Cepheid variable has a period of 20 days. What is its absolute magnitude? If its average apparent magnitude is 5, how far away is it?
2. Which shows a greater range of luminosities, a young cluster or an old cluster? Why? If all members of a cluster form at the same time, why do they not evolve in the same way?
3. If you observed the turnoff point of a globular cluster to be at spectral Type A5 would you suspect something was wrong? Explain. Do the same for an observed turnoff point for a galactic cluster at K0.

4. *Project:* Observe the Pleiades. What color are most of the stars? How many do you see? Observe the Hyades. Compare to the Pleiades.
5. *Project:* Observe M31 in Hercules and the Ring Nebula in Lyra. (See Chapter 3 for their positions.) Describe and discuss what you see.

FURTHER READING

Aller, L. H. "The Planetary Nebula," *Sky and Telescope* (May 1969 through July 1970).

Byrd, D. and Patterson, J. "What Makes a Nova Blow Up?" *Astronomy,* p. 50 (July 1977).

Glasby, J. S. *Variable Stars.* Harvard University Press, Cambridge, Mass. (1960).

Percy, J. "Pulsating Stars," *Scientific American,* p. 66 (June 1975).

part 5

Death of a Star

Stellar death gives rise to some of the strangest animals in the astronomical zoo, perhaps the strangest of all being the black hole, the topic of Chapter 20.

It is important throughout our discussion in the chapters of Part 5 to remember that objects like white dwarfs, pulsars, and black holes are not just objects that "are suddenly there," but are part—the end state—of the evolution of a star. Most important is recognition that evolution is a thread that runs throughout astronomy.

In Chapter 19 of this section we turn to Einstein's theory of relativity. Our object in doing this is to provide background for understanding of black holes (and, later, cosmology). Although it may be tempting to shrug off Einsteinian concepts as far-fetched, it is vital that we recognize that, bizarre as some may at first appear, they are for the most part not speculative; most aspects have now been verified.

Albert Einstein, discoverer of relativity, was a modest, unassuming scientist, perhaps the greatest that has lived. At the 100th anniversary of his birth the press and broadcast media were filled with anecdotes and tributes about Einstein, most of which he would probably have found embarrassing, yet most of which are deserved. His discoveries have touched on almost all areas of physics and astronomy—indeed, on everyday life as well. In this section we will see how true this is.

Aim of the Chapter

To introduce some of the objects and phenomena that occur in a massive star near the end of its life.

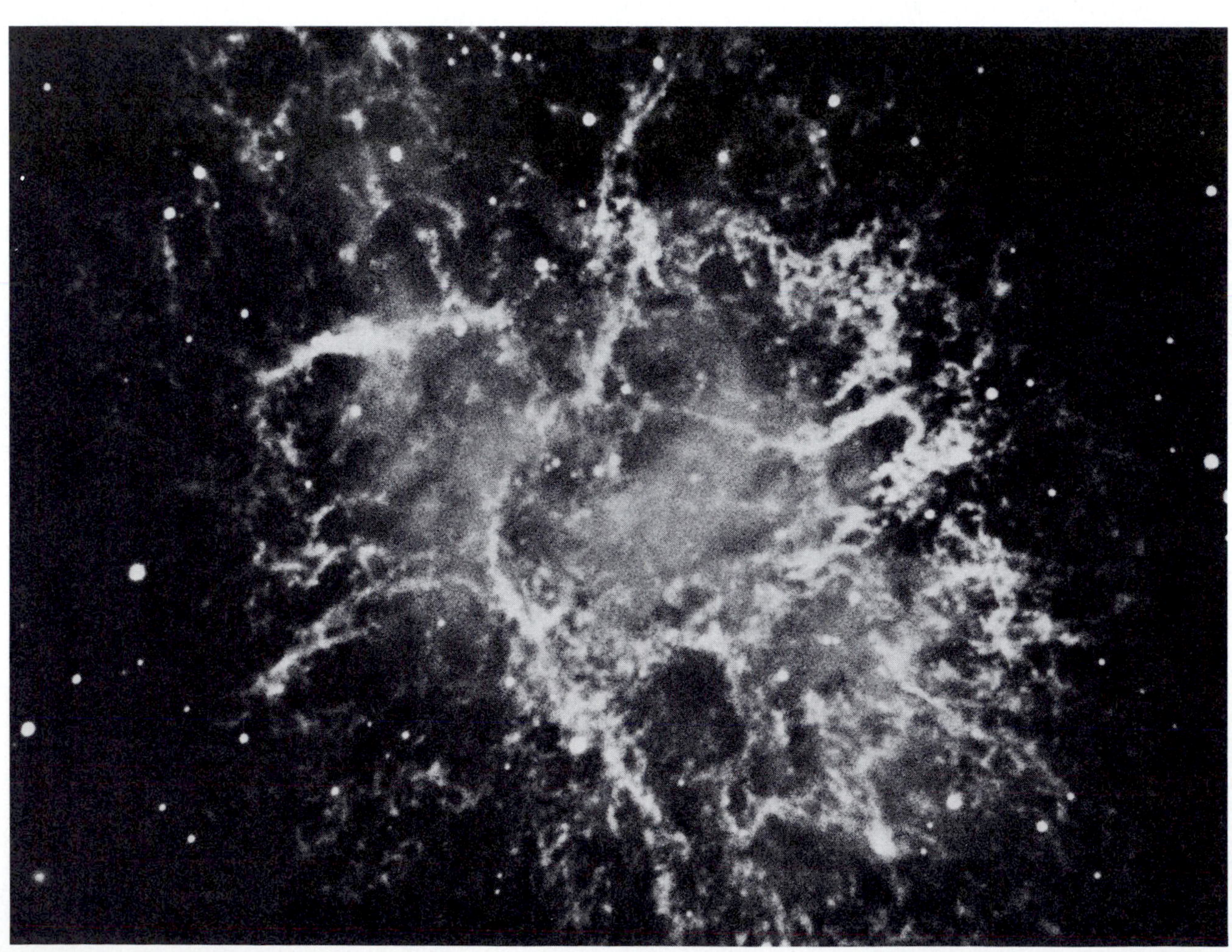

chapter 18

White Dwarfs, Supernovae, Neutron Stars, and Pulsars

In the early nineteenth century, Friedrich Bessel, trying to determine the distance to the bright star Sirius by the parallax method, discovered that Sirius' path had an odd wobble associated with it (Fig. 18.1), rather than the smooth curve he had expected.

In Bessel's view there could be only one reason for the wobble: Sirius had to have a dark companion—possibly a large planet or a dead star; that he could not see it could easily be because it was beyond the limit of his telescope. After determining the distance to Sirius, Bessel turned his attention to another bright star, Procyon, which, to his amazement, also had a wobble—much like that of Sirius; again the cause was probably a dark companion. Thus, two of the brightest stars in the sky seemed to have unseen companions. Twenty-five years later, Alvin Clark, testing a new telescope, noted a tiny speck of light slightly to the side of the main image of Sirius. Sirius did indeed have a tiny companion, and it was not dark; in fact, it was brighter than might be expected, having a magnitude of 8.7.[1] The main reason it had not been seen, was that it was virtually drowned in the light from Sirius.

From its spectrum Walter Adams determined the surface temperature of Sirius' tiny companion to be 8000°K—hardly a dying star. With a temperature almost as high as that of Sirius (10,000°K), and approximately the same distance from us as Sirius, it had to be extremely small to be so difficult to see.

Using the inverse-square law, astronomers easily determined that its surface area had to be about $\frac{1}{2800}$ that of Sirius—only a few times larger than Earth (perhaps as large as the planet Uranus). Since it was part of a binary system, its mass could also be determined; it turned out to be equal to that of our sun!

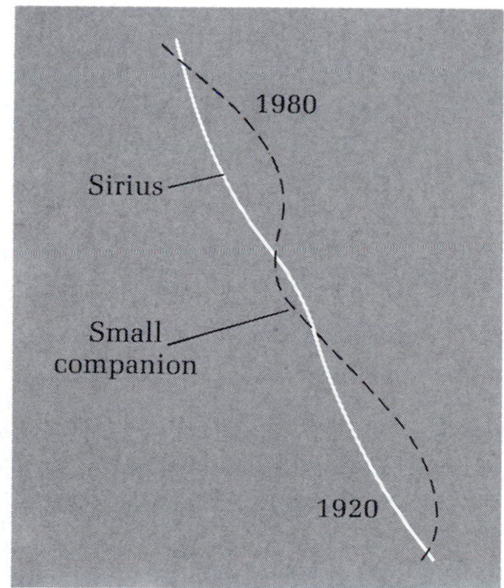

Figure 18.1 Paths of Sirius and its small companion, which revolve around their common center of mass.

[1]We now refer to it as Sirius B, its larger companion being Sirius A.

How could a star become this small and dense? As we saw earlier, a star with a mass about that of the sun will burn hydrogen, and, when its core temperature increases above a certain value, helium reactions will be triggered that leave carbon as "ash." But a star of this mass will never get hot enough to trigger nuclear reactions in the carbon. What happens, then, is that the nuclear furnace simply goes out, and, without an outward pressure to counteract gravity, the star begins to contract. Slowly, over millions of years, it shrinks, and as it shrinks, its density increases. With no energy being generated in the core, its density soon becomes so great that any leftover radiation is trapped, and slowly trickles outward to the surface over billions of years.

WHITE DWARFS

Eventually the star becomes what is called a white dwarf, with a density of about 10^8 gm/cm^3. A handful of its material brought to the earth would be almost impossible to lift or move; a tablespoonful would weigh about 5 metric tons. The mechanism by which such density is produced is as follows. Because of the high temperatures inside the white dwarf, the electrons are stripped from the atoms, leaving ionized atoms. (Of course, if all the electrons are gone, these ionized atoms are bare nuclei.) Thus there is a kind of sea of free electrons intermingled with bare nuclei, and, because of the extreme pressures inside the white dwarf, these nuclei and free electrons will be pushed together.

The volume of an atom, under ordinary conditions, is about 10^{12} times that of its nucleus, which means that an atom is mostly empty space. Consequently when the original components are squeezed together they take up much less room than they did as atoms. However, according to a well-known principle of quantum physics no two electrons with similar velocities can occupy the same small volume. Hence, when a certain volume is reached the electrons can be packed no more tightly. We refer to the electrons at this stage as a degenerate electron gas. It is the degenerate electron gas pressure that keeps the star from collapsing on itself. (Note, though, that the nuclei—because they are much smaller than the "smallest" electron volume—are still a long way apart.)

In the early 1930s, while still a student, Indian astrophysicist Subrahmanyan Chandrasekhar worked out the details of electron degeneracy. He found that, for a star like our sun, it was impossible to compress the electrons beyond the white dwarf stage. But it was later shown that in more massive stars the compressional forces did finally become great enough to squeeze the star further. In these massive stars the protons and the electrons, pushed into one another, form neutrons, a wholly different particle, and the resulting sea of neutrons can be compressed.

The limiting value that he found—now called the Chandrasekhar limit—was 1.4 $M_{\odot}$. Stars larger than this would not become white dwarfs unless they somehow lost some mass before their final collapse.

Properties of White Dwarfs

There are, at the present time, about 200 known white dwarfs. All are about the size of Uranus and have surface temperatures ranging from about 5000° to 50,000°K. Their densities range from about 10^6 gm/cm^3 to 10^8 gm/cm^3. Their mass is always less than 1.4 $M_\odot$, and always greater than about .01 $M_\odot$. (This latter is the approximate minimum allowable mass for stars in general.)

Despite their tremendous density, white dwarfs are still mostly gaseous. They may have a slight crust since their outer layers are not degenerate, and they are likely to have a small atmosphere. Furthermore, most probably spin rapidly and have strong magnetic fields.

With its nuclear furnace inoperative for millions of years, a white dwarf still glows white. This is because the degenerate matter that makes up most of a white dwarf is generally incapable of capturing photons emitted from near its core. This means that the radiation from deep within the star that escapes to the surface loses little energy and, because of this, the surface will be much hotter than it would be in the case of an ordinary star.

White dwarfs remain hot over a long period. It takes about three billion years for one to cool from 10,000°K to 6,000°K, and perhaps another five billion for it to cool to 3,000°K. At this point it is a red dwarf. Finally, it becomes a black dwarf and fades gradually to what we might call cosmic ash.

SUPERNOVAE

In August, 1885, astronomers noticed a sudden brightening in a fuzzy patch in the constellation Andromeda. Some thought this patch was a gas cloud within our galaxy (the Milky Way); others thought it might be another island universe of stars, distinct from ours. Within days after its discovery the spot was visible almost to the naked eye (about seventh magnitude). It remained visible in the telescope for approximately seven months and caused considerable controversy, but eventually interest faded. In the mid-1920s, Edwin Hubble showed that this patch was, indeed, an island universe of stars—a galaxy, known now as the Andromeda galaxy (Fig. 1.8)—at quite a distance from us. Astronomers suddenly realized that the flareup could not have been an ordinary nova (Chapter 17), for it was far too bright.

Zwicky and Baade found that such flareups were common phenomena. Over a few years they located many objects of this type in other galaxies and named them supernovae.

A check of old records showed that a number of bright, apparently new stars that had been seen in our galaxy had probably been supernovae. (Tycho mentions one, and Kepler also, and these have now been established to have been supernovae.) Astronomers had also noticed what looked like the remnants of an exploded star in Taurus.

Figure 18.2 The Crab Nebula, photographed through special filters; it is shown in Fig. 18.4 (color) in a photograph made in the red light of hydrogen.

The Crab Nebula

The remnant in Taurus is now known as the Crab Nebula (Fig. 18.2). Edwin Hubble found that a comparison of old and recent photographs of it indicated that it had expanded at a rate of about 1000 km/sec. Measuring the Crab's overall size, he then determined that the explosion had occurred about 900 years earlier. A search of the most extensive astronomical records from that time (those kept by Chinese astronomers) showed that a bright new star had appeared in Taurus on July 4, 1054—a few years before William the Conqueror invaded England; within days it was the brightest object in the sky (aside from the sun and moon)—so bright, in fact, that it could be seen during the day. It was visible for 653 days. Recent records indicate that the event may have been seen by both the Japanese and the Arabs, and there is some indication that the Pueblo Indians saw the event; drawings have been found in caves in Arizona (Fig. 18.3) depicting a particularly bright object near a crescent of the moon. Calculations show that the Crab supernova may have appeared near the moon at this time.

The remnant, named the "Crab Nebula" by Lord Rosse, is frequently called simply "the Crab," and, as can easily be seen from Fig. 18.2, the name is appropriate; there are many crab-like filaments near its perimeter.

The Crab was also seen by Charles Messier in the 1700s. Although disappointed, since his main interest was in comets, Messier constructed a table of these fuzzy "nuisances" so they could be distinguished from "important objects," like comets; the Crab Nebula was his first entry.

Though rather dull-looking through a telescope, when photographed through special filters (Fig. 18.2; Fig. 18.4, color) the Crab is a truly spectacular object.

Tests of the optical properties of the Crab have shown that there are highly energetic electrons associated with it; these electrons are being accelerated and therefore emit radiation. (Because the mechanism for generating this radiation is similar to that of large accelerators here on earth, called synchrotrons, we usually refer to it as "synchrotron radiation.") In the 1950s I. S. Shklovskii determined that the Crab Nebula was indeed generating energy and was far too "bright," after 900 years, to be just a slowly cooling stellar remnant.

Mechanisms

If electrons were being accelerated in the Crab, there had to be strong magnetic fields somewhere in it. (The synchrotron mechanism requires magnetic fields.) But where were they? What was causing them? In our earlier discussion we saw that a mass of 1.4 $M_{\odot}$ was the upper limit for white dwarfs. We also know that most stars lose considerable mass after they leave the main sequence. Mass is lost, for example, during the planetary nebula stage, and there are several other ways it can be lost. This means, then, that white dwarfs are produced when stars somewhat larger than 1.4 $M_{\odot}$ collapse. Studies have shown that 4 $M_{\odot}$ is a realistic upper limit when mass loss is taken into consideration; in short, most stars of mass 4 $M_\odot$ to ~ .01 $M_\odot$ eventually become white dwarfs.

Events in stars with masses between 4 $M_\odot$ and 8 $M_\odot$ are quite similar to each other. As shown in Fig. 18.5, there are hydrogen- and helium-burning shells, and, inside the helium-burning shell, a core of C–O. When this core reaches 600,000,000°K the carbon in it is triggered, but it is degenerate (p. 320) and so does not expand when heated. Thus, the additional heat stays in the star, causing it to burn even more furiously. Soon the burning is out of control and an explosion occurs. This explosion is similar to the helium flash, but with one difference: The helium flash was contained within the star; in this case, there is a visible manifestation—a supernova. Few, if any, events in the universe are more spectacular. The star is literally ripped apart, its remnants thrown into space at tremendous velocities.

This means, then, that stars in the range 4–8 $M_\odot$ give us supernovae. However, a supernova explosion can occur in more massive stars as well, though the mechanism is different. The carbon in the core of a star with a mass greater than about 8 $M_\odot$ is not degenerate when it is triggered. Although the pressure should be greater the more massive the star, in these stars it is not the case, and consequently they burn without exploding.

First the carbon is ignited, then the oxygen, and if the star is

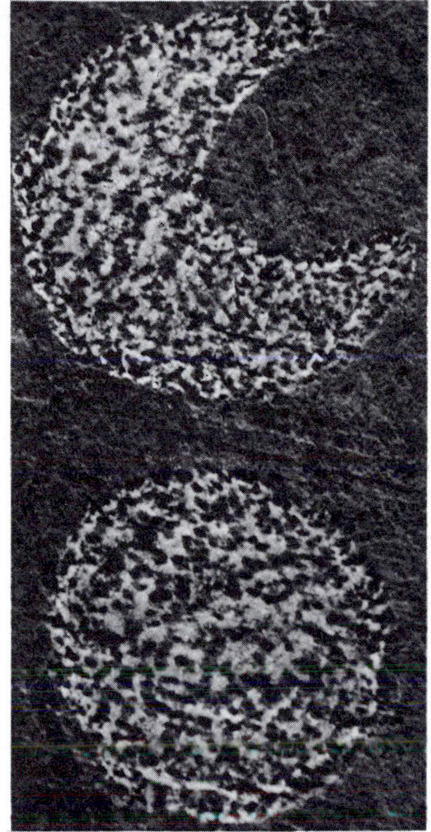

Figure 18.3 (above) Pueblo Indian cave drawings that may depict the Crab supernova of 1054 A.D.

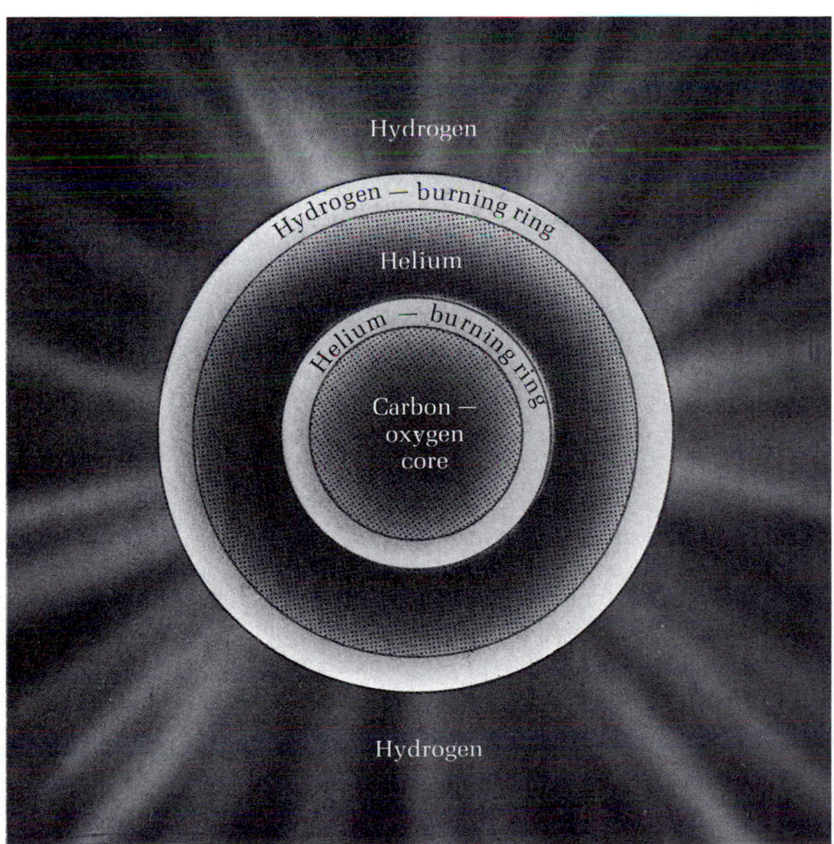

Figure 18.5 The structure of the core of a star in the 4 $M_\odot$ to 8 $M_\odot$ range just before it becomes a supernova.

massive enough this sequence continues through neon and silicon until, finally, the core is iron. In each case when the fusion reactions are triggered energy is given out, except in the case of iron, which absorbs energy.

With no energy being released from the core the star becomes unstable and begins to shrink, and as it shrinks it heats. When the temperature is great enough, the iron nuclei break apart, and protons and neutrons pour out. The collapse continues until, finally, the pressure is so high that the protons and electrons smash into one another with incredible force; this creates new particles: neutrinos. Then immense clouds of neutrinos start to emerge from the star. Until recently it was believed that it was this outrush of neutrinos that was responsible for the blowing off of the outer 80–90 percent of the star. Calculations now show that this may not be true. The neutrinos are delayed in the core by as much as a second (the whole process takes only a few seconds) and, by the time they emerge, the outer layers may already be flying off into space.

If the neutrinos are not responsible for the explosion, "hydrodynamic bounce" may be. In this interpretation the rapid collapse of the core is arrested when the neutrons become degenerate; they can be compressed no further. Suddenly the collapse is reversed, and, like a rebounding ball, the star recoils with an incredible energy. The outer layers are still collapsing, but the rebounding material strikes them with such force that they are immediately blown off into space.

As described above, the heaviest element generated in stars is iron. Calculations show that the elements heavier than iron may be generated in the supernova explosion itself. If true this means that the explosion performs two important functions in the universe: it distributes the elements up to iron (as listed, according to atomic number, in the periodic table) out into space, and it creates elements heavier than iron and distributes them into space. Thus, everything in our solar system, which includes our bodies, contains atoms that were at one time generated and dispersed in a supernova explosion.

Supernovae may also be responsible for the well-known phenomenon of cosmic rays, the showers of particles and radiation that rain down on us when high-speed particles from space hit our atmosphere.

Types

Supernovae are divided into two classes according to how their brightness rises and falls: Type I reach their peak brightness in about 20 days, then fade slowly over a year or so; Type II reach their peak in about 30 days, then fade away in about 150 days. Spectral analysis shows that Type I supernovae have much more helium in their envelopes than do Type II, but no definite relationship between the two types has been found. There is some indication that Type II explosions occur in more massive stars, which would mean they are related to stars with iron cores.

Frequency

Studies of galaxies other than our own show that supernovae occur typically about every 50 years. This means that we should see a supernova in our galaxy every 50 years or so. Yet, in the last 1000 years we have seen only five supernovae. (Of course some may have occurred on the other side of our galaxy, where we could not see them, but five remains a small number for such a long period.) Were one like the Crab to occur tomorrow it would give astronomy a unique opportunity—especially in view of the array of instruments and techniques that astronomers now have—to study it extensively from beginning to end.

Remnants and Influences

There are many remnants of supernovae that occurred prior to the last thousand years. (Slightly over 100, including radio remnants, have been found.) One in the southern hemisphere in the constellation Vela, called the Gum Nebula, is believed to be about 15,000 years old. Another remnant composes the long tenuous coils of luminous gas that form the Cygnus loop in the constellation Cygnus (Fig. 18.6).

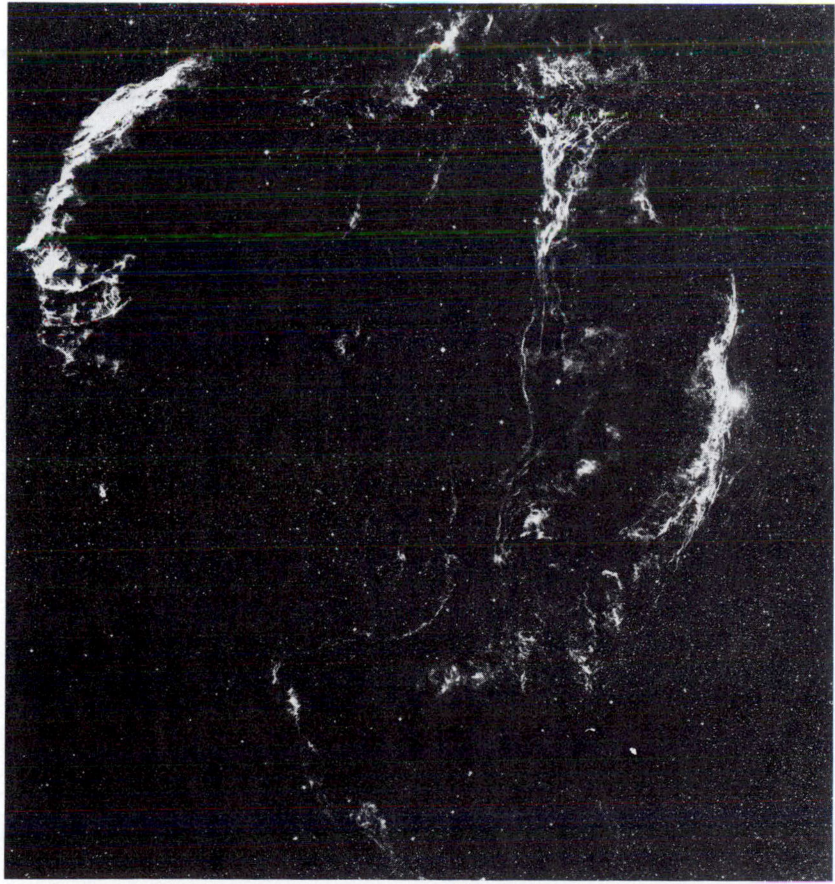

Figure 18.6 A supernova remnant—a loop of nebula in Cygnus.

A tremendous amount of radiation is given off in a supernova explosion; were one to occur quite close to us—say about five light years away, it would be deadly to life on Earth. It is noteworthy that the envelope around the Crab Nebula has now expanded outwards by about seven light years. There is no star nearby at the present time that could explode as a supernova, though there may have been pre-supernova stars relatively close in the past that would have been dangerous to us.

Several astronomers have speculated that various animal and plant species on earth may have been significantly changed by the radiation emitted in supernova explosions. This may or may not have occurred, but had there been a supernova nearby, there is little doubt that numerous mutations would have occurred in the species that were here—enough possibly to give us new species.

Another interesting suggestion related to supernovae has recently been made by David Schramm and Robert Clayton, who believe there is evidence in meteorites to support the idea that our solar system was formed as a result of a nearby supernova explosion.

NEUTRON STARS

We have followed the supernova to the point where the star was blown apart. While there seems no clear-cut mechanism that would leave a remnant core—particularly in the case of the C–O core explosion—there is now considerable evidence that a core is left.

Baade and Zwicky were the first to consider this possibility, in 1933 advancing the view "that supernovae represent the transition from ordinary stars into *neutron* stars, which in their final stages consist of extremely closely packed neutrons."

That a neutron star could actually exist was an intriguing possibility. It would be like a gigantic nucleus, and would no doubt have some particularly strange properties. In 1939 J. R. Oppenheimer and George Volkoff showed that a neutron star would, indeed, be stable and could exist; more important, however was the disclosure that there was another limiting mass: Over a certain critical value even the neutrons could not keep the star from compressing further. (This is discussed further in Chapter 20.)

A neutron star would be extremely dense—much denser than a white dwarf. Where a tablespoon of matter from a white dwarf would weigh five tons, the same amount from a neutron star would weigh a million tons. A neutron star would be about 20 km across, (Fig. 18.7), but like the earth, it would have a crust, and perhaps, some small "mountains" (a few centimeters high). Small as they are, though, the tops of some of these mountains would probably poke through the atmosphere. The outer crust would be about 100 m thick and extremely rigid, perhaps 10^{16} times as rigid as steel; it would, therefore, take a considerable force to disrupt or crack it. (But, as we will see later, this might happen.)

Figure 18.7 A neutron star in relation to Manhattan in New York City. Barely a point compared to Earth, it would be many times heavier than our sun.

A neutron star would also be spinning and would be likely to have a strong magnetic field. In fact, even had the original star had a low spin rate and a weak magnetic field, their values in the neutron star would be extremely high. Both angular momentum (spin) and magnetic flux (field intensity per unit area) are conserved in nature. A skater for example, who is spinning with arms extended, then pulls them in, spins much faster. The same thing is true of a star as it decreases in size. As it gets smaller it spins faster (and its magnetic field gets stronger). When first formed, neutron stars may have spin rates of 100 rev/sec and magnetic fields up to 10^{12} gauss.

PULSARS

History shows that some very important discoveries were made accidentally, while the researcher was looking for something else. Such a case, discussed below, was the discovery of pulsars.

According to calculations it seemed that neutron stars should exist. In fact, many of their properties had been predicted, but, as of the mid-1960s, none had been discovered. A few years earlier, however, what we now call "quasars" had been discovered (Chapter 23). British astronomer Antony Hewish had just developed a new technique for locating quasars. As we know, stars twinkle (because of our atmosphere) and most of the planets do not (primarily because they are, strictly speaking, not point sources). Hewish noticed that a similar phenomenon was associated with radio sources; some "twinkled" (or, rather, "scintillated") and some did not. Hewish believed he could distinguish extended (rather than "point") sources, like radio galaxies (Chapter 22), from quasars by studying their scintillations: Quasars were strong scintillators. To investigate the possibilities, Hewish built a radio telescope that was sensitive to signals of short duration. (Most radio telescopes at that time were not.) In 1967, Jocelyn Bell, while analyzing the data from the telescope, noticed that the taped records included signals that were neither scintillations nor man-made interference. Bell noticed that the pulses were regular, at an unbelievably accurate period of 1.33 sec. Furthermore, the object generating the impulses kept sidereal time, and, therefore, must be associated with an astronomical object.

Someone suggested, perhaps jokingly, that it might be a message from extraterrestrial beings—LGMs (little green men), and others wondered whether this might not be the case. Hewish showed, however, that the pulses could not be coming from a planet revolving around some distant sun since they showed no Doppler shift other than that due to the earth's own motion.

Other such sources were soon discovered, two by Bell.

More and more pulsars (as they came to be known) were discovered and catalogued. The first one, discovered by Jocelyn Bell, is now called CP 1919. ("CP" stands for Cambridge pulsar, and $19^h\ 19^m$ is its right ascension).

Character of the Pulses

Each pulsar has characteristics that easily distinguish it from all others. One of these characteristics is the period; CP 1919, for example, after correction for the Doppler shift due to the earth's rotation and revolution, has a period of 1.33730113 sec, with an accuracy of 1 part in 100 million, comparable to that of atomic clocks.

Pulsars can also be distinguished by the overall "averaged" shape of their pulses. As Fig. 18.8 shows, pulses from the two pulsars CP 0834 and CP 0950 can easily be seen to be quite different. The way in which various frequencies arrive also differs from pulsar to pulsar. This is, in fact, how we determine their distances, for the lower frequencies arrive at the earth slightly later than the higher.

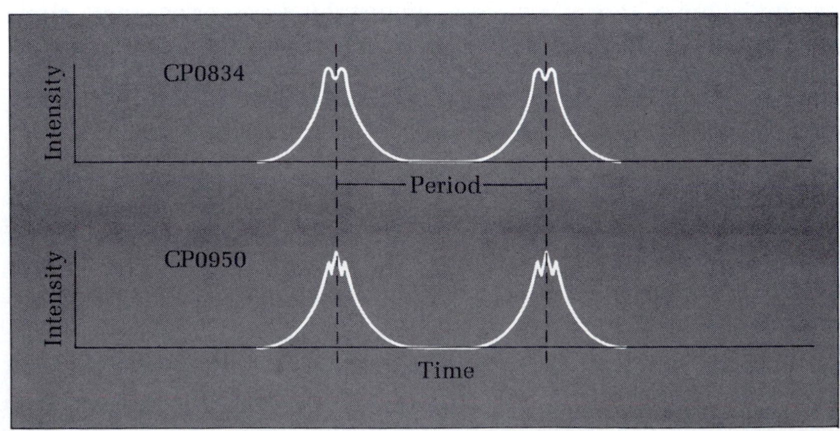

Figure 18.8 Averaged pulses of the pulsars CP 0834 and CP 0950.

This is because, though radio waves travel at the speed of light in empty space—in a perfect vacuum—space is not empty; there is gas (the interstellar medium) between the stars, and the free electrons in this gas slow low-frequency waves more than high frequency, so that measurement of the delay in arrival times (of different frequencies) gives the pulsar's distance. Analysis of known pulsars signals show that they are relatively nearby—in our region of the Milky Way galaxy. Most, in fact, appear to be in the plane of the galactic equator, which is, of course, what would be expected if they are in our galaxy.

Pulsar Models

Because of their short periods, pulsars had to be small—their size limited by aspects of the theory of relativity (Chapter 19)—and the only known stars this small were white dwarfs (or neutron stars—if they existed). Also, since pulsating (alternately expanding and contracting) stars with longer periods were well known, it was assumed the pulsars were pulsating. With a model of this type in mind, astronomers calculated the possible pulsation periods for both white dwarfs and neutron stars, but found that neutron stars pulsed a little too rapidly, and white dwarfs a little too slowly; neither type of star fit the data.

In 1968 Ostriker put forward the model of a spinning star with an associated directed beam somewhat like a lighthouse. He took the rotating star to be a white dwarf; although white dwarfs do not have the stability under high rotational speeds that neutrons stars do, Ostriker felt there was not then enough evidence that neutron stars existed. However, Tommy Gold suggested that a neutron star model was feasible.

As long as pulsars with extremely short periods (very rapid spin) were not found, the white dwarf model seemed to be the most reasonable. Within months, however, a pulsar was discovered within the Crab Nebula that had a period of .03 sec; pulsars had to be neutron stars!

The model, as it stood then, was a tiny spinning neutron star, and further examination soon showed that these neutron stars had to have strong magnetic fields. The pulsar mechanism must somehow be connected with these magnetic fields; perhaps it was a beam (or possibly two beams) directed out along magnetic field lines.

The Crab Pulsar

Pulsars had been "seen" so far only with radio telescopes; all initial attempts to identify them optically failed. Photographs of the regions of CP 0950 and CP 1133 showed no objects whatsoever. In January of 1969, W. J. Cocke, M. J. Disney, and D. J. Taylor made the first sighting; it was of the Crab pulsar, and the object was indeed flashing in the visible region of the spectrum—off, on, off—30 times a second! It had not been noticed before because all previous photographs of the Crab Nebula had been long exposures, so that the pulsar had looked like an ordinary star.

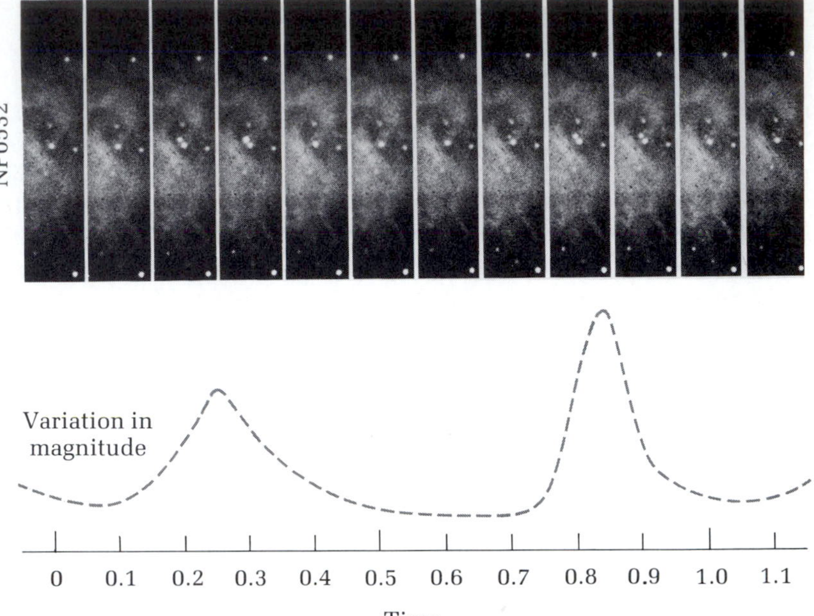

Figure 18.9 At top, sequence of photographs shows the Crab pulsar in its "on" and "off" modes; at bottom is a plot of the pulsar's light intensity over the same period.

Actually, Cocke, Disney, and Taylor had not *seen* the flashing, but had obtained a photomultiplier (p. 91) recording of it. Joseph Wampler and J. Miller were the first to literally see the pulsar flashing on and off (Fig. 18.9). To do this they used a rotating shutter, employing the familiar stroboscope effect, such as may be seen when a rotating spoked wheel, illuminated only at intervals, appears to be moving much slower than it actually is. Setting the shutter at the same speed as the pulses (30/sec) would have left the pulsations too fast to see. (The eye would see 30 pulses per second as a continuous light). With the shutter set at a slightly lower frequency, the pulsar would appear to be slowed.

The light curve of the Crab pulsar showed two peaks per period, with one peak about twice the intensity of the other. In addition, the main peak was narrow and pointed, whereas the smaller was broad. The narrowness of the main peak set an upper limit on the size of the pulsar: It could be no more than about 20 km across.

The Crab pulsar proved to be an astronomical Rosetta stone, study of it leading to a much better understanding of pulsars and our present model of a pulsar.

Formation

Earlier we saw that when the outer layers of the star are blown off into space the core implodes. As the core collapses, spin and density increase rapidly, as does the associated magnetic field; soon it has a density of 10^{14} gm/cm^3 and a magnetic field of 10^{12} gauss. Most of the protons and electrons have disappeared, crushed together to form neutrons. (The positive charge on the proton cancels the negative charge on the electron, creating a neutral particle.) The star is now composed mostly of neutrons.

As with the earth, the spin and magnetic field axes of most pulsars are not aligned; in fact, they may be almost perpendicular to each other (Fig. 18.10). This means that the field lines will be dragged around with the star as it rotates, and according to Ostriker and Gunn, the particles emitted by the star ride out along these field lines, as a surfer rides the crest of a wave. It's easy to see, then, how a stream of particles, accelerating outward, create a beam of radiation. In fact there will be two radiation beams (Fig. 18.11), one at each magnetic pole,[2] and each time one of these beams is pointed towards the earth we see a pulse of light.

That a pulsar can be observed only if it is oriented so that the earth lies near the plane of radiation of its beam explains why astronomers have detected relatively few pulsars.

[2]Beams will be at the north and south magnetic poles because charged particles move along magnetic field lines much more readily than across them.

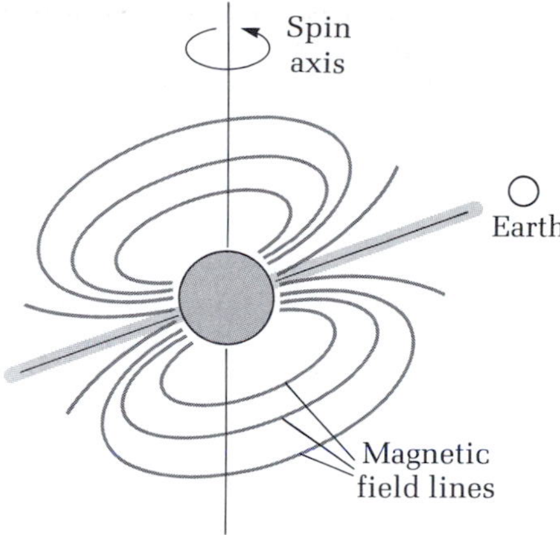

Figure 18.10 Diagram of a spinning pulsar, showing how the axes of the magnetic field and the spin are not aligned.

Figure 18.11 Speculative conception of a pulsar.

The Mystery of SS433

The latest in the list of puzzling objects that have confronted astronomers over the years is called SS433. Because most stars exhibit only absorption spectral lines, stellar objects that show emission lines receive particular attention. One item in a list of such objects compiled by C. Bruce Stephenson and Nicholas Sanduleuk was SS433. Little attention was paid to SS433 until 1978, when David Clark and Paul Murdin showed that there was an apparent connection between SS433 and an X-ray source (4U1908 + 05) that had been discovered by the Uhuru satellite. Bruce Margon and several associates, examining the spectrum of SS433, found three sets of emission lines, each of which acted differently. One set showed almost no motion, but the other two showed motion, strangely, in opposite directions. At any given time one set had a red shift, indicating a recessional velocity, while the other set had a blue shift, indicating an approaching velocity. And the magnitude of the velocities was incredible, the combined velocity being nearly 50,000 km/sec. Continued observation showed that the shifts had a period of 164 days.

What could cause this? Most astronomers would think first of a Doppler shift associated with a binary system, where one of the two stars is approaching and one receding from us. But when the details of this model were worked out it was found that the two objects would have to be exceedingly dense (probably black holes) and astonishingly massive—a billion times the mass of our sun.

Since this was obviously impossible, several other models were soon put forward. One, from Andrew Fabian and Martin Rees, suggested two beams or jets of matter associated with the source. When one of the jets was directed towards us the other would be pointed in the opposite direction. (This would account for the simultaneous red and blue shifts.) But other things were not yet explained. A modification by Bruce Margon and George Abell helped explain the 164-day period. They suggested it was associated with a precession of the object; in other words, like a spinning top, its spin axis traced out a large circle in the sky. This precession moved the jet towards and away from us in a period of 164 days.

But what caused the jet? This has been explained by assuming the "object" is actually a binary system consisting of a large ordinary star and a dense companion—a neutron star or black hole. Matter is assumed to be flowing from the ordinary star into a disk around the dense object. According to J. Katz this material may be overwhelming the object; it may, in effect, be overloading it so much that it is blowing its top, and the jets we see may be the excess that is being blown off.

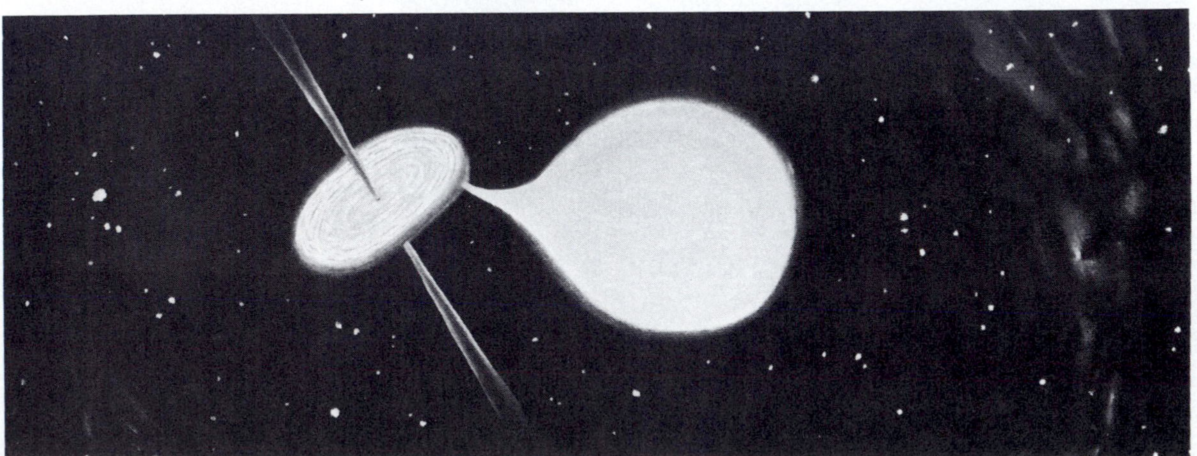

An artist's conception of SS433.

When the Crab pulsar was first formed it probably rotated about 100 times a second. (This is typical of most pulsars.)[3] Gradually, as it lost energy, it slowed to its present 30 times a second. This slowing has been detected; the period of the Crab pulsar has been shown to be increasing at the rate of about a millionth of a second a month.

Although pulsars are slowing down, they are still remarkably precise timekeepers. In 1969, however, there was a sudden small increase in the Vela pulsar's spin rate. At first thought to be a once-in-a-lifetime event, it was repeated in 1971. Later, the same phenomenon—now called a "glitch"—was observed in the Crab pulsar (after which the pulsar continued to slow down as it did before).

The cause, or at least part of it, lies in the rapid spin of the star and in its steel-like crust of packed neutrons. It is spinning so fast it must be quite oblate. On the other hand it is gradually losing energy, and therefore spinning slower and slower. With the force (spin) that produced the oblateness gradually diminishing, the crust must occasionally "relax," much as the earth's crust relaxes built-up strain. When this happens on Earth the result is an earthquake; in a neutron star, then, there should be an equivalent "starquake." Astronomers are now convinced that it is indeed a starquake that is causing the sudden small increases in spin rate of the Crab pulsar. The mechanism in the Vela pulsar, however, is still not known.

[3]In late 1982 a pulsar was discovered that pulses at the incredibly fast rate of 642 times-per-second.

SUMMARY

1. The first white dwarf discovered was the companion of Sirius. It is about the size of Uranus but has a mass equal to that of our sun.
2. A white dwarf is exceedingly dense because the nuclei and electrons of its atoms have been compressed into an extremely small volume (compared to the original volume of the atom). The material is called a degenerate electron gas at this stage.
3. There are about 200 known white dwarfs. They are about the size of Uranus and have temperatures of about 5000°–50,000°K. Their densities range from about 10^6 gm/cm^3 to 10^8 gm/cm^3 and they have masses up to 1.4 $M_\odot$.
4. The supernova phenomenon was discovered as a result of a flareup in the Andromeda galaxy in 1885.
5. The Crab Nebula was the supernova of 1054. Its rate of expansion is about 1000 km/sec. I. S. Shklovskii discovered in the 1950s that it contained a "powerhouse" that accelerated electrons outward.
6. There are two mechanisms for supernova explosions. The first occurs in stars between 4 $M_\odot$ and 8 $M_\odot$. They have C–O cores just prior to the explosion. Explosions can also occur in much more massive stars (ones that have an iron core).

7. Supernova explosions occur in a given galaxy about once every 50 years. In our galaxy they appear to be overdue. Many supernova remnants have been found (e.g., Loop Nebula in Cygnus).
8. Neutron stars were first predicted by Zwicky and Baade in 1933. Oppenheimer and Volkoff showed that collapse could lead to a neutron star.
9. A typical neutron star would be about 20 km across and have a density of 10^{14} gm/cm^3. It would probably be spinning and have a strong magnetic field (10^{12} gauss).
10. Pulsars, discovered in the early sixties by Hewish and Bell, were later shown to be spinning neutron stars.
11. One of the most spectacular pulsars is the Crab pulsar. It has a period of $\frac{1}{30}$ sec, making it the fastest (and therefore youngest) known pulsar. Its optical counterpart has recently been discovered.
12. Although pulsars are precise timekeepers (comparable to atomic clocks) they do gradually lose time (a millionth of a second a month or less). Also, sudden changes in period have been detected in several pulsars. Astronomers believe some of them are caused by "starquakes."

REVIEW QUESTIONS

1. What did Bessel find odd about the path of Sirius? What did this mean?
2. Describe the companion of Sirius. Why was it unusual?
3. What is a degenerate electron gas? Is it actually a gas, or is it a solid? How dense is it?
4. What is the approximate size of a white dwarf?
5. Describe the physical features of a typical white dwarf.
6. What eventually happens to a white dwarf?
7. What is the difference between a supernova and a large nova?
8. How did astronomers find out that the Crab Nebula supernova took place in 1054?
9. Explain the mechanisms for a supernova.
10. Why doesn't the C–O core expand when it reaches 600,000,000°K and nuclear reactions are triggered in it ($4\ M_\odot - 8\ M_\odot$)?
11. Approximately how much of a star is blown off in a supernova explosion?
12. Describe the two types of supernovae. Are they related (as far as we know) to the two known supernova mechanisms?
13. Approximately how many supernovae remnants have astronomers found? Describe a typical one.
14. Discuss the events that led to the prediction of neutron stars.
15. Describe a typical neutron star.
16. Who discovered pulsars? How did the discovery come about?
17. What happened to convince astronomers that pulsars were neutron stars rather than white dwarfs?
18. Were pulsars seen immediately after they were discovered with radio telescopes?
19. Describe the light curve of the Crab Pulsar. What can we learn from it?
20. Describe in detail the presently accepted model of a pulsar. Explain why we see pulses.
21. What is a glitch and what causes it?

THOUGHT AND DISCUSSION QUESTIONS

1. Suppose one of the stars in α Centaurus became a supernova. How bright do you think it would get as seen from Earth? (Make a rough estimate based on what you know about the Crab Nebula Supernova.) Would the expanding envelope be likely to reach us? If it did, how would it affect us?
2. If a star suddenly brightened to magnitude 0.0 (the approximate transition point between supernova and nova magnitudes), how could you determine if it was a nova or supernova?
3. What will the pulsar in the Crab Nebula and the nebula surrounding it be like in 1000 years; in 10,000 years?
4. Explain why a pulsar spins much faster than a white dwarf when first formed. Explain why the magnetic field of a pulsar is also much greater.

FURTHER READING

Bell-Burnell, S. J. "Little Green Men, White Dwarfs, or What?" *Sky and Telescope*, p. 28 (March 1978).

Hewish, A. "Pulsars," *Scientific American*, p. 25 (October 1968).

Hopkins, J. "Supernova!" *Astronomy*, p. 6 (April 1977).

Ostriker, J. "The Nature of Pulsars," *Scientific American*, p. 48 (January 1971).

Rudermann, M. "Solid Stars," *Scientific American*, p. 24 (February 1971).

Wheeler, J. C. "After the Supernova, What?" *American Scientist 61*, p. 42 (1973).

Aim of the Chapter

To describe relativity theory as a background to understanding black holes (Chapter 20) and cosmology (Chapters 24 and 26).

chapter 19

Special and General Relativity

We have seen that ordinary stars collapse near the end of their life to form other types of stars. A star about the mass of the sun gives us a white dwarf; stars with masses between 4 $M_{\odot}$ and 8 $M_{\odot}$ give us a much smaller and denser object called a neutron star, and if we go to even more massive stars we get another type of object with even stranger properties. To understand how these last objects form and why they have such bizarre properties we must first examine the theory of relativity.

The concept of relativity actually embodies two theories—the special and general. The theory of special relativity, introduced in 1905 by Albert Einstein, applies only to straight-line, uniform motion—in other words, to unaccelerated motion. The theory of general relativity is an extension of the special theory to all types of motion including acceleration; it was introduced in 1916.

The major event of those that led to the theory of special relativity was the discovery that light is a wave. This meant that light had to have a propagating medium, and, as we saw in Chapter 4, scientists invented one; they called it the "aether." This aether presumably permeated the entire universe, and had some rather amazing properties: It was unaffected by gravity, gave no resistance to matter, was exceedingly low in density, and was transparent and rigid. In addition, it gave rise to a fixed frame of reference for the universe, which can be explained by an analogy in which the universe is immersed in a bowl of jelly, the jelly representing the aether. All the stars, galaxies, and so on will move through this jelly with certain velocities, and of course we can measure their velocities with respect to the jelly. Note that the jelly is playing a particularly important role here: If there were no jelly, to what would our measurement of the velocities of the stars be relative? Obviously, we could compare them only to one another; in other words we could say a star was moving at such and such a speed relative to another star—but, of course, the other star might also be moving.

The inventors of the idea of the aether were not sure whether the sun was moving through it. It might be sitting still in it, but, since the earth was moving around the sun, there was no question that it (the earth) was moving through the aether.

THE MICHELSON-MORLEY EXPERIMENT

This was the situation about 1880, when physicists E. W. Michelson and A. A. Morley decided to measure the earth's velocity through the aether, and, in the process, to determine if the sun was moving with respect to it. To illustrate the technique they decided to use, we will assume, for simplicity, that the sun is at rest in the aether. The earth is then moving through it at 29 km/sec (its orbital velocity). If a light beam is then projected in the direction of the earth's velocity it should appear (to us on Earth) to be going 299,792 km/sec (the speed of light) minus 29 km/sec (the speed of the earth), or 299,763 km/sec. (Remember that the aether is transmitting the beam.) If, on the other hand, the beam of light is projected in the opposite direction, it should appear to be traveling 299,792 + 29, or 299,821 km/sec.

To clarify this, consider another analogy: Assume you wish to row up a river against the current, which is 2 m/sec. If you normally row in calm water at 5 m/sec, then, in the river, you will row upstream at $5 - 2 = 3$ m/sec (with respect to the shore). This means that the speed of the water relative to you will be 3 meters per second. But if you row with the current you will travel at $5 + 2 = 7$ m/sec and the speed of the water relative to you will be 7 m/sec. From the earth we see the aether as a river with a current (opposite to the direction of Earth's movement) of 29 km/sec. This means that the light beam (the rowboat in our analogy) will be retarded if it is projected "upstream" against this current; similarly it will be assisted if projected in the opposite direction. Michelson and Morley set out to measure the difference between the two cases. Their apparatus was easily capable of detecting it, and they were stunned when the experiment showed that there was no difference; regardless of which way the beam was projected it still traveled at 299,792 km/sec. This would seem to mean that, were the light beam a car traveling at, say 90 km/hr, then regardless of how fast we went in trying to catch it, it would still be traveling 90 km/hr faster than us. We know, of course, that this does not happen at the relatively low velocities of everyday experience. As another example, assume that we set up a large searchlight beside and pointed in the same direction as a rocket ship that is capable of unlimited velocities. The rocket ship blasts off at 299,000 km/sec, then, a few seconds later, we turn on the searchlight. Almost instantaneously the observer in the rocket ship sees the beam pass him, and, if he measures its velocity, he will find that it is 299,792 km/sec greater than his. Strange as it may sound, this is the case. It almost seems that 2 plus 2 is not equal to 4, and, indeed, it is not when we deal with velocities near that of light. A correct answer cannot be arrived at simply by adding the two velocities.

SPECIAL RELATIVITY

Many scientists pondered the Michelson-Morley enigma. Among them were H. A. Lorentz and G. F. Fitzgerald, both of whom arrived at the conclusion that an object must shrink in the direction of its motion, relative to a fixed observer, as its velocity increases.[1] (The effect, now usually referred to as the Lorentz-Fitzgerald contraction, would not be noticeable until velocities near that of light were reached.) Both also arrived independently at a formula for predicting the amount of shrinkage at various velocities, but neither could explain it. The solution was to come from a young man working as a patent examiner in Berne, Switzerland—Albert Einstein. Virtually cut off from the rest of the scientific world, Einstein was examining

[1]Consider the following experiment: Attach one of two meter-sticks of *exactly* the same length to the side of a space ship and keep the other on Earth as a reference. The space ship blasts off and then passes repeatedly overhead at velocities that come closer and closer to that of light. As it does, an observer on Earth compares the meter-sticks. He will find that the one on the space ship will be shorter, and that, indeed, the faster the space ship travels, the shorter the stick attached to it will become relative to the one on Earth (Fig. 19.1). In fact, even the space ship will appear to get shorter. Finally, at a velocity very close to that of light both the space ship and the meter-stick attached to it almost disappear.

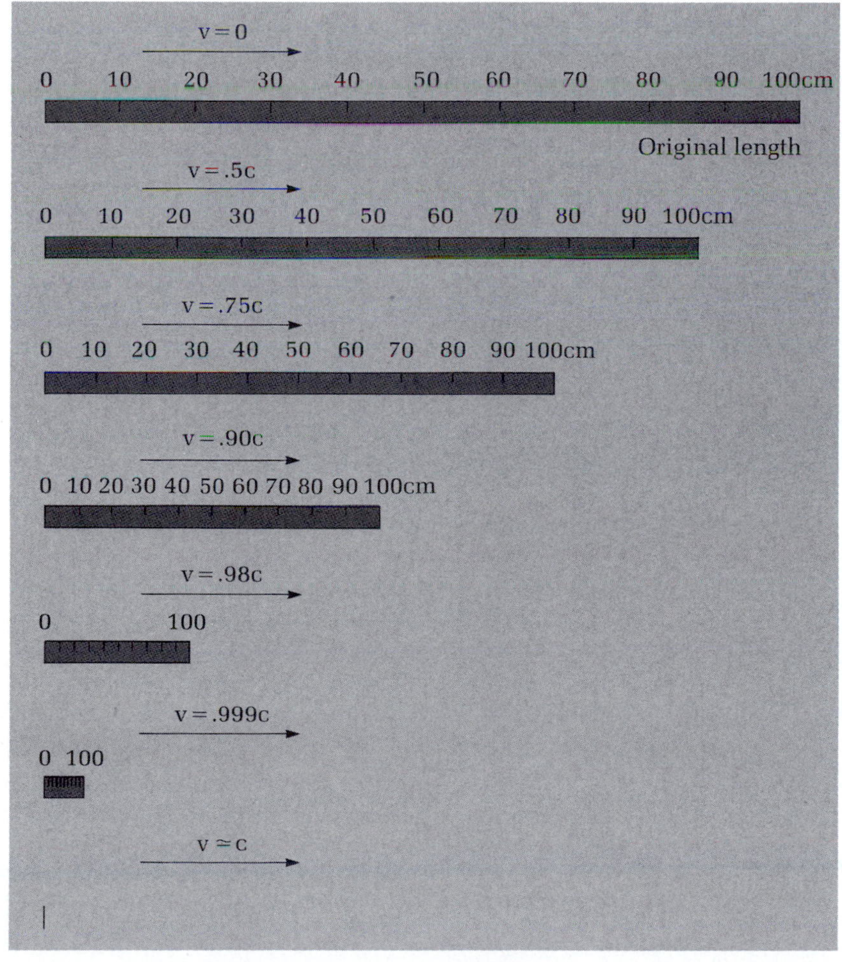

Figure 19.1 Apparent contraction of a meter-stick as its relative velocity approaches that of light.

Boarding a Berlin street car one day, Einstein noted, mistakenly, that the conductor had given him the wrong change. The conductor quickly recounted it, then glaring at him, thrust it back, blurting something like, "It is correct—trouble is, you don't know how to figure."

Obviously the conductor didn't know he was talking to a man who had done some of the most complicated and abstract "figuring" the world had ever seen. Oddly enough, Einstein's parents had also doubted his mental powers. They had been concerned when he didn't talk until he was three. (And even then his speech was slow and awkward.) In school he despised the rote learning pupils had to undergo. Later in life he wrote: "The teachers at the elementary schools were like sergeants, and at gymnasium [high school] they were like lieutenants."

Einstein had not yet finished gymnasium in Munich when his parents moved to Milan. Left behind to finish his schooling, he was soon homesick and trying to find an excuse to leave. Finding one became unnecessary when, one morning, his instructor took him aside and *asked* him to leave.

"Why?" Einstein replied.

"You're a bad influence in class; your presence destroys the respect of the students [for me]."

Einstein, overjoyed, left immediately for Milan, and, for the next year or so was what we now call a "dropout," wandering on foot around Northern Italy and thoroughly enjoying himself. But his freedom suddenly ended when his father's business went bankrupt. Einstein was informed that he had to go back to school to prepare for a vocation. The Swiss Federal Polytech in Zurich was selected, and, in 1895, he took the entrance examinations—and failed them. However, the examiner, noticing that he had done quite well in the mathematics and science sections, advised him to get a high-school diploma in Switzerland, so that he would not have to retake the examinations. A year later Einstein had his diploma, and, in 1896, at the age of 16, entered the Polytech at Zurich.

Einstein was not a particularly conscientious student. He preferred to loiter in the cafés with other students, most of whom remembered him as witty and amiable, but not too studious. Skipping classes came naturally; after all, he only had to pass the final exam, he thought. Fortunately, he had a conscientious roommate, who took excellent notes and passed them on to him; so, with a little help, he managed to graduate. But the extensive cramming he had had to do just before finals left him with a severe distaste for schooling.

He had hoped to get some type of assistantship with one of his professors after graduation, but no one wanted him. Although he was near the top of some of his classes, he had barely managed to get through others. In 1901, however, he got a job examining patents in Berne. It was here that he began an extensive self-learning program, and by 1905 he was hard at work on some of the most difficult problems in physics. In a one-year period he published several papers that set the world of physics on fire. Included in them was one on relativity that is now considered to be one of the most important papers in physics ever written.

By 1909 he was back at Zurich as a professor; from there he went to Prague and then to Berlin. In 1933 he came to the Princeton Institute for Advanced Study in the United States. He was awarded the Nobel Prize in 1921.

Einstein led the type of life he preferred above all others—a simple one. Despite the honors and praise showered on him, he had a distaste for fame and fortune. Money meant little to him. (Someone once found a check for $1500 being used by him as a bookmark.) He shunned publicity, and, from all accounts, was happy. He liked to be comfortable and frequently wore old crumpled clothes—and no socks. He shaved in the bathtub with the same soap he used for bathing. Having several different kinds of soap when one would do was just too complicated, he said.

this problem also, but from a different point of view. (There is still some question as to whether he had even heard of the Michelson-Morley experiment at the time.) Einstein was examining a set of equations that had been published by physicist James Clerk Maxwell—equations that describe the wave properties of all electromagnetic phenomena, including light. What would it be like to travel alongside and catch a light beam, Einstein asked himself. Maxwell's equations gave a strange result: If you could somehow catch up with a light beam and travel alongside it at 299,792 km/sec, the light beam would no longer be a wave—it would, in effect, cease to exist. Like Michelson and Morley, Einstein had discovered seemingly paradoxical phenomena at the velocity of light.

He began his investigation of the matter by making two basic assumptions: 1) The laws of physics are the same in all uniformly moving systems. 2) The speed of light is constant (299,792 km/sec in vacuum) regardless of the motion of the source. With these he could do away with the aether; an absolute frame of reference was no longer necessary, or needed. The first assumption tells us that all motion in the universe must be relative. To illustrate, assume that you are somewhere far out in space, well away from planets or stars, and a chair suddenly passes you at a high velocity. But did it pass you? You may have passed it. According to relativity you can never know which is the case, so the question simply has no meaning: Motion is motion only when it is relative to something (i.e., an object). But light has a special property: it always travels at 299,792 km/sec, regardless of your motion.

With these basic postulates Einstein was immediately able to verify the length-contraction formula derived by Lorentz and Fitzgerald. But, unlike Lorentz and Fitzgerald, who were trying (by the use of a single unexplained formula) to force observation and theory to agree, Einstein had formulated a complete and concise theory. With it he could investigate many other aspects of the problem: How did time, energy, and mass behave? He had the tools to find out.

In addition to verifying the Lorentz-Fitzgerald contraction, Einstein showed that time slows down relative to a fixed observer as we approach the velocity of light. In this case assume we have two clocks; again we put one aboard our space ship (p. 338), and keep the other for reference. As the space ship passes overhead at ever-increasing velocities its clock appears to go slower and slower relative to ours (Fig. 19.2). Finally, very close to the speed of light it almost stops.

If time slows down (relative to Earth's time) as we approach the speed of light, we should be able to make long interstellar flights in a relatively short time if our velocity is high enough. Take as an example a space ship that leaves Earth, and travels to a star and back at velocities such as $.5c$, $.99c$, $.999c$, and $.9999c$, where c is the velocity of light. Assume in each case that one year passes for the people in the space ship. The question is, then, how many years will have passed on Earth during the trips at each of the above-noted velocities. Consider $.5c$ first. A simple calculation gives 1.1 years for the time

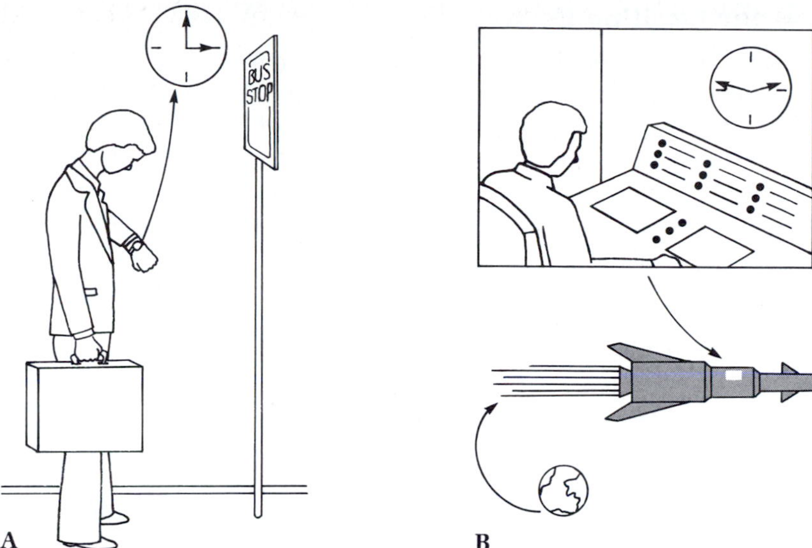

Figure 19.2 As the velocity of light is approached, time slows relative to a fixed observer. Here *A* is on Earth, and *B* is in a rocket moving near the speed of light relative to *A*. Though their clocks were initially synchronized, in these circumstances *B*'s will run slower than *A*'s.

that passes on Earth. At a velocity of .99*c*, 7 years will have elapsed on Earth while the space ship travels to and from the star; at .999*c* 22 years will have passed on Earth. (Thus, the space travelers are only one year older when they return, but their friends are 22 years older.) Finally, at a velocity of .9999*c*, 71 years will have elapsed.

The Twin Paradox

With the discovery of the slowing of time at velocities near that of light (*c*), space flight to the stars took on a new meaning: If your velocity was close enough to *c*, you could travel tremendous distances in only a few days (by your watch). But it wasn't long before someone discovered an apparent flaw in Einstein's discovery—a "paradox." Twins A and B are exactly 25 years old. Twin A makes a trip to a certain star and back in a space ship traveling at .998 *c*. It takes him one year. When he gets back he says to Twin B: "Aha, I'm only 26 now and you're 40."

"Not so," says Twin B. "I'm 26, and you're 40. Remember, all motion in the universe is relative. Actually, it was the earth that moved off into space and returned; your space ship just stood still."

Twin B does, indeed, have a point; after all, motion *is* relative. But surely one of the twins is right and one is wrong. Who is younger? It was ten years before an answer to this question came, and it was general relativity that gave the answer. If we look again at the twins we see that one had to accelerate to get to velocities near *c*, and it is this one who would (because of this acceleration) experience a force. According to general relativity the twin who experiences this force will be the younger.

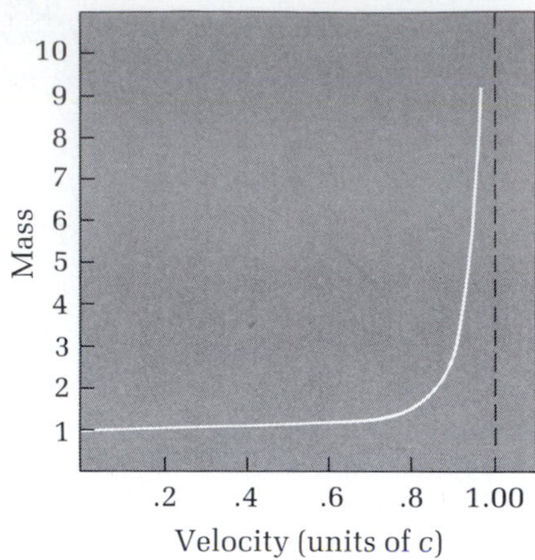

Figure 19.3 Relation of mass and velocity, as latter approaches that of light.

c as the Limiting Velocity

Einstein went on to show that mass is also effected by velocity, increasing with increasing relative velocity (Fig. 19.3). And, again, there is the special phenomenon at the velocity of light: In this case it is that mass becomes infinite. This phenomenon (and the above-described Lorentz-Fitzgerald contraction and the time-slowing phenomena) prompt a question: Is a velocity equal to *c* possible? The answer is, no; matter can travel only at velocities up to (but not including) *c*. It may approach *c* as close as we wish, but it can never equal it; the reason is clear: if the mass were infinite it would take an infinite force to reach *c*.

But what about a case in which, say, two space ships approach each other at $.8c$? Obviously their relative velocity is $.8c + .8c = 1.6c$, which is greater than *c*. But, again things are not as simple as they appear; if we are aboard one of the rocket ships we cannot just add $.8c$ and $.8c$: The velocities must be summed according to the appropriate relativistic formula, and this formula never gives a velocity greater than *c*, regardless of the magnitude of the two velocities being summed.

Thus, it appears that *c* is the limiting velocity in the universe. However, physicist Gerald Feinberg recently postulated particles—he calls them tachyons—that travel only at velocities greater than *c* and cease to exist at *c*. So far no one has observed these tachyons.

We have discussed many of the basic concepts of physics—length, velocity, mass, and we have seen that there are significant changes in their values at velocities close to *c*. Einstein showed that energy also changes in the same ways, but, more importantly, that energy and mass are related: In effect, they are two forms of the same thing, and, in certain nuclear reactions, mass can be changed into energy. As we saw earlier, stars are powered via this conversion.

Four Dimensional Space-Time

Consider a star far out in space. When you look at it you are actually looking back in time. You see it as it appeared many years ago because its light took many years to travel across the empty space between you and the star. For all you know, this particular star may no longer exist; it may have exploded and dissipated off into space thousands of years ago. In another example, an observer on Earth sees two supernova explosions take place simultaneously thousands of light years away and in different parts of the sky. (This is very unlikely in practice.) Far out in space another observer, traveling in a space ship, also sees the two explosions, but to him they are *not* simultaneous. Both observers are correct: One saw the events from a point where their light arrived simultaneously; the other did not.

In setting up his special theory of relativity, Einstein carefully examined such relations of space and time; in particular he noticed that Newton had treated them as separate entities, while it seemed far more natural that they were somehow related. Newton assumed that locating an object in space required three coordinates, as

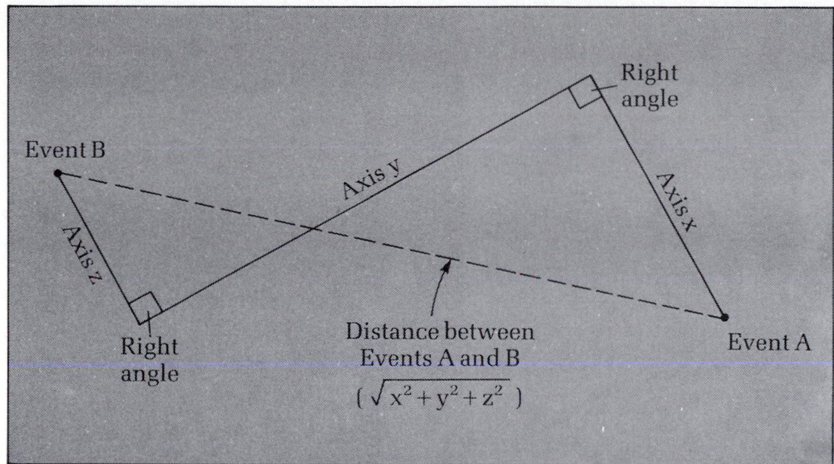

Figure 19.4 Locating an object in space in Newtonian terms. Figure is, in effect, a perspective view, so that lines forming right angles are not perpendicular.

shown in Fig. 19.4 (with distances calculated from the formula $\sqrt{x^2 + y^2 + z^2}$), but Einstein realized that four were needed—three of space and one of time. This space-time is arrived at by multiplying the time interval (in seconds) by *c* (in meters/sec), to get units of length (meters). This product is squared and subtracted (under the square root sign) from the sum of the squares of the coordinates. Intervals of the type obtained in this way are, in fact, the only ones that make sense when we wish to connect events occurring over great distances.

Scientists represent space-time in two dimensions. They let one axis, call it x, represent all three space axes; perpendicular to it is the time axis, and light rays are envisioned as traveling along lines at angles of 45 degrees to the axes (Fig. 19.4). An event is represented by a point, and a sequence of points is a line, called a "world line." However, not all world lines are possible: One starting at "Now" and extending up the "light line" (45 degrees) must represent an object traveling at the speed of light, and we know nothing can travel that fast. (This is referred to as a light-like trip.) One starting at Now and tracing out an angle less than 45 degrees with the x-axis is also impossible; it represents a trip at a velocity *greater* than that of light. (This is referred to as a space-like trip.) Finally, we have lines that represent trips at speeds *less* than the velocity of light and that are, of course, possible. (These are referred to as time-like.)

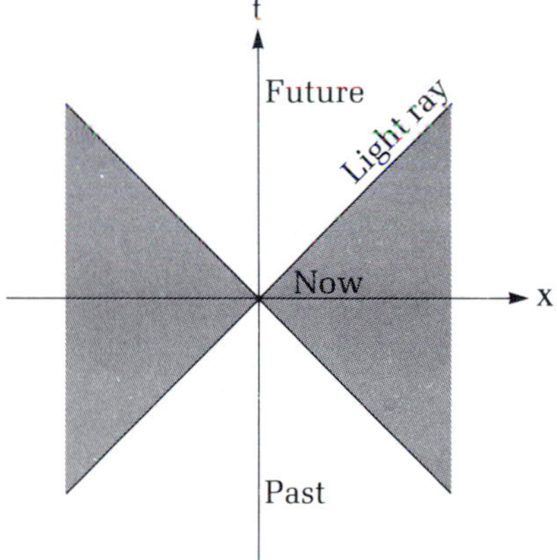

Figure 19.5 A two-dimensional space-time diagram. The vertical axis represents time; the horizontal axis (labeled x) represents all three space axes. The shaded areas correspond to impossible velocities—velocities greater than *c*.

GENERAL RELATIVITY

Einstein's theory, as it was presented in 1905, applied only to straight-line, uniform motion, in other words, to unaccelerated motion. The next step was to generalize the theory to include all types of motion. Although, in common experience, there is a distinct difference between uniform and nonuniform motion—speed up suddenly while

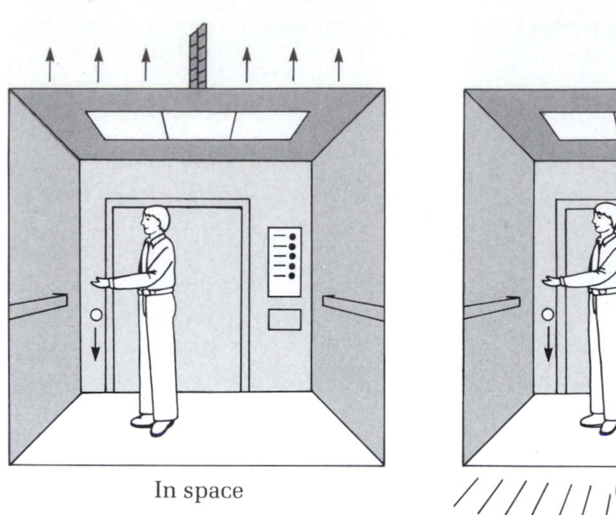

Figure 19.6 Two equivalent situations according to Einstein. The elevator at the left is in space and moves upward at an acceleration of 9.8 m/sec. The elevator at right is on Earth. The ball falls at the same rate in both elevators.

driving, and you feel yourself pressed back against the car seat—Einstein could not accept this. As a first postulate in his generalized theory he stated that no observer, regardless of how he is moving, can determine his state of motion by experiment. He argued that the experience of nonuniform motion might just be the acceleration produced by a gravitational field. It is also possible that, conversely, we could mistake the force of acceleration (which we call the inertial force) for that of gravity: Pilots, when banking steeply in fog, sometimes have this experience, and find it difficult to determine the direction of the surface of the earth.

Einstein saw that, since there are two distinct forces in nature—one associated with acceleration and one with gravity—and since force and mass are related (Newton's second law), then there are also two different but equivalent (this postulate is now called the equivalence principle) kinds of mass: inertial and gravitational. To understand some of the implications of this, consider an observer in an elevator near the top of a high building. The elevator cable breaks, and the elevator begins to fall. The passenger, if he jumps, floats to the ceiling; if he tries to drop something it just hangs there. It is almost as if he were out in space—and, indeed, this is exactly the way things would be were he in space. Now assume he actually is out in space in the same elevator (Fig. 19.6). A cable is attached to the top and the elevator is suddenly accelerated upward at 9.8 m/sec^2 (normal acceleration at the surface of the earth). If the observer now drops an object it falls, apparently, normally, and he therefore believes he is on the surface of the earth. Einstein's equivalence principle seems to be well satisfied in this case.

But according to the equivalence principle, anything that works in the accelerating elevator should also work on the surface of the earth. Knowing this, the observer stops his elevator (sets the acceleration to zero), drills a hole in one side of it and allows a light beam to

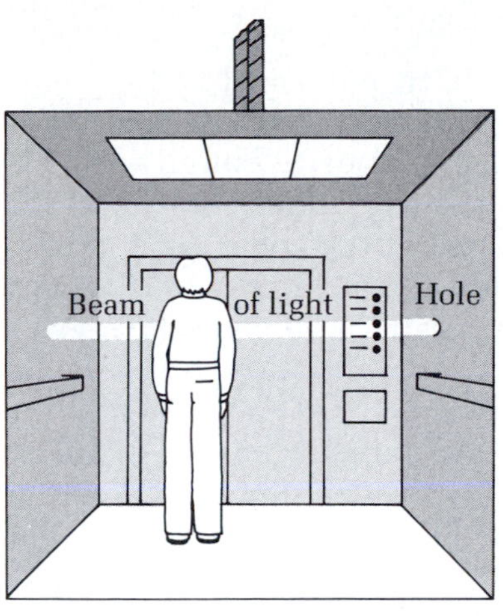

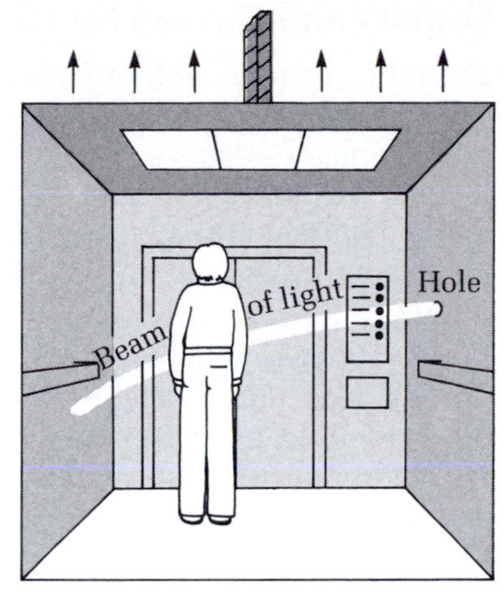

Figure 19.7 The bending of a light beam (exaggerated for illustrative purposes) in an elevator (right) that is being accelerated upward. Elevator at left shows beam when there is no acceleration.

enter and cross to the opposite wall (Fig. 19.7). He marks the spot where the beam hits, then sets his elevator accelerating "upward" again. The light beam is now deflected downward as it crosses the elevator. This is to be expected, of course, since the elevator is accelerating upward as the beam moves across it, and so, by the time the light reaches the opposite wall, that wall has moved "up." According to the equivalence principle, we should expect a similar phenomenon in a gravitational field: A gravitating object should also deflect a light beam (Fig. 19.8).

Einstein calculated such a deflection for the stars near the limb of the sun. At the eclipse (the best time to see a deflection) of 1919, and many times since, Einstein's prediction was verified.

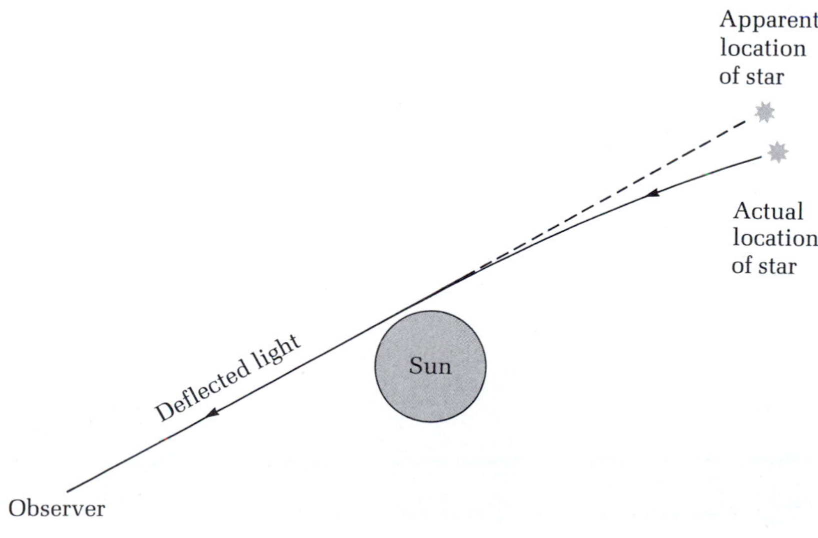

Figure 19.8 A beam of light deflected by the gravitational field of the sun. The position of the star will appear shifted.

Gravity and Curved Space

Einstein's generalized version of relativity, published in 1916, contained many new and different concepts, the strangest of which was a new geometry of curved space. To set up a geometry a number of self-evident axioms is needed. Euclidean geometry (the geometry of flat space, familiar to most high school students) includes such axioms as: through any two points, only one line can be drawn; and, through a point only one line can be drawn parallel to a given line. From these axioms one then proves theorems, such as: the sum of the interior angles of a triangle is 180 degrees.

Figure 19.9 A triangle on a spherical surface. The sum of the interior angles is greater than 180 degrees.

Euclidean geometry is, of course, not the only type of geometry we have. Consider the surface of the earth: The sum of the interior angles of a triangle drawn on it will be greater than 180 degrees (Fig. 19.9). It is, of course, easy for us to see why: the surface is curved, and flat-surface geometry does not apply.

In this case we see the curvature because the surface is two dimensional (in three dimensions) and we live in a three-dimensional space. If, on the other hand, we were two-dimensional creatures, we would not be able to see this curvature. One way to determine whether there was curvature would be to mark off a large triangle on the surface and measure the interior angles. If they add up to 180 degrees we would know the surface was flat; if their sum was greater we could assume it was curved like the earth's surface. (We would have to be sure, though, that we used a large enough triangle to notice the curvature.) But how do we visualize a *curved* three-dimensional space? It's difficult, to say the least; only four-dimensional creatures (if such existed) could do it.

Two curved-space (non-Euclidean) geometries have been set up, one by Russian mathematician N. Lobachevsky and another by German mathematician B. Riemann. In Lobachevsky's, the sum of the interior angles of a triangle is less than 180 degrees, and it can be visualized as the geometry of a saddle-like surface (Fig. 19.10). We have already met Riemannian geometry (Fig. 19.10); in its two-dimensional form it is the geometry of the surface of a sphere. Riemann, of course, developed his theory for three and even more dimensions.

Einstein used the ideas of non-Euclidean geometry to develop his theory of curved space. He assumed that matter curved the space around it and that, when other matter encountered this curved space, it moved through it in a "natural manner," which means that it followed a geodesic. (A geodesic is the shortest distance between two points.) Its path would appear curved to us because the space is curved.

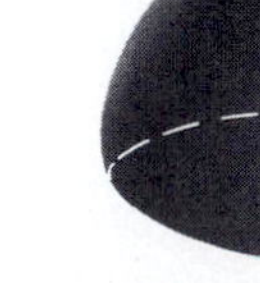

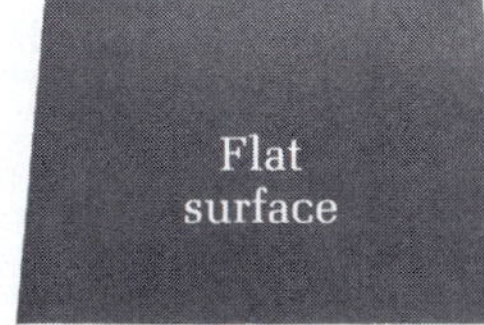

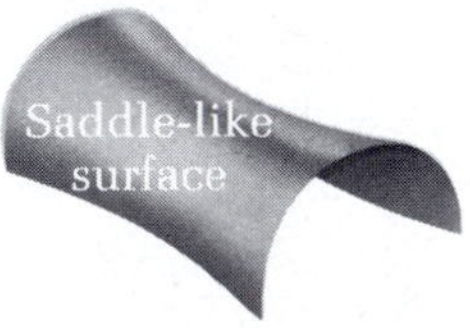

Figure 19.10 The three possible geometries of space.

Using this concept, Einstein derived equations describing the orbits of the planets. To test the validity of his theory he applied his equations to prediction of the orbit of the planet Mercury. A small shift in the direction of the major axis of Mercury's orbit (Fig. 19.11) had been noticed for some years, and this shift was not predicted by Newton's theory; predictions based on Einstein's equations, however, showed no discrepancy from observation. Thus a theory based on concepts completely different from Newton's—curved space and geodesics—seemed to give a more accurate result. Within a few years there was no doubt that it was a better theory.

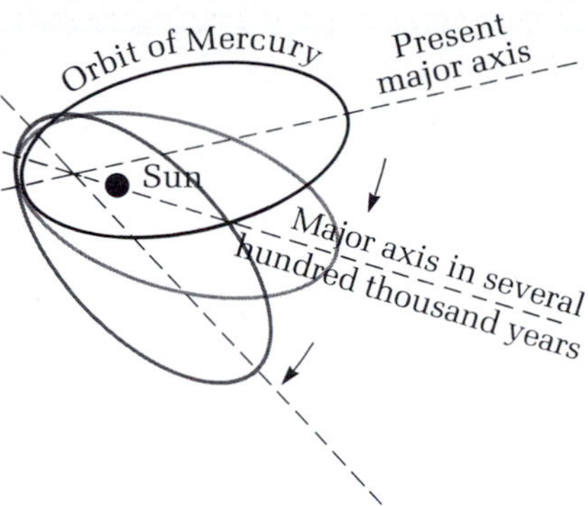

Figure 19.11 Precession of Mercury's orbit. The same effect occurs in a binary system containing a pulsar, and Einstein's theory predicts both.

OTHER THEORIES AND TESTS

Einstein's theory is a thing of beauty. It has a basic simplicity; there are no adjustable parameters and no unnatural or "special-case" assumptions. Based on the idea of a "tensor field," it is therefore referred to as a tensor theory. Although it appears at present to be the best theory of gravity, it is not the only one.

Over the years many theories have been put forward; a few are still considered viable. All are patterned after general relativity in that they are based on the concept of a curved space. The major difference in these other theories lies in their coupling of another field to the tensor field of Einstein. In the best known of these, the Brans-Dicke theory, the field coupled to Einstein's tensor field is associated with a gradual change in the universal gravitational constant.

The predictions of general relativity seem to be closer to the experimental values than those of any of the other theories, but the others—particularly the Brans-Dicke theory—give values that are relatively close, and consequently must be taken seriously.

For example, as we have noted (above) in regard to the precession of Mercury's orbit, general relativity predicts almost exactly the amount observed (43 seconds of arc per century). The Brans-Dicke theory, however, predicts about 5 seconds of arc per century less. In recognition of the fact that if the sun was not exactly spherical this would affect the precession, Dicke and his co-workers measured the sun, and determined that it was oblate to a degree that accounted almost exactly for the difference of 5 seconds of arc. In 1974, another group, headed by H. A. Hill, found almost no oblateness. It is actually quite difficult to make a measurement such as this to a high degree of accuracy, and so the issue remains unresolved.

Other tests of gravity theories have been conducted. We have seen that the light from a distant star is deflected as it grazes the sun (p. 347). But light is only one form of electromagnetic radiation and, since the curvature of space affects all forms, it would also be expected to influence the trajectories of radio waves. For example, a radio beam reflected off a planet just before it disappears behind the sun would be expected to be delayed because of the curvature of space caused by the sun, and the consequent lengthening of the path the beam would have to travel.

Experiments of this type have been conducted by Fomalont and Sramek. Using widely separated radio telescopes (for increased accuracy) they measured the differences (in time of arrival) of signals from two radio sources, one of which was soon to be occulted by the sun. Any change in the difference just before occultation gave a measure of the curvature of space near the sun. Predictions based on Einstein's theory again agreed best with observations. Tests similar to those just described have also been made using quasars as sources.

Another recent test used the laser beam reflector left on the moon by Apollo 11 astronauts. With this reflector extremely accurate determinations of the distance to the moon can be made, and changes of even a few centimeters can be detected. Since the Brans-Dicke theory predicts a gradual change in the universal gravitation constant, the distance to the moon should gradually be changing (note that it is also changing due to other effects, i.e., tides, so these would have to be allowed for.) Results obtained so far again favor general relativity over the Brans-Dicke theory.

In another test, still in the planning stage, a spinning gyroscope will be placed in a satellite sent into orbit around the earth. Part of its precession will be due to the curvature of space, and this will be measured.

SUMMARY

1. The theory of special relativity was introduced in 1905 by Einstein; it applies only to straight-line uniform motion. General relativity was introduced in 1916; it applies to all types of motion.
2. The Michelson-Morley experiment was set up to measure the earth's motion through the aether. It gave a null result; no motion was detected.
3. Lorentz and Fitzgerald, to explain the Michelson-Morley results, postulated (independently) that objects shrink in their direction of motion.
4. Einstein's two basic assumptions in setting up his special theory of relativity were: a) The laws of physics are the same in all uniformly moving systems. b) The speed of light is constant regardless of the motion of the source.
5. According to Einstein's theory, objects shrink in their direction of motion, time slows, and mass increases as our velocity increases relative to a fixed observer. The effects are noticeable only near the velocity of light.
6. It is impossible for matter to travel at the velocity of light. Only photons can travel this fast.
7. Events in the universe should be described using both space and time.
8. One of the basic postulates of general relativity, the equivalence principle, assumes the equivalence of inertial and gravitational mass.
9. Einstein replaced Newton's concept of gravitational force by curved space. Planets move in geodesics in this curved space.
10. Einstein's theory explained the precession of Mercury's orbit.
11. Several other theories of gravity have been introduced, the best known being the Brans-Dicke theory. All are similar to Einstein's theory.

12. Most tests have indicated that Einstein's theory is in best agreement with experimental data.

REVIEW QUESTIONS

1. What is an absolute frame of reference? Would the aether constitute an absolute frame of reference? Why?
2. Why was the Michelson-Morley experiment performed?
3. Explain the significance of the result of the Michelson-Morley experiment.
4. What were the two basic assumptions Einstein made in setting up his special theory of relativity?
5. Explain the twin paradox. How and when was it resolved?
6. Is there a limiting velocity in the universe? Explain why.
7. List all the things that happen (according to special relativity) to an object as it approaches the velocity of light.
8. Define an event in space-time. How do we determine the distance between two events?
9. Explain the difference between space-like, light-like, and time-like trips.
10. What is the principle of equivalence? Explain some of its implications.
11. A light beam that grazes the sun is bent slightly. How did Einstein explain this?
12. Briefly describe the three types of geometry discussed in the chapter.
13. What is a geodesic? Why is it important in general relativity?
14. Explain how the general theory of relativity was tested using the planet Mercury.
15. How does the Brans-Dicke theory differ from Einstein's?
16. How is the shape of the sun important in relation to relativity?
17. Describe some of the recent tests of general relativity. What appears to be the overall conclusion of these tests?

THOUGHT AND DISCUSSION QUESTIONS

1. The equation for length contraction (Fig. 19.1) is $l = l_o\sqrt{1-v^2/c^2}$, where l_o is the original length (at $v = o$ relative to us), l is the length we observe at velocity v, and c is the velocity of light. Using this formula, find out how long a meter stick will appear at velocities $.9c$, $.99c$, $.999c$. What happens at $v = c$?
2. Time slows according to the formula $t = t_o/\sqrt{1-v^2/c^2}$. If it takes us one month by our watch (t_o) to get to a distant star when traveling at $.9c$, how much time (t) will have passed back on earth? Make the same calculations for speeds of $.99c$ and $.999c$. What happens at $v = c$?
3. The principle of equivalence states that a gravitational field equivalent to that on Earth can be created by accelerating upward at 32 ft/sec^2. But if we start from near Earth and accelerate long enough we get to velocities much greater than the velocity of light (which is the uppermost velocity in the universe). Explain this seeming paradox.

FURTHER READING

Barnett, L. *The Universe and Dr. Einstein.* London (1949).
Einstein, A. *Relativity.* Crown, New York (1961).
Kaufmann, W. J. *The Cosmic Frontiers of General Relativity.* Little, Brown & Co., Boston (1977).

Aim of the Chapter

To explain the different types of black holes, and to present observational evidence of them.

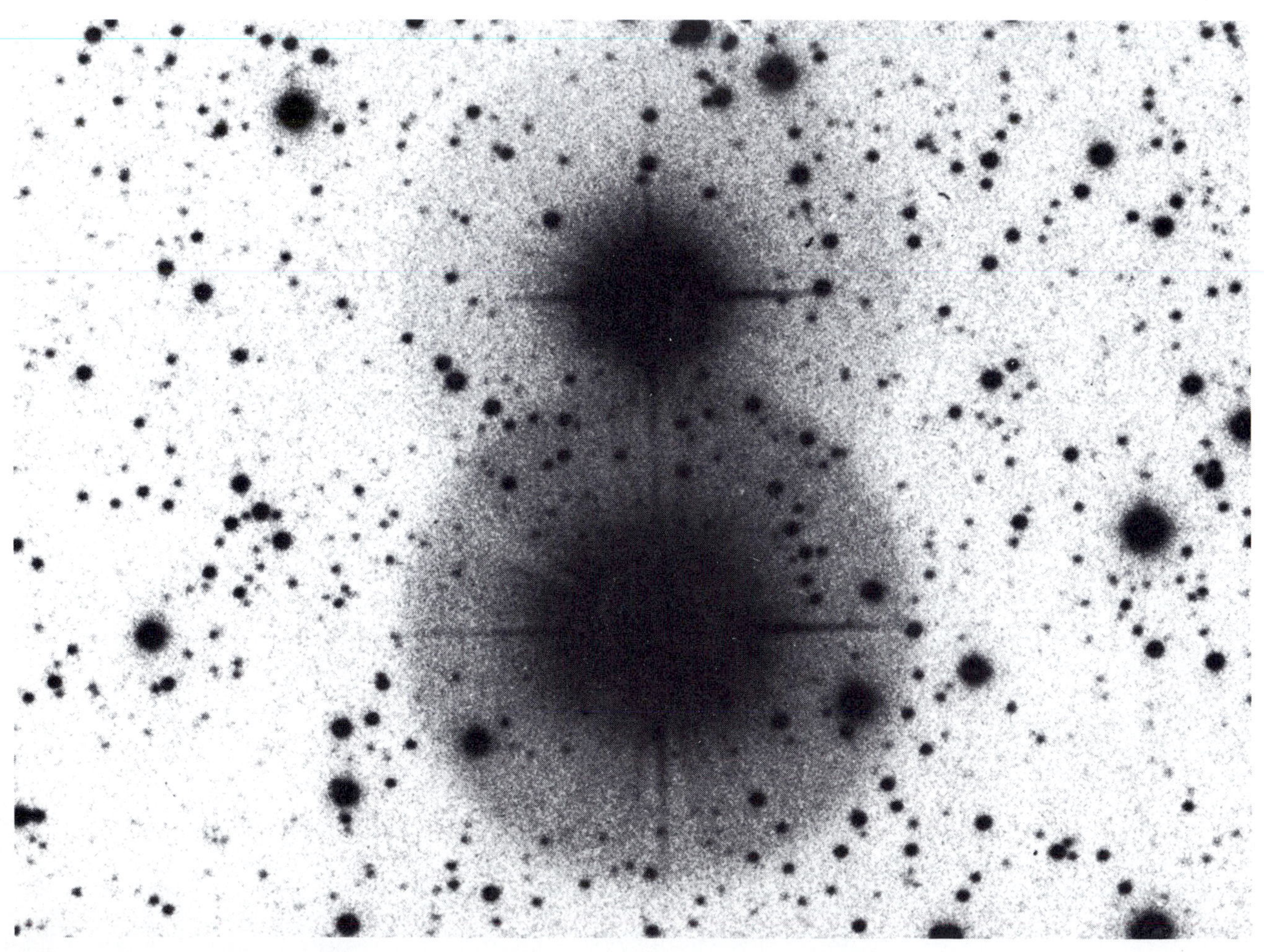

chapter 20
Black Holes

In Chapter 18 we mentioned that Oppenheimer and Volkoff noticed that a neutron star, if sufficiently massive, would also become unstable and collapse. The neutrons could not hold the surface out; they would be crushed and, this time, no subsequent process could stop the collapse. Russian physicist Lev Landau came to the same conclusion, and further study by Oppenheimer and Hartland Snyder came to the same conclusion: If the star had a mass greater than about 3 $M_\odot$, it would collapse forever; it had no stable end state. The limiting mass here of 3 $M_\odot$ is uncertain; some argue that stable neutron stars can exist up to perhaps 8 $M_\odot$. Of course, this is the final mass of the star; its mass while in the main sequence may be much greater than this, since considerable mass is lost before the star collapses.

BLACK HOLES FROM NONROTATING STARS

The theory of these massive ever-collapsing stars developed rapidly. John Wheeler of Princeton dubbed them "black holes," for it was known that, although they collapsed forever, they left a strange black sphere marking the position of the collapse (Fig. 20.1). Although the fundamental tool in the study of black holes is general relativity, black holes are part of all seriously considered theories of gravity. The Brans-Dicke theory, for example, also predicts black holes. In a sense, even Newton's theory of gravity predicts them.

Suppose, for example, the earth is at the present position of the sun and is expanding while its density (mass per unit volume) remains constant. As it expands, its gravitational pull increases, and, therefore, so will its escape velocity (the velocity necessary to escape Earth's gravity). By the time the expanding Earth's surface reaches the orbit of Venus, its escape velocity will have increased enormously—to hundreds of thousands of kilometers per hour. When it approaches Mars, the escape velocity will be 299,792 km/sec—the velocity of light. But, since photons travel at 299,792 km/sec, then any further expansion will result in an escape velocity that prevents even photons from escaping from the surface, and we will not be able to see the object. The object will have become a black hole, though not the type we will be talking about in this chapter; these are much smaller and are explainable not by Newton's theory (as in the case above), but by Einstein's.

From a distance, a black hole appears like its name—a black "hole," or absence of light, outlined by background stars (Fig. 20.2). Examined carefully, it would show a gravitational field unchanged from before its collapse; thus, if a planet revolved around a massive star, and the star suddenly collapsed to a black hole, the planet would probably continue to revolve around it for billions of years.

The black holes generated by collapsing stars (called stellar-collapse black holes) are small, most only a few miles across. They are, indeed, black in appearance; no light whatsoever is reflected from or emitted by them (but see p. 361), and they can definitely be thought of as massive sinks in space. Anything that comes within a certain distance of a black hole is pulled in, and, as the surface (Fig. 20.3) of the black hole is passed, space and time interchange: Just as we cannot control time outside a black hole—it passes at the

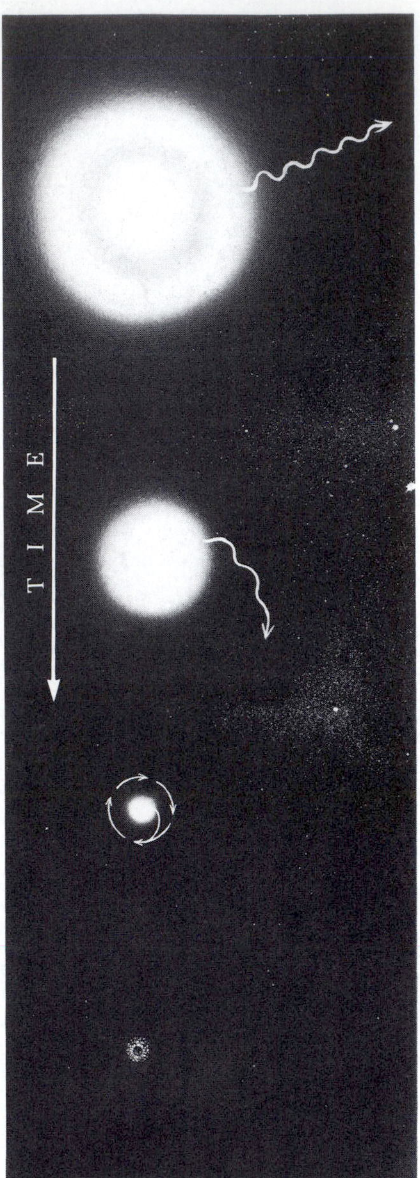

Figure 20.1 Effect on a beam of light (arrow) of the collapse of a massive star and the consequent increase in the curvature of space around the star. Wavelength increases during the collapse, finally becoming infinitely long.

Figure 20.2 How a black hole might look. The large number of stars near its perimeter is an illusion caused by the extreme curvature of space in the vicinity.

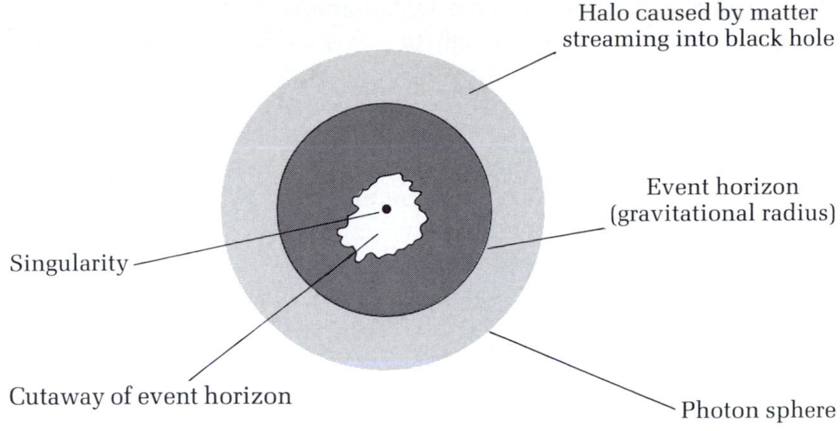

Figure 20.3 Structure of a black hole. The photon sphere is the region where photons have taken up circular orbits around the collapsing matter. The event horizon, or gravitational radius, is the limit below which contact with the exterior is impossible. The singularity is the indefinitely collapsing remnant of collapsing matter.

same rate regardless of what we do—so, too, inside a black hole, we would have no control over space. We would be pulled into its center despite any struggle we might put up; no escape would be possible.

In a star that is becoming a black hole, almost immediately after the collapse begins, surges of photons (X-rays and γ-rays) are given off. Soon, under the influence of the tremendous pull of the star's increasing gravity, those photons leaving the surface at an angle begin to leave in curved paths. The star then passes rapidly in succession through three critical stages. The first is reached when photons that attempt to leave the star parallel to its surface are pulled into orbit around it; for those, escape is no longer possible. The star then passes through what is called its event horizon, shown in Fig. 20.3, and no photons whatsoever can now be emitted. This is the second critical stage; the star is now a black hole.

The matter of the star as it passes inside the event horizon continues collapsing forever, which means that it is crushed to zero volume and infinite density at the center of the black hole (the third stage). This center point is called the singularity (Fig. 20.3).

A star collapses to a black hole in less than .001 sec, and it is possible that the latter stages of the collapse are obscured by a gas cloud. However, if we were able to view the collapse—at some distance—we would see a rapid decrease in size, and, with it, color changes ending in red. This would occur because, according to relativity theory, time slows as you pass through a strong nonuniform gravitational field, and photons are like clocks, vibrating with a very precise period. When time slows down their period lengthens, and this change in wavelength causes the emitted light to change color, until it eventually becomes red.

As the star enters its first critical stage photons are trapped in orbit around it. This is the last we would see of the star, though the photons cause a kind of reddish halo that lingers for a while. Finally the halo turns black, but it grows no smaller. Of course, because of the rapidity of the whole process, in practice we would see only a large bright star change almost instantaneously to a black sphere (assuming you are close enough to see the black sphere outlined against the background stars).

In short, then, the distant observer sees the star turn black and become frozen—fixed—at a certain size. A closer look however, would show a slowing down in the final stages of the collapse; although the star approaches closer and closer to a certain size, it never quite seems to reach it.

To an observer actually on the surface of the collapsing star things would be different: The star would appear to collapse in a finite time; in a fraction of a second the observer would pass through the event horizon and be crushed at the center; for him there would be no "frozen" star. If, early in the collapse, he pointed a light toward the limb of the collapsing star, the beam from it would, as we have said, be bent into a curve. As the star grew smaller the beam would be increasingly bent (assuming the light remains the same distance from the surface), until finally it curved completely around the star. If, as the star became a black hole, the observer pointed the beam away from it, he would find that at a certain distance from the black hole the beam escapes being pulled into it for most directions in space (Fig. 20.4). As he came closer, however, the beam would be increasingly attracted to the black hole, and the cone of light formed by the beam (called the exit cone) would get narrower and narrower (Fig. 20.4), finally closing as he passed the event horizon.

As he approached the event horizon, tidal forces would stretch him until eventually he had the shape of a piece of string. Once passed through the event horizon any messages he attempted to send to the outside would never arrive; although they would leave him normally, the event horizon is receding from him at the speed of light, so that the messages could never reach it.

Another property of black holes is the slowing of time the closer one approaches to the event horizon; this is, of course, the slowing of time predicted by relativity theory.

A star has a certain radius as it passes through its event horizon, and this radius (at which light can no longer escape) can easily be calculated; it depends only on the mass of the star, and is called the gravitational radius or the Schwarzschild radius.

All objects have gravitational radii—even the human body. In the case of the sun it is about 3 km. (For a human body it is microscopic.) This means that if all the matter of the sun were compacted into this 3-km radius we would not be able to see it; it would be a black hole. The more massive the object (giant stars) the larger the gravitational radius and the lower the corresponding average density. Quasars are thought by some to be excessively massive (approximately 100,000,000 times as massive as our sun); if so, they would have a particularly large gravitational radius, perhaps as large as a billion kilometers. In such a case the average density of matter as the quasar collapsed through its event horizon would not be excessive, being equal only to about that of water.[1] Our observer

[1]The assumption here, for sake of illustration, that quasars collapse when they get old, is probably not true, as we will see later.

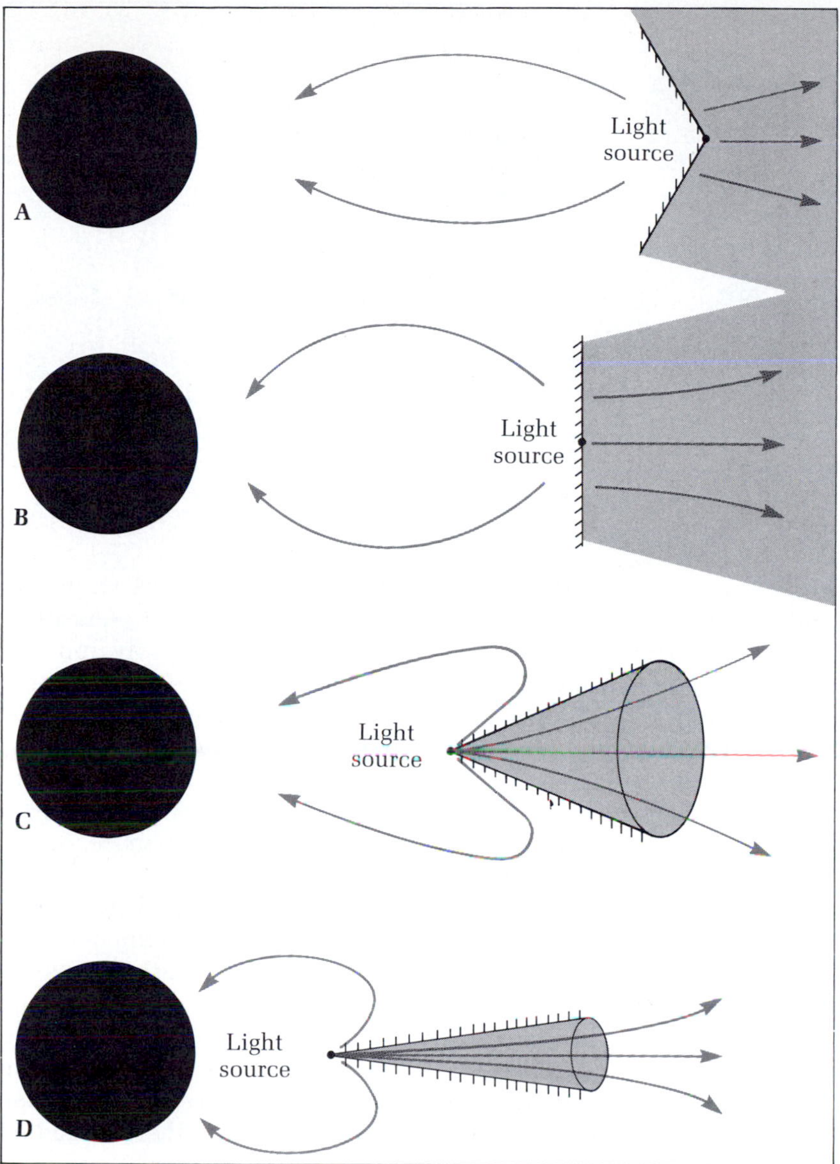

Figure 20.4 **A light source at various distances from a black hole. If the beam is projected in any direction outside of the darkened region (referred to as the exit cone) it will be pulled into the black hole.**

would, then, be able to pass through the event horizon with little effect; he would, indeed, still be alive at that point and would feel no tidal forces. The quasar would, of course, continue to collapse, and eventually he would be overcome.

In Chapter 19 we talked about Einstein's prediction that matter curves space-time. According to his theory, the more concentrated the matter, the greater the curvature. Black holes are extremely concentrated (that is, exceedingly dense), and therefore, the space around them must be severely curved. It is, of course, impossible to see curved space, but as we saw earlier we can project a light beam into it and see the curvature of the beam. An analogy (Fig. 20.5) can be made with a thin rubber membrane (representing a two-

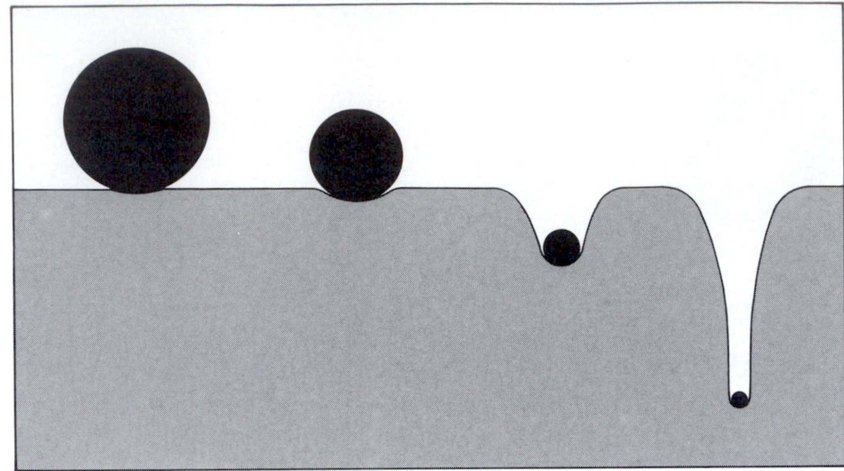

Figure 20.5 Effect on a rubber membrane of balls of approximately the same weight but of increasing density (and therefore decreasing size). The smaller balls make a deeper indentation because they exert a greater force per unit area.

dimensional slice of space through the center of a star) stretched tight and fastened at its edge. A large Styrofoam ball (representing the original star) placed on the membrane only slightly indents (curves) it, just as the space around an ordinary star is only slightly curved. Next, representing the star as it gets smaller and denser, a smaller wooden ball placed on the membrane curves it slightly more than the Styrofoam ball because the force per unit area is greater. Continuing in this way, we represent a later stage by a smaller, denser ball, say of iron. Finally, a still smaller lead ball placed on the membrane seems almost to drop through and disappear.

Just as the rubber membrane curved and stretched more and more as heavier and heavier balls were placed on it, so too does the space curve around a massive collapsing star as the star becomes increasingly dense. As mentioned above, although this curvature cannot be seen, it can be established mathematically, so that, for example, the degree of curvature around a given star at each stage of its collapse can be calculated. Einstein and Rosen were making just such a calculation in the 1930s when they discovered that, instead of the usual single set of solutions to their equations there were two. The first set showed that the space around the collapsing star developed a long, narrow bottleneck (Fig. 20.6), as in the case of the membrane when the lead ball was placed on it. The second set seemed to correspond to another bottleneck in space, somehow attached to the back end of the black hole; in other words, according to their calculations, if the bottleneck in space is followed down far enough (that is, to, and past, the black hole itself), it eventually begins to open up again. What did it open up into? The only answer seemed to be: another universe; black holes led to other universes! Further investigation of these Einstein-Rosen bridges (also known as "tunnels" or "wormholes") showed that they could also open up into distant regions in our own universe. We could not, however, use them to get to distant points in our universe, or possibly, other universes, since a velocity greater than that of light would be needed.

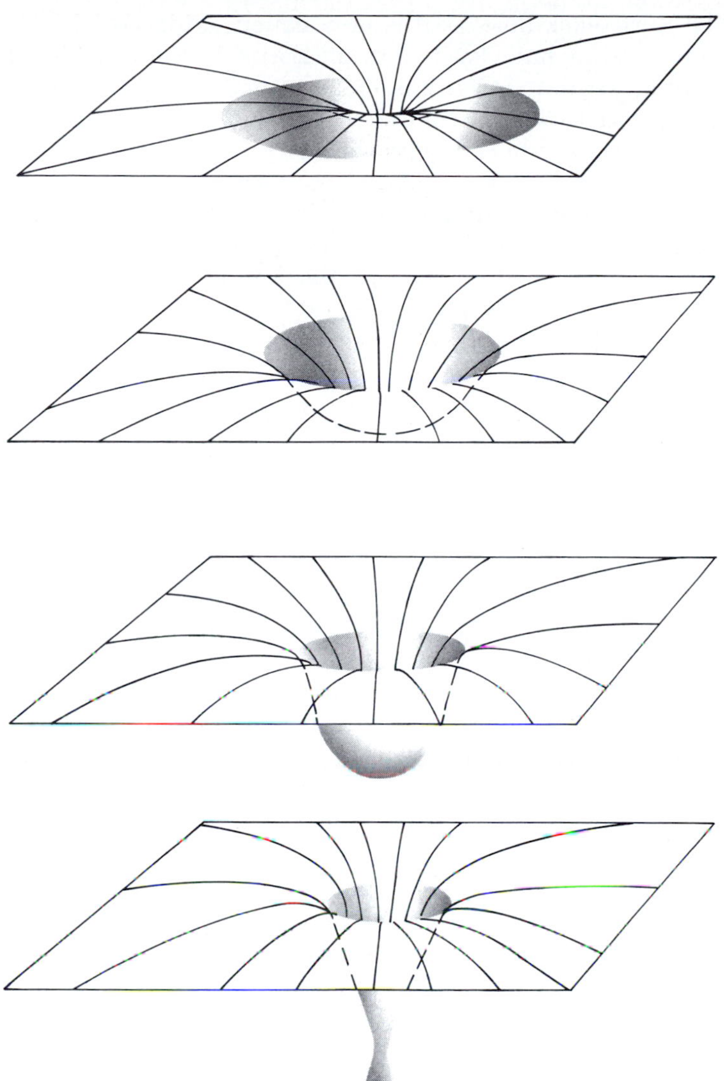

Figure 20.6 Curvature of space around a collapsing star.

BLACK HOLES FROM ROTATING STARS

The Kerr Black Hole

So far we have been considering only spherical, nonrotating stars. A much more likely case is that of rotating stars. Most stars rotate to some degree, but the collapse of rotating stars is much more difficult to deal with than the collapse of nonrotating ones. In 1963, Roy Kerr developed a set of solutions to Einstein's equations that properly incorporated rotation. There were many surprises. A sufficiently massive rotating star also becomes a black hole in space when it collapses, but it differs in many respects from an ordinary black hole.

The Kerr black hole, as it is now called, unlike the ordinary, non-rotating black hole, has two event horizons. Suppose the black hole is initially at rest. As it begins to spin a static limit (and an ergosphere) forms close to the event horizon. The point singularity becomes ring shaped and an inner event horizon forms just around it. As it spins faster the inner event horizon expands outward and the outer event horizon shrinks until they merge. If spin increases the double event horizon shrinks until it disappears, leaving only a "naked singularity," one no longer clothed in a black hole, and, so, consisting only of matter crushed to zero volume.

The possibility that such objects might exist has been considered by physicists as the outcome of the collapse of a star. Clearly, an ordinary collapse would not lead to one, but, were there sufficient asymmetry in the collapse, it might be possible: For example, a spinning object that became sufficiently elongated (like a football) as it collapsed, might end with no event horizon.

An object that encountered a Kerr black hole would be pulled into orbit around it (in the direction of spin), the closer the faster. It would then pass into a region where it could stay at rest only if it could travel at the velocity of light to counter the spin. The surface passed through to enter this region is the static limit. Although it cannot remain at rest here, the object can still escape. If it continued inward it would pass through the event horizon, to where space and time interchange their roles—as in the case of an ordinary black hole. Continuing on towards the center of the black hole the object would encounter the second event horizon, where space and time are again interchanged. Directly ahead is the ring-shaped singularity, which can now be avoided.

As Fig. 20.7 shows, the static limit and the event horizon touch at two points along the spin axis of the star. The surface corresponding to the static limit is oblate, but the event horizon is still spherical. Between the two surfaces is the region called the ergosphere, which is important in discussions of the Kerr black hole.

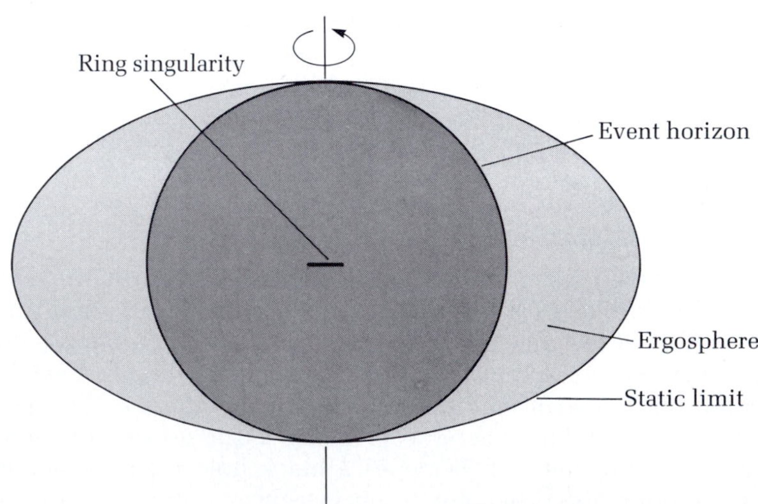

Figure 20.7 A Kerr black hole.

The collapse of a spinning star also shows points of similarity and dissimilarity with the ordinary star. Again, the star gets smaller and changes color, but, as it gets smaller it spins faster, in conservation of angular momentum (p. 327).[2] As it spins faster, it becomes increasingly oblate; finally it becomes a Kerr black hole.

The Kerr black hole, like the ordinary black hole, also has a "wormhole" associated with it. However, not all the matter pulled into it ends in the singularity, crushed to nothing; some of it may pass through the wormhole and emerge at the other end into other universes or distant points in our universe (a velocity greater than that of light is not required in this case).

Energy can be extracted from black holes. For example, were an object to enter the ergosphere of a Kerr black hole and break into two parts, with one falling into the singularity and the other escaping, the escaping particle would be emitted with considerably more energy than the original particle entered with.

This additional energy can come only from the black hole itself. The more energy that is extracted from Kerr black holes, the more slowly they spin. Eventually, after a large amount of energy is extracted they will no longer spin; in effect, they will have become ordinary black holes.

Reissner-Nordström and Kerr-Newman Black Holes

In addition to ordinary and Kerr black holes there are the Reissner-Nordström black hole and the Kerr-Newman black hole. The former has a slight electric charge when it collapses and ends with slightly different properties. Initially, with no charge, there is one event horizon and a point singularity. With a small increase in charge an inner event horizon begins to form near the singularity. With more charge its radius increases and the outer event horizon shrinks. After a large increase in charge the two horizons merge, but the singularity remains a point. If charge continues to be increased, the double event horizon shrinks until it disappears, and leaves a naked singularity. The latter is also a charged star, but one that is spinning. Some black holes in nature may be of this type initially, as most stars spin and some may have at least a small amount of charge. There are two event horizons here which also merge as spin and charge increase. A naked singularity also results if sufficient spin and charge are added.

Mini Black Holes

In the early 1970s Jacob Bekenstein noticed that black holes appeared to have a surface temperature greater than zero. According to his calculations they absorbed and emitted radiation just as any object with a temperature above absolute zero does. But this seems to be in direct contradiction of the evidence that black holes only absorb (pull in) matter.

[2]Angular momentum is defined as (mass) × (velocity of the mass) × (distance to mass). For the total angular momentum of a star you must theoretically divide the star into small masses and sum each of these masses.

According to Stephen Hawking black holes do emit (as well as absorb) radiation by pair production (that is, production of two particles of equal and opposite charge) in the neighborhood of a black hole. According to Hawking if these particles are generated just outside the event horizon one may drop into the black hole and the other may fly off into space. Furthermore, the particle that falls through the event horizon has negative energy, and, therefore, decreases the energy of the black hole. To an observer at a distance from the black hole, it would therefore appear to be giving off energy; and, since all heated objects radiate, or give off energy, it would look like a "hot" object.

However, an antiparticle (one of the two generated just outside the event horizon) can be regarded as a particle moving backward in time. This means, then, that the particle creation described above is equivalent to a particle (moving backward in time) emerging from a black hole, being scattered, then moving forward in time. In other words, particles would be emerging from black holes, which is impossible according to the way we have been looking at them—that is, according to general relativity. However, in this area another theory must be applied—namely, quantum theory, according to which black holes do, indeed, emit particles by a process called particle evaporation.

But quantum theory is of importance only for exceedingly tiny objects; as it turns out, this is also the case for black holes. Particle evaporation from average-size black holes (that is, those resulting from stellar collapse) is negligible; for tiny black holes, though, it is important, and Hawking has shown that tiny black holes may, indeed, exist, and may have been formed in the "Big Bang" explosion (Chapter 25) that created the universe. If this explosion was sufficiently inhomogeneous during the first fraction of a second of its existence, small pockets of matter may have been compressed into tiny black holes under the tremendous pressures present. And these "mini black holes," as they are sometimes called, because they would have only a small mass could lose it all in a relatively short time. According to Hawking, black holes lose mass at a gradually increasing rate as they get smaller, until finally they lose it so fast they literally explode. Since mini black holes form with a range of different masses, this means that all those below a certain limit (shown to be about 10^{15} gm; it is not visible to the naked eye at this mass) have now exploded.

The energy output of a black-hole explosion would be equivalent to the explosion of a million one-megaton hydrogen bombs; so far we have no evidence of explosions of this type. It has been suggested that one might have been responsible for the Tunguska explosion of 1908 (p. 249). It has also been suggested that our sun may have mini black holes at its center, which absorb neutrinos and, thereby, account for the apparent lack of neutrino emissions from it.

There is no reason to believe that larger black holes were not also formed in the Big Bang explosion. In fact, an entire spectrum of black holes—ranging from microscopic ones to those with masses equal to that of galaxies—may have been formed. Many astronomers now feel that some galaxies, and perhaps even quasars, might contain black holes at their cores. We will have more to say about this in later chapters.

WHITE HOLES AND WORMHOLES

Earlier we mentioned that a black hole is, in essence, connected to a wormhole or tunnel in space that may lead, at its other end, to other universes or other parts of ours. Certainly this other end could not be another black hole, since matter is pulled into black holes, and, in this case, matter would be gushing out. Astronomers refer to this other end as a white hole (Fig. 20.8).

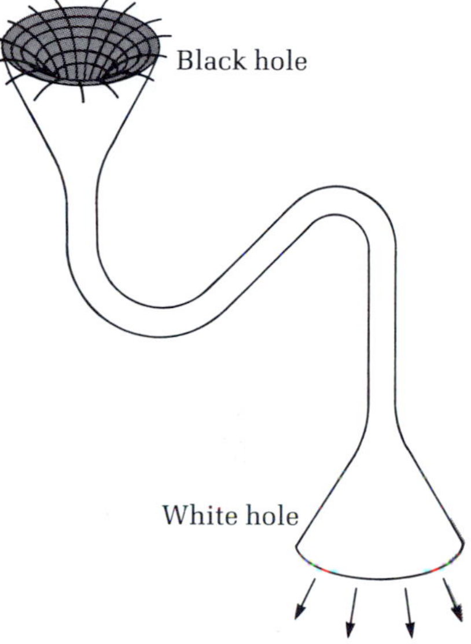

Figure 20.8 A wormhole in space.

Although the concept of black holes is now becoming quite widely accepted (with reservations), the question of the existence of white holes is still well within the realm of speculation. It has been shown that, were a white hole connected to a stellar-collapse black hole, it would be cut off from us forever; matter could not pass through the associated wormhole. However, a white hole produced in the Big Bang explosion along with a black hole might not be cut off. Theoretically, something going into this black hole could pass through and emerge at the white hole. Unfortunately for this position, Douglas Eardley has shown that all white holes will eventually turn into black holes, so that we would have no exits to our wormholes—only entrances. "Mini" white holes might also exist, but Hawking has shown that, if they do, they would be completely indistinguishable from mini black holes!

THE SEARCH FOR BLACK HOLES

Although it is extremely unlikely that we could see a black hole directly (most are only a few kilometers across, and we can barely see Mercury, which has a diameter of 4800 km), there are indirect methods for detecting them.

In the case of a double star system in which one of the stars collapses to a black hole, the other star continues to revolve much as it did before the collapse. Eventually, as its fuel is depleted, it begins to expand, its outer layers approach the black hole, and they are pulled in (Fig. 20.9). As the star continues to grow, its matter spirals rapidly into the black hole, the intense gravitational field compressing and heating it. Just before this matter passes into the black hole, surges of X-rays are emitted—X-rays that can be detected here on earth.

It would appear, then, that binary systems are prime candidates in our search for black holes. While, of course, one of the two stars in the system would be invisible, astronomers are quite familiar with them: spectroscopic binaries are a good example of systems of this type.

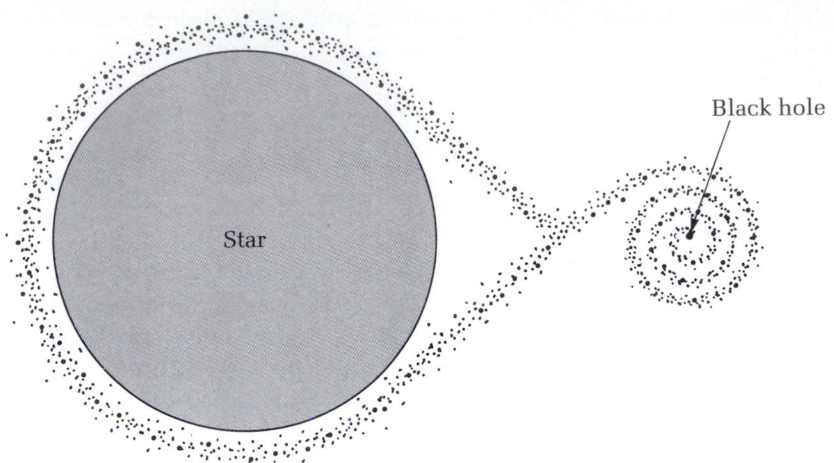

Figure 20.9 A binary system of an ordinary star and a black hole near the end of the life of the former, when its outer layers begin to be pulled into the black hole.

There are also ways of detecting black holes not part of a binary system, for a black hole embedded in a gas cloud, for example, or passing through an interstellar cloud of dust and gas would also be a strong source of X-rays. And Bessel's clue with respect to Sirius (p. 319)—a star with a wobbly path—can disclose a black-hole companion.

Once these and other techniques were established binary systems immediately took the spotlight. Two of the first astronomers to look at systems of this type were Kip Thorne and Virginia Trimble, who published a table (now called the Thorne-Trimble table) of possible candidates—binary systems with massive components.

In 1972 the first UHURU catalogue of X-ray sources was published. Of its more than 100 entries interest eventually centered around CYG X-1 (Figs. 20.10–11; 20.12, color), which emitted rapid pulses of X-rays, the shortest indicating that it was possibly the size of a black hole. The visible component of the system was located (it is approximately 9th magnitude) and the ratio of the two masses was shown to be approximately 1:2. Estimates indicated the mass of the secondary to be about 8 $M_{\odot}$ —easily massive enough to be a black hole.

Although the matter is not yet completely settled, the probability that CYG X-1 is a black hole is considered by most astronomers to be about 90 percent. Another candidate is V861 Sco in the constellation Scorpius (in many respects similar to CYG X-1), and it is thought that the elliptical galaxy M 87 may have a black hole at its core. Indeed, as we mentioned earlier, even quasars may contain black holes.

Gravitational Waves and Gravitational Lenses

There is also some evidence for a black hole at the center of our galaxy. The first clue came from Joseph Weber, who showed that the core of our galaxy appeared to be losing matter at a tremendous rate: up to 300 stars a day were disappearing.

Although Weber's results were not confirmed, many astrono-

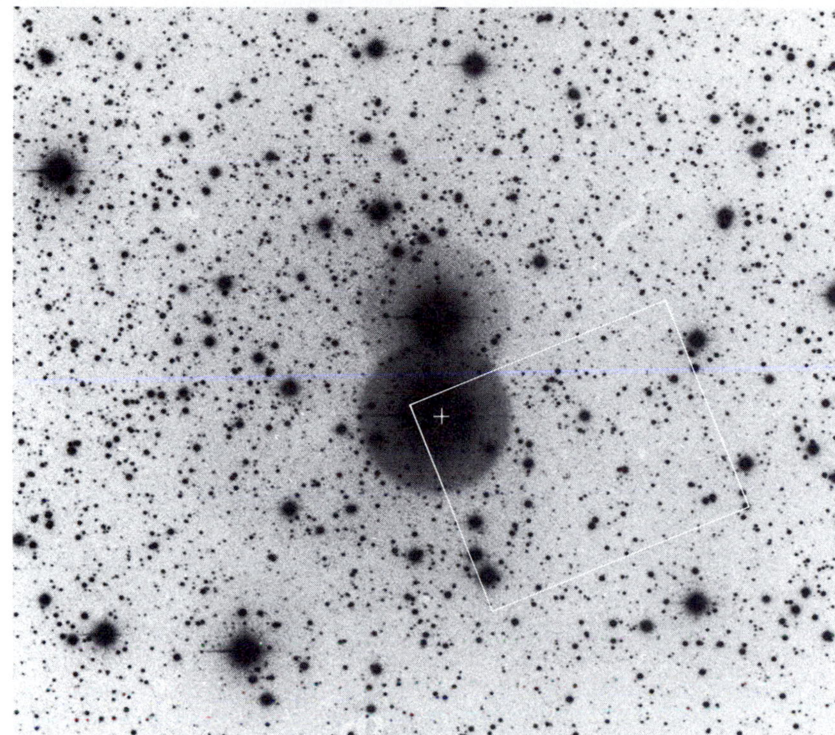

Figure 20.10 **X-ray source Cygnus X-1 (within the inscribed rectangle) is too small to be seen. The large dark object in the center of the photograph is the blue star believed to be associated with Cygnus X-1.**

mers now feel that there may be a gigantic black hole at the center of our galaxy (Chapter 21).

Weber's conclusions were based on a study of gravitational waves ("ripples in the curvature of space-time"), which are predicted by Einstein's general theory of relativity. Einstein showed that they would be given off by accelerating masses, much as electromagnetic waves are given off by accelerating charges. They would, however, be much weaker than electromagnetic waves, and they would be emitted only by asymmetrical systems. A binary star system would have the needed asymmetry, but, according to Einstein's calculations, the intensity of waves from such a system would be far below

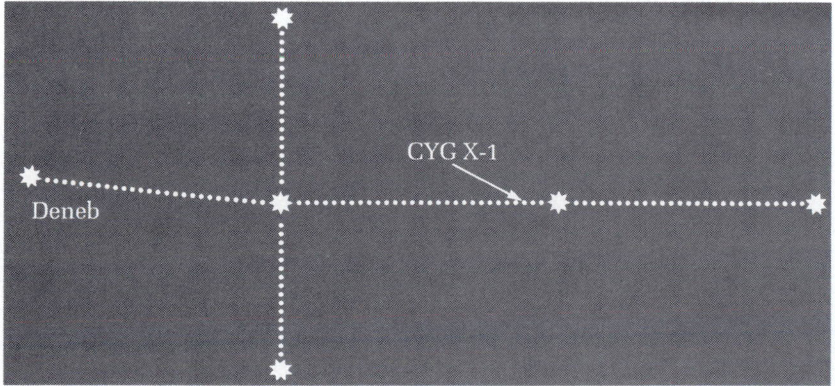

Figure 20.11 **Position of Cygnus X-1 in the constellation of Cygnus (Northern Cross), visible in the summer sky.**

the limits of detectability. The stars would have to be excessively dense and close to one another for the waves to be detectable, and the only dense stars known at that time were white dwarfs, the intensity of which would be too weak.

Most scientists are convinced that gravitational waves exist but that our instruments are not sensitive enough to detect them. (In the very near future a second generation of detectors will be put into operation.) Meanwhile, scientists at the University of Massachusetts have reported indirect evidence of gravitational waves being emitted by a binary system.

If gravitational waves are, indeed, found they may turn out to be a particularly useful tool. At present everything we "see" in the universe is seen via the electromagnetic waves that they emit. But electromagnetic waves are limited; we cannot, for example, see deep into the core of a star, or into the heart of a supernova explosion. In particular, we cannot see past the event horizon of a black hole. Gravitational waves may allow us to do this.

Another important gravitational phenomenon is the gravitational lens. It has been known for many years that stars deflect light; this was, in fact, one of the crucial tests of Einstein's general theory of relativity. It had even been suggested that this effect could be used, under the proper conditions, as a "lens." Using a nearby star or, even better, a black hole (the effect is considerably greater in this case) we could magnify distant objects. But of course we would have to be positioned in exactly the right place (that is, in line with the star and distant object we wish to magnify), and the probability of this was low. It was much more likely in practice that an alignment of this type would be approximate. In this case we would see a double image (Fig. 20.13) of the distant object or, in some cases, three or more images. In May 1979, astronomers D. Walsh, R. Carswell, and R. J. Wyman announced that they believed they had detected the first gravitational lens. They had discovered two side-by-side quasars (Chapter 23) that appeared to have exactly the same properties in the visible region of the spectrum and in the infrared, ultraviolet, and radio regions as well. The object causing the deflection was a galaxy, that, strangely enough, is not midway between the quasars in our line

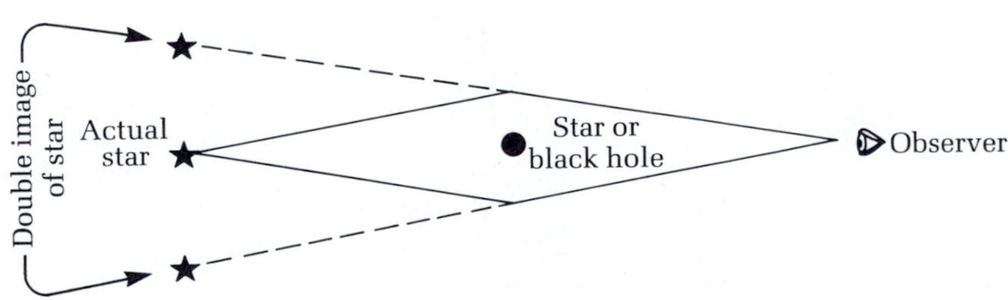

Figure 20.13 Mechanism of the gravitational lens.

of sight. A second probable gravitational lens, discovered by R. J. Wyman and several colleagues in June 1980, showed three identical images. Astronomers are now generally convinced that we are, indeed, seeing the gravitational lens effect.

SUMMARY

1. Black holes are not an outgrowth of general relativity, but are part of all seriously considered theories of gravity.
2. If the earth were at the position of the sun and grew while maintaining constant density it would be a black hole by the time its surface reached the orbit of Mars.
3. The surface of a black hole is called an event horizon. Escape from the black hole is impossible once the event horizon has been passed. At the center of the black hole is a singularity, a point of zero volume and infinite density.
4. If you watched a collapsing star from a distance it would appear to become fixed at a certain size (its gravitational radius). If you were on the surface of the star as it collapsed, however, you would pass rapidly through the event horizon and be crushed at the center.
5. Black holes curve the space around them. If we could see this curvature near the black hole it would look like a wormhole in space. A velocity greater than that of light is needed to pass through the wormhole of a Schwarzschild (ordinary) black hole.
6. Spinning black holes are known as Kerr black holes. Besides an event horizon they have a static limit. The region between the static limit and the event horizon is called the ergosphere. Kerr black holes have ring-shaped singularities. Matter can pass through (at velocities less than c) their wormholes.
7. Two other types of black holes are the Reissner-Nordström black hole (charged) and the Kerr-Newman black hole (charged and spinning).
8. Black holes emit particles; the smaller the black hole the greater the emission rate. The effect is generally negligible in stellar-collapse black holes but is significant in mini black holes. When black holes become tiny enough they explode.
9. Black holes may be connected via space-time wormholes to white holes, which are, in effect, time-reversed black holes. Naked singularities, or singularities not surrounded by an event horizon, may also exist.
10. Our best black-hole candidate at present is CYG X-1. It appears to have all the required properties of a black hole.
11. Einstein's general theory of relativity predicts that oscillating masses (that is, a binary system) should give off gravitational waves. These waves may have been indirectly detected.
12. Astronomers may have discovered a gravitational lens.

REVIEW QUESTIONS

1. Are black holes predicted only by the theory of general relativity?
2. Describe the sequence of events when a star collapses to a black hole.
3. What is an event horizon? What is an exit cone? What is the connection between the exit cone and the event horizon?
4. If there were no tidal forces, would there be any force acting on a person free-falling into a black hole?
5. How many types of black holes are there? Explain how they differ.
6. What is an Einstein-Rosen bridge? Could we travel through one?
7. What is the static limit? How is it related to the ergosphere and event horizon of a Kerr black hole?
8. Do the static limit and event horizon of a Kerr black hole touch? Where?
9. Do all types of black holes have ergospheres? If not, which ones do?
10. Can energy be extracted from all types of black holes? Explain.
11. How many event horizons does a Kerr-Newman black hole have?
12. Explain how particles are emitted from black holes according to the mechanism proposed by Hawking.
13. What is the basis of the argument for the existence of mini black holes?
14. What happens to a black hole as particles are emitted from it?
15. Why is particle creation critical for mini black holes? What eventually happens to them?
16. Explain how naked singularities may be created.
17. What is a white hole? How is it related to black holes?
18. Discuss some of the methods astronomers could use to detect black holes.
19. Discuss the CYG X-1 system.
20. What is a gravitational wave? Do they appear to exist?

THOUGHT AND DISCUSSION QUESTIONS

1. The formula for the gravitational radius is $R_{grav.} = 2\ GM/c^2$, where G is the universal gravitational constant (6.67×10^{-11} m^3/km sec^2) and c is the velocity of light (3×10^8 m/sec). Calculate the gravitational radii of the sun, the earth, and the moon.
2. Suppose Observers A and B have clocks that are synchronized well away from a black hole. If A falls towards a black hole:
   a) How does B see A's clock run (faster or slower than his)?
   b) How does A see B's clock run?
   c) If A has passed through the event horizon, how will B see his clock run? How will A see B's clock run?
   d) If A and B are inside the event horizon but A is closer to the singularity, how will B see A's clock run? How will A see B's clock run?
3. Assume you are out in space in a spaceship and encounter a Schwarzschild black hole. You get caught up in its gravitational field and are pulled into it. Describe in detail what you would experience. (Assume you remain alive until you reach the singularity.) Do the same for a Kerr black hole. (Assume you pass through the singularity.)

FURTHER READING

Block, D. L. "Black Holes and Their Astrophysical Implications," *Sky and Telescope,* pp. 20/87 (July/August 1975).

Hawking, S. "The Quantum Mechanics of Black Holes," *Scientific American,* p. 34 (1977).

Parker, B. "Black and White Holes: Minis, Maxis and Worms," *Star and Sky,* p. 32 (December 1979).

———. "Other Black Hole Candidates," *Star and Sky,* p. 26 (February 1981).

———. "Mini Black Holes," *Astronomy,* p. 26 (February 1977).

———. "Searching for Gravitational Waves," Encyclopedia Britannica Yearbook of Science and the Future (1981).

Penrose, R., "Black Holes," *Scientific American,* p. 38 (May 1972).

Shields, G. "Are Black Holes Really There," *Astronomy,* p. 6 (October 1978).

Shipman, H. *Black Holes, Quasars and the Universe,* Second Ed. Houghton Mifflin, New York (1980).

Sullivan, W. *Black Holes: The Edge of Space, the End of Time.* Doubleday, New York (1979).

part 6

Galaxies and Quasars

In this part we turn to one of the largest units in the universe: galaxies. The amazing distances that we have seen separating the stars are tiny compared with the distances involved in a discussion of galaxies. A traveler moving at the speed of light would take 100,000 years to cross our galaxy. If he then moved out from it in search of other galaxies he might not reach one for 100 *million* years. It is, without a doubt, difficult to comprehend such distances.

Before astronomers knew what the fuzzy objects in the sky were, all were referred to as nebulae. We now know that many of these fuzzy objects are composed of billions of stars—"island universes of stars"—and these are properly called *galaxies*; the term *nebula* is now reserved for gaseous clouds that are *within* galaxies.

The first chapter of this part deals with how the structure of our galaxy—the Milky Way—was determined and with the discovery of 21-cm radiation, which has allowed us to discover parts of our galaxy never before "seen."

In the following chapter—Chapter 22—we look out at the many galaxies in the universe that are like ours and at others that are quite different. We will also see how all these have been classified into three major types. One of the most important unsolved questions in astronomy will be introduced in this chapter: How do galaxies evolve? This chapter will also examine violent galaxies and galaxies that are strong sources of radio waves and in some of which there seem to be gigantic explosions.

Although quasars were discovered over 15 years ago, astronomers are still not certain what they are. In Chapter 23 we will talk about this major enigma of modern astronomy and some of the problems related to it.

Aim of the Chapter

To introduce our galaxy and show how its structure was determined.

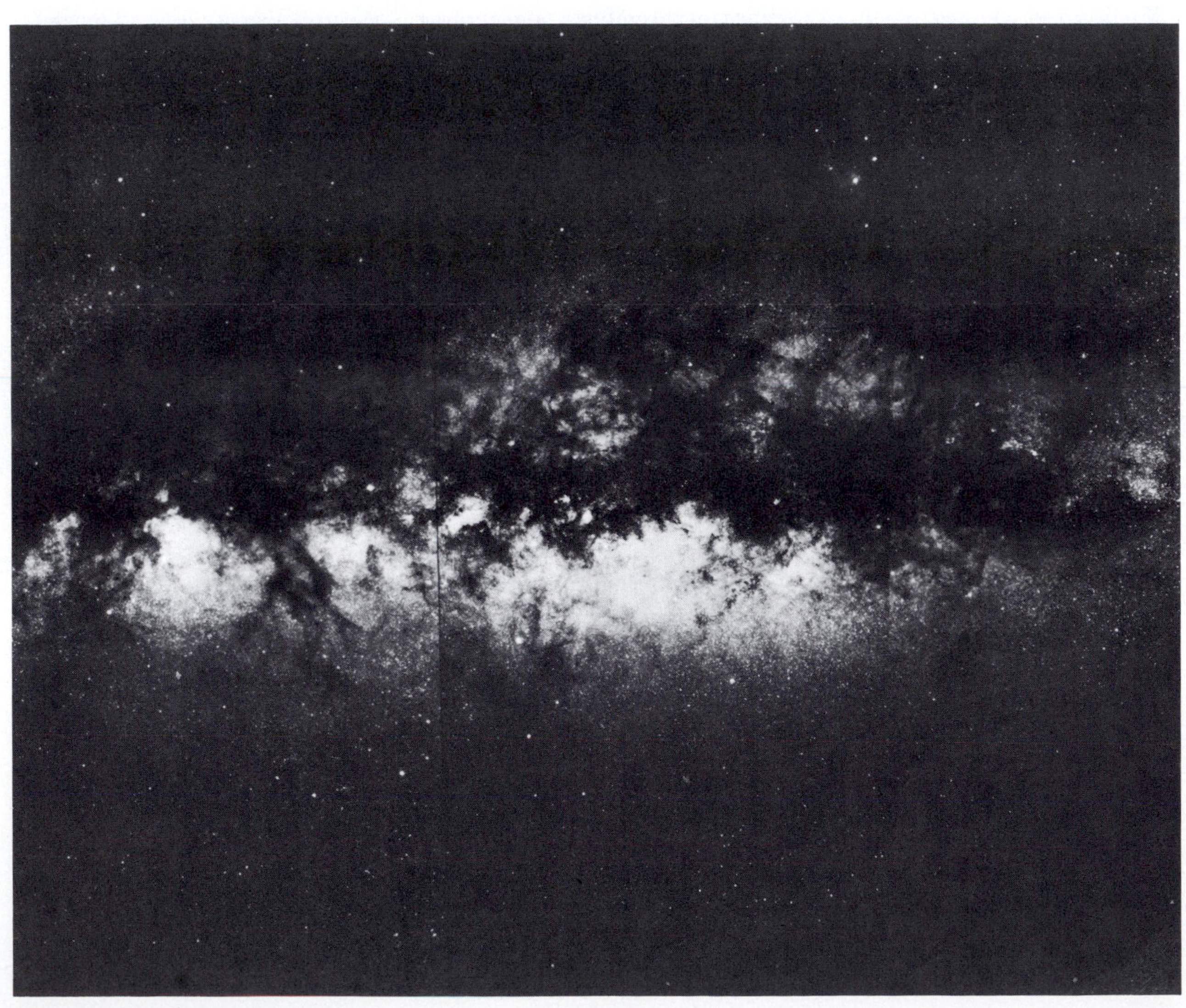

chapter 21

The Milky Way Galaxy

In 1610, when Galileo turned his telescope toward the dimly glowing stream of silvery clouds of our Milky Way (Fig. 21.1), he saw that they were, in reality, thousands of tiny stars (Fig. 21.2). But why were there many stars in this direction and few in a direction generally perpendicular to it?

EARLY MODELS

In 1750, English theologian Thomas Wright suggested an answer: the Milky Way is the gigantic system of stars we live in, and this system is not symmetrical around us. Wright presented what we might call a grinding-wheel model. If we were at the center of this grinding wheel we would see few stars if we looked out through a side; looking out along a radius, however, we would see numerous stars; in fact, we would see an infinite number according to Wright, who assumed that the radius was infinite.

William Herschel, one of the greatest of early astronomers, pondered Wright's model. Herschel was interested in determining the overall structure of our galaxy. Using the technique of star gauging,[1] he deduced that it was a flattened disk about 16,000 light years across (Fig. 21.3) and that we were somewhere near its center. His son John suggested (correctly) that it was disk-shaped because of its rotation. Indeed, the idea of rotation seemed to fit in quite well with the discovery (about 1850) of the first spiral by Lord Rosse of Ireland, who thought they might be gigantic clouds of stars. By 1900, numerous spirals had been seen and photographed (Fig. 21.4), and Cornelius Easton, of Holland, even suggested that our galaxy might be a spiral

[1] Star gauging is akin to determining one's position in a forest by counting trees in various directions. If you are near an edge, the trees in this direction will thin out quite rapidly, while, in the opposite direction, they will seem to go on indefinitely. If you are near the center of the forest, the trees will appear to extend away uniformly in all directions. In star-gauging, the trees are stars.

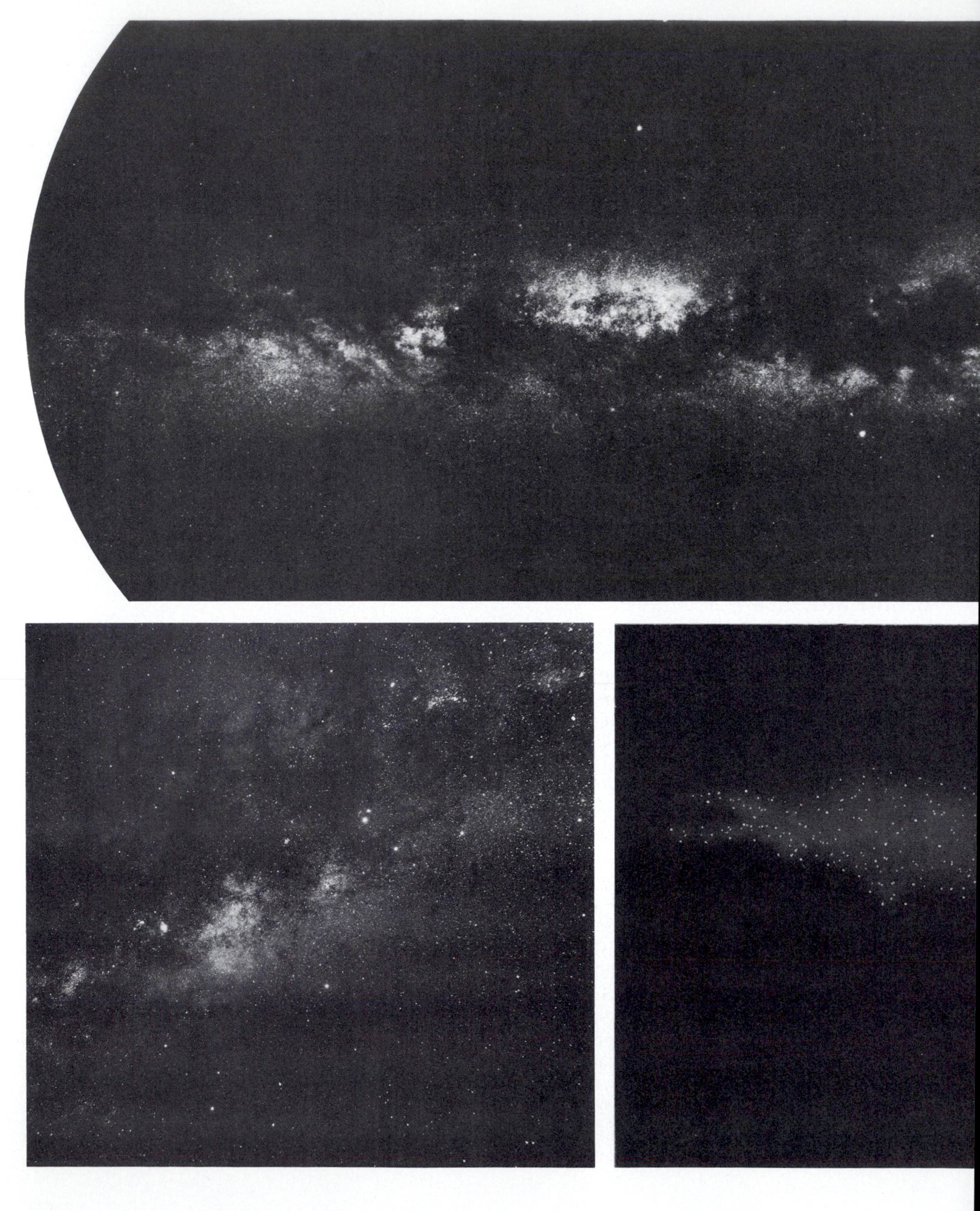

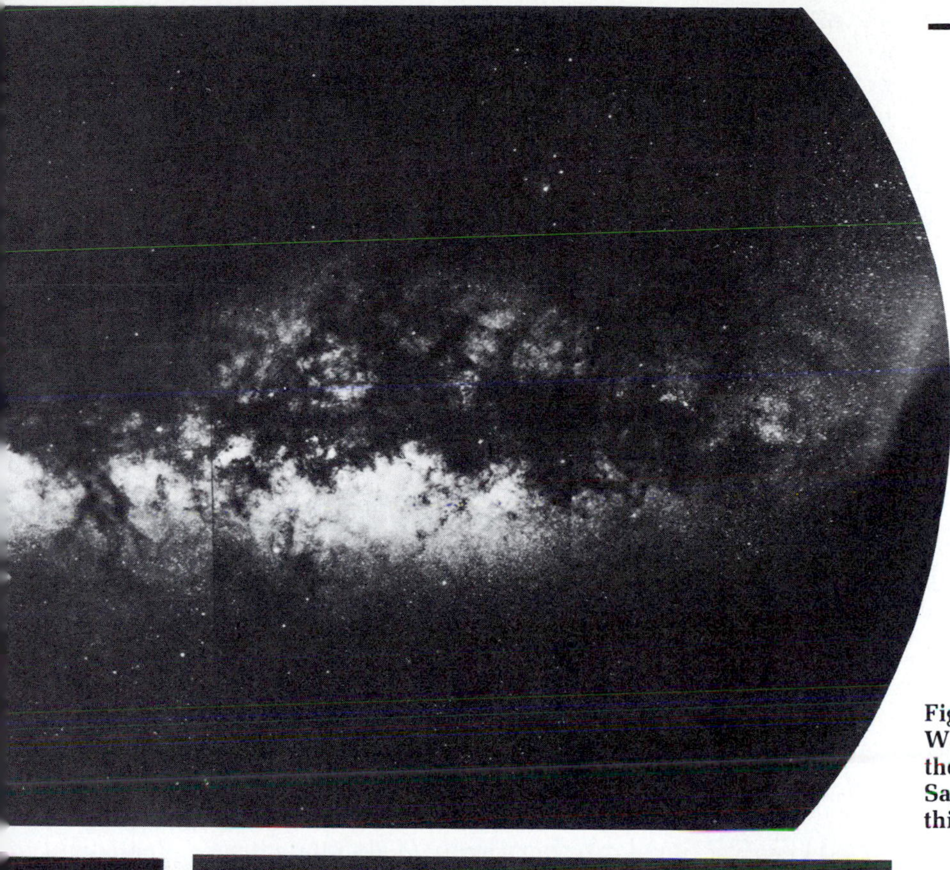

Figure 21.1 Mosaic of the Milky Way. The direction of the center of the Galaxy is in the constellation Sagittarius, which is to the right in this photograph.

Figure 21.2 (far left) A small section of the Milky Way, showing the great density of stars.

Figure 21.3 (center) Herschel's model of our galaxy. The sun was assumed to be at the center.

Figure 21.4 (near left) M51, a representative spiral galaxy.

and proposed a model amazingly close to our presently accepted one. The idea that the spirals might be "island universes" of stars was now being taken seriously. (The concept and name "island universe" originated with Immanuel Kant about 1750.)

KAPTEYN MODEL

Early in the twentieth century, Dutch astronomer Jacobus Kapteyn organized a worldwide effort to look in detail at selected areas of the Milky Way. His resulting model was similar to Herschel's: a flattened disk gradually trailing off towards the edge. It was, however, somewhat larger (approximately 26,000 light years in diameter), and, for most astronomers, it had a serious defect: the sun was at its center. After having been dislodged twice from the center of things—the universe and the solar system—was it possible that Earth was again at the center of something? Kapteyn realized that his model depended on the clarity of space and that, if even a small amount of absorption was taking place (as a result of matter between the stars) his model would be flawed. Distant stars would be so faint they wouldn't be seen.

SHAPLEY MODEL

In 1914 Harlow Shapley began a study of globular clusters. Realizing that Leavitt's period-luminosity relation for the cepheids in the Magellanic Clouds (p. 305) could be an invaluable tool, he began a serious study of it. He set out, first of all, to prove that the relationship was valid for nearby cepheids. Once he had satisfied himself that it was, he turned to the cepheids in globular clusters. Over 100 globular clusters were known, but not all contained cepheids. Eventually he determined the distance to a considerable number and found they were distributed roughly in a sphere centered not on the sun, but a considerable distance away in the direction of the constellation Sagittarius. Shapley realized that this sphere of globular clusters was probably centered on our galaxy, and that this must mean that we were out near an edge (Fig. 1.1K). In short, our sun was just another star out near the edge of a large group of stars.

According to Shapley's calculations the group was gigantic—10 times larger, for example, than Kapteyn's model—and, if the spiral nebulae were also groups of stars, they could be no more than tiny satellites of our system.

OORT AND LINDBLAD MODELS

The idea that our galaxy rotated was fairly well established by the 1920s, but many questions remained. There were stars, with peculiar velocities, that seemed to be flying off in random directions at exceed-

Harlow Shapley

Harlow Shapley was born in Nashville, Missouri, in 1885. By fifteen he was working as a newspaper reporter. Realizing that, because of his educational shortcomings, he was inevitably passed over when it came to important assignments, he enrolled in 1907 at the University of Missouri. His aim was to study journalism, but when he arrived to register, the journalism school had not yet opened, and he was compelled to find substitute courses. Near the beginning of the course listings was *Astronomy*, and Shapley decided to try it; a strange beginning for someone who was eventually to become one of the greatest astronomers of all time.

After graduating in 1911 Shapley went to Princeton to work under Henry Norris Russell, who was struggling with eclipsing binaries at the time and was quite eager to turn some of the problems over to Shapley. Shapley made tremendous progress; within two years he had completed his thesis, now considered to be a masterpiece.

Shortly after he obtained his doctorate he married and accepted a position at the Mt. Wilson Observatory. The route to the observatory was still in primitive condition and a pack train was needed to get to the top. It was a nine-mile trip up 6000 feet, but on the summit was the instrument that Shapley would use to change the course of astronomy—the 60-inch reflector. The next few years were the most productive of his life.

Shapley turned his attention to variables—cepheids and RR Lyraes—and showed that they could be used as "measuring sticks." The ones in globular clusters, for example, could tell him how far away the clusters were. Within four years he had determined the distance to about one quarter of the hundred or so globular clusters known—and soon after astronomers knew that we were not at the center of our galaxy.

Shapley published furiously for several years. After determining the direction and distance to the center of our galaxy, he determined its diameter to be 250,000 ly ($2\frac{1}{2}$ times the presently accepted value). Many astronomers argued that his estimate was at least ten times too large. H. Curtis was among the skeptics. The disagreements culminated in the Curtis-Shapley debate of 1920 (Chapter 22).

Shortly after the debate, Shapley left Mt. Wilson to assume the directorship of Harvard Observatory where he continued to find time to work on variables and galaxies and to make important discoveries.

In his later years Shapley turned to things other than astronomy. Much of his time was spent on humanitarian projects. During World War II, for example, he helped many Jewish scientists to escape from Germany, and, after the war, helped set up UNESCO—all this while continuing to travel and lecture. He died in 1972.

ingly high velocities. Dutch astronomer Jan Oort soon showed that they were actually "intruders" in the Milky Way; they did not follow its overall rotation and, in general, their orbits were not in the plane of the disk. In fact, their velocities were not even as high as they appeared; rather, we were leaving them behind and so it was we who had the high velocity (from the rotation of our galaxy).

The theoretical side of the problem was taken up by Swedish astronomer Bertril Lindblad about the same time. He worked out the dynamics of a gigantic rotating disk of stars and proposed a model that agreed quite well with Oort's observations. Just as the inner planets move faster than the outer ones (Kepler's law), the inner stars in the rotating disk of the galaxy should move faster than the outer ones, and, indeed, Lindblad took this into consideration. In a system as large as this (compared to the solar system), the stars in the outer regions can be expected to follow Kepler's law, but, near the center, most of the mass of the system is outside the orbits of the stars; consequently, Kepler's law will not be satisfied. It turns out, in fact, that the stars near the center move almost like a solid disk. In summary, then, Lindblad and Oort concluded that our system was a gigantic disk that rotated in space. Stars slightly closer to the center than us rotate faster, and those in the nucleus rotate nearly in unison (although Lindblad didn't specify this). Stars farther out from the center than us rotate more slowly and in a first approximation, obey Kepler's law (Fig. 21.5). In essence it was a system with a rather complicated form of differential rotation (i.e., different rotational rates at different distances from the center) and with a surrounding spherical halo of globular clusters and peculiar high-velocity stars.

In 1930 R. J. Trumpler reported that there was an absorbing medium between the stars (see p. 386). Kapteyn's model, then, was incorrect, and Shapley's, although still basically correct, was reduced considerably in size.

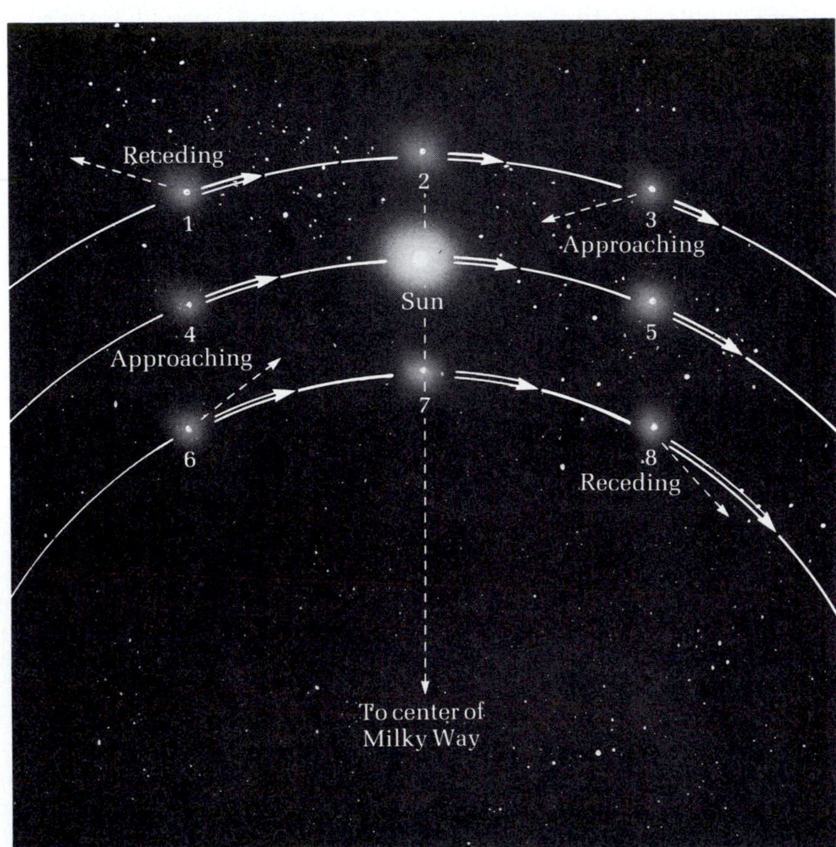

Figure 21.5 Relative velocities of stars in the neighborhood of the sun. Those just inside its orbit in the Galaxy move more quickly than those outside (further from the center of the Galaxy). Thus Star 6 appears to us to be approaching, Star 8 to be receding.

STELLAR POPULATIONS

During World War II, German emigré Walter Baade, using the 2.5-m reflector at Mt. Wilson, and aided by particularly good "seeing" (the lights of Los Angeles were out because of the wartime blackout) was attempting to resolve the inner regions of the Andromeda Nebula (Fig. 1.8) into individual stars. (Hubble had shown that the outer regions could be resolved.) Baade's first few plates, like those that had been taken so many times before, were mushy and diffuse at the center. On the chance the majority of stars here were red, he then tried red-sensitive plates, which proved to be what was needed; the core was soon resolved.

Because Baade's H-R diagram for the region was quite different from the diagram plotted using the stars in the arms, he concluded that there were two types of stars, which he designated Population I and Population II. Those in the spiral arms were generally Population I while those in the core and halo were Population II. Perhaps of more importance was the likelihood that this also applied to our galaxy, which meant that the stars in our core were different from the ones around us. Baade went on to show that globular clusters and the peculiar high-velocity stars mentioned above were also Population II.

Studies have shown that Population II stars formed about 12–15 billion years ago, when our galaxy was very young, and are therefore old and composed almost entirely of hydrogen and helium (initially, all there was in our galaxy). Most of these objects do not orbit in the disk of the Galaxy, but have highly inclined elliptical orbits. Population I stars, on the other hand, formed well after the birth of our galaxy and usually located in the spiral arms, are relatively young, generally hot, and contain many elements in addition to hydrogen and helium. About two percent of their total is carbon, oxygen, neon, silicon, iron, and so on.

Another important result of Baade's study was the discovery that there are two types of cepheids, and that the two have different period-luminosity relations. One of the two types is what we now call W Virginis variables, which Shapley had mistakenly assumed showed the same period-luminosity relation as nearby cepheids. When Baade made the appropriate corrections the estimate of the size of our galaxy was again decreased—to about 100,000 light years across, the presently accepted value.

DISCOVERY OF 21-CM RADIATION

By the early 1940s astronomers knew a considerable amount about the size and shape of our galaxy, and were convinced that it had spiral arms like most of the other spirals that could be seen. However, what the spiral arms actually looked like was obscured by a fog of interstellar dust. To overcome this, astronomers turned to the possibilities of radio astronomy. In Holland, H. C. Van de Hulst, recognizing that most of the matter in our galaxy is hydrogen,

Figure 21.6 Representation of a "spin flip" (change from condition in left-hand diagram to that at right). Left-hand configuration has higher energy, hence the release of a photon when the flip occurs.

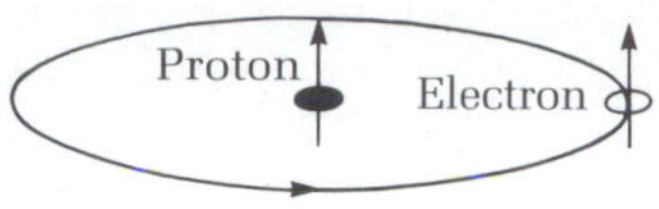

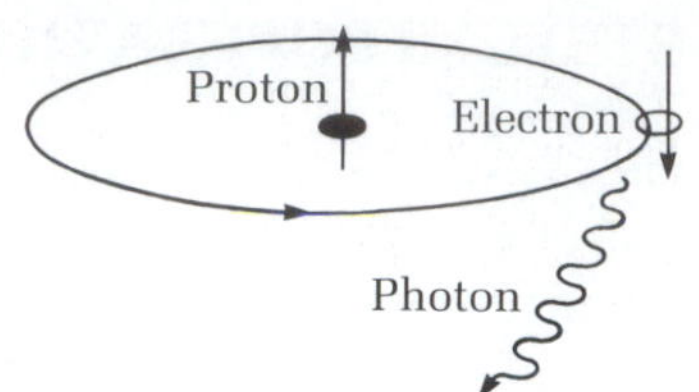

focused on 21-cm radiation, the signal produced by hydrogen. (According to quantum theory, the proton and electron composing the hydrogen atom have a property called spin, as shown in Fig. 21.6. In the high-energy configuration the electron and proton are spinning in the same direction. When the electron changes its direction of spin (called a spin flip) a photon is emitted whose wavelength is approximately 21 cm. Van de Hulst calculated that a spin flip of this nature occurs in a given hydrogen atom once every 11 million years. Moreover, space contains only about one hydrogen atom in each cubic centimeter. Nevertheless, because our galaxy is so immense there is easily enough hydrogen to make the signal observable.)

In 1951, Edward Purcell and E. M. Ewen, using a small, rather crude, instrument constructed mostly of borrowed parts, reported the first detection of this radiation (Figure 21.7); confirmation soon came from Oort and Muller and from F. Bloch in Australia.

Within a few years much was learned about the structure of our galaxy; it was, indeed, a spiral like the Andromeda galaxy (Fig. 21.8).

Figure 21.7 An artistic representation of the structure of our galaxy, based on 21-cm radiation data.

Figure 21.8 (opposite) A conception of the appearance of our galaxy, with edge-on view at top. The small circle indicates position of the sun.

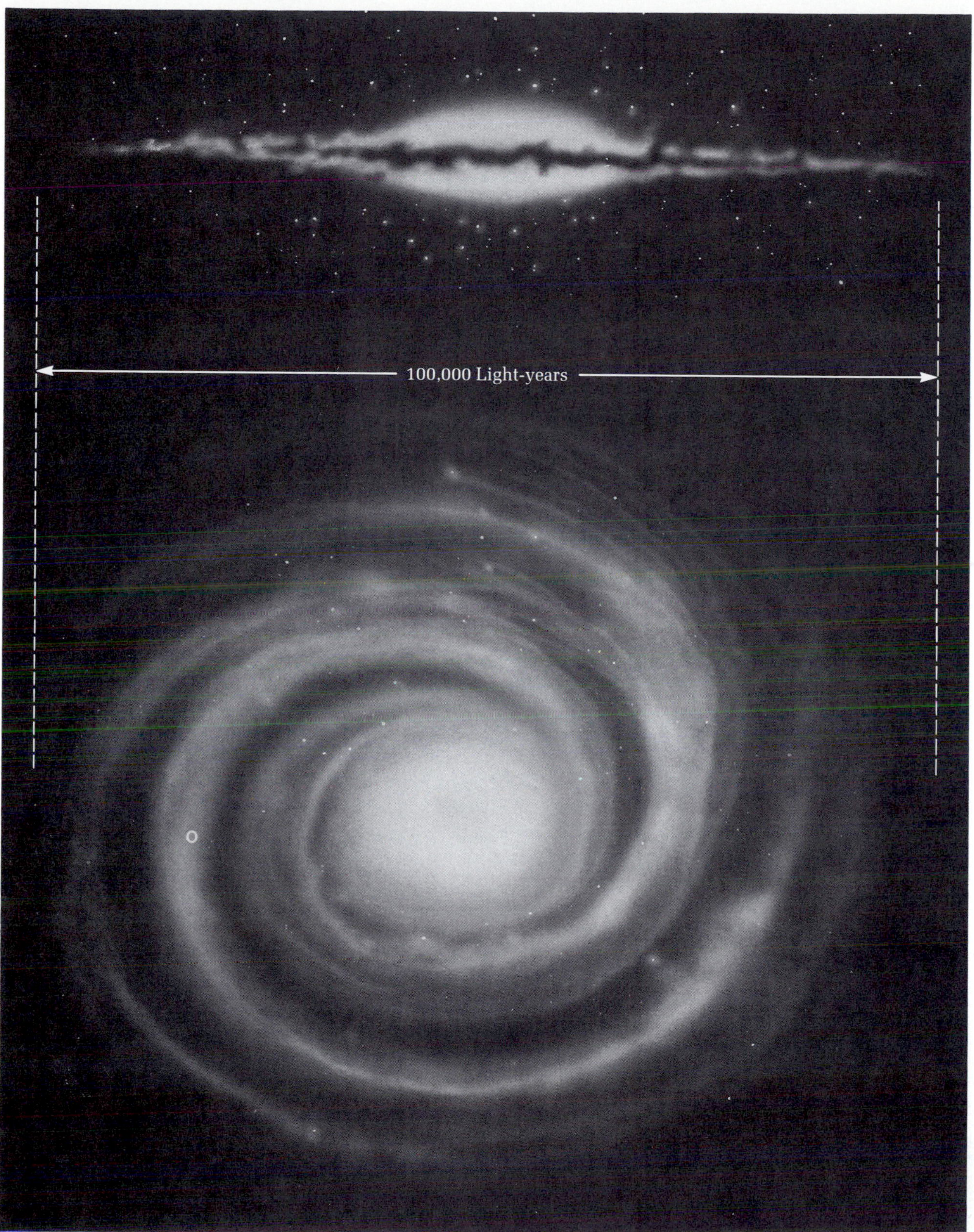
100,000 Light-years

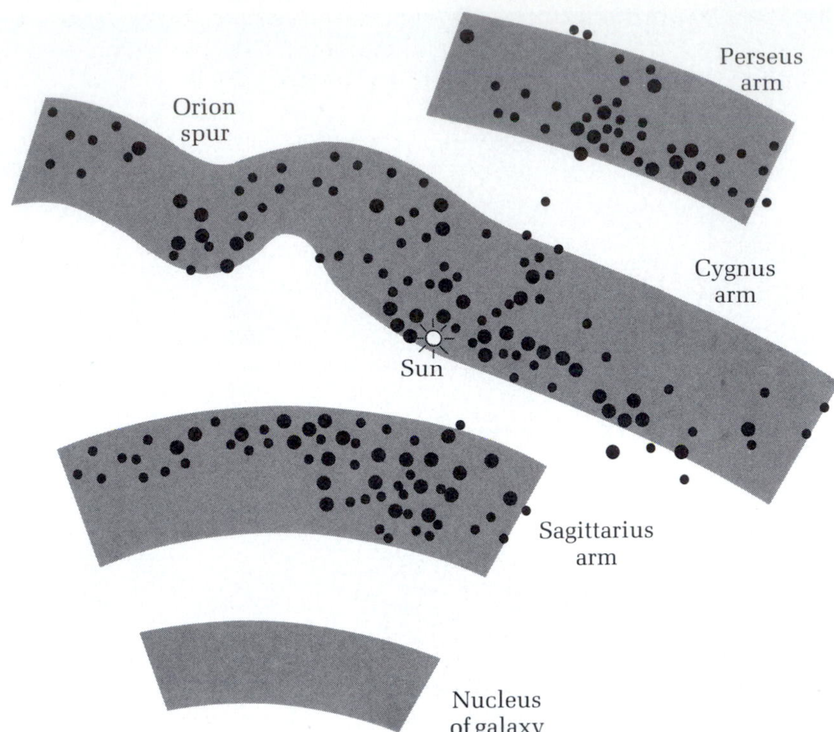

Figure 21.9 Position of the sun relative to nearby arms of the Milky Way.

Several arms were soon identified and named (in general, after the major constellation in their direction), and it was determined that we live on the inner section of one designated the Cygnus Arm, near the Orion Spur (Fig. 21.9). The arm between us and the center is called the Sagittarius Arm, and the arm outward from the center beyond us is the Perseus Arm.

Valuable as 21-cm radiation is, it has a shortcoming: When the density of hydrogen atoms is high, hydrogen molecules begin to form, and these do *not* emit 21-cm radiation. Astronomers began, therefore, to look for signals from other elements and molecules. In 1963, hydroxyl (OH) radiation (with a wavelength of about 18 cm) was discovered, and, within a few years, had been detected in several regions of the Milky Way. Hydroxyl radiation, however, did not conform to the expected spectral pattern, for in addition to absorption lines there were relatively intense emission lines. It was found that certain energy sources, by a kind of "pumping action," were raising the energy of some of the OH molecules to higher levels. When these excited OH molecules returned to the lower state they stimulated emission in nearby molecules, thereby causing them to drop to lower energy states. This is, in fact, exactly what happens in a maser,[2] so that, in essence, astronomers were observing an "interstellar maser."

By the time the OH radiation was finally understood, molecular radio astronomy was well on its way to becoming a major branch

[2] Masers and lasers are basically the same, except that the laser operates in the optical region of the spectrum and the maser operates in the microwave or radio region; laser is the acronym for *L*ight *A*mplification by *S*timulated *E*mission of *R*adiation, while the *M* in maser stands for "microwave."

of the discipline. In a sense, 1968 marks its birth, for that was the year that radiation for water (H_2O) and ammonia (NH_3) were discovered. Rapidly after that came discoveries of formaldehyde (H_2CO), carbon monoxide (CO), hydrogen cyanide (HCN), and cyanoacetylene (HC_3N), and methyl alcohol (CH_2OH), and these constitute only a small fraction of those that have now been discovered.

CURRENT VIEW OF OUR GALAXY

Our galaxy as we know it today has a diameter of about 100,000 light years (about 30 kiloparsecs). The inner 16,000 light years or so (about 5 kiloparsecs), called the nucleus, is a bulging region of Population II stars. Around this nucleus are wrapped two spiral arms, in one of which Earth is situated about 32,000 light years (10 kiloparsecs) from the center. The region in which Earth is located is about 1000 light years thick (Fig. 21.8).

The whole system rotates, with the inner regions rotating faster than the outer. The sun's speed is about 250 km/sec, which means that it takes about 250 million years to make one revolution; since its birth it has done so perhaps 50 times.

The Galaxy has a mass (calculated from Kepler's laws) approximately 100 billion times that of our sun. On the assumption that our sun is an average star, our galaxy contains about 100 billion stars. There is recent evidence, however, that our sun may be twice the average; if this is the case, there are about 200 billion stars in our galaxy.

If we look (in summer) toward the dense star clouds in Sagittarius we are looking directly toward the center of our galaxy (Fig. 21.9). There is, however, so much dust in our way, we cannot see the actual center. Some evidence of this dust is seen in the Trifid and Lagoon Nebulae which lie in this direction (Fig. 21.10, color).

If we turn in the direction of the constellation Cygnus we are looking down the arm in which the sun lies (the Carina Cygnus Arm); this direction is roughly perpendicular to that of the center. The stars in the other two directions shown in the figure can be seen only in winter. The region directly outward from the center is in the direction of the constellation Auriga. Since we are on the inner part of our arm, we are looking across it when we look in this direction. Two well-known objects in this general direction are the Pleiades and Hyades. Finally, in the other direction—down our arm (opposite Cygnus)—we find the constellation Orion.

Nucleus

As we have said, gigantic clouds of dust and gas obscure our view of the Galaxy's nucleus. However, radio waves penetrate gas and dust, and, as it turns out, so do infrared and X-rays. A study of these waves has revealed a particularly strong source at the core, where 60 million stars are packed into a region barely 32 light years across.

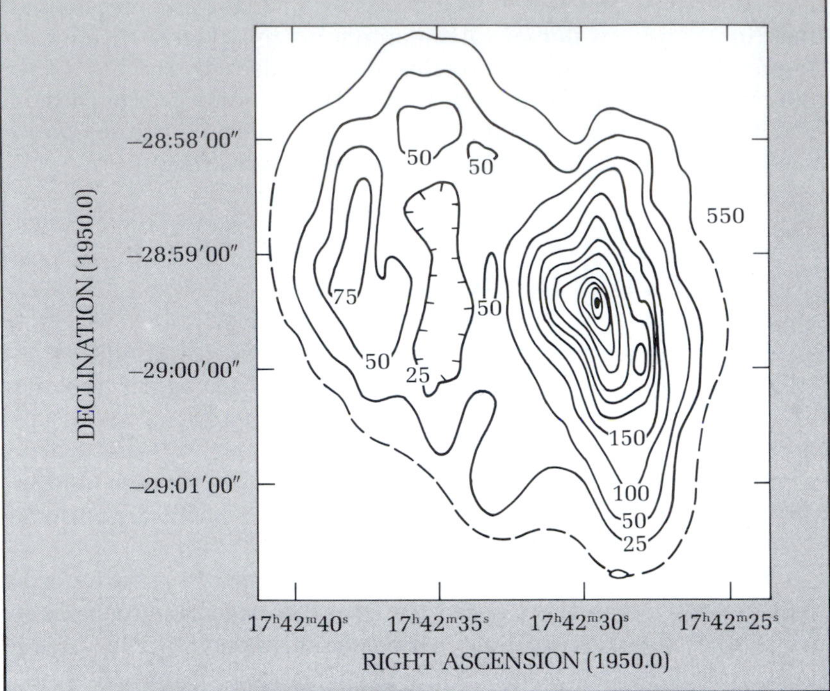

Figure 21.11 Radio map of the central region of our galaxy. Sagittarius A (the strong source) is at the right; Sagittarius B is at the left.

Called Sagittarius A, it is, in fact, one of the strongest radio sources in the sky; nearby is another strong source, Sagittarius B (Fig. 21.11).

In addition to the enormous release of energy in the region, there appears to be a rapid outward motion of gas just outside the nucleus. The cause of these phenomena is still speculative; perhaps it is a chain reaction of supernovae. (Because the stars are so densely packed—and old—one supernova could easily trigger similar cataclysms in nearby stars.) Collisions are also a possibility. Neutron stars and white dwarfs are likely to be common in this region, and collisions between dense stars like this would generate a large amount of energy. (However, stellar collisions are generally considered to be exceedingly rare, even in relatively dense regions.) Finally, there is also the possibility of a gigantic black hole, which may have started from the collapse of a single star and then grown as more and more stars were pulled into it. (Energy would be released as stars passed through the event horizon.)

Gas, Dust, and the Interstellar Medium

Along with most of the gas and dust, the brightest and youngest stars are in the two long trailing arms that radiate from the nucleus. We do not, in general, see the gas and dust; in some regions, however, its density is slightly higher than the average and we see it as nebulae. (Even these regions are an excellent vacuum by earth standards, being many orders of magnitude better than anything we can produce.)

Nebulae are generally classified into three main types: emission, absorption (or dark), and reflection. The emission nebulae are seen as a result of the light they emit. The Great Nebula of Orion (Fig. 21.12, color) is a good example of a nebula of this type. Dark, or absorption, nebulae emit no light, and are detectable (by the light they block) when they lie in front of emission nebulae or dense star fields; the Horsehead Nebula (Fig. 21.13) is an excellent example of one. Reflection nebulae are dust and gas clouds illuminated by the light from nearby stars.

Their glow results from the fact that photons from nearby stars knock the electrons from the atoms of hydrogen (which makes up a large portion of the gas of these nebulae), creating ionized hydrogen (a mixture of free protons and electrons). Regions of this type are called H II regions. Regions in which hydrogen is present in its normal atomic form are referred to as H I. Bright, hot stars such as O and B types (p. 263) are usually surrounded by large H II regions. Because the atoms in and around H II regions are continually being

Figure 21.13 The Horsehead Nebula.

struck by photons, their electrons (when they have electrons) are constantly cascading up and down as they shift energy levels. This leads to the emission of light of many different colors, giving rise to some of the most beautiful objects in the sky (Fig. 21.12, color).

Most of the gas in our galaxy—known as the interstellar medium—is quite invisible to us and lies primarily in the disk. About 90 percent of it is hydrogen, the remaining 10 percent being mainly helium, with traces (less than 1 percent) of such elements as carbon, nitrogen, oxygen, neon, silicon, and iron. Molecules of many different types are also present.

About 1 percent (by weight) of this material is considered to be interstellar dust—tiny grains of ice, rocks, or perhaps carbon that have a diameter about that of smoke particles. An important property of this material is that it reddens starlight. If a beam of light from a distant star passes through a dust cloud on its journey to your eyes, it will interact with the dust in this cloud, and the interaction will cause a scattering of some of the blue light out of the beam. The light that is left will be enriched in red light.

In addition to this reddening there is an overall extinction in all colors—a general obscuring of the light. The effect of this can be seen in many nebulae. (Note the dust lanes in Fig. 21.10.)

The Spiral Arms

The long, trailing arms of our galaxy have been mapped in some detail, one arm being distinguished from the other largely on the basis of the signals they send us: Since, as we saw earlier, stars in the arms rotate at different speeds, the stars in the outermost arm rotate slower than those in the arm just beyond us; this difference produces a Doppler shift in the signal. (We are, of course, also moving, but our motion relative to each arm is still different.)

Why our galaxy has spiral arms is a difficult question, whose answer is still speculative.

Consider a simple experiment. Sprinkle some powdered material into a pail of oil and stir the oil in the center of the pail. The outer regions will also move in a circular path, but at a slower rate than the center (just as the outer regions of our galaxy do). After only a few swirls the powder will form spiral arms around the center very similar to the arms of a spiral galaxy. Is this, then, the mechanism of the formation of our galaxy? If we continue stirring, we see that it cannot be, for the spirals soon get more and more tightly wound, until we can hardly distinguish them. If this were the mechanism, then our galaxy, which has spun on its axis at least fifty times since its birth, should be much more tightly wound.

In a different model—call it the sprinkler model—we assume, in analogy with the common lawn sprinkler, that matter is being shot out of the core, and that, as the galaxy rotates, this matter forms long trailing arms. This time, as long as the rotation rate is constant and there is a continual supply of new matter from the center, the shape will remain constant. Again, the model has deficiencies: First, there is no evidence that matter is moving outward along the arms

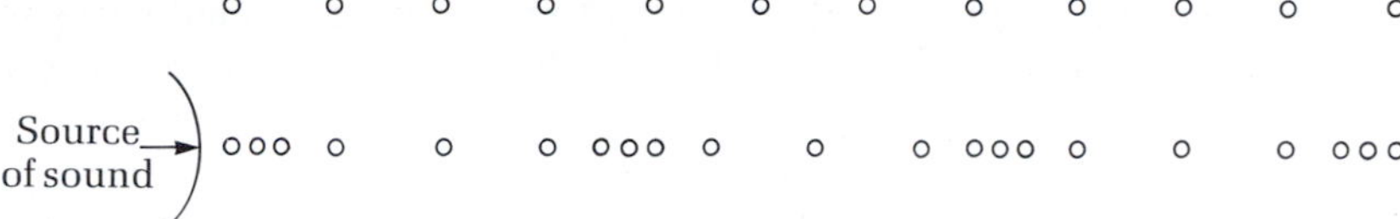

Figure 21.14 Representation of a sound wave. Circles symbolize air molecules; the upper line shows molecular orientation before the wave is transmitted.

and, second, we know nothing about how all this matter could be generated at the core.

There are, of course, other possible and similar models, but a radically different theory seems needed. Such a theory—now called the density wave theory—was introduced by Lindblad and developed mathematically by American astronomers C. C. Lin and Frank Shu in the 1960s. This theory assumes that a wave of matter spreads through our galaxy, and that it is this wave that has produced the spiral arms. An analogy can be made with various waves (on a much smaller scale, of course) common to our experience on Earth; a sound wave is a good example. As Fig. 21.14 shows, when a signal is impressed on an air molecule, it moves slightly; this motion is transferred to the next molecule and so on, until a series of condensations (close-spacing of the molecules) and rarefactions (wide-spacing) is set up. A wave of this type on a galactic scale would also cause condensations and rarefactions to occur in the gas and dust of a galaxy, and stars would form in the condensations. We see these regions as arms.

The theory actually postulates two waves that move through a galaxy to create a shock wave. As the shock front passes, compression and heating occurs, which form new stars.

This theory, too, is not entirely satisfactory. Although it overcomes some of the problems, others remain: For example, what initially caused the shock wave, and what keeps it going?

SUMMARY

1. The first serious attempt to determine the structure of our galaxy was made by Herschel. Using a technique called star gauging he concluded that it was pancake-shaped.
2. A worldwide effort organized by Kapteyn shortly after the turn of the century produced a model similar to Herschel's, but larger, and with the sun at the center.
3. Oort and Lindblad concluded, on the basis of observation and theory, that our system is a gigantic rotating disk of stars with the stars near the center rotating faster than those further out.
4. In 1930 Trumpler discovered that there was an interstellar absorption. Kapteyn's model was overthrown as a result, and Shapley's model was considerably reduced in size.
5. Baade discovered two populations of stars, designated Population I and Population II. Population I stars are mainly in the arms of galaxies; Population II are mainly in the core and halo.

6. In 1951 Purcell and Ewen discovered 21-cm radiation. It had been predicted by Oort and Van de Hulst, and was particularly useful in determining the structure of our galaxy.
7. Molecules of many different substances—including water, ammonia, and alcohol—were discovered in space during the 1960s and 1970s.
8. Our galaxy is approximately 100,000 light years across, and may contain as many as 200 billion stars. We are located about 32,000 light years from the center in the Carina Cygnus Arm.
9. The nucleus of our galaxy is in the direction of Sagittarius. We do not see it because of intervening gas and dust.
10. Nebulae, regions of gas and dust where the density is higher than the average, are classified as to type: emission, absorption, or reflection.
11. The interstellar medium permeates our galaxy. It is generally invisible to us and consists mostly of hydrogen with some helium and trace amounts of other elements.
12. The spiral arms of our galaxy are assumed to have been formed as a result of a density wave that spread through it.

REVIEW QUESTIONS

1. Explain star gauging. Does it always work? Discuss a case where it would not.
2. A number of stars in our galaxy appear to move peculiarly. Explain. How did Oort resolve this problem?
3. Explain the rotational motions of the stars in our galaxy. Are they in exact agreement with Kepler's laws? Why or why not?
4. What were the events that led to Trumpler's discovery of the absorbing medium? What were the consequences of this discovery?
5. What is the difference between Population I and Population II stars? What brought about this difference?
6. Explain the phenomenon that gives rise to the 21-cm radiation. What gas is involved? Do other gases also give off radiation?
7. What is a maser? How could there be one in space?
8. Briefly describe our galaxy. Discuss such aspects as its size, mass, and motion.
9. Describe what part of our galaxy we see as we look at various sections of the Milky Way (e.g., Cygnus, dense star clouds of Sagittarius).
10. Describe the nucleus of our galaxy. What is strange about it?
11. Explain the differences between the three types of nebulae. What is the difference between a nebula and a galaxy?
12. List the components of the interstellar medium.
13. Explain how astronomers distinguish signals received from the various arms of the Milky Way Galaxy.
14. How were the arms of our galaxy formed (according to present theories)? Discuss the deficiencies of these theories.

THOUGHT AND DISCUSSION QUESTIONS AND PROJECTS

1. How many times farther than the nearest star is the core of our galaxy? What is the length of the sun's orbit around the Milky Way (circumference $= 2\pi r$)?
2. How would you go about proving that the Milky Way Galaxy is older than the sun?
3. Describe what a trip into the nucleus of our galaxy would be like. If there is a black hole at the nucleus what would astronomers look for to verify it?
4. *Project:* View the Milky Way with binoculars or a small telescope and answer the following: Where is it most dense? How does the view through binoculars or telescope compare to the naked-eye view? What is the direction of the Milky Way (E-W or N-S) and what are its general naked-eye features?

FURTHER READING

Bok, B. and Bok, P. *The Milky Way,* 5th ed. Harvard University Press, Cambridge, Mass. (1981).

Bok, B. "The Milky Way Galaxy," *Scientific American,* p. 92 (March 1981).

Chaisson, E. "Journey to the Center of the Galaxy," *Astronomy,* p. 6 (August 1980).

GeBalle, T. R. "The Center Parsec of the Galaxy," *Scientific American* (May 1979).

Jaki, S. L. *The Milky Way, an Elusive Road for Science.* Science History Pub., New York (1972).

Turner, B. E. "Molecules in Space," *Scientific American,* p. 51 (March 1973)

Whitney, C. A. *The Discovery of our Galaxy.* Knopf, New York (1971).

Aim of the Chapter

To show how galaxies differ and to introduce the various classifications of them.

chapter 22

Other Galaxies

As the twentieth century approached, astronomers had not yet determined the nature of the "spiral nebulae." Though they had shown that most were in a direction away from the Milky Way, which indicated that they were likely to be beyond the boundary of our galaxy, little was known about their size or their composition. Some of the nonspirals appeared to be gaseous: In the early 1860s Sir William Huggins had discovered that the planetary nebulae gave a spectrum similar to that of a gas, but that the spectra of spirals were like those of stars.

With the discovery of the cepheid period-luminosity relation (p. 305) a new tool became available, though it would be a few years before it would be used to examine spirals. As late as 1920 there was still considerable controversy, with Shapley strongly against the island universe concept, and others strongly for it.

By 1915, A. Van Maanen, using a blink comparator (p. 230), had examined several of the closer spirals and believed that there was some evidence for a small amount of internal motion (though he cautioned that many more plates should be examined before the result could be accepted). He continued working on other spirals, however, and by 1920 was convinced that there was indeed a definite, but slight, rotational movement in several of them.

DISTANCES, SIZES, AND MASSES

In 1919, Shapley published a paper summarizing Van Maanen's results and discussing the various arguments for and against the island universe concept for spirals. He concluded that Van Maanen's measurements showed motion large enough to rule out the island universe hypothesis. (Proper motions of this size would not be possible if these objects were millions of light years away—as they would have to be if they were galaxies.) H. D. Curtis was sure he had sufficient evidence to prove otherwise: He had observed several "novae" in spirals, and their low observed brightness indicated a distance greater than 50,000 light years. The disagreement led to a public debate between the two in 1920. As it turned out, both points of view were based on erroneous assumptions. Shapley had relied on Van Maanen's observations, which were later shown to be in error, and his version of the period-luminosity relation was also later shown to be incorrect. (Intervening matter had not been taken into consideration.) Curtis, on the other hand, considerably underestimated the size of our galaxy, and this led him to underestimate the size and therefore the distance of distant galaxies. He had also assumed that we were at the center of our galaxy.

The final resolution came a few years later with the work of Edwin Hubble, who began a study of galaxies with an examination

Edwin Hubble

Edwin Hubble, born in Missouri in 1889, was a quick, alert student who excelled without studying. Academic achievement was matched in athletics, and, during his undergraduate years at the University of Chicago, he collected letters in boxing, track and field, and basketball. His boxing skills were so good that he was invited to fight an exhibition bout with Georges Carpentier, the then world light-heavyweight champion.

After graduation in 1910 he went to England to study law on a Rhodes scholarship. Three years later he was practicing law, but found that it bored him. Realizing that it was astronomy that really mattered to him, he re-enrolled in the University of Chicago, this time in the Department of Astronomy. Here he began what would become his lifelong study of nebulae. The nature of these diffuse objects was still not known, but by the time he had finished his thesis he had a strong suspicion that they were large groups of stars, perhaps billions of stars, that were outside our system (Milky Way).

Hale offered him a position at Mt. Wilson upon his graduation but the United States had entered World War I and Hubble telegraphed his reply: "Regret cannot accept your invitation. Am off to war." After the war, in which he was wounded, Hubble eagerly accepted Hale's invitation and was soon working again on his favorite subject of nebulae. Within a few years he had proven what he earlier suspected: the spiral diffuse patches were indeed systems of stars beyond the bounds of our galaxy. He continued studying and classifying them. When Slipher discovered that many nebulae were receding from us, Hubble devoted himself to the subject. The culmination of this work was his model of the "expanding universe" (Chapter 24).

Hubble published extensively throughout his life; his papers are considered to be among the most important in astronomy. He was indeed a giant among astronomers. In many ways, however, he was an enigma. Some of those who worked with him described him as cold, overbearing, and arrogant; others found him charming, friendly, and always helpful.

Figure 22.1 M33, a spiral galaxy in Triangulum.

of irregularly shaped nebula NGC 6822.[1] He obtained several photographs, in which individual stars stood out clearly—and some of these were cepheids. This meant that the cepheid period-luminosity relation could be used to determine the distance to the nebula. Hubble's calculations, to his surprise, showed NGC 6822 to be 700,000 light-years away; it had to be well outside our galaxy. Hubble then turned his attention to the spiral nebula M33 in the constellation of Triangulum (Fig. 22.1). Photographs soon resolved the outer arms into stars, and there seemed no doubt that M33 was an island universe of stars, and, as such, was likely to be well beyond the outer limits of the Milky Way. Final proof came when he found cepheids and, from their periods, calculated the distance to M33 to be an

[1]NGC stands for New General Catalogue, which is a standard list of extended objects in the sky (e.g., galaxies) compiled over a hundred years ago by Cambridge astronomers. Another, but much shorter catalogue was compiled by Messier in 1784; the objects in it are also designated with a numeral, but preceded by an M—e.g., M13.

Figure 22.2 (above) The core of M31, the Andromeda galaxy. A closer view (lower right) shows the resolution of its arms into individual stars.

Figure 22.3 Elliptical galaxy NGC 205, one of two small galaxies visible near the Andromeda galaxy.

astonishing 800,000 light years. Finally he turned to the spiral in Andromeda called M31 (Fig. 1.8 and 22.2) and showed that it was at a similar distance. The Shapley-Curtis debate was over: The spirals were indeed galaxies—island universes of stars.

Actually, Hubble's calculations, made before the interstellar medium had been discovered and before it was known that there were two kinds of cepheids, were incorrect; we now believe the distance to M31 to be about 2 million light years. M31 can easily be seen on any clear, moonless, fall night as a faint patch in the constellation Andromeda. What we are seeing, however, is a ghostly image of how it appeared in the distant past; the light that strikes our eyes today, actually left this galaxy 2 million years ago. Thus, it cannot be known for another 2 million years whether it exists today.

As Fig. 22.2 shows, M31 is slightly inclined to our line of sight. It consists of a bright nucleus surrounded by long spiral arms and is no doubt quite similar in appearance to our galaxy, but somewhat larger. Nearby are two much smaller galaxies that are distinctly different in appearance; they are small, elliptical, and have no spiral arms (Fig. 22.3). M33 (Fig. 22.1), in the same region of the sky (as seen from earth) as M31, also has a bright central nucleus (but not as bright as M31's) and long spiral arms. Its arms are much more loosely wound, however, and, because it contains more young stars, it is generally somewhat bluer than M31.

With a telescope, galaxies of many shapes can be found. Just above the handle of the big dipper (Ursa Major), for example, there is the even more loosely wound spiral M101 (Fig. 22.4). In the southern hemisphere the two galaxies called the Magellanic Clouds, which appear to be completely irregular, can easily be seen (Fig. 22.5). At a

Figure 22.4 M101, a loosely wound spiral.

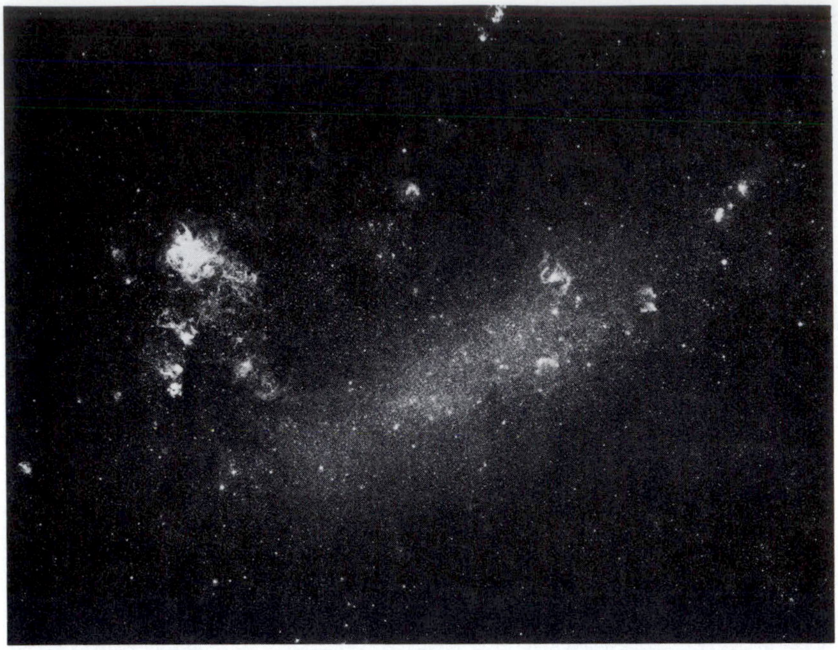

Figure 22.5 The Large Magellanic Cloud. Note the large number of nebulae; the Tarantula can be seen at left.

distance of approximately 150,000 light years are our nearest intergalactic neighbors, thought by many astronomers to be satellites of the Milky Way. They are so close that we can actually see numerous emission nebulae in them; the largest is the Tarantula Nebula (Fig. 22.5).

CLASSIFICATION

We have seen that galaxies are of many different shapes: M31 and M33 are disk-shaped with long spiral arms; the companions of M31 are elliptical; and the Magellanic Clouds appear to have no regular shape. Shortly after completing his work on M33 and M31, Hubble, pondering the possibility of a relationship between the various galactic shapes, divided about 600 galaxies into four major groups: spirals, barred spirals, ellipticals, and irregulars. The Andromeda galaxy (M31) was typical of the ordinary spiral with a bright ellipsoidal nucleus. Spirals that had a bar-like central region, from which the arms trailed off and spiralled back, made up the second group. Both barred spirals and ordinary spirals are further subdivided, the ordinary spirals into Sa, Sb, and Sc (Fig. 22.6), and the barred spirals into SBa, SBb, SBc, where a, b, and c, represent, respectively, tightly wound spiral arms, medium tightly wound spiral arms, and loosely wound spiral arms (Fig. 22.7).

Most of Hubble's 600 galaxies were spirals, probably because they are generally more prominent in photographs then ellipticals. We know today, however, that ellipticals are in the majority. Hubble also subdivided the ellipticals according to roundness, from 0 to 7, the roundest being E0 (Fig. 22.8 and 22.9). Elliptical galaxies also

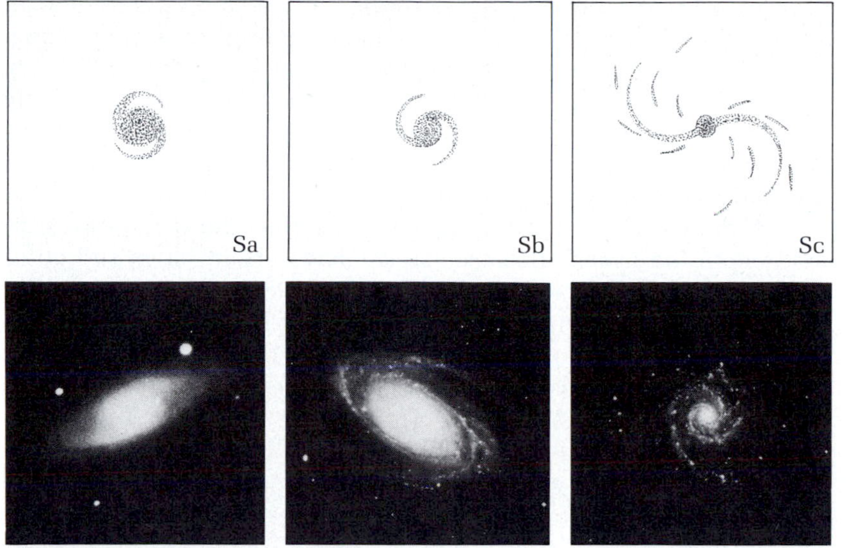

Figure 22.6 Drawings and representative photographs of the three main types of spiral galaxies—Sa, Sb, and Sc.

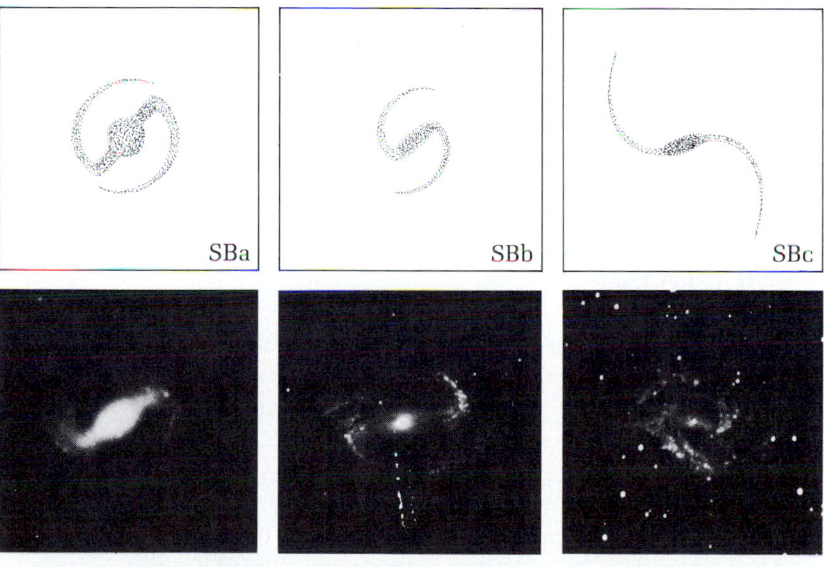

Figure 22.7 Drawings and representative photographs of the three main types of barred-spiral galaxies—SBa, SBb, and SBc.

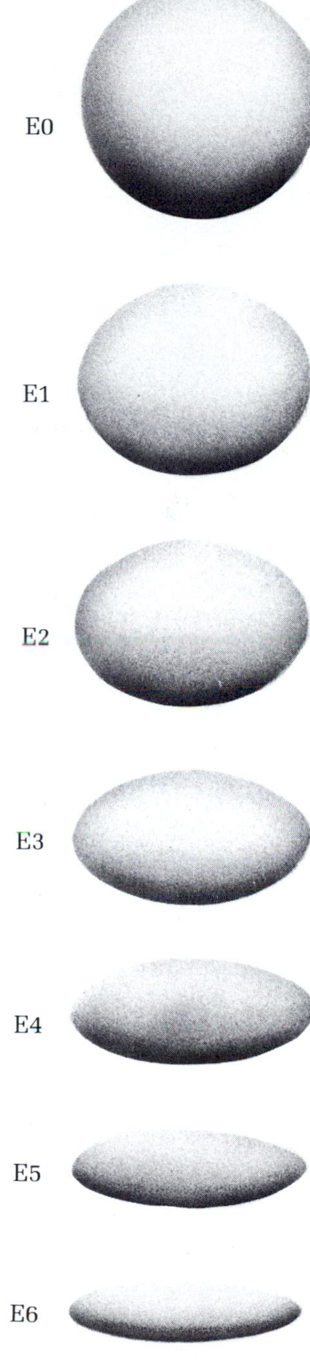

Figure 22.8 The classification scheme for elliptical galaxies, examples of which are shown in Fig. 22.9.

differ from spirals in that, in general, they consist of much older stars and have no gas or dust in them. Some are gigantic, containing as many as 10^{13} stars. But many—perhaps the majority of—ellipticals contain 10^{8} or fewer stars. A typical spiral, by way of comparison, contains about 10^{10} to 10^{11} stars. The irregular galaxies are not subclassified to any degree, and only a few percent of the total fit into this category.

Although Hubble's classification has now been extended, the basic categories of elliptical, spiral, and irregular are still used. Hubble showed that there is a relatively smooth transition between

Figure 22.9 Top: The giant elliptical galaxy M87 (NGC 4486) in Virgo. Type E0. Bottom: The elliptical companion of the Andromeda galaxy (NGC 205). Type E5.

the various types; there is, for example, only a small difference between adjacent types such as, say, E5 and E6 (Fig. 22.8). He summarized his results in a now-famous "tuning fork" diagram (Fig. 22-10), in which the prongs of the fork consist of spirals, while the handle consists of a sequence of ellipticals from E0 to E7. The transitional galaxy between the two sections, called an SO galaxy, has many of the characteristics of a spiral but does not have spiral arms.

Of the various types of galaxies in the diagram the spirals generally spin the fastest. Because, as we saw earlier, inner and outer regions rotate at different speeds (the core fastest, moving like a solid mass; the outer regions slower with distance from the nucleus), we will restrict our comparisons to the speeds at the center regions.

The amount of gas and dust in the galaxies also varies as we go from one end of the tuning-fork diagram to the other. In the ellipticals there is a complete absence of gas and dust; SO spirals also contain little or none; tightly wound spirals contain some but not nearly as much as loosely wound spirals; and, finally, irregular galaxies contain the most.

The distribution in types of stars is also different. Earlier (p. 379) we talked about Population I and Population II stars—our sun being a typical Population I star, and Population II stars being generally very old and metal-deficient. As we have said, in spirals Population II stars are found in the nucleus and in the halo, and Population I stars generally in the arms. In short, spirals contain both Population I and Population II stars. Ellipticals, on the other hand, contain mostly Population II stars, and irregular galaxies consist almost entirely of Population I stars. (They have considerable gas and dust, and many stars are still forming in them.)

EVOLUTION

Hubble, having finished his classification, turned to the evolution of galaxies. It was tempting to speculate that the tuning-fork diagram implied an evolutionary sequence (i.e., that a galaxy, as it ages, moves from one end of the diagram to the other). Though Hubble

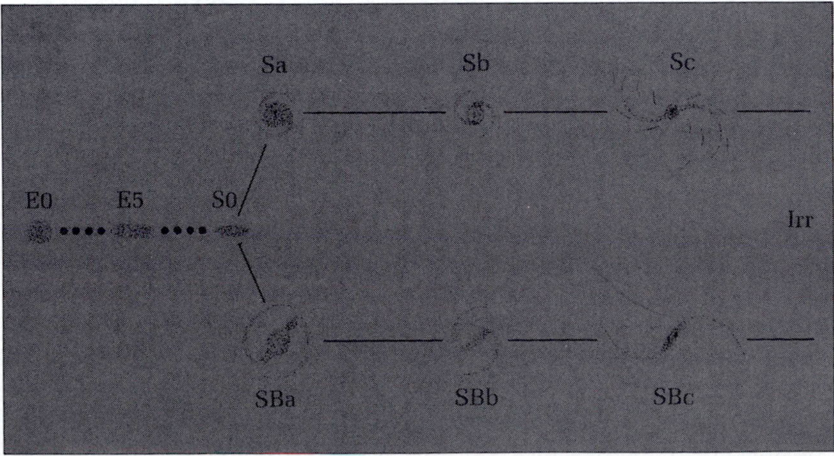

Figure 22.10 Hubble's tuning-fork diagram of the relationships he postulated between galaxies.

recognized that the evidence was inadequate, he expressed the idea privately that galaxies were likely to begin as spherical ellipticals, gradually evolve through Types E2 to E7 to S0, then to barred or ordinary spirals with tightly wound arms. After this the arms would gradually open and finally they would end as irregulars.

However, Hubble lacked one vital piece of information, discovered as galaxies were studied in detail: Ellipticals have no gas or dust, whereas spirals do. Loosely wound spirals and irregulars, in fact, have a considerable amount and, since stars form only where there is gas, this meant that Hubble's ideas had to be incorrect. Many astronomers then began to wonder if perhaps the evolution took place in the opposite direction. According to this scheme, galaxies would begin as irregulars, evolve first to loosely wound spirals, then finally to ellipticals. But even this view had its drawbacks: Although spirals have considerable gas and dust and contain large numbers of very young stars, they also contain old stars. Thus, the possibility of evolution in either direction finally had to be discarded.

It is now generally assumed that all galaxies were formed at approximately the same time and are, therefore, all about the same age. In particular, the view is that there are no "dead" galaxies; the universe just is not old enough. Their shapes are determined by their initial properties at birth: mass, rotational speed, turbulence, and so on. Thus, one type does not evolve from another; it is different because it started with a different set of initial properties.

Of course galaxies do change as they age, and astronomers now feel they have a reasonably good idea how these changes take place. Unlike the study of stellar evolution, though, the study of galactic evolution is still in its infancy, but progress is being made.

In the present view of the formation of galaxies a central role is played by the two basic galactic forms: spherical and disk. Assumed is the Big Bang theory of origin (discussed in Chapter 24), which postulates that the universe was uniformly filled with gas (hydrogen and helium) in its early stages. Eventually fluctuations in density manifested themselves. Whether these fluctuations had their origin in the Big Bang or in later developments is not known. In any case, it is generally believed that these accumulations of matter grew until they were exceedingly massive, when gravity pulled each inward. As they slowly collapsed, their average density increased and, finally, they became protogalaxies (Fig. 22.11), massive spherical clouds of rotating gas in which stars had not yet formed. Some rotated rapidly, others slowly. (The origin of the rotation is uncertain; it may have been the Big Bang explosion itself. Another difficulty was pointed out many years ago by Sir James Jeans, who showed that fluctuations of the type just discussed could not possibly lead to structures as large as galaxies. Later modifications of the theory met most of the objections, but difficulties remain.) As the protogalaxies continued to collapse inward, stars began to form in some (presumably because, as their rotation slowed, the turbulence in them diminished). Because the resulting galaxies developed early they remained spherical or elliptical in shape. The stars in these galaxies (the ellipticals we see today) are now very old—entirely Population II.

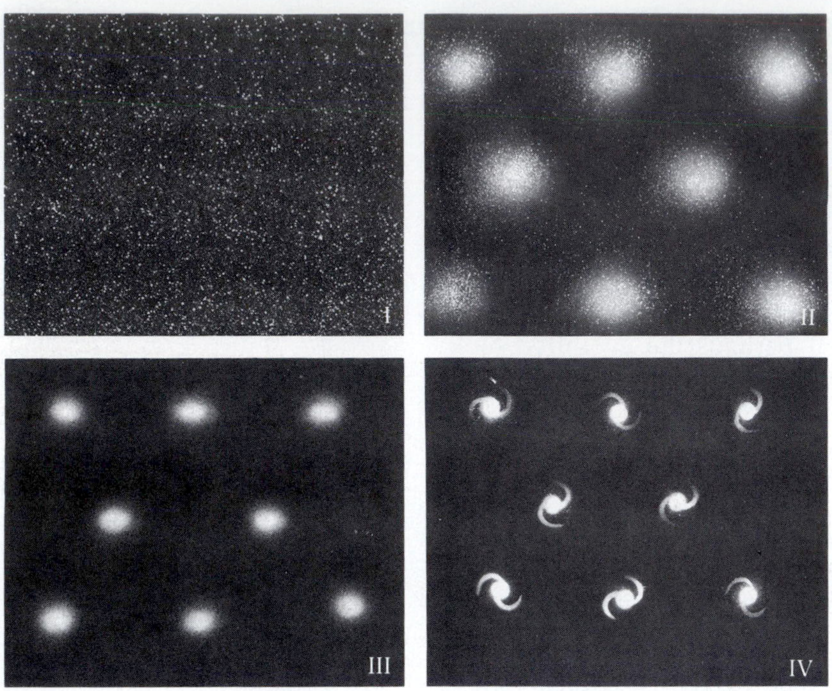

Figure 22.11 Schematic representation of the formation of galaxies: I, the universe uniformly filled with hydrogen and helium; II, break-up of the gas as density fluctuations occur; III, formation of protogalaxies; and IV, galaxies.

Figure 22.12(opposite) Probable method of formation of barred spirals. Stars moving outward along the rotating bar pass into space and are left behind.

In other protogalaxies, star formation is assumed to have begun in the core. The development of such cores was therefore similar to that of ellipticals and so they began to resemble elliptical galaxies. In the outer regions of these protogalaxies, however, the gas remained diffuse and below the critical density needed for star formation; the random motion of the gas molecules being converted mostly into heat and radiation. During the long delay in the formation of stars the outer regions gradually flattened (as a result of rotation) and the galaxy eventually began to take on the appearance of a sphere surrounded by a disk. Finally, stars began to form in the disk; stars (mostly Population I) still form there in the spirals we see today.

Our theory also addresses the question of how the spiral arms formed. As we saw in Chapter 20, the spiral arms in our galaxy are assumed to have developed as a result of (and to be maintained by) a density wave that spread throughout it. It is likely that this is also the way they formed in other galaxies. How this wave started is not known, but once it developed it would create a shock front as it traveled, and conditions just behind this shock wave would probably be ideal for star formation. Galaxies have weak magnetic fields (about 10^{-6} times that of the earth's magnetic field) along their spiral arms, and it has been suggested that magnetic fields may play an important (but still unknown) role in spiral arm formation and stability.

The density wave mechanism applies only to ordinary spirals. In the case of barred spirals (Fig. 22.12), astronomers feel stars may move outward along the bar to its end, then off into space. Assuming that the bar rotates, they will then be left behind it in a spiralling arc as the bar moves away.

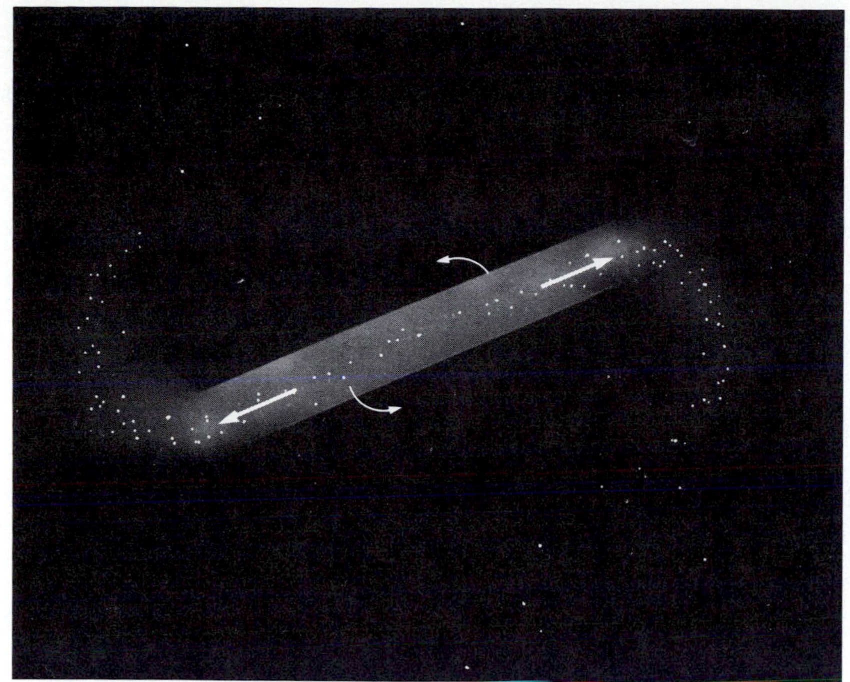

Figure 22.13(above) Peculiar galaxy NGC 4038-9. Note antenna.

Collisions may also play a role in the development of arms in a small percentage of galaxies (called peculiar galaxies). As Fig. 22.13 shows, tidal forces produced by two galaxies approaching one another would draw out long strands of gas, or antennae, as they are sometimes called. In recent years, several computer simulations of collisions of this type (Fig. 22.14) have given what appears to be a reasonably accurate picture of what may have happened.

We have established that, in general, galaxies do not evolve from one type to another, but that eventually there will be *some* change in type. Spirals, for example, have only so much gas and dust and, when it is used up, their arms (made up mostly of gas, dust, and short-lived stars) will probably disappear. Some astronomers have suggested that spirals may eventually evolve into S0-type galaxies (the galaxy transitional between the spirals and ellipticals in Hubble's tuning-fork diagram), which contain no gas or dust, have no arms, but have the characteristic disk shape of a spiral.

Many astronomers now believe that surrounding objects may play a role in the development of galaxies. As we just saw, collisions, and probably even near collisions, severely perturb their shape. Furthermore, galaxies generally occur in clusters (below), some of which contain mostly spirals, some mostly ellipticals, and still others, both. Compact clusters, for example, contain mainly elliptical galaxies; loosely bound and irregular clusters, on the other hand, contain mainly spirals. Many clusters also have, near their centers, several giant ellipticals, which some astronomers believe derive from gas that has been "ripped" from the outer regions of nearby galaxies by tidal forces.

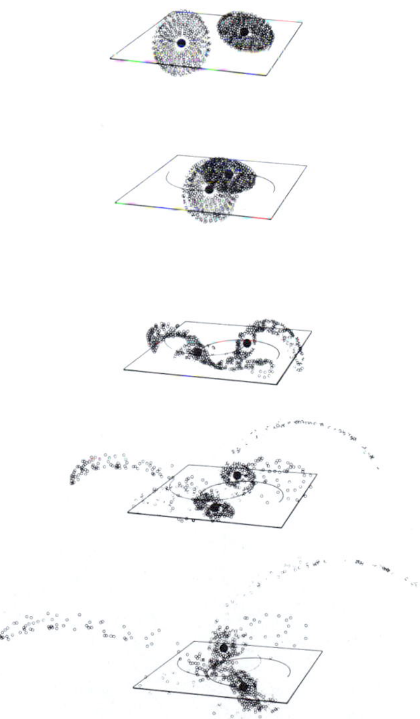

Figure 22.14 Computer simulation of the outcome of a collision between two galaxies. Each stage is separated from the preceding one by 200 million years.

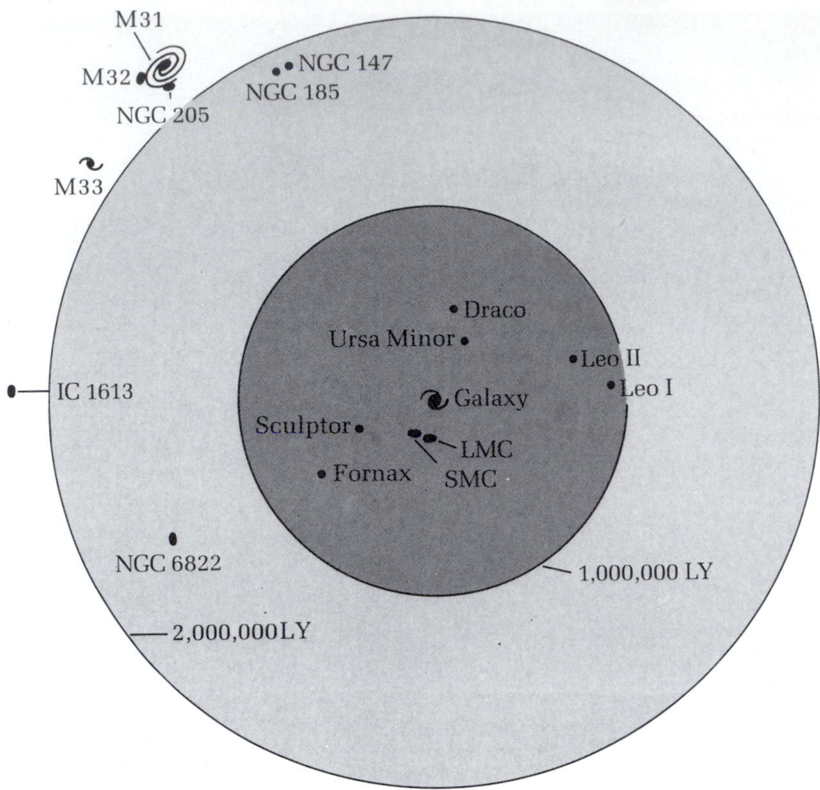

Figure 22.15 The local group of galaxies. Our galaxy is at the center.

CLUSTERS

Gigantic as they are, galaxies are not the largest units in the universe; there are clusters of galaxies, and there may even be clusters of clusters—called superclusters.[2] Our galaxy is the second largest member of what is called the local group, which consists of about 30 galaxies (Fig. 22.15); M31, the Andromeda galaxy, is the largest. Our two nearest neighbors, the Magellanic Clouds, are irregular galaxies, as are two others. Apart from these and three spirals, all are ellipticals—many, dwarf ellipticals.

What appeared to be two additions to our group (found by Italian astronomer Paola Maffei in 1968) are now believed to be farther away than originally thought—about 16 million light years—and consequently not likely to be part of our local group.

Our group is, of course, one among known thousands. Nearby is the Hercules cluster (Fig. 22.16), a sphere-shaped group of galaxies that astronomers refer to as regular (because of its regular shape). It is about 300 million light years away and is approximately 10 million light years across. At its center are two gigantic elliptical galaxies. The Virgo cluster, a good example of an irregular cluster, is about 65 million light years away and about 6.5 million light years across.

Figure 22.16 The Hercules cluster, embracing several different types of galaxy.

[2]Our cluster may belong to a supercluster that has its center in the direction of the constellation Virgo and is called the Virgo supercluster. It may contain as many as 100 members, including our group and the Virgo and Coma clusters.

The various galaxies would appear to stay together for the same reason that stars are frequently found in groups—because they are gravitationally bound to one another. It is, in fact, quite easy to calculate how much mass is needed to bind a system of galaxies gravitationally. When calculations of this type were performed, however, astronomers were astonished to find that many clusters did not have enough mass to hold them together. Indeed, some groups have been found that need 5 to 50 times the mass they have if they are to be bound gravitationally; one has been found that needs 400 times as much.

This mass discrepancy is now referred to as the "missing mass" problem. According to calculations, the mass must be there, but in what form? Dust and small chunks of matter can easily be ruled out (they are detectable), and so can most other possibilities. Black holes large enough to supply the missing mass would have to be gigantic (more massive than galaxies), and, were they this large there should be some indication of their presence—distortions, for example, in nearby galaxies.

S. Van den Bergh, looking for evidence of black holes in the Virgo cluster, found that four or more members of this cluster exhibit distortion. But Van den Bergh has shown that these distortions can be explained in other ways. He concludes that the Virgo and similar clusters are not likely to contain black holes of galactic mass; but not all astronomers agree.

RADIO GALAXIES

In 1948 a strong radio source was discovered in Cygnus. Within a couple of years it was narrowed down to an area subtending about one minute of arc. Optical telescopes were then used to locate the source—two fuzzy masses. Cygnus A, as this object is now known, is one of the strongest radio sources in the sky (Fig. 22.17), emitting a million times as much energy as the Milky Way. It was later shown to be a galaxy apparently undergoing an explosion of some sort.

Many radio sources of this type are known (see Figs. 22.18–20). M82 in Ursa Major, for example, also appears to be undergoing an explosion. In most cases the major sources of radiation are two lobes on either side of the galaxy (Fig. 22.19). The lobes are thought to be clouds of plasma (charged particles) that have been blown out of the galaxy in an explosion. In a few cases—the giant elliptical M87 is a good example—jets are seen emanating from the core.

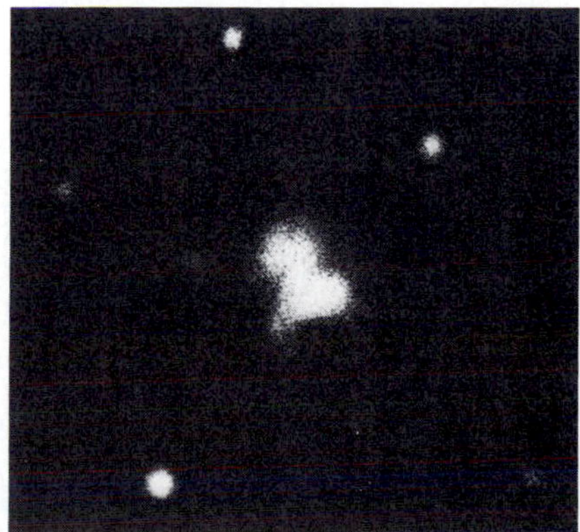

Figure 22.17 Cygnus A, one of the strongest radio galaxies.

What causes these explosions is uncertain. Perhaps there is a chain reaction of supernova explosions, or maybe gravitational energy is being released as a result of stellar collisions. (The density of stars in the cores of such galaxies is high.) Several astronomers believe there may be gigantic black holes in the cores of radio galaxies, and that, as stars pass nearby their outer layers may be ripped off by tidal forces, with radiant energy given off as this gas falls into the black hole.

Figure 22.18 M82, a galaxy in Ursa Major that appears to have an exploding core.

Figure 22.19 The two zones (or lobes) of radio emission of Cygnus A (shown in a close-up view in Fig. 22.17.)

Figure 22.20 The radio source Centaurus A.

Although most strong radio sources are now known to be a result of explosions in galaxies, it is not impossible that some are due to galactic collisions. Collisions of this type do occur occasionally. As two galaxies pass through one another their associated hydrogen gas clouds could interact and emit radio waves.

SEYFERTS AND N GALAXIES

In 1943 Carl Seyfert discovered a number of galaxies with particularly bright, dense nuclei. These galaxies, now called Seyfert galaxies (Fig. 22.21), also exhibit broad emission lines characteristic of rapid movement of hot gases. Thus considerable amounts of gas may be pouring out of their cores. Their energy output is tremendous: about 100 times that of our galaxy. Its source is uncertain, but many of the bright nuclei are strongly variable at optical, radio, and even X-ray frequencies, indicating some process not familiar to us. Closely associated with Seyfert galaxies are N galaxies, which also have bright compact cores. Some astronomers believe they may be elliptical galaxies with particularly bright nuclei.

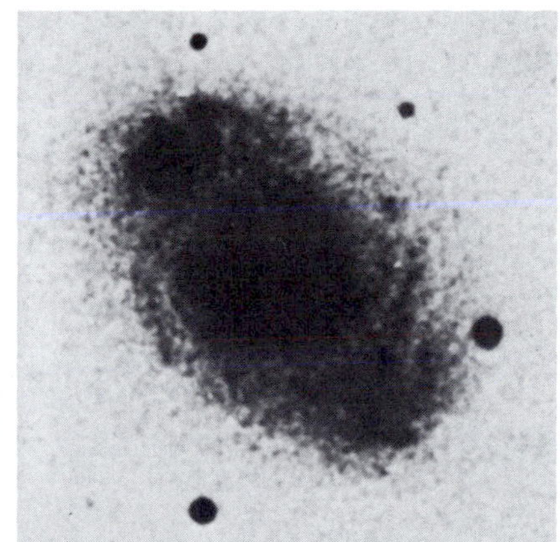

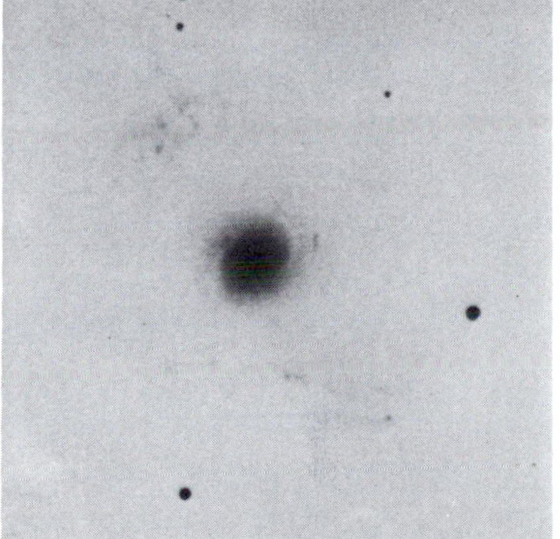

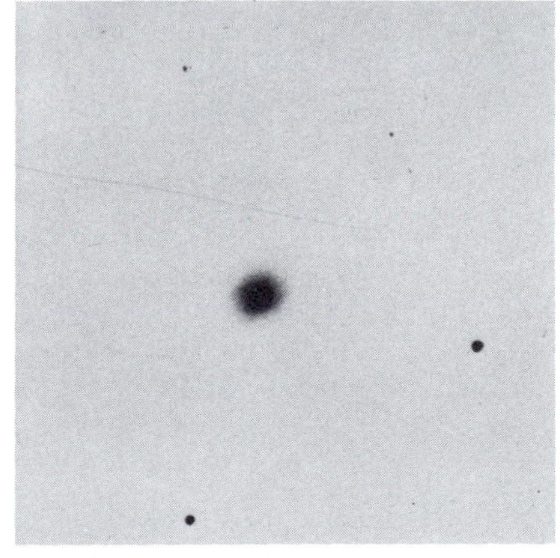

Figure 22.21 A Seyfert galaxy in three exposures, the longest at the top and the shortest at the bottom.

SUMMARY

1. The Curtis-Shapley debate, which took place in 1920, helped clarify the various arguments for and against the "island universe" concept of galaxies.
2. The controversy was resolved a few years later when Hubble determined the distance to several nearby galaxies. The "spiral nebulae" were shown to be systems similar to the Milky Way system, but well outside it.
3. The Andromeda galaxy, M31, is approximately 2 million light years away. It is quite similar in appearance to the Milky Way galaxy, but somewhat larger. Another nearby spiral, M33, is in this same region of the sky (as seen from Earth).
4. Our two nearest intergalactic neighbors, the Magellanic Clouds, are at a distance of 150,000 light years.
5. Galaxies are classified into four main groups: spirals, barred spirals, ellipticals, and irregulars. The spirals and barred spirals are classified into a, b and c subgroups according to how tightly their arms are wound. Ellipticals are classified as E0 to E7 according to their appearance (roundness).
6. Hubble believed there was a relationship between the various types of galaxies. He summarized his results in his "tuning-fork" diagram.
7. Astronomers later showed that galaxies do not evolve in either direction along the "tuning fork." They are different because they started with a different set of initial parameters.
8. The form of barred spirals can be explained by assuming an outward motion of stars along the central elongated (barred) region.
9. We are in what is called the local group of galaxies, which consists of about 30 galaxies. The Milky Way is the second biggest in the group. Most of the galaxies in this group are dwarfs.
10. There are numerous clusters of galaxies spread throughout the universe. The Coma and Virgo clusters are two of the better-known ones. Many appear to have "missing mass"—less mass than they should have to bind them gravitationally.
11. Many galaxies are strong emitters of radio waves. In some cases these galaxies appear to be undergoing explosions.
12. Seyferts and N galaxies are galaxies with particularly bright nuclei.

REVIEW QUESTIONS

1. Why was Shapley convinced that the "island universe" concept was wrong?
2. Summarize some of the arguments presented by Shapley and by Curtis in their debate. Which ones were in error?
3. How did Hubble resolve the debate?
4. Compare and contrast the galaxies M31 and M33 with our galaxy.
5. What are the four major classifications of galaxies? How are they subdivided? Sketch examples of each type.
6. What is the "tuning-fork" diagram? Do galaxies evolve in either direction along it? What is the evidence for or against this.
7. Describe the population of stars (Populations I or II) in the various types of galaxies.

8. Describe briefly how galaxies were formed. How are they likely to change as they age?
9. How do "peculiar" galaxies differ from other galaxies? Can we explain their appearance? How have we attempted to do this?
10. Describe our local group. How many galaxies are in it? Which is the largest?
11. Are Maffei I and II considered to be members of our group? What is their distance from us?
12. How are clusters of galaxies classified?
13. Discuss the "missing mass" problem. Why are we sure there is some missing mass?
14. What is different about Cygnus A? Describe it. Are there other galaxies with this same property? If so, list and discuss them.
15. What is a Seyfert galaxy? How does it differ from an ordinary galaxy?

THOUGHT AND DISCUSSION QUESTIONS AND PROJECTS

1. What evidence do we have that a) elliptical galaxies do not evolve into spirals; b) spirals do not evolve into ellipticals; c) spirals do not evolve into barred spirals; d) barred spirals do not evolve into irregulars?
2. What is the ratio between the size of the Milky Way Galaxy and the distance to the Andromeda galaxy?
3. Suppose we wanted to find the velocity of the center of the Andromeda galaxy relative to the center of our galaxy using the Doppler effect. What corrections would have to be made before we could accept the result?
4. Assuming the Big Bang theory (Chapter 25) to be correct, describe how you think galaxies formed.
5. *Project:* Observe the Andromeda galaxy. Sketch what you see. What is its approximate angle of tilt? Can you see individual stars?

FURTHER READING

Arp, H. C. "The Evolution of Galaxies," *Scientific American,* (January 1963).
Hodge, P. W. *Galaxies and Cosmology,* McGraw-Hill, New York (1966).
Parker, B. "The First Galaxies," *Astronomy,* p. 94 (November, 1981).
Sandage, A. *The Hubble Atlas of Galaxies.* Carnegie Inst. of Washington (1961).
Shapley, H. *Galaxies.* Harvard University Press, Cambridge, Mass. (1972).
Strom, S. E. and K. M. Strom. "The Evolution of Disc Galaxies," *Scientific American* (April 1977).

Aim of the Chapter

To describe the discovery of quasars and to indicate the problems that now surround them.

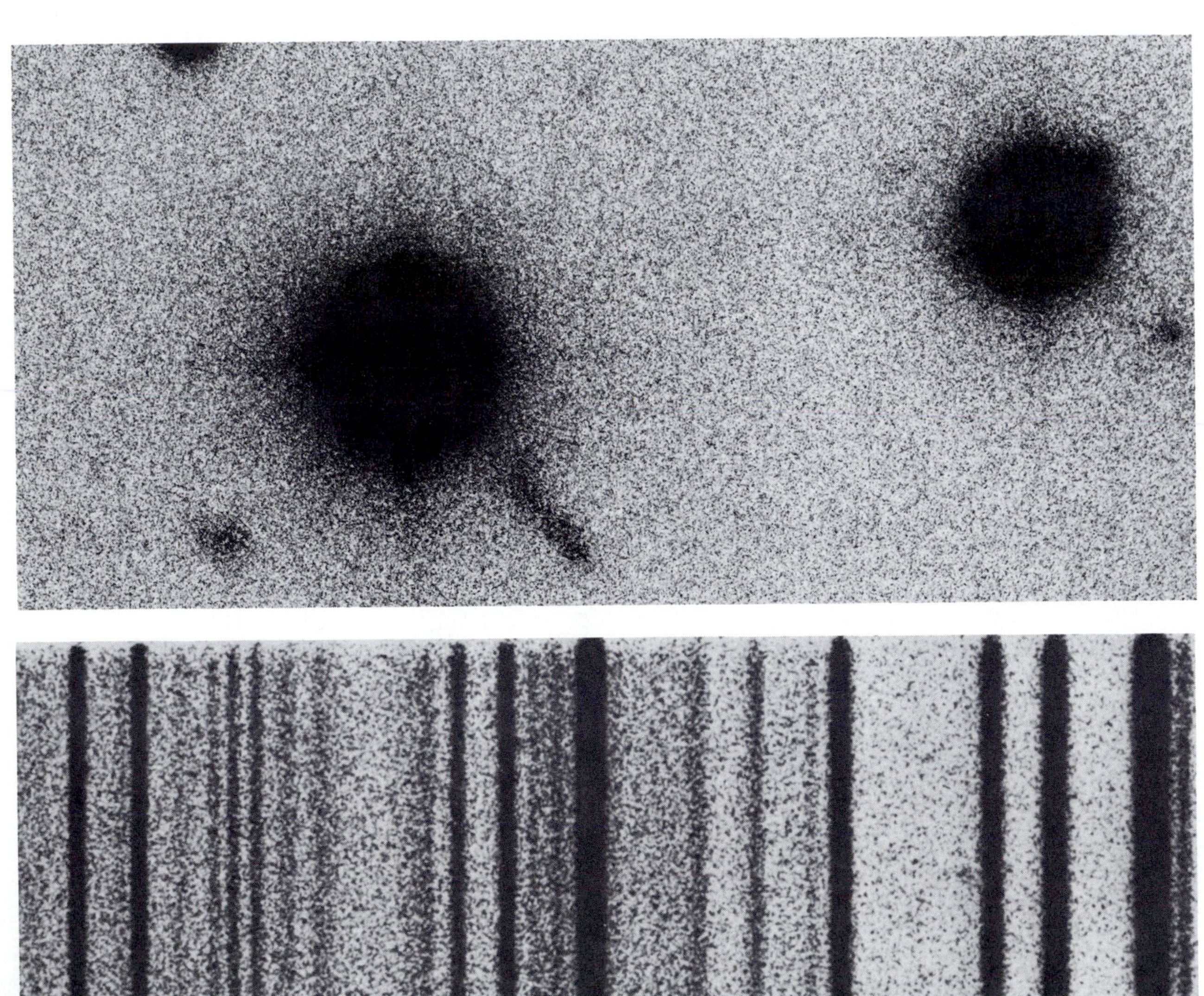

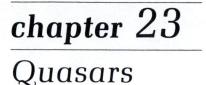

chapter 23
Quasars

Considerable progress was made in radio astronomy during the 1950s; several large radio telescopes went into operation, including Jodrell Bank's Mark I. Cambridge astronomers published their first catalog of known radio sources in 1950, a relatively brief listing containing about 50 entries. In 1955 they undertook a much more extensive survey, which resulted in the Second Cambridge Catalog. As there were many problems with false entries and improperly resolved sources, a third survey was initiated a few years later. It was completed about 1960, and the Third Cambridge (3C) Catalog was published soon after. This catalog was by far the most successful of the three; it eventually turned out to be particularly useful to radio astronomers.

For the most part, though, the coordinates in the 3C Catalog were only approximate. Radio sources have exceedingly long wavelengths compared to light waves, and objects in the sky could not be pinpointed as they could with optical telescopes. The resolution of radio telescopes during the 1950s was poor. Some of the brighter, more extended objects in the 3C Catalog were soon identified optically, but many remained unidentified.

The resolution of radio telescopes was not significantly improved until quite recently. The breakthrough was the use of radio interferometers with exceedingly long baselines, a technique we discussed in Chapter 5. Briefly, it consists of linking and coordinating two well-separated radio telescopes (the farther apart the better). Today, with computers and atomic clocks, there is no need for a physical link; computer tapes can be brought together and compared later.

DISCOVERY

Even during the early 1960s, however, resolution was poor. But since most of the radio sources were extended objects such as gal-

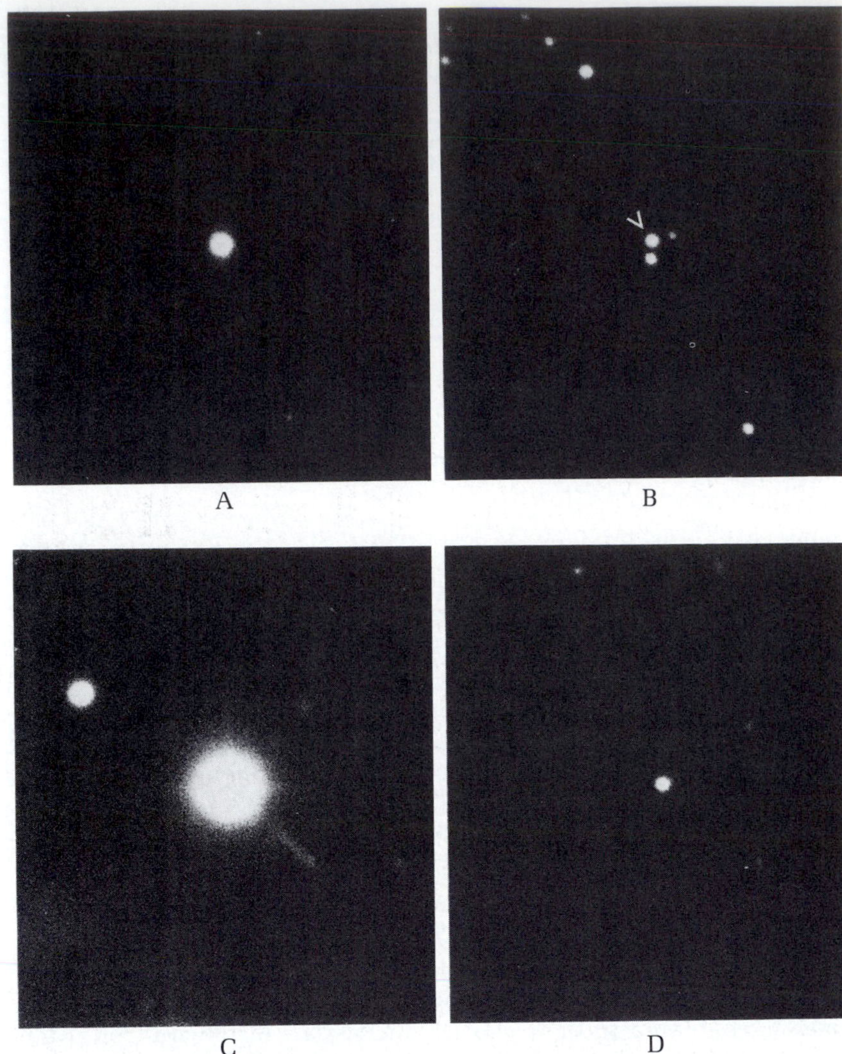

Figure 23.1 Four quasars: (*A*) 3C 48, (*B*) 3C 147, (*C*) 3C 273, (*D*) 3C 196. Note their starlike appearance.

axies, their optical counterparts were easily found. In addition, most of them were peculiar in some way (some, for instance, appeared to be exploding), so it was not surprising that they were radio sources. In a number of cases, though, the angular diameter of the source appeared to be exceedingly small. Radio astronomer Tom Matthews selected several for detailed study and, about 1960, was able to fix the source 3C 48 (Fig. 23.1) accurately enough for optical identification. He passed the information on to Allan Sandage at Palomar, and Sandage photographed the region using the 5-m reflector. The results were puzzling: The only object with which the source could possibly be associated was a 16th-magnitude bluish star. This was particularly strange in that no star, with the exception of our sun (because it is so close), had ever been identified as a radio source.

Sandage obtained the spectrum of 3C 48 soon afterward, and it was even more puzzling. He could not identify any of the lines; indeed, they were emission lines rather than absorption lines. Emission lines are usually seen only in the spectrum of gaseous clouds or nebulae such as the Orion Nebula; most objects such as stars and normal galaxies exhibit only absorption lines. In addition, the lines were broad, indicating some sort of motion associated with the source, perhaps an outpouring of gas.

Sandage studied the lines for weeks but could make no sense of them. Eventually he passed them on to a colleague, Jesse Greenstein. Greenstein was also puzzled; he thought the object might be the result of a supernova, but there were problems with this suggestion. One of the major difficulties was the dearth of lines—there were only four.

Almost two years passed before a similar spectrum with more lines was obtained. (It was not necessarily the *additional* lines that led to the final identification.) Noticing that the source 3C 273 (Fig. 23.1) was to be occulted by the moon during 1962, Cyril Hazard and two colleagues decided to use the event to obtain a more accurate position for it. They expected a gradual cutoff of the radio waves as the lunar edge approached the source, but the cutoff was exceedingly abrupt. In fact, there were apparently two closely associated sources connected with 3C 273. They obtained their positions accurately and passed them on to Maarten Schmidt at Palomar.

Schmidt photographed the region with the 5-m reflector, and as in the case of 3C 48 there was only a dim, bluish star near the position of the source. This time, though, the star was a little brighter (13th magnitude), and there appeared to be a jet emanating from it (Fig. 23.2). This jet, it turned out, was the second source noticed by Hazard; actually, most of the radio waves come from it rather than from the "star" itself. Schmidt then obtained the spectrum; it was similar to that of 3C 48, but there were more lines.

Schmidt showed the spectrum to Greenstein, and the two tried without success to analyze it. Several months had passed when an editor from *Nature* asked Schmidt to write a report on 3C 273. Looking again at the spectrum (Fig. 23.3), Schmidt noticed that several of the lines looked strangely familiar. They were similar to the

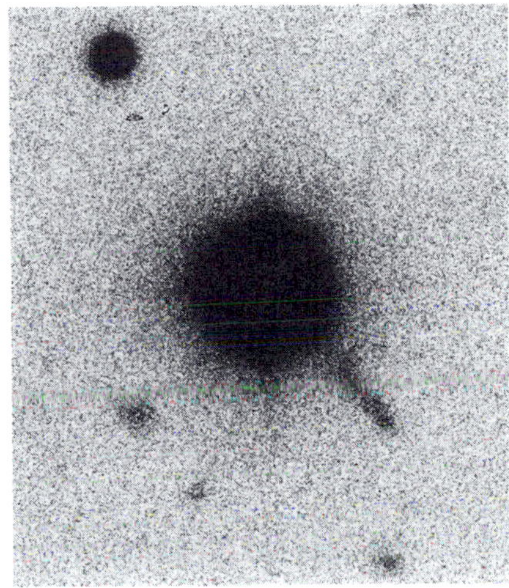

Figure 23.2 Quasar 3C 273, with emanating jet. The object looks like any other 13th magnitude star, except for the faint narrow jet, visible out to about 20 seconds of arc from the quasi-stellar object.

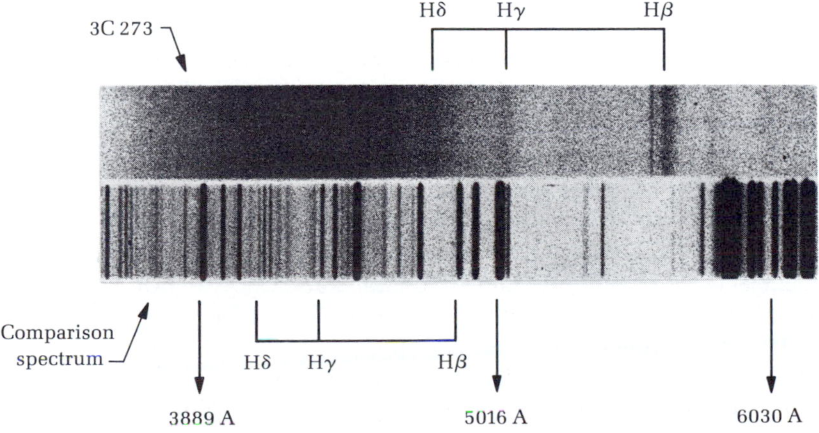

Figure 23.3 Spectrum of the quasar 3C 273, with comparison spectrum below. The comparison spectrum consists of hydrogen and helium lines. The Balmer lines Hβ, Hγ, and Hδ in the quasar spectrum are at longer wavelengths than in the comparison spectrum. The red shift of 16 per cent corresponds to a distance of two billion light years in the expanding universe.

Balmer series of hydrogen, except that they were in the wrong place and at least one line was missing. The spectral lines of galaxies exhibit red shifts, but this source, a star, would not be expected to show a large red shift. However, Schmidt became convinced that these lines must be hydrogen lines shifted toward the red end of the spectrum. He made the appropriate calculations, and they did, indeed, fit a red shift of 15.8%.

He showed his calculation to Greenstein, who performed a similar one for the spectrum of 3C 48. This source also seemed to have a red shift, in fact a much larger one, 37%. How could a star be red-shifted this much?

As we saw in Chapter 6, a shift of spectral lines toward the red end of the spectrum is usually attributed to the Doppler effect. According to this interpretation, a red shift means that the object is traveling away from us. But a 37% shift would correspond to a speed nearly 37% that of light, an incredibly high speed for a star.

Many galaxies have spectra with large red shifts. As we saw in Chapter 1, this is because such galaxies are speeding away from us; the more distant the galaxy, the faster it appears to be traveling. But the objects being studied by Schmidt and Greenstein were starlike. Could they be participating in the expansion of the universe as galaxies do? Although this view presented numerous problems, it seemed, at first, to be the only possible one. Astronomers refer to it as the cosmological interpretation of the red shift. We will see later that there are other interpretations, but for now let us assume that this one is valid.

Although the sources were starlike, they were quite different from stars in many respects; they were therefore called quasi-stellar sources (QSS). The term *quasar* later came into wider use.

In succeeding years many more quasars were discovered (the count is now more than 1500). A useful technique for locating them was devised by Allan Sandage and Martin Ryle in 1964. Color studies had shown that quasars had an *ultraviolet excess;* they are brighter in the ultraviolet region of the spectrum than they are in the blue region. Since very few other known objects have an ultraviolet excess, Sandage and Ryle were able to use this property to locate potential quasars. They would select a region of the sky that had a good chance of containing quasars, photograph this region using an ultraviolet filter, then shift the plate a millimeter or so to one side and rephotograph it using a blue filter; the result would be a group of double images. Examining these images—looking in particular for ones where the UV image was the brighter of the pair—they could locate quasar candidates. They would then take the spectrum of the candidate for a positive identification.

Sandage soon discovered, though, that many of the objects with excess UV were apparently not associated with radio sources. Their spectra were similar to that of QSSs, but they did not emit radio waves. Since they seemed to be somehow related to quasars, they were called quasi-stellar objects (QSO) to distinguish them from QSSs. Eventually it was decided to lump QSOs and QSSs together and call them both quasars. We will adhere to this convention.

Astronomers continued to discover more quasars. Recessional velocities of 70% and 80% of the velocity of light soon became common, and if we assume they are cosmological, the majority of them are farther away than the most distant known galaxies. At present the farthest known quasar has a recessional velocity slightly over 90% of the velocity of light, which means that it is almost seven times as far from us as the most distant galaxy.

It is important to remember that quasars are generally much older than the galaxies we now see. (However, we may be seeing only quasars in the outer regions of the universe because they are so much brighter than the galaxies that are here.) As we look outward, we see galaxies as they appeared in the distant past—hundreds of millions of years ago in some cases. But we are seeing quasars as they appeared many billions of years ago, almost as long ago as the beginning of the universe.

PROPERTIES OF QUASARS

Energy Output

A galaxy about 1 billion light years away would appear to be little more than a fuzzy point of light. It might consist of perhaps 100 billion stars, and we see the galaxy because of their combined absolute luminosity. Yet a quasar, which looks like a single star and seems to be at approximately the same distance, appears brighter than the galaxy. Such an object would have to be extremely bright, in both the visible and also the radio regions of the spectrum (since some quasars, QSSs, are strong radio sources).

Not only are they bright in the visible and radio regions of the electromagnetic spectrum, but Frank Low and Harold Johnson have shown that they are also exceedingly bright in the infrared region, and many are now known to be X-ray sources. Their total energy output is greater than that of an entire galaxy, greater even than that of a radio galaxy. A typical quasar (QSS) gives off about 100 times as much energy as our galaxy, and yet it appears starlike.

Variability

It seems unlikely that a quasar is a gigantic star with a mass equal to or greater than that of our galaxy. There are several things that limit the size of a star. As we saw in Chapter 16, a star with a mass greater than about 80 $M_\odot$ cannot remain in equilibrium; the outward radiation pressure is always greater than the inward gravitational pull.

The size of any astronomical object is also related to its variability. A large object such as a galaxy cannot change its magnitude in a few days or even a few months. If such an object is to pulse or change in intensity, its distant parts must "communicate" with one another, because the pulse or change must be correlated across the object. But since the uppermost speed in the universe is the speed of

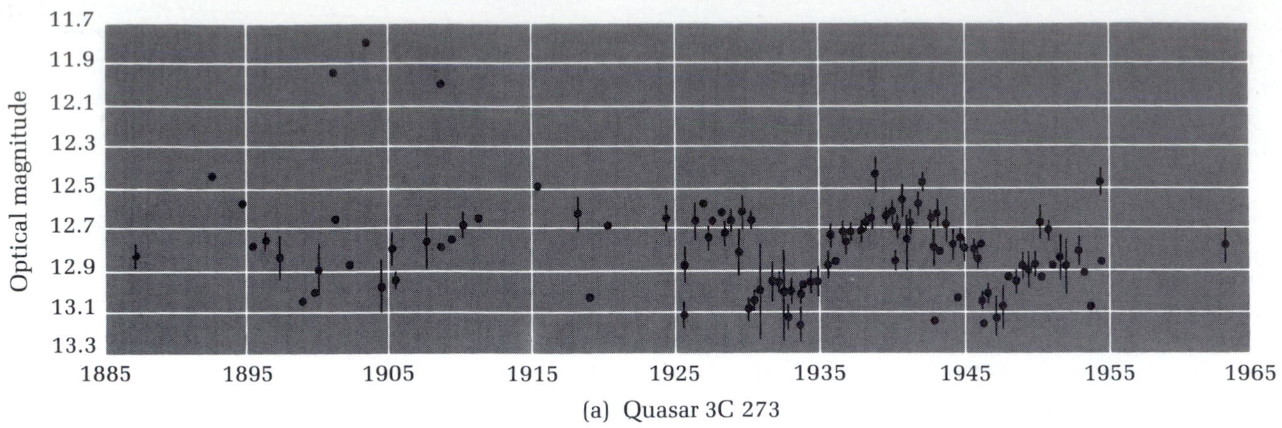

Figure 23.4 Variation in brightness of quasar 3C 273.

light, the distant parts of the galaxy cannot communicate faster than this speed. Therefore, the variability of a source gives a good indication of its size. If its intensity changes rapidly, say, in a few days, it cannot be more than a few light days across.[1]

Shortly after 3C 273 was identified optically, astronomers examining old survey plates at Harvard found that it had undergone changes over the years (Fig. 23.4). It was found to vary erratically, sometimes over a period as short as a week or so. Other quasars were also found to change considerably in only a few days. The conclusion was inescapable: *Quasars could not be galactic in scale.* Although they had the luminosity of perhaps a hundred galaxies (each containing hundreds of millions of stars), they could be no larger than our solar system.

Spectra

The spectral lines in the first few quasars that were discovered were all emission lines. In 1966, however, a quasar was discovered that also had absorption lines, and astronomers measuring the red shift of these absorption lines found that it was different from that of the emission lines. This meant that the quasar, or parts of it, somehow had to be traveling away from us at two different velocities. (In general, it has turned out that the absorption lines usually give a velocity *less* than that of the emission lines.)

Looking more closely at the absorption lines, astronomers found another problem: They were not all shifted by the same amount. Indeed, some quasars exhibited five or six different shifts. Astronomers are still not certain what causes this, but they have proposed a model that seems to resolve most of the difficulties. Recall that absorption lines are caused by light passing through a cool, absorbing cloud of gas. The absorption lines in the spectrum of a quasar could therefore be due to clouds of gas around a central bright core (the emission lines would be due to the core itself). For example, there may

[1]One light day, the distance light travels in one day, is approximately three times the distance from the sun to Pluto.

be several layers of these clouds at different distances from the core. If these clouds were expanding away from the core at several different speeds, the differences in red shift would be explained.

INTERPRETATION

So far we have assumed that quasars are cosmological. This implies two things: that the shift is due to the Doppler effect and that the quasars are participating in the expansion of the universe. If this is true, they are indeed billions of light years away (the range is from 400 million light years to almost 12 billion). But it is possible that the Doppler interpretation is valid and the objects are not cosmological. According to James Terrell, they may have been blown out of the core of our galaxy in a gigantic explosion that took place some time in the past (or perhaps is still going on). In this case they could have velocities approaching that of light but would not be as distant or as luminous as we now believe they are; they would be within our galaxy or just outside it (see Fig. 23.5).

At first glance this seems an excellent solution to many of the problems, but when examined in detail it creates more problems than it overcomes. If quasars were indeed blown out of the core of our galaxy, some should be coming toward us and would thus exhibit blue shifts; yet all quasars exhibit red shifts. It could be argued

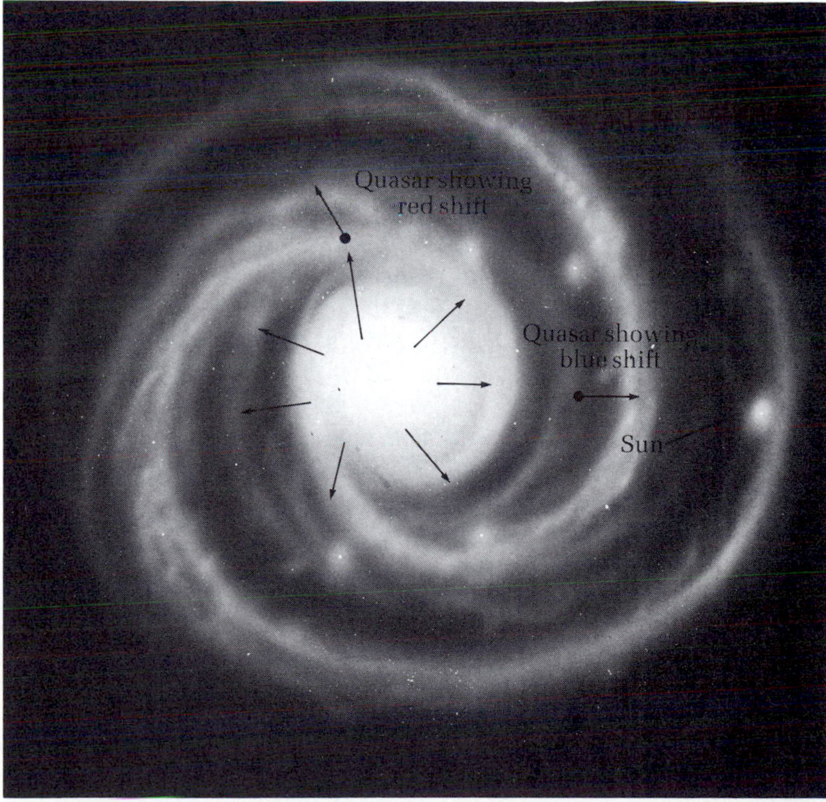

Figure 23.5 If quasars were blown out of the core of our galaxy, some of them would be approaching us and would therefore show a blue shift.

that the explosion occurred so long ago that they have now all passed us, but even this leaves problems.

First, if quasars were blown out of the core of our galaxy, then quasars were likely also blown out of the cores of other galaxies, and some of these would certainly be approaching us. So far, though, none that are approaching have been found. There is another problem that may be even more significant: When Terrell made his suggestion, it appeared that there were only a small number of quasars in the universe, but this number increased drastically when radio-quiet quasars were discovered. If we assume that the radio-quiet ones are quasars that are now depleted of energy, then there may have been several million quasars in the universe since it began. And if each of them had a mass of, say, 10,000 $M_{\odot}$ and if all of them were expelled from the core of our galaxy, there would be little, if anything, left of it now. They would have carried off *all* its energy.

There are two additional arguments in favor of the cosmological interpretation. James Gunn has found a quasar (PKS 2251+11) that appears to be associated with a cluster of galaxies in the direction of the constellation Pegasus. It has the same red shift as the other galaxies in the group, and thus it is logical to conclude that it is within the cluster. The second argument is related to the observation of absorption of radio waves by intervening galaxies. This also seems to confirm the cosmological interpretation.

On the other hand, H. Arp has shown that certain quasars appear to be associated with clusters of galaxies but do not have the same red shift as the galaxies. So again there is controversy.

If we asume, then, that the shift is due to the Doppler effect, a cosmological interpretation seems to work best. But it is possible that the shift is not due to the Doppler effect. In Chapter 20, for example, we saw that an extremely massive collapsing object also reddens light that is emitted from it. The spectrum of such an object exhibits a gravitational red shift.

It might seem possible, then, that quasars are massive collapsing objects and that we are seeing a gravitational red shift. However, in the mid-1960s Schmidt and Greenstein published a paper that ended most speculation along these lines. They calculated the mass that such an object would have to have, assuming it was at various distances from the earth. If it were 30,000 light years away, for example, its mass would have to be 10^{11} $M_{\odot}$, as great as that of our galaxy. And yet, at 30,000 light years, it would still be inside our galaxy (it would be outside only if it were generally outward from us, or out of the plane of the Galaxy). Furthermore, a million or so objects of this mass would make the average mass density of the universe considerably greater than it is. Selection of distances greater or less than 30,000 light years does little to resolve the problem.

Our entire discussion assumes, of course, that the known laws of physics apply to quasars. Even if quasars are subject to different physical principles, we have no idea what form such alternative laws might take. For the present, at least, we must conclude that the Doppler interpretation is valid and that quasars are cosmological.

MODELS EXPLAINING ENERGY OUTPUT

If quasars are cosmological, though, it is difficult to explain their great energy. Many models have been suggested, most of them quite speculative. One of the more frequently mentioned possibilities is a star millions of times as massive as our sun. As we have seen, however, an object this massive is not stable. Alternatively, if millions of ordinary stars were packed into a relatively small region of space—a superdense globular cluster—and if the frequency of collisions were high enough, the energy output would be sufficient. Even better would be a densely packed sphere of pulsars. If the frequency of collision were high (and if there were enough pulsars), the energy output would easily be sufficient.

A quite different model is based on the interaction of matter and antimatter. When matter and antimatter meet, they annihilate one another with the release of a tremendous amount of energy. However, large quantities of matter and antimatter cannot be brought together and completely annihilated; the explosion blows them apart before they can completely mix.

Clearly, most of these models are not convincing. This may be why the most acceptable at present are "black hole" models, even though we are still not sure black holes exist. (See Box, p. 418)

POSSIBLE RELATIONSHIP WITH GALAXIES

In general, quasars do not look like ordinary galaxies, but certain galaxies are themselves quite different from ordinary ones. These are the Seyfert galaxies and N galaxies (Chapter 22). Many astronomers now believe that quasars may be associated with these objects.

Seyferts and N galaxies have especially bright cores. Short exposures (Fig. 23.6) show only their cores, which look curiously like quasars. Because of this similarity, it has been suggested that quasars may be related to Seyferts and N galaxies; in fact, they may even be related—more distantly—to the cores of ordinary galaxies.

According to one theory, ordinary galaxies, Seyferts, N galaxies, and quasars may form an evolutionary sequence. Perhaps quasars formed first, then gradually changed into Seyferts (or N galaxies), which in turn evolved into ordinary galaxies. There is some evidence for this: Nearby (out to the farthest observable galaxies) there are few quasars, but farther and farther out (and consequently further and further back in time), more and more quasars are found. Their number density increases rapidly with distance (and time) until suddenly, at a distance corresponding to about 90% of the speed of light, there are no more quasars.

This apparent cutoff seems to indicate that there were no quasars in the universe before this time. They suddenly appeared at this point in time and then, because of their high energy output, evolved rapidly, changing relatively soon into radio-quiet quasars.

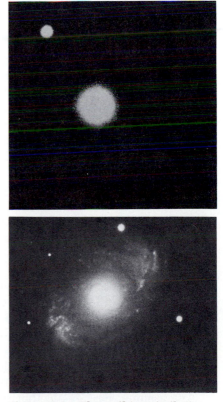

Figure 23.6 The striking similarity between quasars and the cores of Seyfert galaxies is exemplified by quasar 3C 273 (upper) and Seyfert galaxy MGC 4151.

Black Hole Models of Quasars and Radio Galaxies

Quasars and radio galaxies have extremely high energy outputs. A typical quasar, for example, emits a hundred times as much energy as an entire galaxy. Astronomers still do not have a completely satisfactory explanation of where this energy comes from. A clue may be found in the fact that a number of quasars and radio galaxies have jetlike streams of gas shooting outward from their core. One of the first quasars discovered, 3C 273, has a jet of this type, as does the elliptical galaxy M87. The jets are obviously being caused by explosions deep within these objects.

But what is causing the explosions? In the case of M87 there is considerable evidence that the core is small and exceedingly dense. In addition, it is abnormally bright, and the stars in the region seem to be moving much more rapidly than would be expected. Quasars also have small light-emitting regions. Some quasars change their brightness in only a day or so. This means they must be small (compared to galaxies), perhaps no larger than our solar system.

How do we combine this evidence for small, compact cores with the tremendous energy outputs of these objects? According to Martin Rees of Cambridge University, the best candidate is a gigantic black hole. Rees and others believe that there may be massive black holes in the cores of many, and perhaps all, quasars and radio galaxies. To fit the observed data, however, they would have to be extremely massive—a billion times as massive as our sun. And, of course, something else would be needed; a large black hole by itself is not an energy source. The energy is presumably being released as gas pours into the black hole. This gas is believed to be in the form of an accretion disk that surrounds it. This disk would be much like the one postulated for Cygnus X-1, only much larger.

There are at least three different ways in which a disk of this type could be produced. We assume, first of all, that there are stars in orbit around the black hole. This would be expected in the case of a radio galaxy, but we assume that it is valid for quasars as well. Near the core these stars will likely be close together. As they age and become red giants, much of the gas in their outer shells will be thrown off into space, and most of this gas will likely end up in the accretion ring.

Galaxy M87, with emanating jet.

Stellar collisions may also account for some of the gas of the ring. If the density of stars is high, collisions will be common. Much of the debris and gas from such collisions will no doubt find its way into the accretion disk. The third and perhaps most important source of gas is tidal force. Some of the stars orbiting the black hole would no doubt come quite close to it; tidal forces would then rip off their outer layers and pull them inward.

The gas and debris in the accretion disk will gradually be pulled toward the black hole. Finally, in a burst of released radiation, the gas closest to it will plunge through the event horizon and be lost forever. However, the nature of the radiation in the case of gigantic black holes is quite different from that released by smaller systems such as Cygnus X-1. Whereas Cygnus X-1 gives off mostly X-rays, systems of this type will give off mostly UV and visible radiation, but in some cases X-rays are released as well.

There are two possible explanations of how extremely massive black holes could be formed. As we mentioned in Chapter 20, exceedingly massive black holes may have been produced by the Big Bang. But even if a black hole was small initially, it could grow as it swallows up gas and debris. As it gets larger, its sphere of influence becomes larger and it grows even faster, until finally it is massive enough to pull in entire stars.

While there are still problems with black hole models, they seem to supply many of the needed answers. They will likely play an increasing role in explanations of exotic objects.

There is other evidence that suggests a relationship between quasars and galaxies. The object BL Lacertae has been observed for years, but its nature could not at first be determined. It is surrounded by a small, faint nebulosity and varies considerably in luminosity. Furthermore, its spectrum was unusual in that it was continuous; no discrete spectral features could be seen on it. This meant that no red shift could be determined and therefore that there was no way of determining how far away it was.

In 1974, however, J. Gunn and J. B. Oke reported they had found a dim discrete feature, which seemed to indicate a 7% red shift. In 1975 J. Miller and S. Hawley of Lick Observatory, using a ring aperture to block out the central region, obtained clear spectral features. The 7% red shift found by Gunn and Oke was correct. Even more interesting, the spectrum of the outer nebulous region was almost identical to that of a normal elliptical galaxy. It was the core that was strange. Since this discovery, several other similar objects have been found. They are now called BL Lacertae objects, or *blazers* for short.

M. J. Rees and R. Bladford of Cambridge have presented a model that appears to explain what has been observed. They believe that quasars and blazers are the same type of object seen from different directions. In their model a dense, doughnut-shaped cloud surrounds a dense core—perhaps a black hole. Two powerful nonthermal radiation beams emanate from the core (in a direction perpendicular to the plane of the gas ring). When we see this object face-on, the radiation is overpowering and we see no discrete spectral features; it looks like a blazer. When we see it at some angle or edge-on, however, there are discrete spectral features and it appears to us to be a quasar.

OTHER PROBLEMS

Although some progress in understanding them has been made, quasars are still an enigma. We are not sure of their energy source; we do not know how large or how massive they are. It is most widely believed that they are cosmological, but there are problems with this interpretation. A quasar has been discovered that is apparently connected via a luminous bridge to a galaxy of very different red shift, and there are other such pairs (see Chapter 26). Moreover, in 1971 a quasar was discovered that appeared to consist of two components, one apparently moving away from the other. The red shift of the objects indicated that they were in the outer reaches of our universe. Four months later they were again checked. Astronomers were amazed by how much they had separated in such a short time. How fast were they traveling? Calculations seemed to indicate that they were separating at several times the speed of light, which according to the theory of relativity is impossible.

Several of these so called "superluminous sources" have now been discovered. M. J. Rees and others have shown that they are probably optical illusions caused by the way we are seeing them.

Several models have been put forward. One is called the "flashlight" model. In this case it is assumed that a central quasar is surrounded by an invisible cloud of gas. A beam, or perhaps two beams, of light (or other radiation) from the quasar sweep across this gas, illuminating it. We see these illuminated spots moving away from one another. If the rotational velocity is high enough, or if the gas cloud is far enough away, velocities greater than that of light will be seen.

Dr. Lynden-Bell has given this theory an interesting twist. According to his model (in which, again, two radiation beams illuminate a gas cloud), the velocity of separation will be less than the velocity of light only if the age of the universe is about half that which we now accept (p. 438).

SUMMARY

1. Shortly after the Third Cambridge Catalog was published, Matthews pinpointed one of the radio sources with small angular diameter (3C 48) accurately enough so that a photograph could be taken of it. The photograph showed a starlike object. The emission lines in the spectrum of this object could not be identified.
2. A photograph of the source 3C 273 showed a few more lines (as compared to 3C 48), but they also could not be identified. A jet emanated from 3C 273.
3. Schmidt showed that the spectral lines had undergone an extremely large red shift.
4. The objects were called quasi-stellar sources (QSSs) and quasi-stellar objects (QSOs). Both types of object are now referred to as quasars.
5. The recessional velocities of quasars are much greater, in general, than those of galaxies. Recessional velocities greater than 90% of the velocity of light have been observed.
6. Quasars are extremely strong sources of energy. A typical quasar gives off as much energy as 100 galaxies like ours (assuming that quasars are cosmological).
7. Many quasars are variable. Some, in fact, have time variations as small as a day or so. This indicates that they are relatively small compared to galaxies.
8. Some quasars exhibit both emission and absorption lines. In some cases the red shifts of these lines are different for the same quasar.
9. Although there have been arguments for quasars as local phenomena (objects blown out of the core of our galaxy), the evidence is now strong that they are at cosmological distances.
10. Numerous models have been put forward to explain the energy output and various other properties of quasars. So far none of them have been completely successful.
11. There may be an evolutionary relationship between quasars and Seyfert galaxies. Ordinary galaxies may also be distantly related.

REVIEW QUESTIONS

1. Why was it difficult during the 1960s to locate and photograph quasars? How has this problem been overcome?
2. Why were Sandage and others puzzled by the spectra of 3C 48 and 3C 273?
3. Why did astronomers not expect large red shifts for starlike objects at that time?
4. What is the difference between a QSS and a QSO?
5. Explain the technique that Sandage and Ryle developed for locating quasars.
6. Why is the variability of a quasar important? What kind of limit does it impose? Why?
7. Discuss the arguments against quasars as "local" phenomena.
8. Briefly discuss some of the models that have been proposed to explain the energy output of quasars. Has this problem been resolved?
9. Summarize the problems that still exist in relation to quasars.

THOUGHT AND DISCUSSION QUESTIONS

1. It has recently been suggested that quasars may be an early form of galaxies. List features that suggest this. What are the problems with this suggestion? Discuss.
2. Suppose you obtained the spectrum of a quasar and it exhibited both emission and absorption lines. Suppose further that the absorption lines showed two different Doppler shifts and that the emission lines showed three different shifts. Explain all features of this spectrum in terms of a physical model.
3. Many pairs of quasars appear to be superluminous sources (separation velocity greater than c). Can you think of any explanation for this phenomenon other than the one given in this chapter?

FURTHER READING

Burbidge, G. and M. Burbidge. *Quasi-Stellar Objects.* W. H. Freeman, San Francisco (1967).

Darling, D. "The Quasar Connection," *Astronomy,* p. 6 (December 1979).

Golden, F. *Quasars, Pulsars and Black Holes.* Scribners, New York (1976).

Kahn, F. D. and H. P. Palmer. *Quasars.* Harvard University Press, Cambridge (1968).

Schendel, J. R. "Are Blazers Quasars?" *Astronomy,* p. 67 (February 1980).

Schmidt, M. and F. Bello. "The Evolution of Quasars," *Scientific American,* p. 54 (1971).

part 7

Cosmology and the Universe

Our study of the universe began with familiar things such as planets, then moved to stars and finally to galaxies. At each step the object dealt with was larger and, in many ways, stranger. In this part we take a step to the spatial limits of astronomy—to the universe itself. The answers are much less certain here.

Questions of importance in this section will be of the nature: What is the structure of the universe? and how big is it? To many, questions like these may seem impossible. How, for example, do we deal with concepts like an "infinite" universe? What does it really mean to say that the universe goes on forever, and, of course, if it doesn't go on forever, where does it end?

As in all of science there are two approaches to the answers—through theory and through observation. The theories are relatively well established, and have been for some time, but showing which particular theory (or which version of a theory) is correct has not proven to be easy, for we have literally to look to the ends of the universe for the answers. Observations are difficult, to say the least, and sometimes many years are involved in taking a single measurement. Despite the problems, a number of important breakthroughs have been made in the last few years, but many problems remain.

The first chapter in this part—Chapter 24—outlines the history of the discovery of the expanding universe. In particular it shows what observational evidence we now have for this belief. The heart of this part is perhaps Chapter 25, in which the Big Bang theory is introduced and the evidence we now have for it is presented. Major recent advances, such as the discovery of the cosmic background radiation, will be discussed.

Chapter 26 offers convincing reasons that cosmology is far from a closed science: Even where the evidence is relatively strong (for example, in the case of the Big Bang theory) there is controversy.

Aim of the Chapter

To describe some early models of the universe and to trace the discovery of its expansion.

chapter 24

The Expanding Universe

Questions about the size, age, and overall structure of the universe are the concern of the branch of astronomy called *cosmology*. Cosmologists may also consider how the universe began, whether it indeed had a beginning, and how it has evolved.

Strictly speaking, the latter questions are the domain of *cosmogony*, the theory of the origin of the universe. But since it is sometimes difficult to differentiate these fields clearly, we will deal with both without distinguishing them.

Newton was the first to try to formulate a mathematical theory of the structure of the universe. He began with two postulates: that the Newtonian law of gravity is valid throughout the universe, and that Euclidean geometry (the only geometry known at that time) applies throughout it. The major difficulty that Newton faced was the question of whether the universe has a boundary. It seemed that it could not, for if it had an end, one would immediately ask, "What is beyond this?" The possibility of a finite universe of matter contained in an infinite space was also unsatisfactory; gravity would eventually pull the matter inward, and the universe would collapse. The only way around this seemed to be a universe in which the stars and matter extended to infinity. However, an infinite universe would give rise to infinite forces at each point, so that for equilibrium to be maintained, an infinite force in one direction would have to be balanced by an infinite force in the opposite direction. With complications like this it was difficult to explain fully the motions of the solar system. Moreover, an infinite universe led to a perplexing quandary, later known as Olbers's paradox.

OLBERS'S PARADOX

The paradox named after the nineteenth-century amateur astronomer Wilhelm Olbers, who carefully described it, was actually

Figure 24.1 Olbers's paradox as Kepler visualized it: If the universe is infinite, every line of sight should eventually strike the surface of a star. Since this does not occur, the universe must be finite.

known many years before Olbers was born. One of the first to consider it was Kepler. He saw it as follows:

> If we look far enough into the universe, every line of sight must eventually intercept the surface of a star. Thus, if the universe is infinite, the night sky should be bright.

Since the night sky is not bright, Kepler reasoned that the universe was finite (Fig. 24.1).

However, there is more to the paradox. To understand it more fully, let us divide the space around the earth into an infinite number of shells of equal thickness (Fig. 24.2). Assume that the stars are uniformly distributed and, on the average, all of about the same brightness. As we look outward through these concentric rings, the distant stars will appear dimmer than those nearby because the intensity of light drops off with distance. But from Fig. 24.2 we see that each concentric shell has a larger volume than the one just inside it and therefore contains more stars. In fact, it contains just enough more to offset the loss by diminishing brightness. This means that each shell gives off the same amount of light as every other one, and if there are an infinite number of shells there will be an infinite amount of light.

Although Kepler realized that problems existed, it was Edmund Halley who brought the paradox to the attention of the scientific world. In 1720 he published two short papers on it. In one he described the comments of someone who had recently suggested that the only way around the paradox was to assume a finite universe. Halley did not agree; he was convinced that the universe was infinite, but he could not give a compelling argument to support this belief.

In 1744 Swiss astronomer J. P. Cheseaux suggested that the universe was filled with a slightly opaque gas that interrupted light. This would be like looking at street lights through a slight fog; the distant ones could not be seen. Olbers offered a similar solution in 1826. In 1848, however, William Herschel showed that both men were wrong; he pointed out that an intervening gas may indeed absorb some of the light, but eventually it would have to emit it.

The first person to find a valid way out of the paradox was C. V. I. Charlier. We will see later, however, that although his argument is valid, it does not apply to our universe as far as we know. Charlier showed that the night sky need not be bright if there was a hierarchy of clustering among the stars—in other words, if the universe was made up of stars, clusters of stars, clusters of clusters of stars, and so on indefinitely. There is some evidence for such a hierarchy, but it ends after only a few steps. At present astronomers are familiar with stars, star clusters, galaxies, clusters of galaxies, and clusters of clusters of galaxies (superclusters), but that is as far as the chain appears to extend.

Fortunately, we need not depend on this approach to resolve the paradox. Before we look at the solution, let us return to Fig. 24.2. We concluded that since each of the concentric shells gave off an equal amount of light, an infinite number of shells would give off an infinite amount of light. But since stars actually subtend a small but finite area, it would only take a finite number of stars to cover the entire sky. Hence a finite universe would also give us a bright night sky. Why, then, is the night sky not bright?

According to E. R. Harrison of the University of Massachusetts, the universe simply does not contain enough energy. Even if all the matter in the universe were converted to starlight, there would not

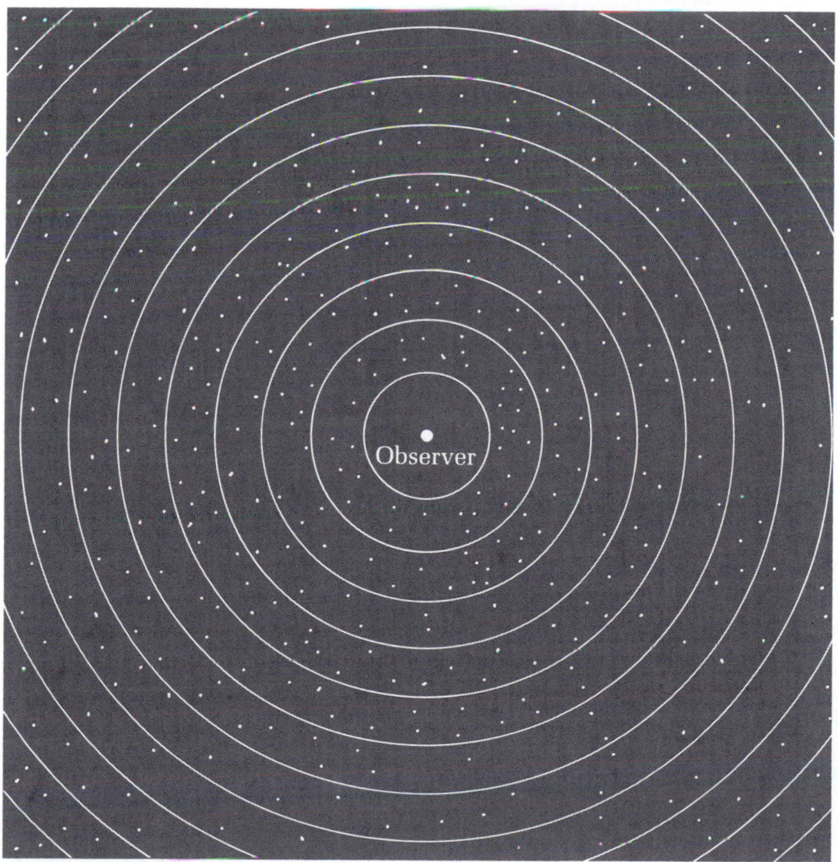

Figure 24.2 A representation of Olbers's paradox: If the shells are infinite in number and each gives rise to the same amount of light, the night sky should be bright.

be enough to make a bright night sky. Harrison has shown that most of the light that would in theory give us a bright sky would come from the very distant parts of the universe (more than 10^{24} light years away). But the universe cannot be this large; to phrase it another way, since we are also looking back in time, it is not old enough. As we look outward, we see younger and younger stars (or galaxies) until finally, beyond about 10^{10} light years, we see no stars (see Chapter 25).

Let's turn now to another explanation of Olbers's paradox that is frequently given. It has been stated that the expansion of the universe completely resolved Olbers's paradox. It might seem, since the light from the most distant galaxies has the greatest red shift, that eventually the dropoff would become great enough to prevent a substantial fraction of the light from reaching us. Harrison has shown, however, that the dropoff rate is not sufficient. Expansion would have to reduce starlight much more than it does for this argument to apply. This does not mean that there is no contribution from the red shift; there is. It is generally assumed that both the red shift and Harrison's argument contribute to the solution of Olbers's paradox.

EINSTEIN AND DE SITTER MODELS

When Einstein first turned his attention to cosmology, he was faced with the same problems that had challenged Newton many years before, but he had just developed a new and powerful tool, the general theory of relativity. Einstein asked whether the universe would necessarily be Euclidian (flat) on a large scale. Perhaps, he suggested, it appears flat to us only because we see such a small section of it. Indeed, according to general relativity, space is curved by matter on a small scale (i.e., around stars). Perhaps it is also curved on a universal scale. Einstein theorized that the universe has an overall positive curvature, and he made two additional assumptions:

1. That the matter in the universe is randomly distributed and static but that on a large scale it is distributed homogeneously (Fig. 24.3).
2. That matter is distributed isotropically everywhere (the same in all directions from any point).

These two assumptions, now referred to as the *Cosmological Principle,* are still basic to almost all cosmologies.

The matter in Einstein's model universe gave it a positive spherical curvature. The more matter it contained, the greater the curvature. Using Hubble's calculations of the amount of matter in the universe, Einstein concluded that a ray of light starting out in a given direction would trace out a gigantic circle, returning after 200 billion years to the point from which it started. There was, in effect, no boundary to this universe; it was finite but unbounded. However, it could contain an infinite number of galaxies (see problems at end of chapter). A universe that folds back on itself in this way is said to be *closed.*

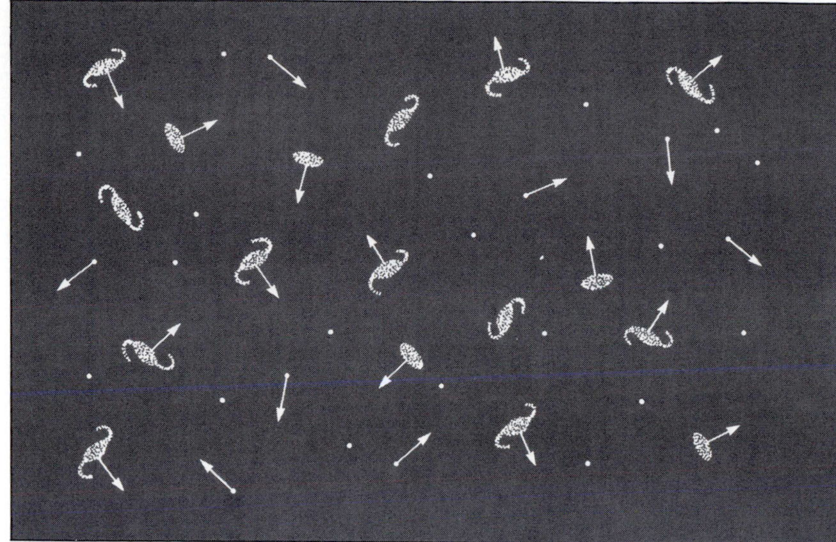

Figue 24.3 Einstein's model of an isotropic, homogeneous universe. The galaxies could move, but their motions would be random. Einstein assumed that aside from local fluctuations the universe would appear the same regardless of an observer's position.

There were still problems with Einstein's model. His general relativity equations implied that the universe either expands or contracts. To avoid this difficulty, he modified the basic equations of general relativity, by introducing a term he called the "cosmological constant"; however, he was never satisfied with this solution, which he finally abandoned in 1935. Indeed, it is ironic that he introduced it in order to obtain a static universe, for it was later discovered that the universe is not static. Thus, if he had accepted the results as they were, he could perhaps have predicted the expansion of the universe.

Einstein's cosmology was the first mathematical theory based on a solid physical foundation and, as such, is a model for cosmological theories even today. We now know, of course, that it is incorrect; Einstein did not know of the expansion of the universe. Despite this, the theory is a milestone in cosmology.

In the same year that Einstein published his cosmology (1917), Dutch astronomer de Sitter introduced a theory that also fit the equations of general relativity. In many ways it was more interesting than Einstein's. The strangest thing about it was its prediction that the universe has an average density of zero. We now know that our universe is quite close to being empty. Hermann Weyl noted that the theory predicted a red shift; this seemed to indicate that the model was not static, as first assumed. In fact, it was later shown that two particles placed in it would move away from one another, and the farther apart they traveled, the faster they separated. Thus, even with a cosmological constant, de Sitter's theory predicted an expanding universe (a contracting version is also possible). This prediction was actually made several years before Hubble put forward his theory of an expanding universe.

De Sitter's universe can be thought of as a special case of Einstein's. It expands because it does not contain enough matter to hold it together. If more and more matter were poured into it, it would eventually attain a rather "shaky" state of equilibrium. From another viewpoint, if Einstein's universe could be made to expand, it would become de Sitter's universe when the distances between galaxies became infinite.

DISCOVERY OF THE EXPANDING UNIVERSE

About 1912, V. M. Slipher of Lowell Observatory in Arizona began a study of the spectra of some of the nearby galaxies. One of the first spectra he obtained was that of M31; a close examination showed that the lines were blue-shifted. When Slipher examined the spectra of other galaxies, he found that most of them were red-shifted.

According to the Doppler interpretation of these shifts, M31 was moving toward us. (Of course, corrections must be made for the rotational motion of our galaxy, but even after this correction M31 is still moving toward us.) However, most of the other galaxies that Slipher studied were apparently moving away from us. He continued observing more and more galaxies; all were red shifted. By 1919 he had encountered recessional velocities up to 1800 km/sec. Hubble's discovery that these objects were outside our galaxy made velocities of this magnitude easier to accept, but they were still very surprising.

Hubble soon became interested in the problem. Using the 2.5-m reflector at Mt. Wilson, he not only verified Slipher's discoveries but went on to look at fainter and fainter galaxies. Each was red-shifted, and the fainter the galaxy, the greater the shift (Fig. 24.4). Hubble then turned his attention to determining the distance to these galaxies. Some of the nearby ones had cepheid variables in them that could be resolved; their distance was therefore easily obtained. In the galaxies farther out, however, cepheids were not bright enough to be seen. Fortunately, it was still possible to resolve some stars—the supergiants—and these were all of approximately the same absolute brightness.

Using this as a guide, Hubble estimated the distance to many more galaxies, but there were even fainter ones—usually in clusters of galaxies—in which no stars could be resolved. As a final step, Hubble assumed that the brightest galaxies in these clusters were of approximately equal absolute brightness. This enabled him to estimate the distances to galaxies hundreds of millions of light years away.

When he announced his expanding universe theory in 1929, Hubble had explored galaxies out to 6 million light years. (Later, when the distance scale was adjusted, all distances were considerably increased.) By 1936, when he published his book *Realm of the Nebulae,* he had explored out to 240 million light years. In this book he showed that as the distance to a galaxy increases, its recessional velocity increases linearly. This means that a galaxy twice as far away is moving away from us twice as fast (Fig. 24.5). The slope

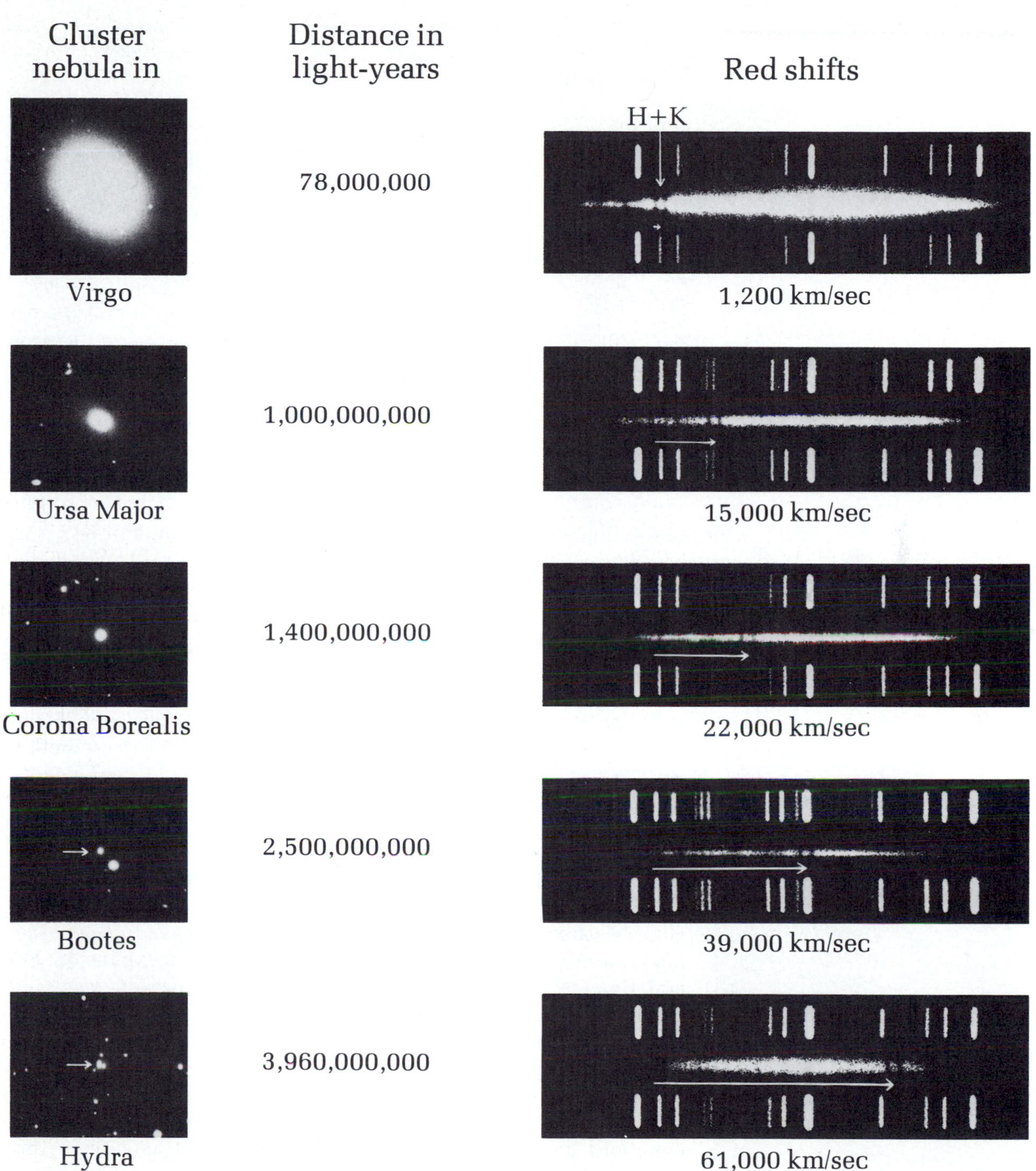

Figure 24.4 Relation betwen red shift and distance for galaxies. The position of the H+K line moves toward the right (the red end of the spectrum) as the distance to the galaxy increases.

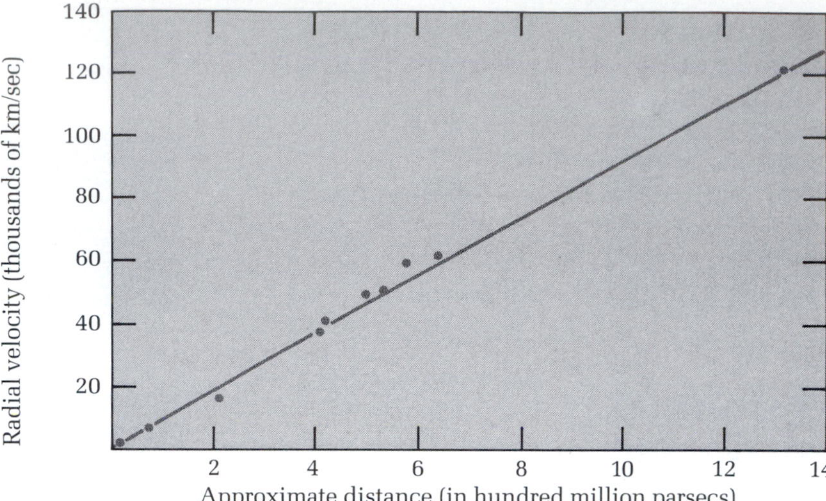

Figure 24.5 An early Hubble plot. The slope of the line is known as Hubble's constant.

of the line in the graph is called Hubble's constant, and this number plays a particularly important role in cosmology.

When Hubble died in 1953, Allan Sandage collected and published many of his best photographs. This collection is now known as the *Hubble Atlas*. Sandage and others have used the 5-m reflector at Mt. Palomar to carry Hubble's work further. They have shown that although his linear relation is well satisfied to the limit of the 2.5-m reflector, this does not seem to hold true for galaxies brought into view by the 5-m reflector. Near the limits of our universe something strange seems to be happening.

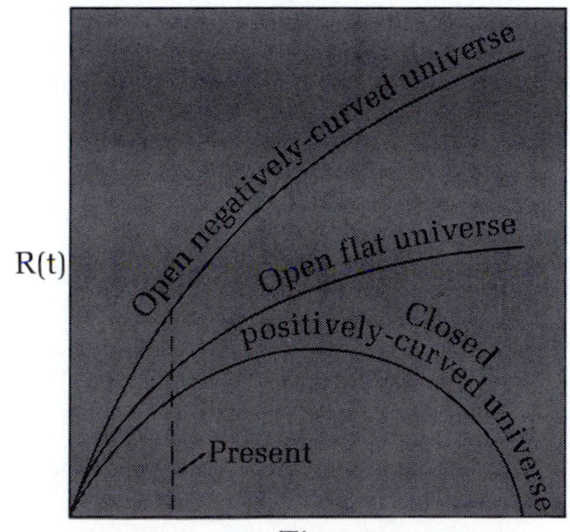

Figure 24.6 The three possible universes within Friedmann's cosmological model. The factor *R* is a measure of the distance between galaxies. An open universe will expand forever; a closed one will eventually collapse back on itself.

EVOLUTIONARY MODELS OF THE UNIVERSE

Hubble's announcement that the universe was expanding generated new interest in theories that could account for this expansion. Several time-varying or evolutionary models had already been published but had received little attention. The first was due to the Russian A. Friedmann, who published in 1922 a paper describing an evolutionary cosmology that did not include the cosmological constant. In discarding the constant Friedmann found that his universe either expanded or contracted. Upon developing the resulting equations, he found that within his framework there were three possibilities:

1. An open, negatively curved universe.
2. An open, flat universe.
3. A closed, positively curved universe.

They are represented schematically in Fig. 24.6 (also see Fig. 25.10). Note that the two open models will expand forever; the closed model, on the other hand, will eventually stop and collapse back on itself.

Friedmann's 1922 paper, as well as one published in 1924, was generally ignored at the time. Because the red shift relation for galaxies had not yet been discovered, there was little reason to consider evolutionary models.

In 1927 the Belgian Georges Lemaitre, completely unaware of Friedmann's work, also published a paper outlining an expanding universe model. His model was different from Friedmann's in that he retained the cosmological constant. This paper also received little attention.

After Hubble's announcement, scientists directed their attention to evolutionary models. Among them was Lemaitre's paper (and later Friedmann's), which soon came to the attention of Sir Arthur Eddington. Realizing its importance, he arranged to have it republished in the English journal *Monthly Notices of the Royal Astronomical Society,* where it would receive much wider circulation. Eddington also discovered that Einstein's static model was not as static as previously believed. Although it was in equilibrium, the equilibrium was unstable. The slightest disturbance would topple it into a state of expansion or possibly contraction.

Like Einstein's model, Friedmann's was theoretical. In essence it was a set of mathematical equations; for completeness it needed a physical counterpart. This was supplied by Lemaitre. In his original (1927) paper he assumed a smooth expansion from an Einstein state to a de Sitter state, but by 1931 he had modified his ideas. In his new theory the initial state was a compact and extremely dense "primordial nucleus." According to Lemaitre this nucleus suddenly became unstable and exploded; large, heavy atomic nuclei were generated and gradually broke down to lighter and lighter nuclei. Eventually the expansion stopped and the universe approximated the Einstein static model. It was during this time, according to Lemaitre, that stars and galaxies formed. But Einstein's universe is unstable, and eventually expansion began again.

George Gamow was largely responsible for the acceptance and popularization of the Lemaitre model. He made many contributions to the theory, wrote several popular books on it, and is generally credited with the fanciful name it now bears—the "Big Bang" theory.

SUMMARY

1. Newton made the first attempts to formulate a mathematical theory of the universe. His model was not stable, and he found no satisfactory answer as to whether the universe has a boundary.
2. Olbers's paradox can be stated as follows: If the universe is infinite and contains an infinite number of stars, every line of sight will eventually intercept the surface of a star. The night sky should therefore be bright. Yet in fact it is black.

3. Two solutions have been given to Olbers's paradox. First, Harrison has shown that there is not enough energy in the universe to give a bright night sky. Second, the light from distant galaxies is red-shifted and becomes redder as we examine increasingly distant galaxies. Eventually no light reaches us. Both arguments contribute to the resolution of the paradox.
4. Einstein introduced the Cosmological Principle, which states that the universe is homogeneous and isotropic at all points.
5. Einstein's cosmology was flawed because he did not know of the expansion of the universe and therefore never included it in his theory. The model proposed by de Sitter predicted an expansion.
6. The first clue that the universe may be expanding came from Slipher's discovery that many of the nearby galaxies exhibit red shifts. Hubble examined more galaxies and eventually showed that most of them are receding from us (exceptions are those in our local group).
7. Friedmann's theoretical model of the universe allows for three possibilities: a closed, positively curved universe, an open, flat universe, and an open, negatively curved universe.

REVIEW QUESTIONS

1. Explain the problems that Newton encountered when he tried to formulate a cosmology.
2. State Olbers's paradox in your own words.
3. How is Olbers's paradox now resolved? Are there other ways of resolving it? Explain.
4. How was Einstein's static universe curved? What eventually happens to a beam of light traveling in it?
5. What is the Cosmological Principle? Is it possible to have a universe that is isotropic but not homogeneous? Homogeneous but not isotropic?
6. Why did Einstein introduce the cosmological constant?
7. How did Einstein's cosmology differ from de Sitter's? Are they related in any way?
8. Why did the discoveries of Slipher and Hubble spur interest in evolutionary models?

THOUGHT AND DISCUSSION QUESTIONS

1. Remembering that galaxies flatten more and more as they approach the velocity of light (relative to us) at the edge of our observable universe, explain how an infinite number of galaxies could fit into our universe.
2. If it were discovered that the red shift of spectral lines of galaxies were due to something other than expansion, how would our view of the universe change? Would there be conflicting evidence? Discuss.
3. Give a model of a universe that is isotropic but not homogeneous. Is it possible that the universe is homogeneous but not isotropic?
4. According to the Friedmann theory there are three possible models of the universe: positively curved, negatively curved, or flat. Sketch a simple two-dimensional representation of each of these (see Fig. 25.10 after you have tried). Show what happens to two lines that are initially parallel in each of them. What are the sums of the interior angles of a triangle for each of the three cases?

FURTHER READING

Berendzen, R., Hart, R. and Seeley, D. *Man Discovers the Galaxies,* Neale Watson Academic Publishers, New York (1976).

Charon, J. *Cosmology.* McGraw-Hill, New York (1970).

Harrison, E. R. "Why is the Sky Dark at Night?" *Physics Today,* p. 30 (February 1974).

Hubble, E. *The Realm of the Nebulae.* Dover, New York (1958).

Wilson, L. A. "The Dark Sky Paradox and Origin of the Universe," *Astronomy,* p. 52 (September 1978).

Aim of the Chapter

To examine the evidence for the leading theories of how the universe began and how it may end.

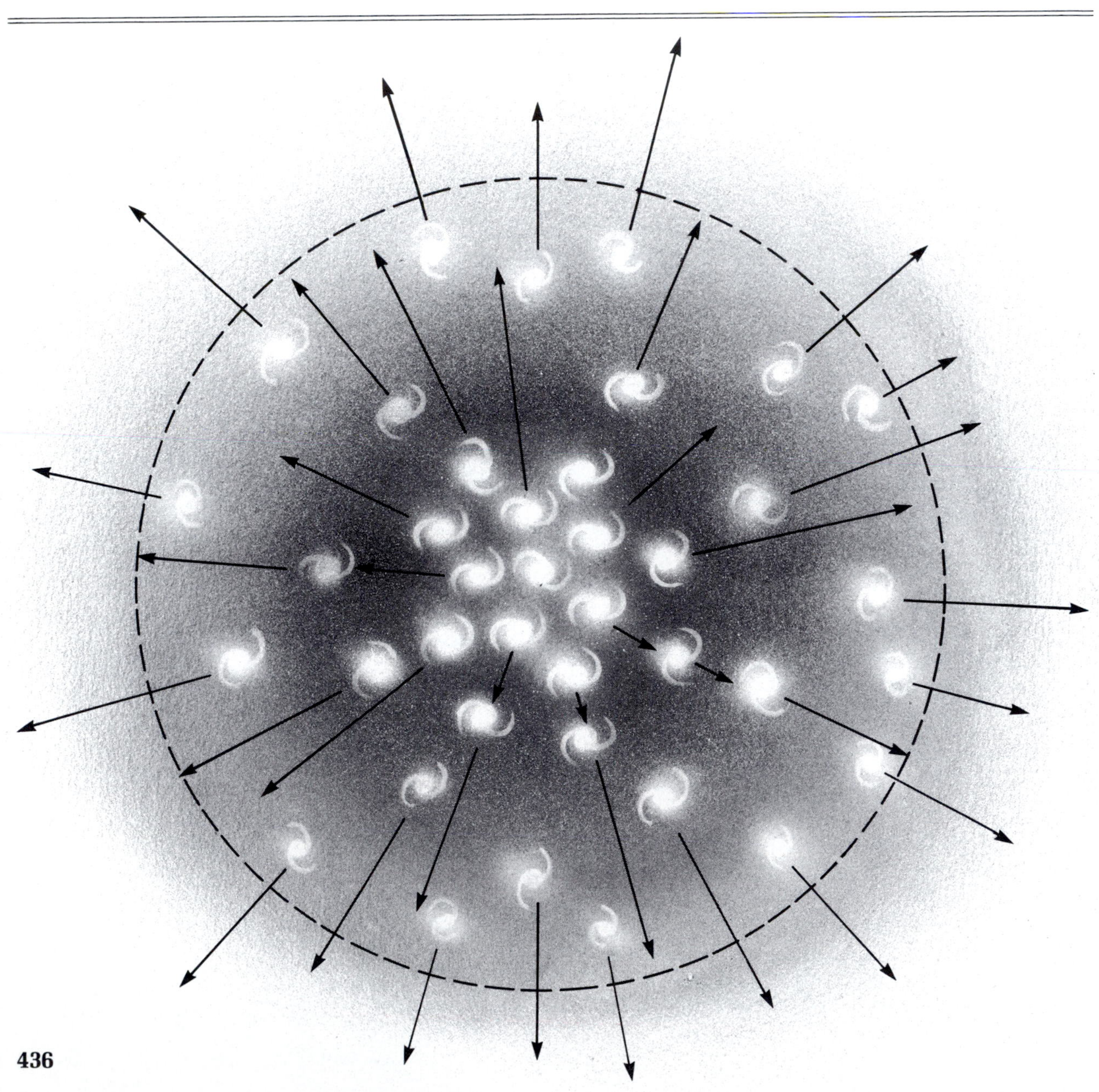

chapter 25

The Beginning and the End

If the universe is expanding now, it is reasonable to believe that it had a beginning. To see why, imagine taking a time-lapse movie of its expansion (with a billion years passing in, say, a minute). If we run this movie backwards, the galaxies will appear to move back together; the universe will contract. And eventually the galaxies will all come together into a fiery nucleus—what Lemaitre called the "primordial atom."

People sometimes ask where in the universe this nucleus was. This is roughly equivalent to asking where the center of the universe is, but the universe does not have a center. It would be incorrect to picture the primordial nucleus as a gigantic sphere in an endless space. The primordial nucleus *was* the universe; there was nothing else.

It is important to realize that it is the space between the galaxies that increases; the galaxies themselves do not increase in size. A useful two-dimensional analogy is provided by a balloon with small patches glued to it; as the balloon is blown up, the distance between each of the patches increases, but the patches themselves stay the same (Fig. 25.1). For a three-dimensional analogy, think of the raisins in a loaf of bread that is being baked; as the bread expands, the raisins separate without changing size.

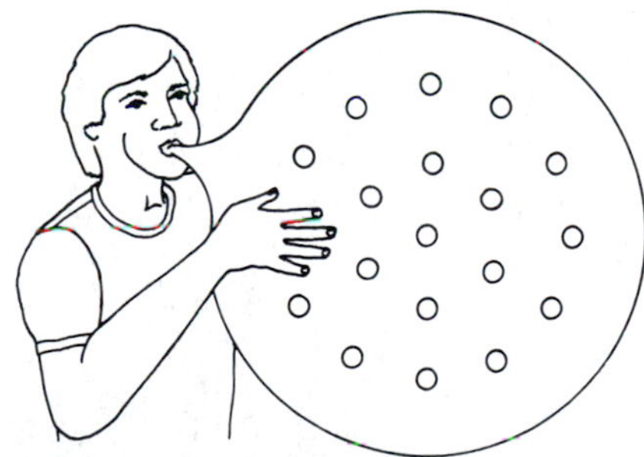

Figure 25.1 The expansion of the universe is analogous to that of a balloon with patches glued to it. The space between the patches increases, but the patches themselves stay the same. Similarly, the distance between galaxies increases, but the galaxies remain unchanged in size.

It might seem, since we find galaxy after galaxy receding from us in all directions, that we are at the center of the universe's expansion. But return to our balloon analogy. To an observer on one of the patches, watching all the others, they will recede as the balloon is blown up; but if the observer moves to a different patch, all the other patches will still appear to move away from him, because the distance between *all* patches increases. Thus we are not in a privileged position in the universe; regardless of where we happen to be, all galaxies will appear to be moving away from us.

THE BIG BANG THEORY

The Age of the Universe

If the universe had a beginning, it makes sense to ask how old it is. Recall that Hubble plotted red shift against distance for a large number of galaxies and found a linear relationship (p. 430). The shape of his curve was constant, and we call the magnitude of this slope H. The age of the universe can easily be obtained from H; in fact, to a first (crude) approximation it is just the reciprocal of H (i.e., $1/H$).[1]

To obtain a better approximation, let us compare the Big Bang to an explosion on Earth. The debris in a terrestrial explosion eventually slows down as it loses energy; when all its energy is used up, it stops. Similarly, we would expect galaxies at a great distance from us (thus very distant in time) to be *decelerating*. Hubble's constant is, therefore, not a constant after all; the straight line will eventually begin to curve for galaxies far enough away. Hence the age obtained from the value $1/H$, which we call the *Hubble age*, is too long (Fig. 25.2). The true age can be determined from the Hubble age by taking deceleration into account. Throughout the 1970s the accepted age was based on a study done over many years by Allan Sandage and G. A. Tammann. Their results indicate that the universe is approximately 20 billion years old.

But not all astronomers agree with this estimate. For years G. de Vaucouleurs consistently obtained a value close to 10 billion years. And in 1979 Marc Aaronson, John Huchra, and Jeremy Mould arrived at a similar value. All these astronomers feel that their results indicate that we live in an anomalous region of the universe.

In Chapter 22 we mentioned that our local group of galaxies belongs to a supercluster that is dominated by the gigantic Virgo cluster (it lies near the center of this supercluster; we are assumed to be near the edge). According to de Vaucouleurs, and to Aaronson, Huchra, and Mould, we are being gravitationally pulled toward the center of our supercluster, that is, in the direction of the constellation Virgo. This means that the galaxies in this direction do not recede from us as fast as they otherwise would. Thus using them to determine the age of the universe would give an overestimate.

We do, in fact, have evidence that we live in an anomolous region: the anisotropy of the cosmic background radiation (see p. 440). But it indicates that we are moving in the direction of the constellation Leo, not in the direction of Virgo. Sandage and Tammann, in contrast, find little evidence for any anomalous motion in their data.

However, there are two other, indirect methods for determining the age of the universe; they give the approximate age of our galaxy, which is a good lower limit on that of the universe. The first method

[1] H is the slope of a plot of velocity vs distance, so

$$H = \frac{\text{velocity}}{\text{distance}} = \frac{\text{distance}}{\text{time} \times \text{distance}} = \frac{1}{\text{time}} = \frac{1}{\text{age}}.$$

Therefore age $= 1/H$.

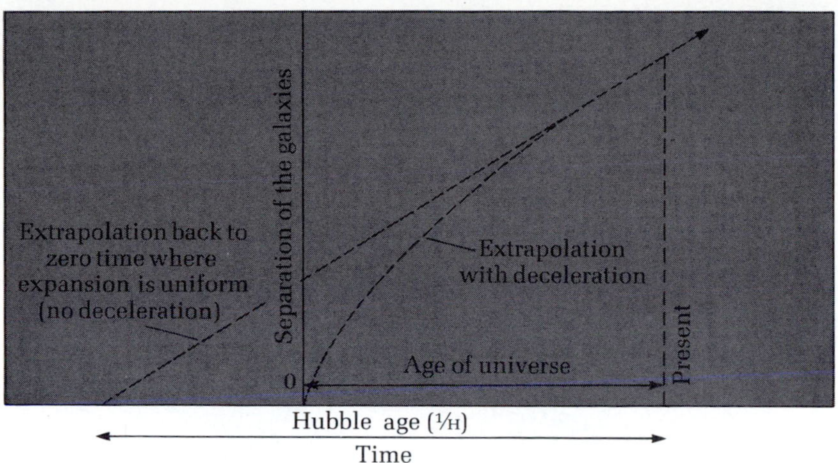

Figure 25.2 **Relationship between the Hubble age and the actual age of the universe. They are different because there is a deceleration in the expansion. The deceleration is a result of the mutual gravitational attraction of the galaxies.**

makes use of globular clusters; as we saw in Chapter 17, a turnoff point is obtained along the main sequence when the H-R diagram of a cluster is plotted. This turnoff point gives the approximate age of the cluster. A study of a large number of globular clusters indicates that their age ranges from about 8 to 18 billion years, which obviously contradicts the recent suggestions that the universe may only be 10 billion years old.

The second method of determining the age is based on the radioactive decay of various nuclei. All radioactive elements have a *half-life,* the time for half of the nuclei present at any time to decay to a different element. A study of the abundance of these elements also indicates that the universe is older than 10 billion years.

A somewhat different approach to the age problem has recently been taken by several groups. They have looked at how the various techniques depend on such things as the average density of matter in the universe and the abundance of helium in the universe. They have found that the age most consistent with all the available data is from about 15 to 18 billion years. At present we are reasonably certain that sometime between 10 and 20 billion years ago the Big Bang explosion took place and the universe began. For convenience we will use 18 billion years in most of the succeeding discussions.

As we look farther and farther out, we see galaxies receding from us at ever-increasing speeds. If we look far enough, we will find galaxies that are receding at near the speed of light (Fig. 25.3). Just beyond this galaxies will, in theory, be traveling at and even beyond the speed of light (relative to us), which according to relativity theory is impossible. In effect, then, this is the end of our observable universe. We will never be able to observe anything beyond the distance at which the recessional velocity is *c*. (If we were at a different position, we would have a different observable universe.)

As we will see, the Big Bang theory is really a *class* of theories. There are many variations on it, and we will discuss some of them in Chapter 26. We will also discuss a theory that takes a completely different point of view, the Steady-State theory. According to it, the

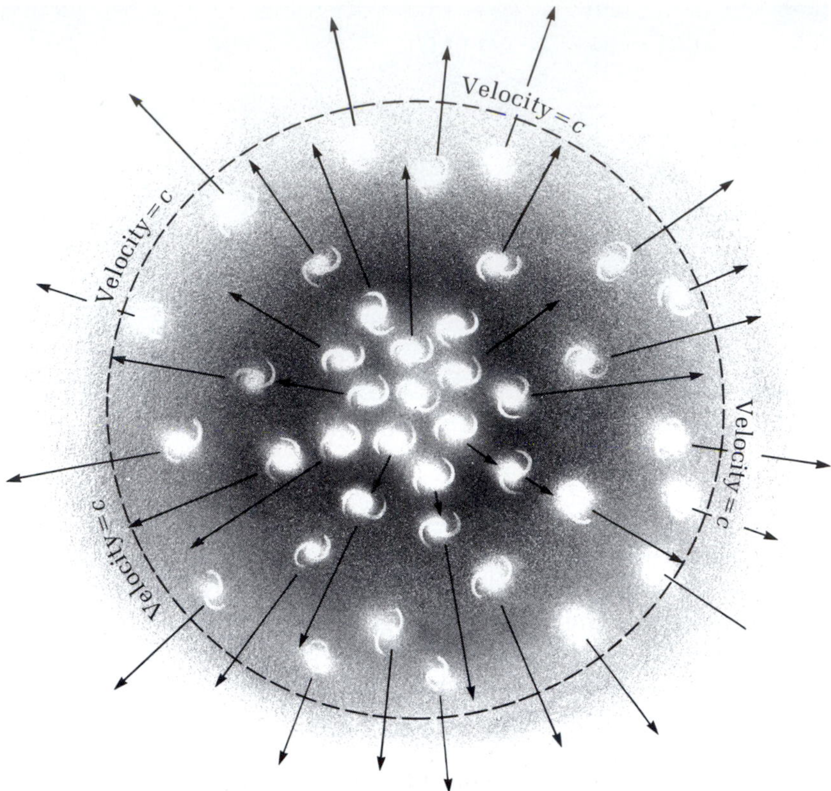

Figure 25.3 Expansion of the universe. The farther away a galaxy is, the faster it moves away from us. Eventually the speed relative to us is *c* (represented by the dashed circle). This is the end of our observable universe.

universe has always been the same, had no beginning, and will have no ending. During the 1950s there was considerable controversy as to whether the Big Bang theory or the Steady-State theory was correct. Later evidence gave more and more support to the Big Bang theory, and today most astronomers are convinced that it is correct (but see Chapter 26 for other possibilities).

The Cosmic Microwave Background Radiation

The most damaging evidence against the Steady-State theory came with the discovery of the cosmic microwave background radiation. In the late 1940's George Gamow, Ralph Alpher, and Robert Herman tried to determine how elements were born in the Big Bang explosion. Their calculations indicated that radiation should have been generated in the early stages of the universe. Originally it would have been exceedingly hot, but gradually it should have cooled to a temperature of perhaps 5°K. It would now be microwave radiation and should still permeate the entire universe. But Gamow and his colleagues did not think that the technology for detecting such radiation was available and so did not look for it.

Fifteen years later Robert Dicke, working on a problem related to the beginning of the universe, concluded, as Gamow, Alpher, and Herman had, that radiation must have been generated at an early stage; P. J. Peebles showed that the radiation would have an effective temperature of about 3°K and should be spread uniformly throughout the universe. Two colleagues, P. G. Roll and D. T. Wilkinson, began to look for it.

But unknown to these astronomers, it had already been found. In 1964 Arno Penzias and Robert Wilson encountered a low-level "noise" in a radio telescope with which they were working near Holmdel, New Jersey (Fig. 25.4). At first they thought it was coming from the antenna. When all attempts to eliminate it failed, they decided that it must be coming from beyond the antenna, from somewhere in the universe. They began looking for the source, but it seemed to be perfectly isotropic. It was coming from everywhere.

When Dicke and Peebles were invited to examine this "noise," they concluded that it was caused by the radiation they had predicted. It had all the required properties.

To determine what caused this radiation and why it is so isotropic, we must examine the events that took place just after the Big Bang. For hundreds of thousands of years the universe consisted of only particles and radiation. There was light, but the universe was opaque, for the radiation was, in a sense, trapped; almost immediately after it was emitted, it was absorbed or scattered again. It could not break free from the matter; we say that it was in thermal equilibrium with it. But gradually the universe cooled, and finally at about 3000°K hydrogen atoms began to form. This signaled the end of the equilibrium; the photons of light broke free (decoupled) and spread out into the universe, which was still expanding, pulling the wavelength of the photons out as it did so. And as their wavelength increased, the photons lost energy and cooled, until today they are microwaves with an effective temperature of only 3°K.

For a better understanding of what the 3°K temperature means, recall that all objects with temperatures above 0°K radiate (i.e., give off photons of all wavelengths) and that each has a characteristic emission curve. If the object is a black body, the curve will be a black-

Figure 25.4 The horn-shaped antenna at Holmdel, New Jersey, used by Penzias and Wilson to discover the background microwave radiation.

body curve (see Chapter 6). Suppose, then, that we place our radio telescope antenna in a black box (and wait for equilibrium conditions) and it detects photons that exhibit a 3°K emission curve. We say that this radiation has an effective temperature of 3°K.

The properties of the radiation found by Penzias and Wilson correspond well to the properties predicted for the radiation from the Big Bang. As we have seen, its effective temperature of about 3°K fits the predictions. Moreover, it should be black-body radiation (Fig. 25.5). When we compare this observed radiation over its range of frequencies (Fig. 25.6) to the well-known 3°K black-body curve (Fig. 25.5), we find an excellent match between the theoretical and observed curves. The radiation is, indeed, from the primordial fireball. (Recently several scientists have noticed that there is still a very slight discrepancy. What this means no one yet knows.)

Once they were certain the radiation was cosmic background radiation, scientists began to examine its properties in detail. All the early measurements showed that to a high degree of accuracy it was isotropic. However, some anisotropy would be expected if we are moving relative to the background radiation. To see why, imagine being stationary in a fog, having some device for measuring the fog's physical density, which shows that the fog is uniform in all directions. If we begin to move in a particular direction, however, we find that the fog appears to be more dense in that direction and less

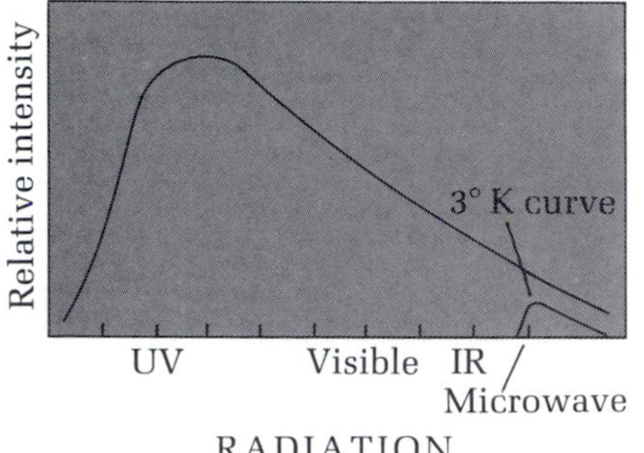

Figure 25.5 A black-body radiation curve for a relatively hot body. The 3°K microwave curve is similar to the larger one but smaller and closer to the right-hand end of the scale.

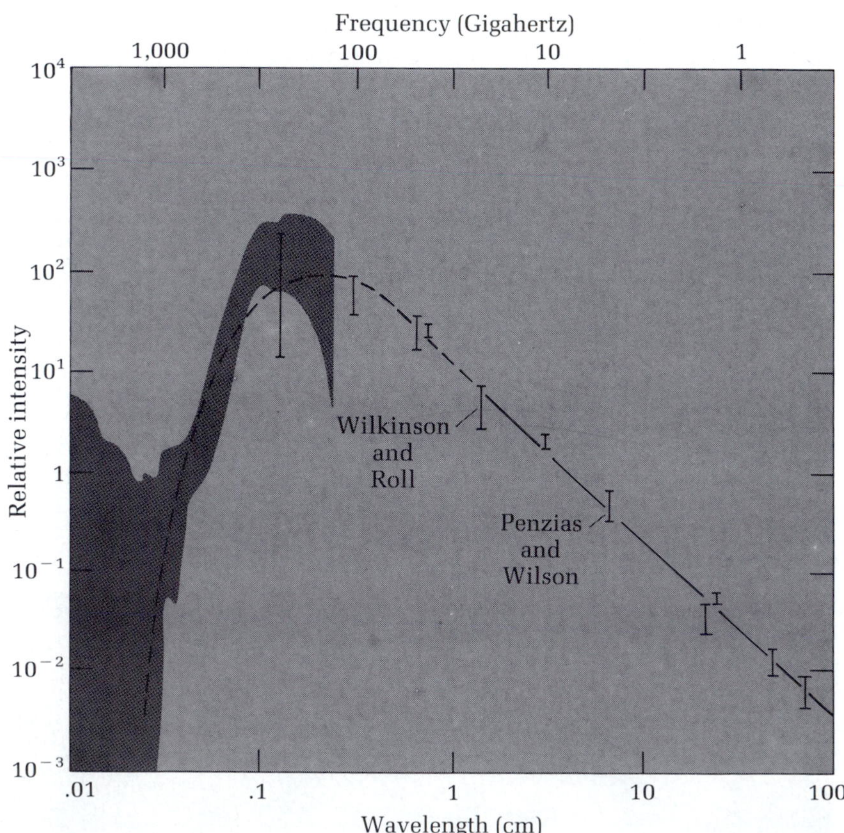

Figure 25.6 The black-body curve for the cosmic background radiation. Several experimental points, including the one obtained by Penzias and Wilson, are shown. The lengths of the bars represent the possible error in the measurements.

dense in the opposite direction, since we will measure a larger number of particles per unit volume in the direction we are traveling.

Similarly, the microwave radiation will not appear isotropic if we are moving through it, and since we know the earth is moving, it is reasonable to assume that we would find some anisotropy. This idea is similar to that behind the Michelson-Morley experiment (Chapter 19). Michelson and Morley were trying to measure our velocity relative to the aether; in this case we are trying to measure our velocity relative to the background radiation. If we are stationary relative to it, the temperature in all directions will be exactly the same. If we are moving in a particular direction, however, the temperature will be slightly higher in this direction and slightly lower in the opposite direction. With this in mind, then, scientists began looking for slight deviations in temperature.

Two groups set out to perform this "new aether experiment," one consisting of R. A. Muller, G. F. Smoot, and M. V. Gorenstein, the other of D. Wilkinson and B. Corey. Within a few years both groups showed that there is, indeed, anisotropy. We are apparently traveling at a speed of about 600 km/sec in the direction of the constellation Leo. In fact, our entire galaxy and local group of galaxies are moving in this direction (Fig. 25.7).

Since this radiation was formed early in the history of the universe, it is interesting to look for evidence of "clumping" (or inhomogeneities) in it. Inhomogeneities likely occurred as galaxies, quasars, and so on formed. So far such inhomogeneities have not been found, but during the next few years, as extremely delicate instruments for checking this radiation are put into satellites, it should be possible to determine whether they exist.

Figure 25.7 Measurement of anisotropy of the background radiation indicates that we are moving in the direction of the cross shown in the diagram. Darker portions indicate higher background radiation.

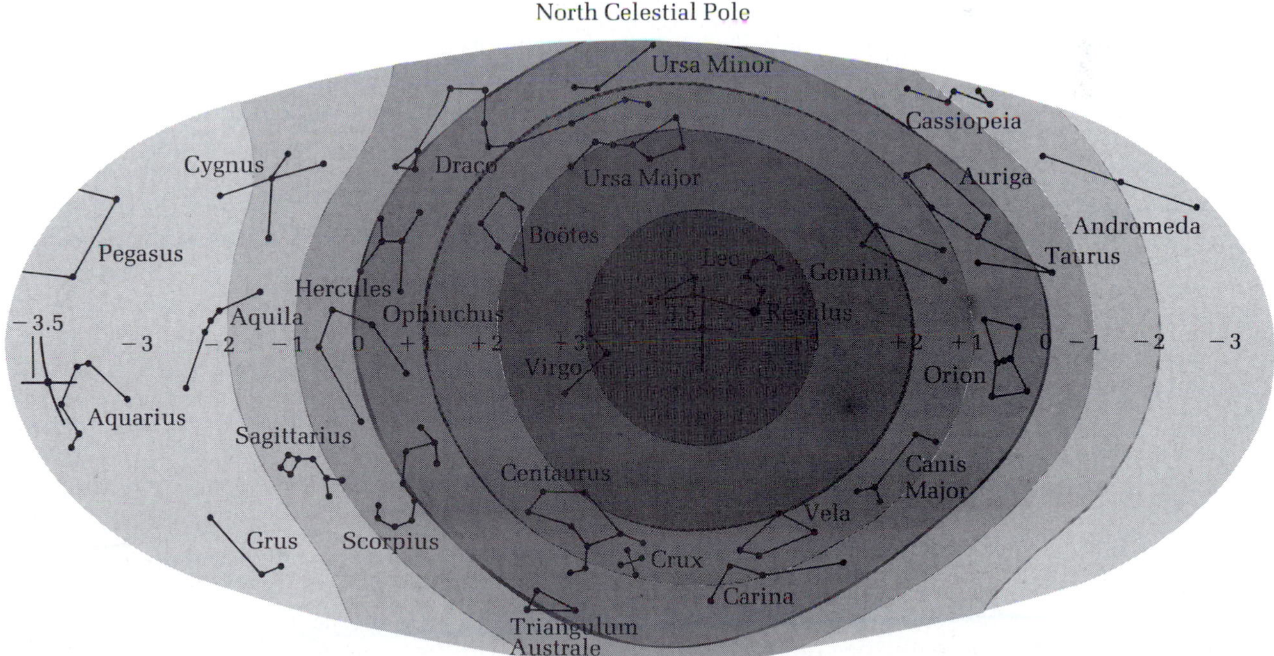

The Beginning of Time

With the discovery of the cosmic background radiation the Steady-State theory soon gave way to the Big Bang theory. Assuming, then, that the latter is correct, let us look at it in more detail, beginning with the explosion itself. In Chapter 24 (p. 432) we saw that Lemaitre and Gamow believed the primordial nucleus was extremely dense and perhaps a few hundred million light years in diameter. Today we have a quite different view of this nucleus. Most astronomers believe that it was initially a "singularity"; in other words, its initial density was infinite and its diameter was zero.

Suddenly, about 18 billion years ago, this singularity exploded and the universe began. Studies have shown that the expansion was relatively uniform and isotropic as early as 1 sec after the explosion, but astronomers are interested in times much earlier than this. They theorize about what happened beginning at 10^{-43} sec after the explosion (at earlier times Einstein's general theory of relativity breaks down). Much of what can be said about this time is quite speculative, but nevertheless it is accepted by many astronomers. And although we are discussing exceedingly short time intervals, so many things were happening at this time that infinitesimal fractions of a second were of tremendous importance.

We begin, then, with the four basic forces of nature (Chapter 15): the gravitational field, the electromagnetic field, and the strong and weak nuclear fields. A number of astronomers now believe that during this early era there was only one basic force. Gradually, however, the forces we now know broke free and changed significantly in strength. One might think of them as phases; for example, as water vapor cools, several phases condense out. The forces of nature may have "condensed out" in the same way (the gravitational field is assumed to have condensed out shortly after 10^{-43} sec, the others later).

At about 10^{-23} sec the universe entered what we now call the *hadron era* (Figure 25.8). It was still in a state of unimaginable chaos. The temperature may have been as high as 10^{32} °K, and only heavy particles (hadrons such as protons and neutrons) and radiation were present. Two different models have been suggested for this period: the elementary particle model (EPM) and the composite particle model (CPM). In the elementary particle model it is assumed that all particles are made up of a small number of elementary particles (quarks, antiquarks, leptons). Temperatures of up to about 10^{32} °K are possible in this description (more accurately, this temperature is generally assumed for the era just before the hadron era), and because of this it is sometimes called the "hot Big Bang" model. It is, incidentally, the more generally accepted of the two models.

In the composite particle model it is assumed that there are no true elementary particles. R. Hagedorn has suggested a particularly interesting theory of this type; he assumes an exponentially increasing spectrum of "resonances" (or fireballs, as he calls them). Basically they are just "composite" particles. One of the important consequences of his theory is the limiting temperature of 10^{12} °K, considerably lower than that allowed in the elementary particle model. Be-

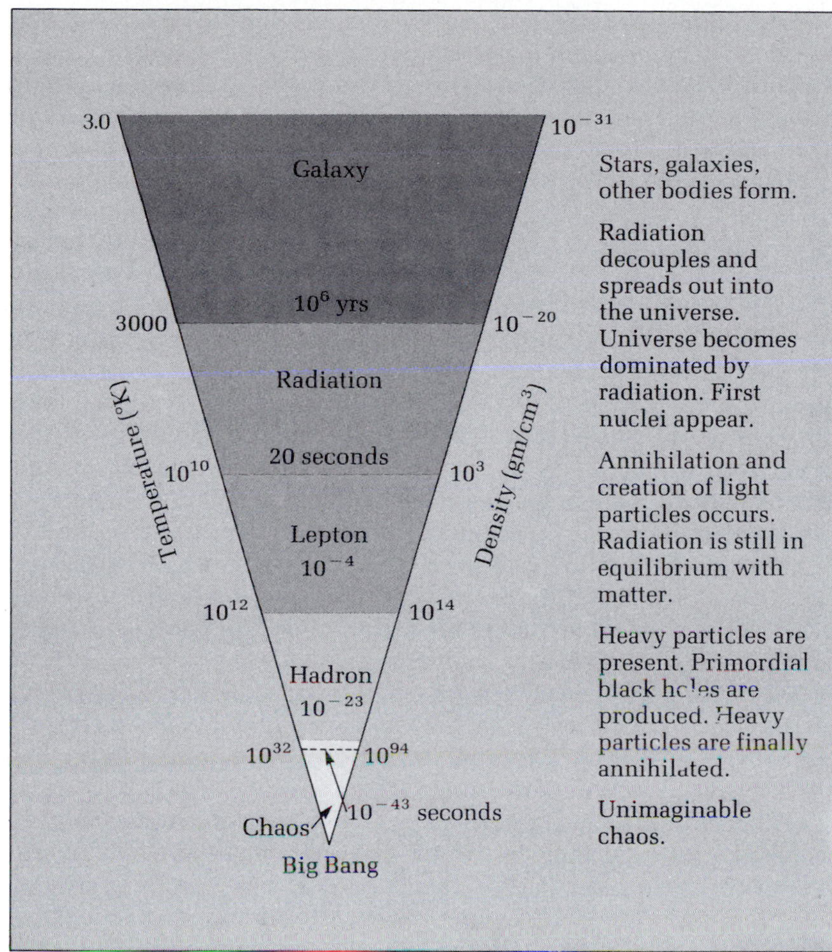

Figure 25.8 Graphic representation of the stages of evolution of the universe, beginning with the Big Bang and ending with the present.

cause of this the Hagedorn model is sometimes called the "warm Big Bang" model.

The photons that were present during the hadron era had considerable energy, easily enough to create proton–antiproton pairs and neutron–antineutron pairs (and likely even heavier particles and their antiparticles). But these pairs quickly annihilated one another after they were generated, so that there was no net change in the number present. The universe was in a state of equilibrium.

About 10^{-4} sec after the explosion the hadron era ended. As the universe expanded and cooled, the photon energies decreased until finally they were not energetic enough to create heavy particles; only light particles such as electrons and positrons could be generated. The universe then entered the *lepton era* (an era of light particles). There were still a few heavy particles in the universe at this stage, but they were rapidly annihilating one another, and some were decaying to other particles. For example, a free neutron (a neutron not in an atom) decays in a short time to an electron, a proton, and an antineutrino.

An important feature of the universe during this time was that it had a slight excess of particles over antiparticles. Indeed, if this had not been the case, the matter–antimatter annihilations would have left only radiation and the universe as we know it could not have evolved. Finally, when the temperature was low enough, the neutrinos in the universe decoupled from the matter and expanded freely.

About 20 sec after the Big Bang the temperature was below 10^{10} °K, and the photons no longer produced electron–positron pairs. The universe then entered the *radiation era.* An important event in its history would soon take place: the generation of the first nuclei. The generation of these nuclei began about three minutes after the explosion and lasted perhaps fifteen minutes. Since that time their abundance has remained approximately constant (more about this later); they are therefore of considerable value in our study of this era, usually called the era of *nucleosynthesis.* The helium and deuterium abundances, for example, depended on the rate of expansion of the universe during this time and on the average density of matter.

Temperatures during the radiation era remained high for many years. All the atoms in the universe were therefore ionized. Finally, after about 10,000 years, the temperature had fallen to about 3000°K and neutral atoms began to form. As the particles recombined into atoms, the radiation in the universe decoupled and began a free expansion. This is the background radiation we examined in the last section (the present temperature is 3°K).

The matter in the universe continued to spread out. By 1 million years after the Big Bang, the average density of matter in the universe had decreased to about 10^{-20} gm/cm^3. The universe then entered the *galaxy era,* in which galaxy and star formation dominated. Indeed, the existence of stars and galaxies implies that the universe could not have been completely uniform during this early period. Small clumps must have formed and gradually grown into stars, and on a larger scale, into galaxies. This clumping could not have occurred before the radiation era; the radiation pressure was so great this time that gravitational collapse would not have been possible.

OPEN OR CLOSED?

For evidence as to whether the universe will eventually stop expanding or will expand forever, we must look at its farthest reaches (which are also far back in time). As we saw earlier, the galaxies in this region should be decelerating because of their mutual gravitational attraction. The important question, then, is: Are they decelerating at a sufficient rate to stop them eventually? Since their gravitational attraction (and consequently their deceleration) depends on their mass, we can rephrase this question as follows: Is the average density of matter of the universe sufficient to stop its expansion? This is now a key question in cosmology.

If the average density is sufficient to stop it, the present expansion will be followed by a contraction phase. The universe will collapse in on itself, ending in a fiery nucleus, a singularity like the one from which it began. It is even possible that the universe could emerge from this singularity and go through another cycle; in fact, it could continue going through cycle after cycle. The model corresponding to this situation is called the *pulsating* or *oscillating* universe model (Fig. 25.9).

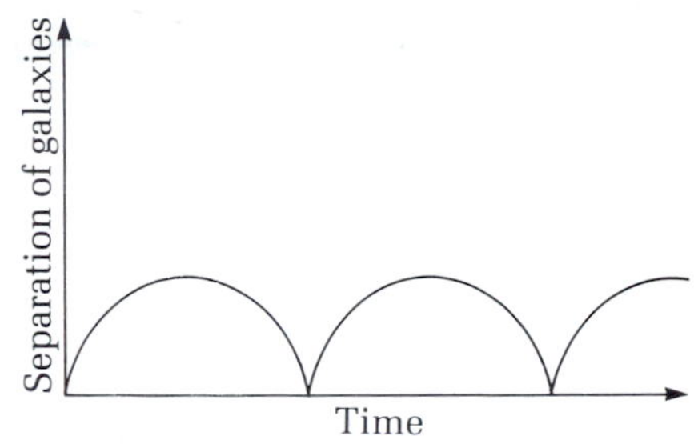

Figure 25.9 Simple representation of a pulsating, or oscillating, universe.

If the universe does collapse back on itself, it is said to be *closed* (it is positively curved in this case; see Fig. 25.10). On the other hand, if it expands forever, it is *open* (it is negatively curved or flat in this case). There are a number of tests that give some indication which case actually holds. The two major quantities that cosmologists measure for this purpose are H (Hubble's constant) and ρ (the average density of matter in the universe). Another constant called the *deceleration parameter, q,* is a measure of how rapidly the universe is slowing down and is formed from H and ρ. We will assume the Friedmann models of the universe throughout our discussion.

As we saw earlier, H is obtained from a Hubble plot. It is particularly important, though, that we extend this plot to the farthest regions of the universe, to see how H changes. It is this change that is significant. The average density of matter, ρ, is obtained by counting galaxies and other forms of matter and radiation in a large volume, estimating their total mass, and dividing by the volume. According to Friedmann's theory, if the actual density of matter is greater than a critical value of about 6×10^{-30} gm/cm^3, the expansion will be stopped and the universe will eventually collapse; if the density is less than this, the universe will expand forever. Hence an accurate value for ρ can tell us whether the universe is open or closed. Unfortunately, it is not easy to get an accurate value.

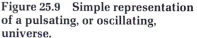

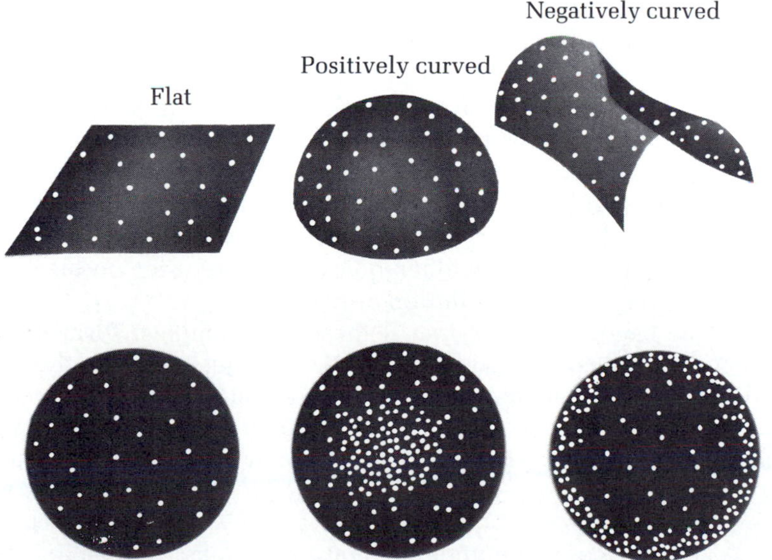

Figure 25.10 Plots of galaxies in three differently curved spaces. The bottom row shows what happens in each case when the plot is done on flat paper. If the universe is positively curved, the plot should also be made on a positively curved surface; when this is done for a uniform distribution and the curved paper is flattened, there is an excess at the center. Similarly, if the plot is made on a negatively curved surface and then flattened, there is an excess near the edge.

Astronomers have also shown that if the deceleration parameter, q, is greater than $\frac{1}{2}$, the universe is closed; if it is less than this value, the universe is open. So it too is important to the question of the universe's end.

Besides the determination of these parameters there are various counts of radio sources and galaxies, although tests of this type have so far failed to yield consistent results. The basic idea is to count the number of galaxies (or radio sources) of various magnitudes in a given direction and compare the numbers obtained to those obtained from other directions. These numbers indicate (in theory) how the universe is curved, and hence whether it is open or closed. The reason is as follows. The points corresponding to the numbers obtained are plotted on a piece of *flat* paper. This means that if the points are uniformly distributed, a uniform plot will be obtained, but if the universe is curved, a uniform plot will not be obtained by plotting on a flat piece of paper. If there appears to be a deficiency of dim galaxies as compared to bright ones, then the universe is positively curved. If there is an excess of dim galaxies, then the universe is negatively curved.

Now, let us look at the results of the tests. Consider q first. Present data indicate that q is less than the critical value of $\frac{1}{2}$; this suggests that the universe is open. Second, consider the average density of the universe, ρ. Is it greater or less than the critical density, 6×10^{-30} gm/cm^3? From galaxies alone (without any allowances for invisible matter) we get about 5×10^{-32} gm/cm^3. But galaxies do not constitute the total mass of the universe; there may be particles or clumps of matter we cannot see. If all visible matter is included in our result, and appropriate allowances are made for matter we know likely exists but cannot see, we get approximately 3×10^{-31} gm/cm^3, which is still far short of the critical value. Thus it seems that this test also gives us an open universe.

The Missing Mass

But what if there is a considerable amount of mass in the universe that we have not yet detected (or even guessed might exist)—enough to close it? Astronomers have, indeed, considered this. The needed mass is frequently called the "missing mass." So, just as many galactic clusters seem to have some missing mass, so too does the universe. If there is indeed a considerable amount of undetected matter, we immediately ask what form it could have. Astronomers now generally agree that, excluding black holes, there are three possibilities: discrete matter, gas, and radiation or neutrinos.

Let us begin with the first of these. The simplest form of discrete matter is dust. Is it possible that the missing mass is in the form of dust? It turns out that an incredible amount would be needed, and it is very unlikely that this much could go undetected. What about larger objects such as dead or extremely faint stars? It has been suggested, for example, that galaxies may be surrounded by extensive halos of faint stars. According to J. P. Ostriker, P. J. Peebles, and A. Yahil, we may be underestimating the masses of galaxies by a factor of 10 or more because of this. If so, there would be sufficient mass in the universe to close it.

Turning now to the second possibility, namely, gas, we find hydrogen to be the most likely candidate. Can we determine how much hydrogen there is in the universe? To a good approximation we can. As we saw in Chapter 21, hydrogen absorbs and emits radiation at a wavelength of 21 cm. But again, studies of this absorption tell us that hydrogen is also not a good candidate for the missing mass.

The third possibility is radiation. We know from our discussion of the early universe that at one time all the mass was in this form. How much still is? Astronomers are uncertain; studies of ionized hydrogen indicate that there may be a background of X-rays throughout the universe, similar to the microwave background, but it seems unlikely that there is enough to close it. We also believe that there is a neutrino background (as yet undetected). In fact, there has been a recent suggestion that the neutrino might not be a massless particle as previously assumed; it might have a small but finite mass. If this mass were in the correct range, it would help close the universe. It would also clear up another problem: the missing mass in clusters of galaxies. As we mentioned in Chapter 22, many clusters of galaxies do not have nearly enough matter to keep them bound. If the neutrino has a small mass in the right range, it would be attracted to large clusters and could account for their missing mass.

Finally, black holes should be mentioned. There would have to be a large number of them to provide the needed mass; nevertheless, they are a relatively good candidate. Primordial black holes are the best candidates, as many of them may have been generated in the early stages of the Big Bang. But if there are this many, where are they? So far we do not know.

In summarizing our data we find relatively little to indicate a closed universe. The evidence that there is missing mass is not strong—neutrinos (if they have mass) are perhaps our best hope. Thus, unless this is the case or unless there is much matter in some completely unfamiliar form, it appears that there is not enough mass in the universe to stop its expansion.

Deuterium

There is another approach to determining whether the universe is open or closed. During the era of nucleosynthesis (p. 446), several light nuclei were formed; these included: H^2, H^3, He^3, He^4, and Li^7. From the Big Bang theory we can predict the abundance of each of these nuclei. Figure 25.11 shows the results of a calculation that gives these abundances as a function of density. One of the first things we notice in this graph is that the abundance of He^4 is generally independent of density. This means that there is little difference in the abundance predicted for an open as compared to a closed universe, and therefore it is of little use in determining whether the universe is open or closed. The helium abundance does depend critically, though, on the expansion rate of the universe during the period when

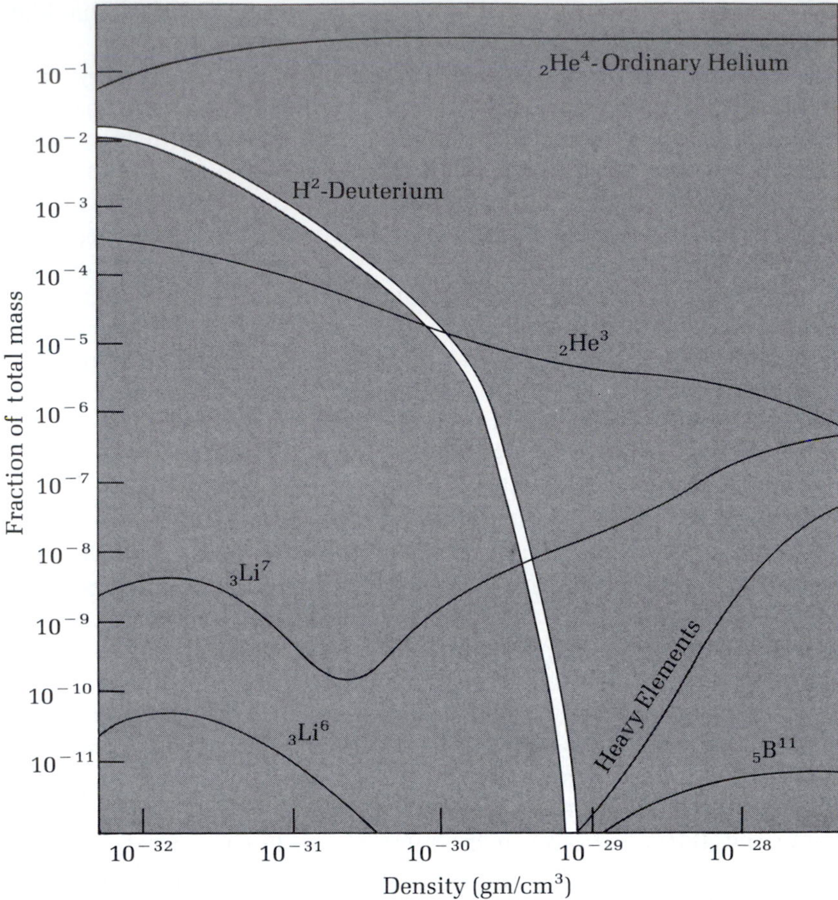

Figure 25.11 Abundances of various elements in the early universe as a function of density. Note the sensitive dependence of deuterium on density.

it was produced, and this allows us to deduce how many different kinds of particles were present at this time.

If we now look at the curve corresponding to deuterium, we see that the deuterium abundance is very sensitive to density (i.e., large changes result from small changes in density). This means that if we can measure the present amount of deuterium in the universe (in practice the deuterium-hydrogen ratio is usually measured), we can determine the present average density of matter.

To see why the deuterium abundance is so sensitive to density, let us look at what happens to deuterium when it is hit by a neutron; as we saw earlier, it becomes tritium. And when this tritium reacts with a proton, it becomes helium. This tells us that if the universe was dense (filled with many neutrons) during this era, it would be easy for a deuterium nucleus to meet a neutron. Most of the deuterium would therefore have become helium. On the other hand, if the density was low, most of the original deuterium would have survived.

Unfortunately, it is difficult to determine accurately how much deuterium has survived. The first accurate measurements, which were made from the orbiting satellite Copernicus, tell us that there is about one deuterium atom to about 3000 hydrogen atoms. This indicates that the universe is open.

Although this does appear to be irrefutable proof, we have to be careful in accepting it. Like so many other methods in astronomy, it is not without its problems. One of these problems is that we do not know for sure how much deuterium has been burnt up in stars.

THE END OF TIME

Let us see what will eventually happen if the universe does expand forever. In this case all galaxies will continue to expand away from any observer. Within each of the galaxies, stellar evolution will go on until all the available fuel is used up. The number of tiny red dwarfs will gradually increase; the number of black holes will also increase, and as time goes on the black holes will continue to grow as they cannibalize everything nearby, until they are exceedingly massive. Many of them will collide and coalesce. And finally most of the matter of the universe will be tied up in supermassive black holes.

But we saw earlier that black holes evaporate. Matter and antimatter is given off in this evaporation, and it will likely suffer mutual annihilation. Over an almost unimaginably long time all of the supermassive black holes will evaporate, leaving only naked singularities. The universe will be an exceedingly strange place: no matter, no life, only radiation and a chaotic twisting of space and time.

On the other hand, it is possible that new evidence will be found in favor of a closed universe. In this case the outward rush of galaxies would stop and contraction would begin. Eventually, after billions of years, an observer's sky would begin to redden as the background radiation became more intense; the red glow would gradually change to yellow, orange, then finally to a blinding blue-white. Long before this final stage, of course, all life would have disappeared.

And again there would likely be numerous black holes present. This brings to mind an interesting question: Will the universe eventually become a black hole? Oddly enough, the entire universe cannot collapse through its gravitational radius and become a black hole; it will only collapse if it is already a black hole.

But how could the universe be a black hole when by Earth standards it is almost a perfect vacuum? When something as large as the universe is considered, a high density of matter is not needed. To see why, let us consider a number of black holes of different mass. For example, if a star of mass 20 $M_{\odot}$ collapses to a black hole, its density of matter as it passes the gravitational radius will be considerably less than that of a star of mass 10 $M_{\odot}$ as it passes its gravitational radius. If we went to an even more massive star, say 40 $M_{\odot}$, the density at the gravitational radius would again be less. Quasars are examples of objects that are extremely massive. If a quasar suddenly collapsed and became a black hole, the average density of matter as it passed its gravitational radius would be about equal to that of water. A man could live there. Eventually, though, assuming he stayed on the surface of the collapsing object, he would be crushed at the center of the resulting black hole. But at the event horizon he would be in no danger. If we continue to examine objects of greater and greater size and mass, we find that in each case the density at the gravitational radius is less. For an object (if it can be called that) as large as the universe, the average density of matter needed for it to be a black hole is extremely small, about 10^{-30} gm/cm^3.

When we talk about the collapse of the universe, then, the important thing is the singularity at the center of this "universal" black hole. All the matter of the universe will collapse into it. What finally happens to this matter? So far we do not know; Einstein's theory of relativity breaks down and we have no theory that can tell us, but even without the use of an adequate theory scientists have speculated on what may happen.

John Wheeler and others have looked carefully at the breakdown point (and beyond) and have obtained some interesting results. Wheeler believes that if the collapsing matter passes a certain critical density, a "fuzziness" develops. His idea is based on one of the fundamental concepts of quantum theory, the Uncertainty Principle. According to this principle we cannot simultaneously distinguish quantities such as momentum and position of a particle to an arbitrary degree of accuracy; as we close in on one and pinpoint it, the other becomes fuzzy. Wheeler (and also Hawking of Cambridge) has shown that as a result of this principle, if collapse occurs beyond a certain point ($\rho \sim 10^{40}$ gm/cm^3), our present fundamental constants of the universe (mass and charge of the electron, gravitational constant, and so on) may be lost.

If the universe should expand again, all the matter would emerge out of the singularity. We do not know for certain if this will happen, of course, but if it does—and if all the previous fundamental constants are lost—it will emerge with new constants, and because of this it will likely have a different expansion rate and, consequently, a different radius. If it continues to oscillate through cycle after cycle, some of the cycles will likely be quite short, perhaps as short as 1 million years; others will be exceedingly long, perhaps hundreds of billions of years.

THE SINGULARITY AND LIFE

If the universe goes through cycles of various lengths, the question arises whether life is possible in all these cycles. Some of them presumably last only a few million years, and we know that life took at least 3 billion years to evolve here on Earth. On the basis of this, then, life would not be possible in the shorter cycles.

And this is not all. A few years ago B. Carter of Cambridge University was examining the effects of changes in the fundamental constants on stars. He found that if they were adjusted slightly one

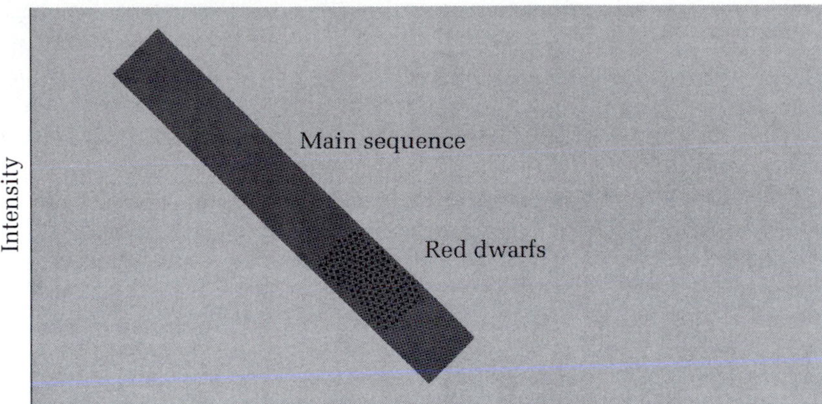

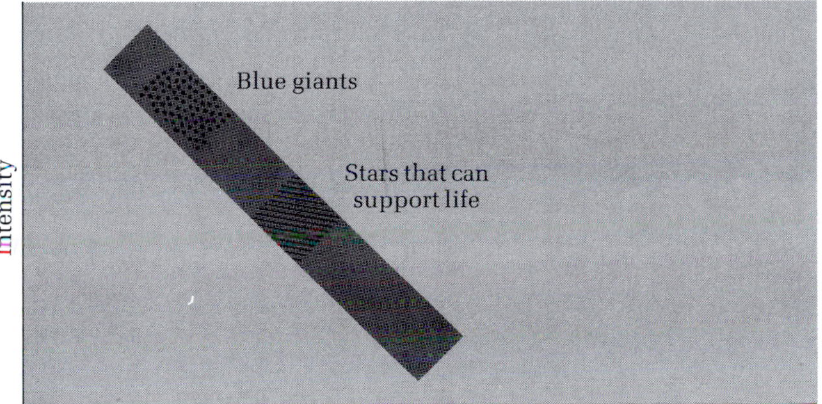

Figure 25.12 The consequences of slight changes in the fundamental constants of the universe. When there is a slight change in one direction (top), only small red dwarfs will form. When there is a slight change in the other direction (bottom), only blue giants will form.

way, most types of stars could not exist—there would only be small red stars. And if they were adjusted slightly in the other direction, only large blue stars could exist (Fig. 25.12). According to this reasoning, if the universe did collapse back on itself and begin a new cycle with changed fundamental constants, the change would have a considerable effect on life. The resulting universe would not likely have the large array of different types of stars our universe has, and it needs this array if life is to exist. As we will see in Part VIII, the best candidates for life-supporting systems are stars near the middle of the main sequence (those like our sun). If the fundamental constants were very different from what they are now, these stars probably would not exist. This means life would be impossible in that universe. Even if they were only slightly different, there would be problems, since massive blue stars play a particularly important role in the universe: They produce the heavy nuclei, and without heavy atoms, worlds like ours could not exist.

According to this theory, then, the nature of our universe is strongly attuned to the fundamental constants it now has. If they were changed even slightly, there would be no life in the universe. This idea is summed up in what is known as the Anthropic Principle. It states that if the universe were different in any way from the way it is, we would not be here.

SUMMARY

1. All clusters of galaxies appear to be moving away from us; actually, all are moving away from one another.
2. According to the Big Bang theory, the universe began with an explosion that took place about 18 billion years ago.
3. The farther we look into space, the faster galaxies appear to be receding from us. There is a theoretical sphere surrounding us that corresponds to a recessional velocity of c. This is the end of our observable universe.
4. An alternative to the Big Bang theory the Steady-State theory, which assumes that the universe is always the same. There is now considerable evidence against this theory.
5. The cosmic background radiation is radiation that is assumed to have been left over by the Big Bang. It now has a temperature of approximately 3°K.
6. Anisotropy in the background radiation has recently been discovered. It indicates we are traveling in the direction of the constellation Leo.
7. It is now possible in theory to trace the universe from very early times. According to recent theories it passed in succession through several eras: hadron era, lepton era, radiation era, and galaxy era.
8. Heavy particles were produced during the hadron era. Light particles were produced during the lepton era. The fireball radiation spread out into the universe during the radiation era; the first nuclei were also produced. Stars, galaxies, and so on began to appear during the galaxy era.
9. There are two options open for the future of the universe: It can expand forever (open), or it can eventually stop expanding and collapse in on itself (closed).
10. Several tests can be performed to see whether the universe is open or closed. The two important quantities for this purpose are H and ρ. Most of the tests indicate that the universe is open.
11. If the universe is open, it will expand forever; eventually only the galaxies in our group will be visible from Earth. Much of the matter of the universe will likely become black holes in the extremely distant future. Eventually these black holes will evaporate.
12. If our universe were very different from the way it now is (i.e., if the fundamental constants were different), life probably would not be possible.

REVIEW QUESTIONS

1. Describe the primordial nucleus.
2. All galaxies appear to be moving away from us. Are we therefore at the center of the universe?
3. What is the difference between the Hubble age of the universe and the actual age? Which is greater?
4. Explain what the 3°K cosmic background radiation is. Where did it come from? Why is its temperature now so low?

5. Is the background radiation perfectly isotropic? If not, what does the anisotropy mean?
6. Show that there is a parallel between the Michelson-Morley experiment and the experiments to measure our velocity relative to the cosmic background radiation.
7. Why can we go back no further in time (theoretically) than 10^{-43} seconds after the Big Bang occurred?
8. Outline the events that occurred just after the Big Bang.
9. What is needed for the universe eventually to collapse in on itself?
10. Discuss and explain some of the basic cosmological tests. What are the important parameters of cosmology? What do they tell us?
11. What are some of the possible sources of the mass required to make the universe collapse in on itself? Which is most likely?
12. Describe the end of time in (a) an open universe; (b) a closed universe.
13. What is the Anthropic Principle?

THOUGHT AND DISCUSSION QUESTIONS

1. Explain the confusion (inconsistencies) that would result if we assumed that the universe had a center (that the primordial nucleus was located at some particular point).
2. When the primordial nucleus exploded, creating the universe, what did it expand into? Discuss.
3. The age of the universe, T, in years can be calculated from Hubble's constant using $T = 10^{12}/H$. If Hubble's constant is 50 km/sec/Mpc, what is the age of the universe? Perform the same calculation for $H = 100$ km/sec/Mpc.
4. If the universe is open, what will eventually happen to the cosmic background radiation? If it is closed, what will happen to it?
5. If you were located in a galaxy near the boundary of our observable universe, would the galaxies in the direction of the Milky Way appear to be approaching or receding from you? Explain.

FURTHER READING

Ferris, T. *The Red Limit.* Morrow, New York (1977).
Gott, J. R., J. E. Gunn, D. N. Schramm, and B. M. Tinsley. "Will the Universe Expand Forever?" *Scientific American,* p. 62 (March 1976).
Laurie, J. (ed.) *Cosmology Now.* Taplinger, New York (1976).
Motz, L. *The Universe: Its Beginning and End.* Scribners, New York (1975).
Parker, B. "The First Second of Time," *Astronomy,* p. 6 (August 1979).
———. "The End of Time," *Astronomy,* p. 6 (May 1977).
———. "The Cooling Fire of Creation," *Astronomy,* p. 14 (March 1980).
———. "The Age of the Universe," *Astronomy,* p. 67 (July 1981).
Sciama, D. W. *Modern Cosmology.* Cambridge, London (1971).
Silk, J. *The Big Bang.* W. H. Freeman, San Francisco (1980).
Weinberg, S. *The First Three Minutes.* Basic Books, New York (1977).

Aim of the Chapter

To introduce some of the alternative cosmological theories.

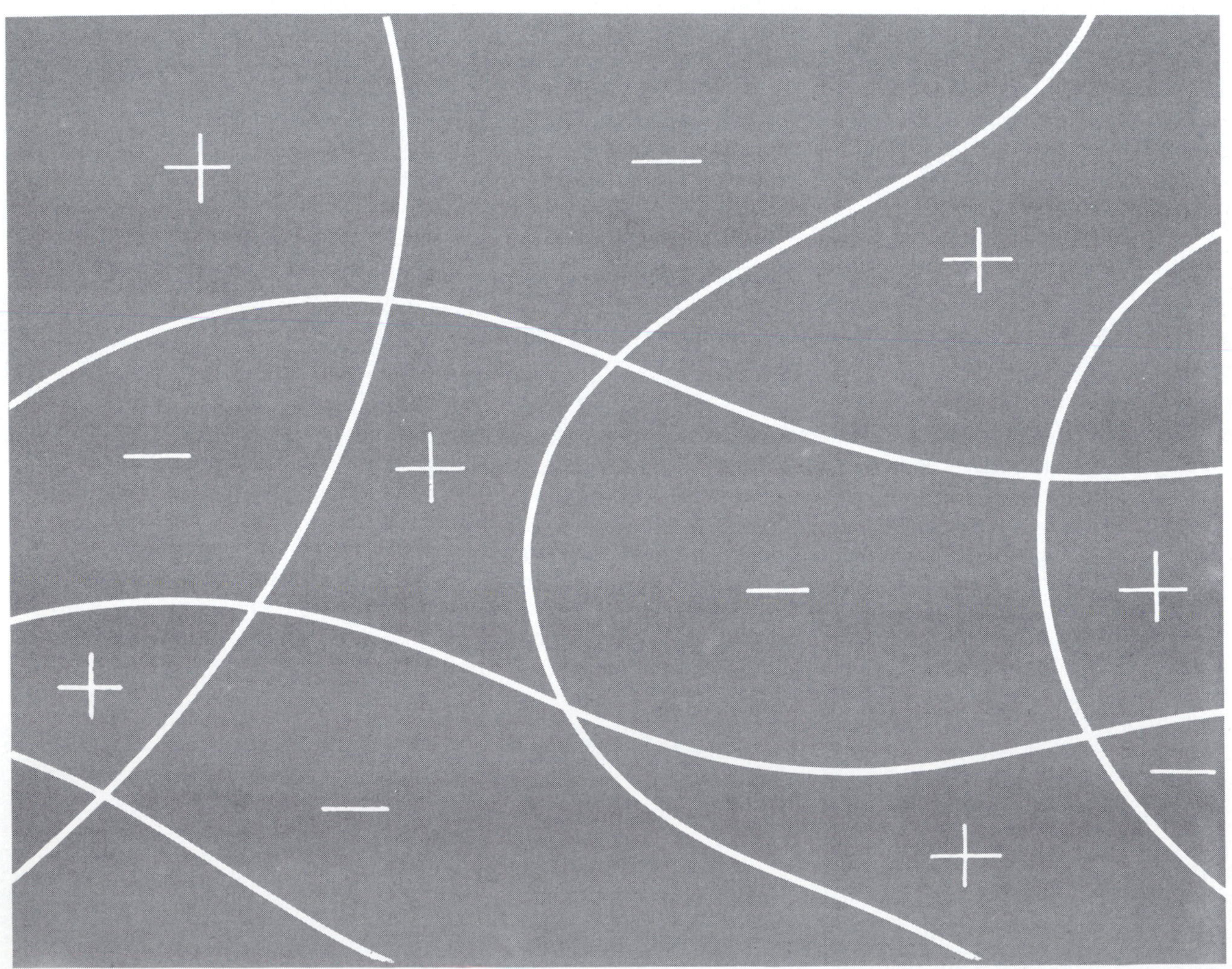

chapter 26

The Rival Cosmologies

In Chapter 25 we saw that the Big Bang theory is at present the most widely accepted cosmology. But it does have a number of important rivals. We have already briefly met one of them, namely, Steady-State. In this chapter we will examine it and some of the others in more detail. Let us begin with a brief overview.

First, in addition to the Steady State theory, there are the varying-G cosmologies. In these theories the gravitational constant, G, is assumed to change slowly over exceedingly long times. Second, we have antimatter cosmologies, the most important of which is the one by O. Klein and H. Alfvén. They assume that the universe is composed of approximately equal amounts of matter and antimatter. Third, there are the tired-light theories; they assume that the red shift is not due to the Doppler effect but rather to a gradual loss of energy as light travels millions of light years across space. Fourth, we will talk about an interesting theory by Fred Hoyle, which he calls a "compartmentized" theory. And finally, we will briefly touch on the hierarchy theory and the charged-universe theory.

THE STEADY-STATE THEORY

During the latter part of World War II, three men stationed at an Air Force base just outside London formed a discussion group. They talked about Lemaitre's model, Gamow's ideas, and Hubble's observations. The three men were Herman Bondi, Fred Hoyle, and Thomas Gold. When the war ended, they continued their discussions at Cambridge.

One day during a lively argument about the beginning of the universe, Gold, the youngest of the three, asked: What if the universe had no beginning and no end but is always the same? Bondi and Hoyle were skeptical at first. After all, they countered, as the galaxies get further and further apart, the average density of matter decreases; so it will not always be the same.

Gold suggested that matter could be created in the vast regions between the galaxies. If the rate of creation were equal to the amount lost at the boundary, there would be a steady state. Only a small amount would be needed—no more than about one hydrogen atom per year in a volume approximately equal to that of the Empire State Building. Although this violates fundamental scientific laws, the small amount created would be impossible to detect.

Bondi and Hoyle were interested but still skeptical; they worked out the mathematics of the idea, however, and soon found that it was indeed viable. Hoyle showed that the theory was consistent with general relativity. In addition, it could be made consistent with a variation, or generalization, of Einstein's Cosmological Principle. This variation is now called the *Perfect Cosmological Principle.* According to it the universe is not only homogeneous and isotropic at all points but also the same at all *times*—it is in a steady state.

In 1948 two papers appeared describing the theory; one was by Bondi and Gold, and the other by Hoyle. The Big Bang theory suddenly appeared to have a worthy opponent. Did the universe begin with a bang, or has it always been the same? This question preoccupied many astronomers during the 1950s. Little by little, though, the foundations of the Steady-State theory were whittled away. First it was noticed that radio galaxies were not distributed uniformly in space. As astronomers probed farther out, their number appeared to increase; if true, this meant there was evolution, or change, in the universe. Then came the discovery of quasars, and soon there could be little doubt that something was different in the outer reaches (early stages) of our universe. The evidence for change was overwhelming. In 1965, with the discovery of the cosmic background radiation, the Steady-State theory suffered its most devastating blow.

Hoyle and J. V. Narlikar made several modifications in an attempt to save the theory. One of them allowed for localized sources of matter, another for variations in the rate of output from the sources. But it soon appeared to be a losing battle, and now even Hoyle is no longer a strong supporter of the theory.

VARYING-G COSMOLOGIES

In the mid-1930s P. A. M. Dirac of Cambridge University discovered a number of interesting coincidences while working with some of the fundamental constants of nature. In particular he noticed that a dimensionless ratio formed from one of the fundamental constants of the microworld (e.g., the radius of the electron) and one from the macroworld (e.g., the gravitational constant, G) always gave the value 10^{40} or some power thereof. He was sure this had some significance. In 1937 he published two papers outlining a theory of "cosmic ratios." Central to his theory was the following postulate:

> Any of the very large dimensionless ratios occurring in nature are connected by a simple mathematical relation in which the coefficient is of the order unity (and the powers are $(N)^n$, where $n = 0, 1, 2, \ldots$ and $N = 10^{40}$). Moreover, if one of these relations varies, they must all vary.

Do any of these relations vary? There are many of them, but for brevity let us consider the three that should be simplest to understand:

$$\frac{\text{Electric force } \mathrm{p} \leftrightarrow \mathrm{e}}{\text{Gravitational force } \mathrm{p} \leftrightarrow \mathrm{e}} \approx 10^{40}$$

$$\frac{\text{Radius of universe}}{\text{Radius of electron}} \approx 10^{40}$$

$$\frac{\text{Mass of visible universe}}{\text{Mass of fund. particle of universe (proton)}} \approx (10^{40})^2$$

It is easy to see that some of the fundamental constants must indeed vary. Consider the second relation; we know that the radius of the universe is varying and therefore, according to Dirac's postulate, others must also vary. As it turns out, we are forced to vary either the charge of the fundamental particle or the fundamental gravitational constant, G.

Dirac selected the second of these and formulated a "varying-G" cosmology (magnitude of G slowly decreasing). It was soon discovered, though, that his theory was not in accord with observation, and it had to be discarded.

But this was by no means the end of varying-G cosmologies. In 1947 a modification of Dirac's theory was published by P. Jordan. Jordan got around some of the problems encountered by Dirac by allowing particle creation, and by introducing some specialized mathematical techniques for dealing with the variation of G. But again, when the theory was examined in detail, it was found to be flawed.

Although Jordan's theory was never completely satisfactory, it did set the stage for a third varying-G cosmology by Brans and Dicke. We talked about the Brans-Dicke theory of relativity in Chapter 19, and their cosmology is based on this theory. At the present time it is in fair agreement with observation.

But is there any evidence that G actually does vary? Routine observations certainly show no such variation, but of course it could vary so slowly that extremely delicate tests would be required to notice the change. This is what is assumed.

To see whether G does vary slowly, we must consider what would change as a result of the variation. It turns out that there are several things. Stars would gradually decrease in luminosity; this means that we would overestimate their lifetime. And, since our sun is a star, it would have had a higher luminosity in the past than it now does, and this in turn means that the earth must have had a higher temperature in the past than supposed. In addition, the earth would have been smaller in the past if the gravitational force was greater on it then. As this force weakened, it would have expanded. Some astronomers believe there is evidence for such an expansion; they point to the large cracks and ridges in the floor of the oceans. An expansion of this type could also explain why continental drift may have taken place in the past (this view is not taken very seriously by most geologists).

Thomas Van Flandern of the U.S. Naval Observatory has recently looked at a number of other effects of a variation in G. According to the theory a gradual decrease in both the earth's rotational rate and the moon's revolutionary rate around the earth should be taking place. Using lunar occultation data that goes back almost 20 years (to when atomic clocks were first used for recording data), Van Flandern has found what he believes is a significant discrepancy in the deceleration of the moon, compared to what would be expected from tides and so on. The amount he finds is almost twice what is predicted. He also finds an apparent discrepancy in the earth's rotational rate. Van Flandern's results are interesting but will require careful checking before being accepted.

ANTIMATTER COSMOLOGIES

Before we talk about antimatter, we must explain what it is. We saw in Chapter 15 that to each type of particle (e.g., an electron) there corresponds an antiparticle (an anti-electron, or positron). When the two collide, they both disappear and are replaced by one or more photons.

Antimatter is of considerable interest to scientists, and there are still many unanswered questions about it. We do not even know, for example, whether matter and antimatter attract or repel one another. It might seem that we could merely project a beam of antimatter along the surface of the earth (which is made of matter) and observe whether it is deflected upward or downward. It turns out, though, that the lifetime of antimatter is too short for such an attempt.

Another interesting question related to antimatter is how much there is in the universe. It seems reasonable that there should be half matter and half antimatter, but most astronomers do not believe that this is the case. Current evidence indicates that matter predominates.

With this brief introduction, let us turn to antimatter cosmologies. The most important one at present (and the only one we will discuss) is that of O. Klein and H. Alfvén. They assume a universe that contains equal amounts of matter and antimatter. Their primordial state consists of an immense sphere (10^{12} light years across) containing a homogeneous distribution of protons, electrons, and their antiparticles (this mixture is usually called an ambiplasma). The density of this primordial ambiplasma is assumed to have been extremely low, so low that there was initially no interaction between the particles and antiparticles. But gradually, over trillions of years, the ambiplasma collapsed inward toward its center of gravity (note

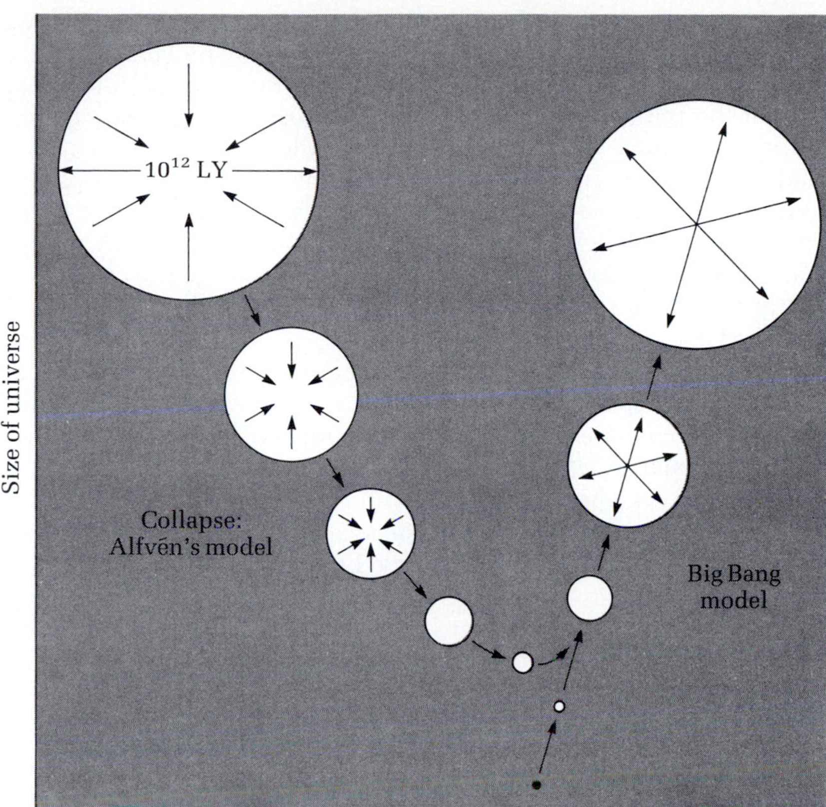

Figure 26.1 Schematic representation of Alfvén's matter-antimatter model of the universe. It can be seen that Alfvén's model fits into the Big Bang model, represented by the straight line at the right.

that Klein and Alfvén are assuming that gravity acts the same way for both matter and antimatter). The inward-falling matter and antimatter picked up speed; proton-antiproton collisions began, and radiation flooded out into the universe. Eventually a substantial outward pressure developed, and as the ambiplasma became more dense, the pressure continued to build. Suddenly it overcame the inward gravitational pull and an explosive expansion began (Fig. 26.1).

If conservation principles are to be obeyed, this expansion should be the same as the inward contraction but opposite in direction. This means that the universe expanded rapidly at first but is gradually slowing down. If this is the case, then the outward expansion will continue at an ever-decreasing rate until the particles and antiparticles are well separated—far enough so that no radiation is generated. At this time it may begin to collapse again; in fact, it may repeat this cycle over and over.

TIRED-LIGHT COSMOLOGIES

According to the usual interpretation of the red shift, galaxies are receding from one another; the farther we look outward, the faster they appear to be receding from us. This is, of course, the way most astronomers picture the universe, and it is based on the well-known Doppler effect, which we know is valid for nearby objects.

The evidence at present strongly favors the Doppler interpretation of the red shift, but there are other possibilities. One of these is that light "tires" as it traverses extremely long distances. We know that light is made up of photons of various wavelengths. A photon corresponding to, say, yellow light has more energy than one corresponding to red light. This means that if the yellow-light photon (the term is used loosely here) gradually lost energy, it would eventually become a red-light photon; in effect, a red shift would have occurred.

But how could it lose energy? One possibility is that in its long flight through space it interacts in some way with the interstellar dust. We know, for example, that when a high-velocity charged particle is deflected by another charged particle, it radiates some of its energy away. A photon undergoes an analogous energy loss; however, the odds against it happening in space are high. Although a light beam travels on the average many light years through space, it actually encounters very little matter enroute. Space is an exceptionally good vacuum. A light beam that travels only a few miles on Earth is likely to encounter as much or more matter, and we know it does not tire. So it does not seem that the light beam in space would tire. As an alternative we could consider an interaction between space itself and light. This again is a possibility, but for the moment it must be considered very speculative, since we still do not know whether or how it happens.

Y. B. Zel'dovich of the U.S.S.R. has shown that there are also other problems for tired-light theories. He notes that if there is an energy loss by the photon, there will also be a change of momentum. This in turn will cause a change in the direction of photon's travel, and hence we will see only "smeared" versions of the original images. Distant stars would appear as small disks; galaxies would appear as large, indistinct disks. Furthermore, because of this smearing the spectral lines would appear broadened. Since neither of these occur, it seems unlikely that a tired-light cosmology is possible.

HOYLE'S "COMPARTMENTIZED" COSMOLOGY

In 1893 Germany physicist Ernest Mach put forward a hypothesis that has since become known as Mach's principle. It can be simply stated as follows:

> Local phenomena (e.g., planets in their orbits) depend not only on local objects but on every bit of matter in the universe, no matter how distant.

Einstein was intrigued with the idea and made a serious attempt to incorporate it in his theory of general relativity. It is generally acknowledged, however, that he failed in this attempt. Others have also formulated theories with this principle in mind. Dicke and Brans consider Mach's principle an integral part of their theory, and recently Fred Hoyle has formulated a new and rather different cosmology that incorporates Mach's principle directly. The central idea of Hoyle's theory is, in fact, that the mass of a body depends on the rest of the mass in the universe. This leads to some rather strange predictions.

Hoyle theorizes that the mass of an object varies according to its position in space-time (with the time part playing a particularly important role). We know that as we look out into the universe we are looking backward in time, but according to Hoyle's hypothesis we are also looking at objects that have much less mass than they once had. In fact, if we look far enough, the mass becomes zero—this is the boundary of our universe.

But Hoyle did not accept this boundary as the end of the universe. He wanted to know what was on the other side. Setting up the appropriate field equations for his theory, he found that there were both positive and negative contributions to the mass in his universe. This meant that mass was like charge with its positive and negative counterparts. Hoyle went on to show that if all the mass on one side of the boundary was positive, then all the mass on the other side would be negative. In fact, the universe was made up of an infinite number of aggregates of opposite types, each cut off from the other by a boundary.

It might seem at first that a mass of one type, say positive, would decrease in magnitude as it approached a boundary until finally it reached zero, and then it would appear on the other side as negative mass. But there is also a coupling constant that does not allow a direct conversion from one sign to the other. Nevertheless, a mass does become zero at the boundary, giving rise to a multiply connected universe, one section cannot see or communicate with another (Fig. 26.2).

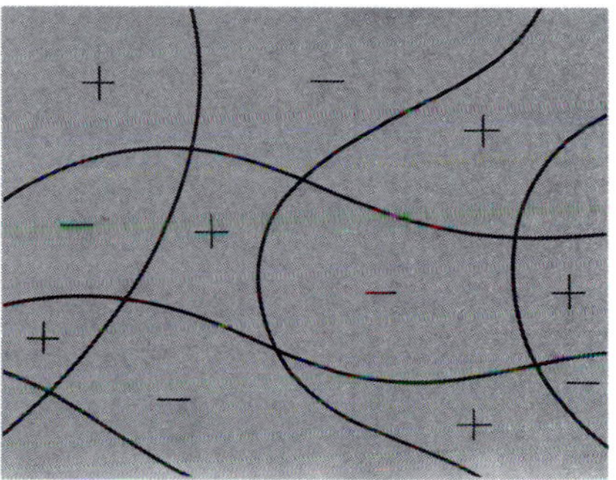

Figure 26.2 Hoyle's compartmentized universe.

As we have seen, the central concept in Hoyle's theory is mass. Indeed, he finds that he can express all the variables of physics (velocity, energy, etc.) in terms of mass, or perhaps we should say in units of mass. But from what we have seen so far, it is obvious that there are problems. First, where does Einstein's theory of general relativity fit in? It is the accepted gravitational theory in our universe. Furthermore, what about the well-known background radiation? We know it exists and is assumed to have come from the Big Bang explosion. There is no such explosion in this theory.

Both those objections are answered by Hoyle. Einstein's general relativity equations are valid within the theory under restricted conditions. According to Hoyle, Einstein's theory is a special case of his much more general theory. It cannot be used to explain things at or across a boundary, but it is valid as long as we remain relatively close, astronomically speaking, to a boundary. And this must mean that we live relatively close to a boundary, since we know that Einstein's equations are valid in our region of the universe.

Hoyle's explanation of the background radiation is as follows: Starlight from the other side approaches the boundary of our universe. As it passes through, it changes in character, emerging on our side as microwave radiation of temperature 3°K.

Hoyle's theory is mind-boggling, but also stimulating. It may be a long time before the theory is accepted, if it ever is, but a theory of this nature is not merely a fanciful concoction of ideas that just happened to occur to a cosmologist. It is a consistent theory with a certain number of basic postulates that are based on well-known laws of mathematics and physics.

OTHER COSMOLOGIES

Somewhat like the positive and negative masses of Hoyle's theory, positive and negative charges play an important role in the charged-universe cosmology of Lyttleton and Bondi, put forward in 1960. Lyttleton and Bondi were concerned with the expansion of the universe: Could it be explained without resorting to a Big Bang? What if the charge of the proton were not *exactly* equal in magnitude to that of the electron? This would mean that each hydrogen atom would have an extremely small residual charge, and with most of the matter of the universe being hydrogen there would be an expansion as a result of the repulsion between the like charges. When laboratory experiments were performed, however, it was found that if a residual charge did exist it was below that detectable by our present technology, and as such too small to cause the expansion of the universe.

Finally, we will briefly discuss the *hierarchy model*. We mentioned this model in relation to Olbers's paradox. Basically, it assumes that the universe is made up of stars, star clusters, galaxies, clusters of galaxies, clusters of clusters of galaxies, and so on. We saw earlier that if this is the case, Olbers's paradox is resolved. However, there is a peculiar consequence of this theory: The concept of mean density (as we have been using it in relation to the universe) has no meaning, or at least does not have the meaning we have assumed it does. Let us see why. We know, first of all, that we calculate the mean density by counting galaxies and other objects in a certain volume, determining their mass, and then dividing by the volume (i.e., $\rho =$ mass/volume). In a hierarchy universe, however, as we consider larger and larger volumes, the average density decreases (volume increases much more rapidly than mass). In fact, as the volume approaches infinity, the average density approaches zero. This means, then, that if this model is valid, we cannot use our simple definition of average density. At present, however, there is little evidence to support this model.

THE RED SHIFT CONTROVERSY

According to the Doppler interpretation of the red shift, the universe is expanding, and most astronomers accept this view. In recent years, however, some startling anomalies have been discovered. Several cases have been presented that seem to contradict the Doppler interpretation. If this interpretation is correct, then, it is important that it be verified.

The best way to verify it would perhaps be to check the red shifts of galaxies in clusters and see if they are all the same. This has been done in many cases, and with few exceptions the red shifts are the same. But what about the more distant quasars? Are they also found in groups? It turns out that quasars do not have a tendency to cluster, so it is best to look for them in association with clusters of galaxies. A few such groups have been found.

Another check on the red shift would be to find two visibly interacting objects and compare their red shifts. Luminous bridges or filaments between the two objects are a good indication that they are at the same distance. Many objects of this type have been found.

In 1966 Halton Arp began an extensive study of peculiar galaxies. Between 1966 and 1970 he was virtually alone in his search for interactions between peculiar galaxies and other objects of different red shift, but as he pointed out case after case, other astronomers began to take interest.

The first evidence for *discordant objects* (interacting or closely associated objects of different red shift) came from the discovery that radio sources, some of them quasars, appeared to be associated with peculiar galaxies. It was almost as if they had been ejected, and on the basis of this Arp formulated an ejection theory for their origin. He then went on to show that most interacting galaxies with high red shifts are associated with elliptical galaxies, while most quasars that are interacting are associated with spiral galaxies.

Arp's hypothesis was given a considerable boost in 1971 when G. Burbidge and his colleagues discovered that quasars are apparently closer (on the average) to bright galaxies than they would be if they were randomly distributed. He found four out of forty-seven to be within 7 arc minutes of bright galaxies. According to Burbidge's calculations, this has only one chance in 250 of being accidental.

But if quasars are generally associated with bright spirals, where are the spirals for the "lone" quasars? Burbidge has suggested that they have moved sufficiently far from their parent spiral that we no longer associate them with it.

Now let us look at comparisons of the red shifts of interacting objects. For example, if two objects are clearly connected by a bridge or filament of luminous material, then it would be revealing to see if their red shifts are equal. Arp has recently discovered several objects of this type. The best-known of them is the Markarian 205–NGC 4319 combination (Fig. 26.3). Markarian 205 is a quasar with a relatively large red shift; NGC 4319 is a spiral with a low red shift. Yet, although these two objects have drastically different red shifts, they appear on photographic plates to be connected by a luminous bridge. If there actually is a bridge between them, this would be disastrous to the cosmological interpretation of the red shift. But is it really a bridge, or just an optical effect, or possibly a background object? Astronomers are not yet certain.

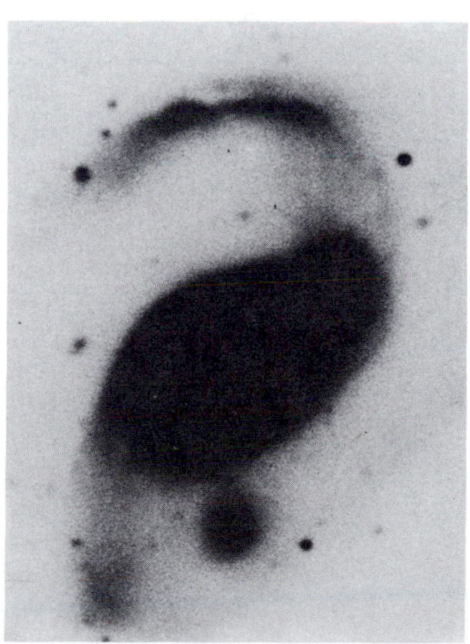

Figure 26.3 The combination Markarian 205–NGC 4319. Note the apparent bridge between the objects.

Arp and others have gathered an impressive amount of evidence indicating that there are discordant objects, but not everyone agrees with his interpretation. Most vocal among this group is John Bahcall.

He and others have looked closely at each of the cases that Arp has presented and have interpreted most of the results differently.

One of Arp's first discoveries was that radio objects were apparently associated with peculiar galaxies and that in many cases they were paired across them. An extensive study of this same problem was made by H. Van der Laan and F. N. Bash. They carefully examined the neighborhoods of all of Arp's peculiar galaxies, compared each of these areas to six other areas in the sky, and concluded that there was no more of a tendency for alignment in Arp's areas than there was in the arbitrarily selected ones. The only case that seemed to be special was the particular one Arp selected, that is, 6° circles centered on his peculiar galaxies.

Evidence for a possible association between quasars and spirals was presented as early as 1966 by Arp. Then when Burbidge and his colleagues found in 1971 that four quasars out of forty-seven were extremely close to spirals, it seemed that there was a particularly strong case for believing they were connected. But Burbidge had examined only the forty-seven objects listed in the Third Cambridge Catalog. Bahcall, McKee, and Bahcall used a more complete list of 222 quasars, and they found these quasars to be randomly distributed relative to spirals. Only in the case of the Third Cambridge Catalog did they find any significant association between the two types of objects. Bahcall and his associates also consider the Markarian 205–NGC 4319 combination to be a spurious photographic effect.

Who is correct? It may be a few years before we know. At any rate, there is little doubt that there is a problem. Whereas only a few years ago there was almost unanimous agreement among astronomers as to the validity of the Doppler interpretation of the red shift, there is now some doubt.

SUMMARY

1. The Steady-State cosmology assumes that the universe had no beginning or no end; it is always the same.
2. Dirac noticed that certain dimensionless ratios formed from the fundamental constants always gave 10^{40}, or some power thereof. These ratios are now called cosmic ratios.
3. In varying-G cosmologies, G is assumed to vary over extremely long times.
4. In the Klein-Alfvén antimatter cosmology the universe began as an immense sphere composed of equal amounts of matter and antimatter. The sphere slowly collapsed inward on itself until finally an explosive expansion took place. This expansion is still going on.
5. Tired-light cosmologies assume that light loses energy as it travels through space. The universe would not necessarily be expanding in this model. There is considerable evidence against this point of view.
6. Hoyle's compartmentized cosmology incorporates Mach's principle. According to this theory the mass of an object varies according to its position in space-time. As we look further and further outward, we see objects with less and less mass; finally, the mass is zero. Beyond this boundary are other "compartments" of the universe that contain objects of negative mass and that cannot be seen from our part of the universe.

7. The hierarchy cosmology assumes that the universe is composed of stars, clusters, galaxies, clusters of galaxies, and so on, ad infinitum. Average density has no meaning in this theory.
8. Arp has found several cases of interacting or closely associated objects that appear to have different red shifts. Such objects are said to be discordant. On the basis of them Arp has formulated an ejection theory.

REVIEW QUESTIONS

1. How does the Perfect Cosmological Principle differ from the Cosmological Principle?
2. How is a cosmic ratio formed? Give an example.
3. What did Dirac postulate on the basis of his cosmic ratios?
4. What are some of the tests for a varying *G*? Is *G* assumed to increasing or decreasing?
5. Do matter and antimatter attract or repel? How would we test this?
6. What is an ambiplasma? What role does it play in antimatter cosmologies?
7. What are some of the arguments for and against tired-light theories?
8. State Mach's principle in your own words. Discuss some of the implications of Mach's principle for a person spinning on a piano stool.
9. Hoyle's compartmentized cosmology assumes that there are several universes. Explain. What happens at the boundaries of these universes? How does Hoyle deal with the problem of the background radiation?
10. Explain why the concept of an average density has no meaning in the hierarchy cosmology.
11. Arp discovered that many radio sources were associated with galaxies. What did he postulate on the basis of this and other evidence?
12. Why are Markarian 205 and NGC 4319 of particular interest? What does Arp believe we are seeing?
13. Outline some of Bahcall's arguments against Arp's conclusions in relation to the various discordant objects.

THOUGHT AND DISCUSSION QUESTIONS

1. Is Einstein's original static-universe model inconsistent with the Steady-State theory? Is de Sitter's theory inconsistent with Steady State? Discuss.
2. Which of the rival cosmologies (also include the Big Bang theory in the first three):
   (a) assumes that the universe is the oldest?
   (b) assumes that the universe is the youngest?
   (c) assumes that the universe is not expanding?
   (d) are generally consistent with the Big Bang theory?
3. List the most positive points (or arguments) that favor each of the rival cosmologies. What is the strongest argument against each?

FURTHER READING

Alfvén, H. *Worlds—Antiworlds*. W. H. Freeman, San Francisco (1966).
Field, G. B., H. Arp, and J. N. Bahcall. *The Redshift Controversy*. W. A. Benjamin, Reading, Mass. (1973).
Hoyle, F. *The Nature of the Universe*. Oxford (1960).
Kilminster, C. *The Nature of the Universe*. Dutton, New York (1971).
Parker, B. "The Redshift Problem," *Astronomy*, p. 6 (September 1978).
——. "Rival Cosmologies," *Astronomy*, p. 18 (March 1978).
Van Flandern, T. "Is Gravity Getting Weaker?" *Scientific American* (February 1976).

part 8

Life Among the Stars

Are we the only civilization in all of space? Chapter 27 addresses the intriguing questions of whether there is life other than our own in the solar system and beyond, as well as the related subject of how life developed on our planet. Many scientists believe the chances are good that there are other civilizations like ours, but there is no direct evidence that this is the case. We have found no sign of extraterrestrial life within the solar system. We have carefully sifted through the lunar soil for bacteria, viruses, or perhaps just hydrocarbons—one of the basic blocks of life—and there was nothing. With the Viking explorations, we tried again, and again there appeared to be nothing.

Understanding how life arose here on Earth might help us in our quest for life elsewhere, and much of Chapter 27 examines that topic. Many believe that life on Earth is a result of a supernatural event, and many scientists do not challenge this view; rather, they take the position that the Big Bang explosion and the creation of the laws of nature can be taken to have been that supernatural event.

In the epilogue, we turn to a discussion of the searches now in progress for extraterrestrial life, as well as possible space exploration projects of the future.

Aim of the Chapter

To discuss the nature and development of life as we know it and to show how scientists go about determining whether there is life beyond Earth.

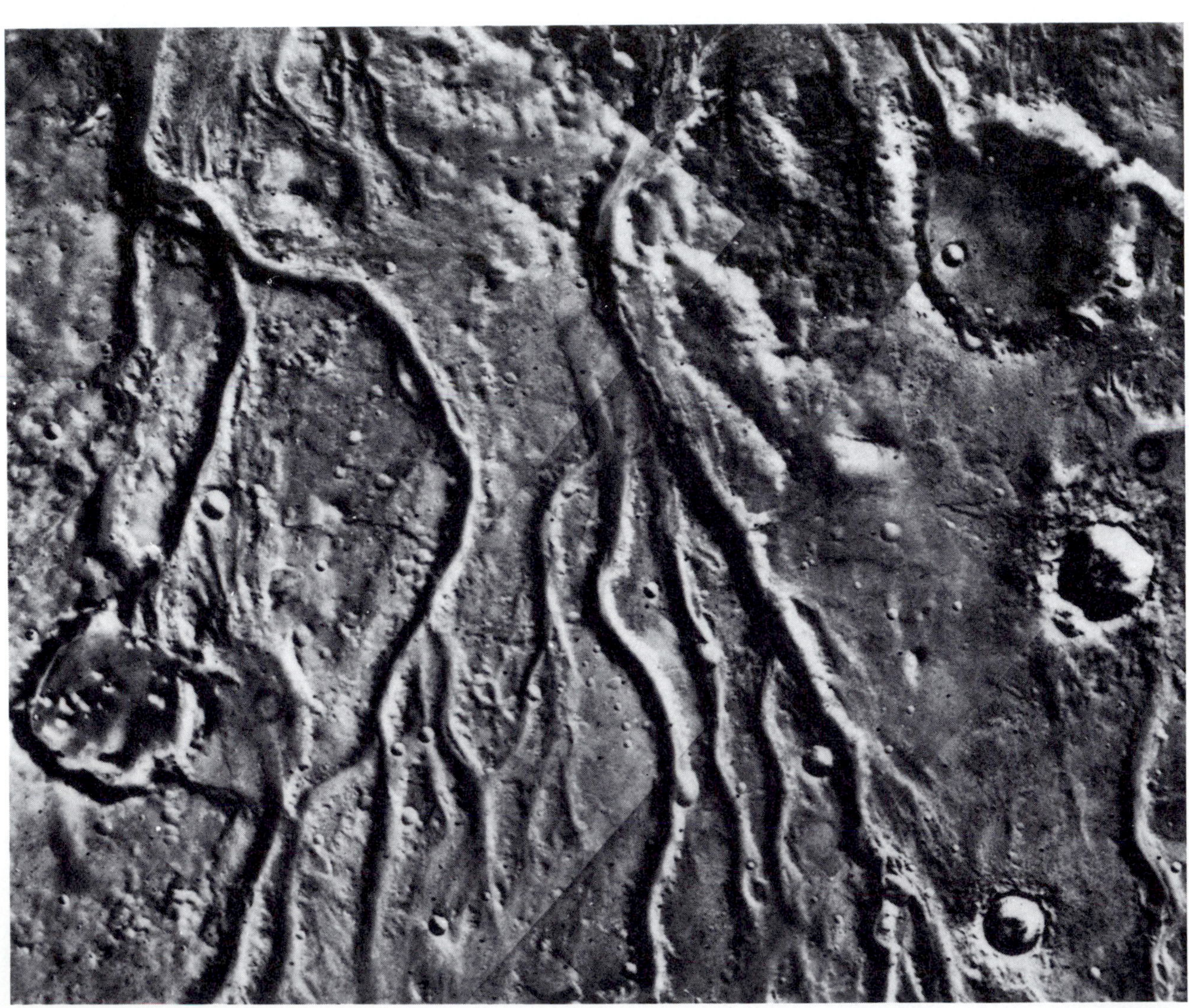

chapter 27

Are We Alone?

Looking at the countless stars that dot the heavens, we wonder whether there is intelligent life somewhere among them. We know we are the only intelligent life in the solar system, but perhaps among the billions of stars beyond our sun there are some with planets that support civilizations. It might be too much to expect the inhabitants to look anything like us, but if they are technically advanced, we could at least communicate with them. In this chapter we will take a serious look at this question.

The quest for extraterrestrial life is by no means new. At one time astronomers believed that many, and perhaps all, of the planets in our solar system might be inhabited. Serious consideration was even given to signaling them with gigantic torches, perhaps huge ditches filled with flaming oil. Although we now realize that such an endeavor would be fruitless, we are still intrigued with the possibility of extraterrestrial life. The search for it has become an important branch of astronomy called *exobiology*.

Exobiology is quite different from other branches of astronomy in that much of it is experimental. Planetary atmospheres are simulated in the laboratory to see whether various organisms can survive in them, or perhaps even evolve naturally (an energy source is generally needed in this case; see p. 475). Rocks are dated, fossils are examined, and radio telescopes are pointed toward potential candidates listening for signals that might mean another civilization. Many different branches of science (biology, chemistry, geology) are involved in the study.

If extraterrestrial life were found, it would indeed be the most significant discovery that humankind has made. But one wonders what the impact of such a discovery would be. Several scientists have stated that if we receive a message from space, we should ignore it. After all, the civilization responsible for it is likely considerably more advanced than we are, and the senders may be looking for new worlds to colonize. Others argue, however, that the likelihood of this is small and we should answer as quickly as possible. We may be able to learn a considerable amount from such a civilization, such as how to overcome our energy problems and how to conquer disease. On the other hand, sudden advances in knowledge or in technology could have destructive effects, and making discoveries on our own is certainly more satisfying than having them handed to us.

Yet we would still like to know: Is there life out there? What scientists already know about how life began on Earth can help us make a much better estimate of whether life exists beyond it. There are, in general, three major hypotheses for the origin of life on Earth:

1. It occurred as a result of a supernatural event that is beyond scientific description.
2. It was produced by a natural evolutionary process.
3. It came from beyond Earth in an exceedingly rare event: the transport of a spore.

The first of these is not within the realm of science, and we will not discuss it. Most of the chapter will be devoted to the second hypothesis, as it is the one most scientists now accept. Before turning to it, let us look briefly at the third possibility, usually referred to as the *panspermia hypothesis*. It assumes that somewhere in the universe there is (or was) another planet containing life. A few billion years ago live spores were ejected from it into space, either as a result of an explosion (perhaps a volcanic eruption) or by the deliberate action of another civilization, and these spores drifted through space until finally they reached Earth. If this is, indeed, the way life came to Earth, our study of the evolution of life would be of little help. It is easy to calculate how much radiation a spore would receive in a journey through space, however, and it is many orders of magnitude greater than the amount needed to kill a terrestrial spore. Because of this problem there is little support for this hypothesis.

WHAT IS LIFE?

Let us turn, then, to the second hypothesis. We begin by asking what life is. It might seem, at first, that we could define it in terms of *change*. All life is in a continual state of change: Cells die, and new ones replace them. In fact, your body now is completely different from the one you had only eight years ago; all the cells have renewed themselves during this time. Of course, your personality and physical appearance have not changed to a large degree, so it is obvious that there must be a master plan somewhere in your body that allows duplication.

However, attempts to define life in terms of change soon encounter difficulties. Many things, such as a stream or a flame, change continually and yet are not alive. Perhaps a better way to define life would be in terms of *reproduction*. All life reproduces itself. But even here we have to be careful: Crystals duplicate themselves, as do viruses, and we know that crystals are not alive. Whether viruses are alive is difficult to say; they seem to be very close to the border between the living and the nonliving.

Reproduction is, nonetheless, the key. In the past few decades tremendous breakthroughs have been made in understanding the life process. The basic unit, the cell, has of course been known since the invention of the microscope, but as scientists have probed deeper and deeper into the cell, they have made some fascinating discoveries. Just as an atom has a nucleus, so has a cell (Fig. 27.1). And within this nucleus are the units that control all growth and development of the cell. The controlling mechanism is in the form of two giant molecules: deoxyribonucleic acid (DNA) and ribonucleic acid (RNA). Of these, DNA is much the larger; it contains the master plan of the cell. It directs the growth and development of the cell with the help of RNA, which acts as a sort of "worker" molecule. RNA plays many important roles in the cell and, unlike DNA, is found both inside the nucleus and outside (i.e., in cytoplasm; see Fig. 27.1).

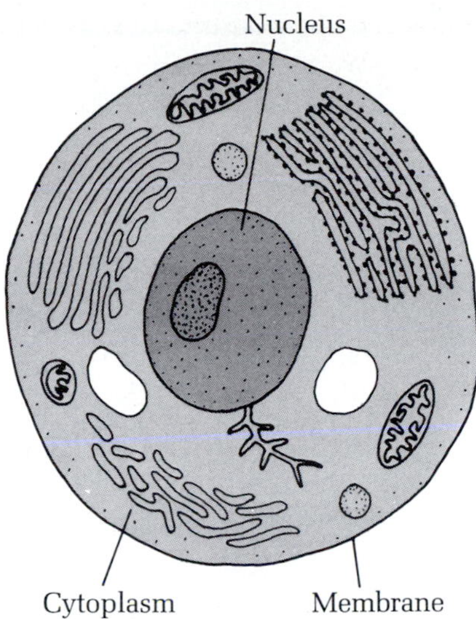

Figure 27.1 Cutaway diagram of a cell showing its components. The nucleus is near the center; the region outside the nucleus is the cytoplasm. DNA is found in the nucleus, RNA in both the nucleus and the cytoplasm. Protein is produced in the cytoplasm.

Because they are so large, biological molecules appear to be complex; however, they are actually quite simple. For example, DNA is composed of two helical strands of sugar and phosphate bonded together by pairs of basic molecules—adenine (A), cytosine (C), guanine (G), and thymine (T). RNA is similar; the two differences are the replacement of T by U (uracil) and the replacement of deoxyribose sugar (in the strands) by ribose sugar. Another important type of biological molecule is protein. It is quite different in structure from DNA, but parts of it are usually spiral in structure, and it is built of relatively simple units. The units in this case are amino acids.

The bases in DNA are not joined randomly; adenine joins only with thymine, guanine only with cytosine. This is important, for it gives what scientists refer to as the code of life. The sequence of bases contains the blueprint for the fabrication of everything needed by the cell (including a duplicate). The preservation of this code is therefore critical, and nature sees to it that it is well protected. Only at replication time (Fig. 27.2), when the molecule unwinds and duplicates itself, and when it passes information to RNA, is this code exposed.

DNA, RNA, and protein play closely interrelated roles. Information is passed from DNA to RNA; RNA in turn assembles protein. Some of this protein is then fed back to the DNA controlling the production of RNA. A simple representation of this scheme is the following:

$$\text{DNA} \rightarrow \text{RNA} \rightarrow \text{protein} \rightarrow \text{DNA}$$

One of the main elements in the molecules we have been discussing is carbon. It is, in fact, one of two elements that has the ability (when combined with hydrogen) to form such large molecules. Because of this we say that life on Earth is based on carbon.

One other element, silicon, can form large molecules, and perhaps somewhere in the universe life is based on it. Silicon-based life would be strange, to say the least; living beings would be made of stone, and the respiratory medium would likely be sand. Such creatures have been described in science fiction, and perhaps we should leave them there.

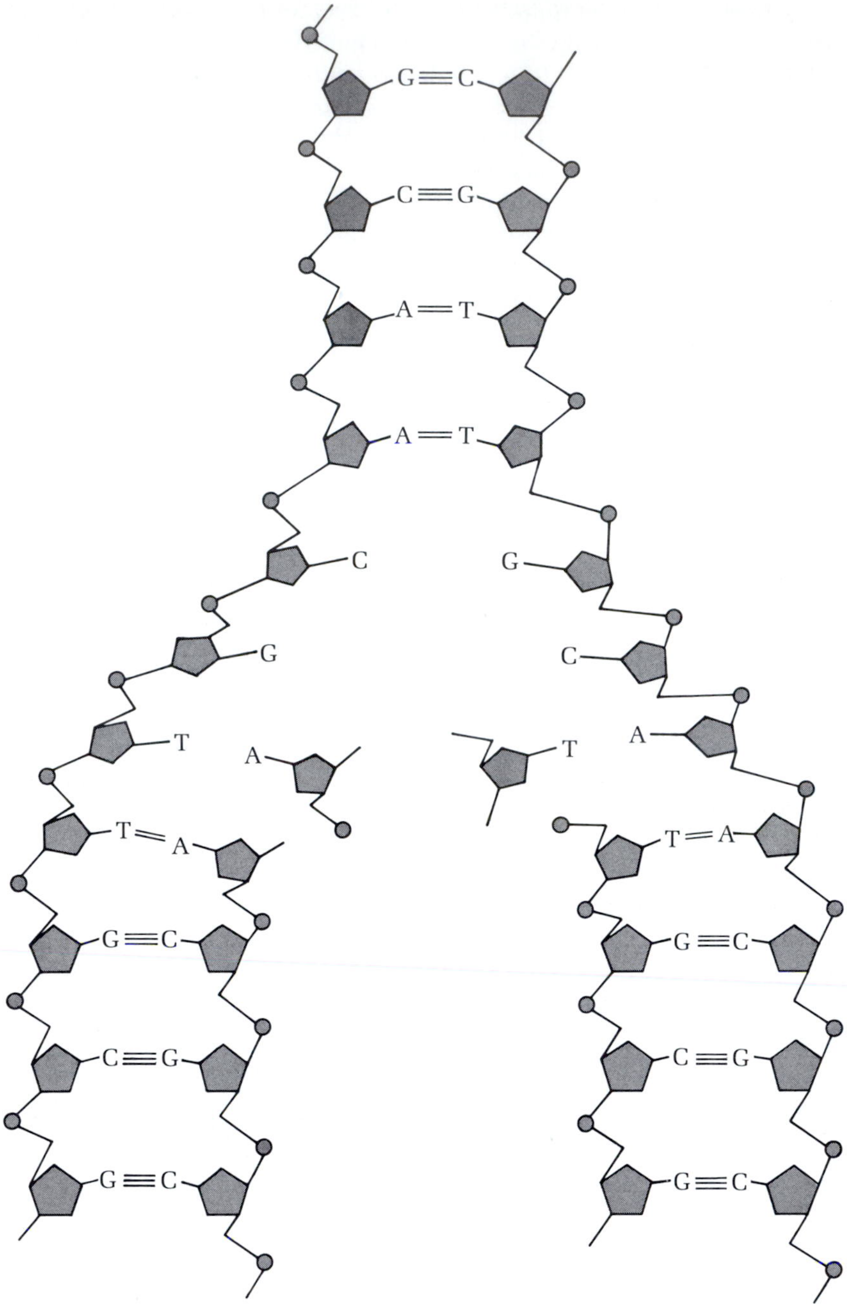

Figure 27.2 Replication of DNA. Notice that A joins only with T, and G only with C.

THE MILLER-UREY EXPERIMENT

With this background on the basic molecules of life, let us turn to how these molecules, and consequently life itself, may have come into being. Among the first to suggest (independently) that life may have evolved naturally on Earth were A. I. Oparin, in Russia, and J. B. S. Haldane, in England. They hypothesized that during its early stages Earth had a very different atmosphere from the one it now has. There was no oxygen; rather, there was a mixture of such gases as methane, ammonia, and perhaps nitrogen. They went on to suggest that if appropriate energy sources (e.g., lightning) were present, life may have evolved in such an atmosphere. (Many years earlier, in a series of cleverly devised experiments, Louis Pasteur had shown that life could not evolve from nonlife in our present atmosphere.)

But for years no one took the suggestion seriously. It was not until almost thirty years later, in 1953, that Harold Urey decided to test the idea's viability. If life had, indeed, formed in this primitive atmosphere, we should be able to duplicate the process in the laboratory. Graduate student Stanley Miller was given the task of performing the experiment. He simulated the primitive atmosphere, using methane, ammonia, water vapor, and hydrogen; a mixture of these gases was circulated through his apparatus (Fig. 27.3). Miller selected a sparking device as his energy source; this would be equivalent to lightning on the early Earth. After a few days the liquid in the apparatus began to darken. Miller stopped and analyzed it, and as Urey (and, earlier, Oparin and Haldane) had predicted, biological molecules were present. The dark liquid contained several types of amino acids, the building blocks of protein. Within a few years the experiment was duplicated by numerous other groups, and the race was on for other "keys" to the origin of life.

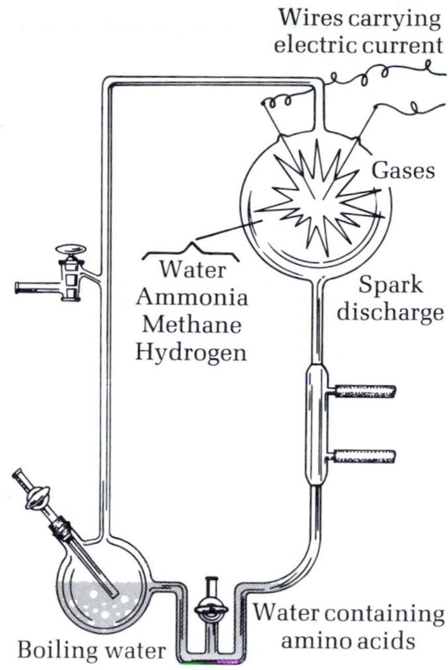

Figure 27.3 In the Miller-Urey experiment a simulated primitive atmosphere was circulated in an apparatus like the one shown. Within a few days amino acids appeared.

The next step was the production of the building blocks of DNA: sugars, phosphates, and the bases A, T, G, and C. With the use of a variety of energy sources (e.g., UV radiation, ordinary sunlight) and catalysts, all of these units were eventually produced naturally. But would the building blocks form DNA if they were mixed together? Again it was shown that with the help of appropriate catalysts the building blocks would form themselves into a "working" DNA molecule. It has also been shown that mixtures of amino acids produce single proteins called proteinoids.

As complex as these molecules are, we cannot really call them life. A molecule by itself is not alive. We still have a long way to go before we have actually created life in the laboratory, but some of the numerous experiments now in process may soon do so.

THE DEVELOPMENT OF LIFE ON EARTH

As we saw in Chapter 7, when the earth condensed out of the solar nebula about 5 billion years ago, it was surrounded by an atmosphere composed mostly of hydrogen and helium. As these gases cooled, others such as methane, ammonia, water vapor, and nitrogen began to form. Being lighter, the hydrogen floated to the top and began leaking slowly off into space. When nuclear reactions were triggered in the core of the sun and radiation was given off into space (Chapter 7), the resulting solar gale pushed the remaining hydrogen (and likely some of the other gases) into space, leaving a planet almost void of an atmosphere.

The earth itself was, at this stage, a roughly homogeneous mixture of silicates and iron, with traces of other materials. But among these other materials were radioactive atoms, and heat from these atoms gradually increased the interior temperature of the earth until it was a molten ball. Iron and other metals sank rapidly to the core;

silicates and lighter materials floated to the top. Then, as cooling occurred, a crust began to form. At first it was little more than a seething, skinlike scum covering a ball of molten lava and metal, but as cooling continued the crust thickened. However, it could not contain the tremendous pressures that were building up beneath it, and gigantic volcanoes began to appear. Eruptions were soon widespread. Initially, mostly nitrogen, methane, and hydrogen were emitted, but later this changed to predominantly carbon dioxide (CO_2) and water vapor. As a result of this outgassing, the earth's atmosphere began to build again. The various gases broke down, reacted, and so on, until finally the atmosphere consisted mostly of methane, ammonia, nitrogen, water vapor, and perhaps some hydrogen, most of which would leak off rapidly into space. (Most of the CO_2 was absorbed in ocean water. Some of it went into the formation of calcium carbonate, and we find it today in the form of shells and limestone.) We refer to this as the "primitive atmosphere" of the earth. Beneath this strange atmosphere vast oceans had begun to appear.

The first life forms to appear were no doubt anaerobic (independent of need of oxygen). They were probably vegetablelike and likely lived in the vast oceans that covered the earth at the time. Over a relatively short period—a few thousand years—this life, according to Oparin, significantly changed the atmosphere. We know that vegetation takes in CO_2 and gives off oxygen. This primitive life therefore supplied considerable oxygen to the atmosphere, after which most of the anaerobic life forms disappeared and were replaced by oxygen-using life forms.

Ultraviolet radiation was also extremely important in the formation of life, but soon after the oxygen formed, a layer of ozone appeared, cutting off the ultraviolet radiation (UV radiation acting on oxygen creates ozone). Since UV radiation is detrimental to life, this blockage was needed if higher forms of life were to evolve. As higher forms developed, however, they had to adapt to the new atmosphere.

Over millions of years the simple forms of life evolved, changed, and gradually grew more complex. Life gradually spread over the earth; different species arose in different environments, until today we have an amazing variety of life forms, adapted to many different conditions.

OTHER LIFE IN THE SOLAR SYSTEM

Did life form on the other planets of the solar system in the same way it formed on Earth? And if not, why? We know that the other planets support no advanced forms of life, but have there ever been lower forms? Since life as we know it requires moderate temperatures (temperatures that are between the freezing and boiling points of water most of the time), we can eliminate such inhospitable planets as Mercury and Pluto as candidates for life.

As we move outward from the sun, temperatures decrease. Far inside Earth's orbit temperatures are too high for life; far outside it temperatures are too low. The region between these two extremes is called the *ecosphere,* and Earth's orbit lies almost in the center of this region. This ecosphere extends from just inside the orbit of Venus to just outside that of Mars. Thus, with the right conditions, life may have evolved on Venus and Mars. Unfortunately, the conditions are not satisfactory in either case; both have atmospheres composed almost entirely of CO_2.

Venus, in particular, has a heavy, dense atmosphere of CO_2 that (as we saw in Chapter 11) acts like a greenhouse and traps radiation. This causes temperatures much too high for life as we know it. Still, there is the possibility of life in the clouds. Although temperatures in the upper layers are well below the freezing point of water, there is a region between the upper layers and the surface where moderate temperatures prevail. A floating form of life could exist here; it would have to be small and light and be able to survive under trying conditions, but this is possible.

Mars has long seemed the most interesting planet as far as extraterrestrial life is concerned. It has an atmosphere, there is evidence of water, and temperatures are reasonable (only a small greenhouse effect operates). Nevertheless, Mars has almost no oxygen (less than 1%) and consequently no ozone layer. This means that its surface is continually bombarded by UV radiation, and consequently, any life there would need protection from this radiation.

It also appears that there is no surface water in the liquid state. Pressures are so low, in fact, that according to calculations water could not exist in the liquid state. There is considerable evidence, however, that there is frozen water under a thin layer of solid CO_2 at the polar caps, and it has been suggested that there may be a frozen layer just beneath the surface over much of the planet. Indeed, if Mars underwent volcanic activity early in its history, as Earth did, it no doubt had a denser atmosphere at that time. It also therefore had a greater atmospheric pressure that would have allowed water to flow on its surface (and, as we have seen, there is considerable visual evidence for this flow; see Fig. 27.4). Perhaps this water is now frozen beneath the surface.

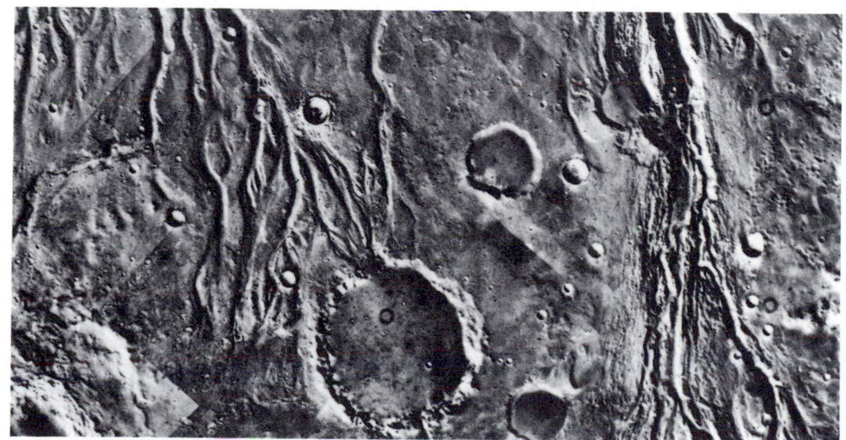

Figure 27.4 Numerous ravines and gulleys on the surface of Mars appear to be old riverbeds.

We know, of course, that there are no higher forms of life on Mars, and now, with the landing of Viking I and II, there is strong evidence that there are no lower forms. Several experiments were performed during these visits, and in one instance several gases (CO_2 and H_2O) were expelled from the soil, indicating the possibility of life. But further examination showed that this was likely just a normal chemical reaction. No large molecules were found at either site, and there was, in general, no indication of life forms. Of course, this does not necessarily rule out life.

At best, then, our solar system harbors only the most primitive forms of extraterrestrial life. But when we turn to the vast regions beyond the solar systems, a different picture emerges.

LIFE BEYOND THE SOLAR SYSTEM

We have not seen a single planet in the region beyond the solar system; yet, according to most astronomers, the prospects for life here are relatively good. Why? One of the major reasons is the multitude of stars. In our galaxy alone there are two hundred billion, and many of them are like our sun. It is not surprising that no planets have been directly observed outside the solar system. At the distance of even the nearest stars, our most powerful telescopes probably could not discern even a very large planet the size of Jupiter.

Fortunately, there are other ways of going about the search. For example, we can make a number of reasonable assumptions about the universality of life (as we know it)[1] and determine the probability that there is life beyond the solar system. We will do so in the next section. Before we can discuss probabilities, though, we must lay the ground rules and talk about what is required of life.

Requirements for Life

To begin with, we will assume that our system is typical of other solar systems in our galaxy. Carl Sagan calls this the *assumption of mediocrity*. We cannot be certain that it is a valid assumption, but it does seem reasonable.

It is important to make the exact meaning of this assumption clear. It does not imply that life will necessarily arise in a system like ours, and it does not imply that if life arises it will look like the life forms on Earth. It only says that our solar system is a typical or average one. With this in mind, then, let us look at what is required for life as we know it.

First, *temperatures* must be moderate—between the freezing and boiling points of water most of the time. There are species of bacteria (and viruses) that manage to survive outside this range, but higher forms of life are not nearly so hardy and would quickly disappear if the average temperature were near one of these extremes.

[1]In the rest of this chapter, "life" always means life as we know it.

Second, *water* must be available. Most scientists feel that it is the only adequate host solvent for life. Certainly, all life on Earth requires water, and it is difficult to imagine life forms that would not.

Third, there must be a satisfactory *atmosphere*. It need not be the same as ours, and in all probability it would not be. Oxygen is a vital component of our atmosphere, and most forms of life require it, but again, there are bacteria here on Earth that survive in an oxygenless environment.

Fourth, an adequate *energy supply* is important. Plants and vegetables (and some algae and bacteria) obtain their energy from sunlight via photosynthesis and by absorbing various chemicals (fertilizers) from the soil. All animals, on the other hand, require a direct food supply of some sort.

Fifth, there must be *protection from radiation*. We are protected here on Earth by our atmosphere; in particular we are protected from UV radiation by the ozone layer. Mars, as we just saw, is continually bathed in UV radiation, and any life forms there would need protection from it.

If these five conditions are to be satisfied, the planet will, as a minimum requirement, have to be in the ecosphere of the star (the region where temperatures are neither too high nor too low for life). The ecosphere would have to be as wide as possible. And since small stars have small ecospheres, there is obviously a minimum brightness requirement for our candidate.

Stability over a long period of time is also important, as life must develop and be sustained. We know that lower forms of life on Earth took about 1.5 billion years to evolve, higher forms perhaps 3 billion years, and civilization about 4.5 billion years; thus, if we are typical, the ecosphere must be stable for at least this long (for corresponding forms). Looking at the main sequence (see Fig. 14.5), we see that this condition forces us to eliminate many stars from consideration. The giants at the top of the main sequence evolve much too fast for life to evolve; in general, they reach old age in only a few million years.

Interestingly, though, these stars play an exceedingly important role in relation to life in the universe. Life requires heavy elements and metals, and they must be made available in relatively short times, astronomically speaking. And, as we saw in Chapter 18, heavy elements and metals are produced and distributed into the universe via supernova explosions by these short-lived stars. If instead, heavy elements and metals were produced, for example, in red dwarfs, which live for 40 to 50 billion years and longer and do not explode, life would not be possible.

We see, then, that we can eliminate all O, B, A, and perhaps a few of the F stars from consideration. Turning now to the stars near the bottom of the main sequence, we find another problem. Their ecospheres are stable over long periods of time—easily long enough for life to develop—but they are, unfortunately, narrow and close to the star (Fig. 27.5). This leads to two difficulties. First, most planets have elliptical orbits, and the probability that the orbit would be entirely within this narrow ecosphere is small. But even if the orbit did remain within the ecosphere, its closeness to the star would sub-

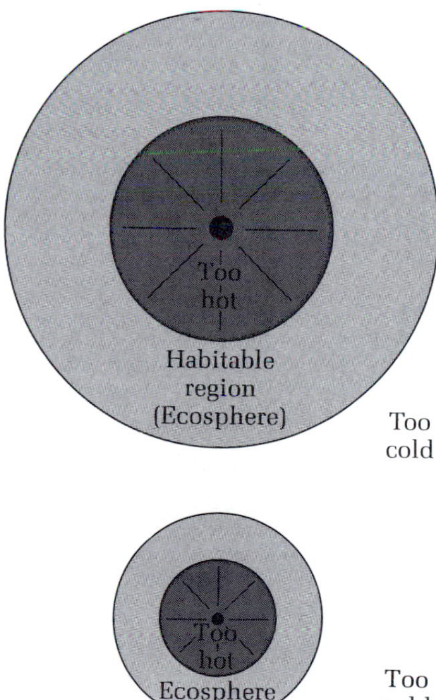

Figure 27.5 Habitable regions (ecospheres) around two stars of different size. The larger star has a relatively large ecosphere; the smaller has a small ecosphere.

ject the surface of the planet to strong tidal forces as it moved in its orbit. Eventually these forces would slow down the planet's rotation; it would then be tide-locked to the star, so that one side of the planet would be fiercely hot and the other extremely cold—both unsatisfactory for life. On the basis of this we can eliminate all M stars and about half of the K stars. This leaves the stars from about F5 to perhaps K5 as the best candidates (some astronomers argue for a slightly larger range). As you might expect, our sun (type G2) is roughly in the middle of this range.

But even if a star is ideal, there can be problems. Consider binary systems. The ecosphere of such a system is not only complicated in shape but also continually changing. In the region outside the system of stars it may be reasonable, but between them it will be narrow or perhaps nonexistent. It is unlikely that a planet in such a system could remain within the ecosphere at all times.

Conditions would likely be even worse in triple and higher-order systems. Surprisingly, our nearest neighbor, α Centauri—a triple system—has at least one reasonable ecosphere, but it is certainly exceptional; to be on the safe side, we eliminate all multiple systems as good candidates.

Probability of Life

Now that we are familiar with the conditions required for life, we can examine the probability that it exists elsewhere. First, let us illustrate the idea with some simple examples. The probability of getting a head when a coin is flipped is $\frac{1}{2}$ (one head, two sides to the coin). To find the probability of getting two heads with two coins, simply take the probability of a head on one coin, $\frac{1}{2}$, and multiply it by the probability of a head on the other, $\frac{1}{2}$. This gives $\frac{1}{4}$, or a probability of one in four.

To generalize, assume you also have a pair of dice. Since there are 6 sides on a die, the probability of rolling a 1 is $\frac{1}{6}$; similarly, the probability of rolling two 1's with two dice is $\frac{1}{6} \times \frac{1}{6}$, or $\frac{1}{36}$. To find the probability of two heads (with coins) and two 1's (with dice), simply multiply $\frac{1}{36}$ by $\frac{1}{4}$, which gives $\frac{1}{144}$, or one chance in 144.

Now let us use these ideas to calculate the probability of extraterrestrial life. We will start with a certain number of stars that are good candidates and make estimates of the probabilities that they have planets, that the planets are in the ecosphere, and that life evolves on the planets. Keep in mind, though, that the scheme is very crude—an approximation at best—and that some of the estimates we will make are only guesses.

First, let us consider all the stars of our galaxy. As we saw in Chapter 21, there are about 200 billion of them. How many of these are suitable suns? To be suitable, they must be in the proper spectral range and must be single. However, there is another limitation that we have not yet mentioned: When the Galaxy was born about 16 billion years ago, some of the stars that formed were massive and evolved rapidly; many of them exploded, spewing their metals and heavy elements out into space. New stars were then formed from this material, and some of the heavy elements went into the making of planetary systems around these stars. We call a star of this type *second-generation;* if this second-generation star then explodes, a third-generation star may arise from its debris, and so on. It is obvious that our candidates must be at least second-generation stars; otherwise, they would have no metals or heavy elements. When we eliminate first-generation stars as potential candidates and take into consideration the other requirements, we are left with about 4%, or 8 billion, of our original 200 billion stars as good candidates.

Next, we would like to know how many of these stars have planetary systems. There is little to go on here, but, as we will see in the next section, there is indirect evidence that several nearby stars have dark objects orbiting them. A number of astronomers, in fact, believe that all stars may have planetary systems, but this is perhaps too optimistic. As a compromise we will assume that half of them do. This gives us 4 billion stars with planets.

Now, in what fraction of their systems is there at least one planet in the ecosphere? Again we only have our system to go by; it contains one (Earth) at the center of the ecosphere and two near the edges. We are not sure this is typical, but let us be optimistic and assume it is. This means that all of our 4 billion systems have at least one planet in the ecosphere.

Finally, on what fraction of these planets will life actually develop? Needless to say, this is a difficult question. It depends, of course, on whether we are interested in advanced forms of life or in primitive forms such as bacteria. Let us consider only advanced forms (preferably with a technology). In almost total ignorance we assume a proportion of 0.1; this gives about 400 million civilizations in our galaxy.

But how many of these could we actually communicate with *now?* Certainly not all of them: The number we just calculated is the number that have existed at one time or another since our galaxy began. That brings us to the question of *longevity.* How long does a typical civilization last? We have had technology for about 100 to perhaps a few hundred years, depending on how one defines it. Many civilizations will no doubt annihilate themselves soon after they develop advanced technology (such as nuclear weapons). Because of this we should take a relatively small number as our lower limit—say, 1000 years. As an upper limit let us take 1 million years.

From these numbers we can easily calculate the number of civilizations that should currently exist in our galaxy. Taking its age as 10^{10} years, we find $10^3/10^{10} \times 400$ million = 40 as in the minimum number, and $10^6/10^{10} \times 400$ million gives 40,000 as the maximum number. We should exercise some care in using 10^{10} years in our denominator; allowance should be made for the fact that the first civilization would not appear for several billion years. But again, within our approximation this can be neglected.

Estimates of the probability of advanced life elsewhere in the Galaxy use numbers that differ considerably, but I believe the numbers presented here compare favorably to most in print. At a 1971 conference on extraterrestrial life, for example, the number agreed on was one civilization per 10^5 stars, or 10^5 civilizations currently in our galaxy (which is close to our upper limit). Stephen Dole, who has spent several years estimating probabilities of this nature, also obtains numbers comparable to ours.

Problems

The estimate of thousands of civilizations in the Galaxy may give us a feeling of optimism. In fact, however, several well-known scientists have begun to wonder if we are not the *only* intelligent life in our galaxy. One scientist, Michael Hart, who has done an extensive computer study of the problem and finds that the traditional way of estimating the width of ecospheres around stars omits many significant factors. He showed that if Earth were a mere 5% closer to the sun than it is (about 8 million kilometers), a runaway greenhouse effect like that on Venus would have occurred long ago and no advanced forms of life would have evolved. On the other hand, if Earth were only 1% farther away (about 1.6 million kilometers), a runaway glaciation period would have occurred about 2 billion years ago and Earth would now be covered with ice; again, no life would have evolved.

Extending his results to stars in general, Hart shows that the range (F5–G–K5) of stars usually assumed to have satisfactory ecospheres is probably too large. According to his results, a K1 star has an ecosphere of effective width zero, and stars greater than about 1.1 $M_{\odot}$ emit so much UV radiation that any planet in their ecosphere would require considerable protection for life to evolve. He concludes that a more reasonable range for life to evolve and last a few billion years is F7–G–K0. But if life is to exist longer than a few billion years (long enough for advanced civilization to evolve), even further restrictions are required. In this case it may be that only G stars are satisfactory.

There are a number of other limiting factors that we have not yet discussed. First, there are the spin of the planet and the tilt of the axis; they give us our day and our seasons. They are not critical in the case of the earth, but if they were changed significantly, disastrous weather changes would no doubt occur and only small sections of the earth might be habitable.

The atmosphere of a planet is also important. It would not necessarily have to be the same as ours, or have evolved as ours did, but some evolution would likely be needed. In our solar system two planets, Mars and Venus, are similar to Earth in many respects, yet their atmospheres are quite different and they do not appear to be satisfactory for higher forms of life. Would most planets in our galaxy follow the evolution of Earth, or would they evolve as Venus and Mars did and develop atmospheres of carbon dioxide? This is an important but unanswered question.

Another problem is related to the number of different species on Earth. Since the earth began, there have been as many as 10 billion different species of animals, insects, and plants on it. Of these only

Figure 27.6 A section of the Milky Way, showing a portion of its billions of stars.

one has evolved to higher intelligence (we could perhaps also include dolphins, making the count two). Again, is this typical of all planets where life evolves? Most scientists would agree that if we wiped the slate clean and started over again on Earth, humankind—as we know it—would not evolve. Our bodies, our brains, and so on are the result of millions of beneficial mutations; it is very unlikely that the same sequence of mutations, the same evolution, would occur again.

On the other hand, many astronomers are very optimistic that there is, indeed, extraterrestrial life. Frank Drake, for example, believes that intelligent civilizations are being created at the rate of about one per year in our galaxy. So perhaps we are not the only intelligent life in our galaxy; maybe there are millions of other civilizations. And beyond the Milky Way (Fig. 27.6) are billions of other galaxies; the probability that there is some form of life in some of them is certainly high.

SUMMARY

1. The search for life beyond Earth is now an important branch of astronomy called exobiology.
2. Scientists generally assume that life developed on Earth as a natural evolutionary process. An alternative possibility is the panspermia hypothesis, according to which life first reached Earth in the form of spores from another solar system.
3. The basic molecules of life are DNA, RNA, and proteins. The master code is contained in the DNA.
4. The Miller-Urey Experiment demonstrated that the building blocks of protein (amino acids) could be produced naturally in a primitive atmosphere.
5. It is believed that the earth condensed out of the solar nebula about 5 billion years ago. Initially it had an atmosphere composed mostly of hydrogen and helium, which was driven off by the solar gale. Volcanoes later gave rise to the *primitive atmosphere.* Life is assumed to have arisen in this atmosphere.
6. In our solar system the best candidate for life is Mars. However, Viking results indicate that the probability that there is life is small.
7. The prospects for life beyond the solar system appear to be relatively good. We have little to go on, however, aside from arguments based on probability estimates.
8. Scientists assume that our solar system is typical of others in our galaxy (the assumption of mediocrity).
9. There are certain requirements for life as we know it: moderate temperatures, water in liquid state, satisfactory atmosphere, energy supply, and protection from radiation.
10. The best candidates for life are G-type stars similar to our sun. Massive stars evolve too fast; small stars have small ecospheres.
11. A probability argument for extraterrestrial life can be given. It predicts that from 40 to 40,000 civilizations could exist in our galaxy.
12. Several scientists have suggested that the usual probability calculations do not take many important factors into consideration. If these are considered, the probability is reduced considerably.

REVIEW QUESTIONS

1. What are the three major hypotheses for the origin of life on Earth?
2. Can we define life in terms of change? Why? If not, how can we define it?
3. What are the three basic molecules of life? Describe each in detail. Where is each found, and what function does each have in the cell?
4. Describe the Miller-Urey experiment. What did it demonstrate?
5. Briefly describe the development of early Earth. How many different atmospheres has Earth had? Why is there not as much CO_2 in its atmosphere as in the atmospheres of Mars and Venus?
6. What is an ecosphere? Describe the ecosphere of the sun. What planets are in it? What types of stars have the most satisfactory ecospheres for life?
7. Why is it more probable that there is life beyond the solar system than within it (aside from Earth)?

8. What is the assumption of mediocrity?
9. What spectral classes of stars are the best candidates for life-supporting systems? Explain why.
10. Outline and discuss the calculation of the probability that there is intelligent life elsewhere in our galaxy.
11. What are some of the problems with this calculation?

THOUGHT AND DISCUSSION QUESTIONS

1. What do you think would be the long-range (and short-range) consequences of proof that life does exist elsewhere in the universe? Of communication with an alien civilization?
2. It has been suggested that a nearby supernova explosion may have eliminated many species, including the dinosaurs. Discuss this possibility.
3. Give several arguments that we are the only intelligent life in our galaxy (see the Reading List). Discuss. Also take the opposite point of view and discuss.
4. Using the probability argument given in this chapter as a model, construct a similar argument using reasonable estimates for the *maximum* number in each case. What total number of civilizations would you expect from this argument?

FURTHER READING

Asimov, I. *Extraterrestrial Civilizations.* Crown, New York (1979).

Dole, S. H. *Habitable Planets for Man.* American Elsevier, New York (1970).

Oparin, A. I. *Origin of Life.* Dover, New York (1953).

Ponnamperuma, C. *The Origin of Life.* Dutton, New York (1972).

Parker, B. "Are We the Only Intelligent Life in Our Galaxy?" *Astronomy,* p. 6 (January 1979).

Sagan, C. *The Cosmic Connection.* Dell, New York (1973).

Shklovskii, I. S. and Sagan, C. *Intelligent Life in the Universe.* Holden Day, San Francisco (1966).

Sullivan, W. *We Are Not Alone.* McGraw-Hill, New York (1973).

A final summary, and a brief look into the future.

Epilogue

In our journey through the universe we have seen an amazing array of astronomical objects: stars in a profusion of types, spinning wheels of stars—galaxies, tiny stellar lighthouses called pulsars, quasars—still enigmatic after twenty years, awe-inspiring black holes, and bleak but sometimes colorful planets and moons. Strangely, even though we have perhaps gained an appreciation for the size and numbers involved we are still no closer to being able to truly comprehend them. The universe is gigantic beyond imagination; no matter how we try to visualize distances such as a million light years, times such as a billion years, or the immense explosion—the Big Bang—that gave us our universe, we cannot: they are beyond human comprehension. But reflecting on them, and other aspects of what we have learned about the universe, can help us toward a recognition of our place in the cosmos, a better understanding of nature in general, and, perhaps more importantly, an appreciation of our own planet. For we now know that, despite the size of the universe and the incredible number of objects in it, there is strong evidence that worlds such as ours are rare. We are privileged to live on a planet that is immensely fertile and rich in resources. And, perhaps, we have gained a vital insight: Our resources are not limitless and our ecology is fragile, so, if we are to preserve our existence, we must protect and conserve what we have.

Some of the greatest astronomical insights in the next few years may come from the telescope (2.4-m mirror) that is to be put into space using the space shuttle. In orbit it will be high above the earth's atmosphere, where the skies are black and highly penetrable. With this telescope astronomers expect to penetrate seven times further into the universe than they have. They hope to see almost all the way back to its earliest events—the birth of galaxies and quasars. Other telescopes will also be put into orbit using the space shuttle. Plans are already underway to loft a cryogenically cooled .85-m infrared telescope in the European Spacelab, which should be operating by 1987. Also planned is the sending aloft of a large solar optical telescope in the late eighties.

Although numerous ventures planned for the United States space exploration program—a Galileo Jupiter orbiter and atmospheric probe, a Mars rover vehicle and sample returner, a Saturn atmospheric probe, and a Venus orbiting image radar vehicle—have been cut back or cancelled, corresponding programs in Europe and Japan are gaining momentum. The European Space Agency is planning a close pass of Halley's comet in 1986 and a Kepler mission to Mars a few years later.

Farther into the future is the prospect of space stations. Small ones, equipped for astronomical observation and other scientific research, will no doubt come first, but later, more elaborate space stations that can support colonies may be established. For several years, Gerald O'Neill and others have been working out the details of how, using present technology, we could bring space colonization to reality. O'Neill visualizes large, ring-shaped cylinders in space, that simulate gravity on their inner curved surfaces by spinning. Their sides would be made up of alternating strips of land (landscaped to resemble Earth) and windows, and shutters on the windows would allow the day-night cycle (and perhaps even seasons) to be simulated. Up to 200,000 people could live in these stations. While many future astronomical programs will be conducted from the vantage point of space, Earth-based programs will not be at a standstill. As we saw in Chapter 5, astronomers are working on arrays of NGTs. These, in conjunction with new electronic techniques and more sophisticated computers, will greatly assist ground-based astronomers.

Gerald O'Neill's conception of a space station of the future. An interior view is shown on page 486.

The search for intelligent life among the stars will probably become increasingly important also. Although we cannot see them directly, there is strong evidence that there are planets around many nearby stars. This might mean that planets are common in space. Most, no doubt, will be barren of life, but if only a tiny fraction contain intelligent life, it is worth the search. Peter van de Kamp has shown, by studying proper motions of stars over long periods, that our second nearest stellar neighbor, Barnard's Star, may have two dark companions. Barnard's Star is, unfortunately, a small red dwarf with a narrow ecosphere, and neither of the suspected planets is predicted to be in this ecosphere. Other nearby stars also appear to have dark companions but, like Barnard's Star, are poor candidates as life-supporting systems. Indeed, if we search the stars out to twenty light years, we find only two stars worth serious consideration—and neither can be nearly as hospitable to life as our sun. They are ϵ Eridani and τ Ceti. τ Ceti resembles our sun more than any other star in our neighborhood: It is single, has a mass of .82 $M_{\odot}$, and is of spectral type G4; it is 11.8 light years away. Unfortunately, it does not appear to have a dark companion. ϵ Eridani (mass .7$M_{\odot}$, spectral type K2, 10.8 light years distant), on the other hand, has a dark companion, but it is not like Earth: It has a mass about six times that of Jupiter, and is likely to be a dim (or perhaps dead) star. Although our nearest stellar neighbor, α Centauri, is a triple system, it is worth considering. The two larger stars of the system are separated, at nearest approach, by about 12 A.U., and it is possible that both have reasonable ecospheres, but so far we have no evidence that there are planets within them.

The object of finding a nearby civilization is, of course, communication. A civilization on α Centauri, for example, would be easy to detect and communicate with (although nine years would be required for an answer to any message). Even a civilization at a distance of 30 light years would be relatively easy to detect, and communication would be possible. But beyond this distance difficulties compound. The number of candidates increases drastically as we go further and further out, but the possibility of carrying on a two-way conversation diminishes. A civilization at, say, 100 light years would be virtually impossible to communicate with.

The first step (before communication) is, of course, detection. What, in fact, would we listen for? Messages that have been purposely directed at us? *We* have sent messages into space: A coded message was beamed in the direction of the globular cluster M13 in 1974 during the ceremony when the power of the giant Arecibo dish was increased. It will take about 24,000 years to reach its destination, and, if anybody is listening, we could get a reply about 50,000 A.D. (Messages in another form—plaques aboard Pioneer 10 and 11—were sent out in 1972 and 1973.) It is easy to see that the probability that we would intercept a message purposely sent us by another civilization is very low. First, there is the problem of what frequency they would use. The sender would no doubt realize that the message would have to penetrate a planetary atmosphere and would have to be in a relatively "quiet" region of the electromagnetic spectrum if it were to be detected. Giuseppe Cocconi and Philip Morrison pointed out several years ago that there is such a region: 21 cm or 1220 mHz. Most of the searching that has been done so far has been done at or near this frequency. Frank Drake showed in 1977, however, that ionized clouds in our galaxy give rise to a slight Doppler shift in 21-cm waves, so that any message received at this frequency would be likely to be incomprehensible.

Is there any other way we could detect a civilization? Fortunately, there is. Beginning about 30 years ago the earth became a noisy, bright radio source, second only to the sun in our neighborhood. This was brought about by the introduction of VHF-TV, UHF-TV, and FM radio; the waves from these sources easily penetrate our atmosphere and many of them radiate into space. Like ripples on a pond, they are spreading out in a sphere around us. Frank Drake estimates that there are now 400 stars within this sphere, and, if any of them were listening, they would know we are here.

If there were another civilization in space it would also be likely to give off radiation such as this. Some of these civilizations may, in fact, be surrounded by radio spheres hundreds, or even thousands, of light years in radius, and if they are, we could detect them. Even now, however, we are turning to underground cables and satellites for transmitting our television and radio programs, so that in perhaps 50 years we will be radio-silent. This means that most civilizations (assuming they are like ours) will pass through a phase a few decades long in which they will transmit radio waves into space, and, after that, become radio-silent. Stars with advanced civilizations orbiting them could, therefore, be surrounded by expanding radio "shells" only a few tens of light years thick.

A listener in space tuned into our stray radiation would detect a variable radio source; two peaks, one from the U.S. and one from Europe would be noticeable over a period of 24 hours. Without extensive study they might not even suspect the message was from a technological civilization. Of course, if they finally did realize that it was, they could—in time—determine a considerable amount about the earth.

Many searches for such radiation have been made in recent years, but so far none has been detected. The problems associated with large-scale searches are, unfortunately, enormous. First of all, if we included all the stars out to, say, 1000 light years in our search, there would be hundreds of thousands of candidates, and each would have to be scanned at, perhaps, a million different frequencies. It would be far more difficult than looking for a needle in a haystack. Confidence, however, is so high that a program called Cyclops has been envisioned that will consist of 1500 radio telescopes, each 100 m across, grouped in a circle 10 km in diameter. The cost will be enormous—about 10 billion dollars today. It could, however, be built in stages, spreading the cost over many years. With such an array our chances of detecting an advanced civilization—assuming one exists—would be greatly enhanced. We could easily tune into one as far away as 300 light years.

If the search is finally successful and a civilization is found, we will want to visit it. But, even were we so fortunate as to discover a civilization around α Centauri, it would take us 225,000 years at a velocity of 40,000 km/hr to visit it and return.

The generations of time required for interstellar space flight are, at the present time, an insurmountable barrier. Furthermore,

An artist's conception of the proposed Cyclops project, which would consist of as many as 1500 radio telescopes.

A Hollywood version of future space travel, from the movie *Star Trek II*.

the amount of fuel necessary would present an even more serious problem. Even were the energy source nuclear fusion (assuming we had mastered controlled fusion) the rocket would have to be larger than the Empire State Building to store all the hydrogen needed for a flight to α Centauri. Engineers use the term "mass ratio" to describe this situation; it is the ratio between the takeoff mass and the payload mass (the mass that arrives at the destination). The mass ratio for a trip to α Centauri is about one billion. This means that if we wanted to deliver 10,000 tons to Centauri we would have to start with 10,000 billion tons.

An ingenious way around the fuel problem was proposed in 1960 by R. W. Buzzard. Realizing that interstellar space is filled with hydrogen (about 1 atom per 5 cm^3), he suggested that a large scoop be mounted on the front of the rocketship to collect this fuel. If the ship is to be self-sustaining, large magnetic fields would also probably be needed to help guide the particles toward the scoop. The gas that is scooped up could then easily be stored until needed. With this technique, reasonable velocities—up to about $.10c$—could be obtained. At this rate we could get to α Centauri in about 40 years. Any way you look at it, though, unless there is a tremendous breakthrough of some sort, travel to the stars is a long way off. However, at that or some other far-off time mankind will have no choice. As resources dwindle and as our ultimate resource, the sun, runs down, we will be forced to leave the solar system and venture into interstellar space in search of other worlds suitable for colonization.

In the meantime, the drama of the universe continues to unfold, with us as sometimes comprehending witnesses of the awesome and wonderful spectacle. With your newly acquired knowledge you will be among those to whom its workings will be somewhat less mystifying, but even more wonderful.

ESSAYS

Amateur Astronomy and Astrophotography

Now that you have learned about the planets, stars, and galaxies you may want to study them first-hand with your own telescope. In the last few years thousands of people have taken up astronomy as a hobby and have found it to be fascinating.

It is best to start simply. A good pair of binoculars or a spotting scope will allow you to see more than you might think. The moons of Jupiter, the rings of Saturn, the craters of the moon, and many deep sky objects will be visible. Eventually, though, you will want to buy an astronomical telescope. The two main types on the market are reflectors and refractors, or, for a little more money, you can get a small Schmidt-Cassegrain. The advertising pages of *Astronomy* or *Sky and Telescope* will give you a good idea of what you can get for your money. You may also want to visit some of your local stores; many department stores now carry astronomical telescopes.

The feature of the telescope that is most important in determining what you will see is the size of the objective (mirror in a reflector; large lens in a refractor). The larger the objective, the more light the telescope will gather and the better your view (assuming the oculars themselves are of high quality). Don't be misled by claims of high power—it's the size of the objective that counts.

Inch-for-inch, refractors cost more than reflectors; this is one reason why most amateurs use reflectors. If you can afford it, a 6-inch reflector is a good starting instrument; it is, in fact, unwise to select anything less than about 4 inches in a reflector or 2.5 inches in a refractor.

After you have decided on a size, look carefully at the instrument's stability. If it jiggles easily or seems nonrigid in any way, don't buy it. With the high powers you will be using, these jiggles will be considerably magnified—and very frustrating. Look at how well the telescope is built and how easy it is to use. Do the slow-motion controls work smoothly? Do the clamps hold the telescope rigidly? Check each of these items carefully, as large telescopes (particularly reflectors) are difficult to move from position to position and must be held solidly in place. Their unwieldyness is one of the reasons why small, compact Schmidt-Cassegrain telescopes have become so popular in recent years. They are expensive, but because they are easy to use, they are worthwhile to many people.

A clock drive is desirable, but not essential. The earth is in continual motion and any object you are viewing will move across the field of view in a minute or so (depending on the power you are using). Frequent adjustments are frustrating, and a clock drive will do away with them once the telescope is equatorially mounted. Finally, you will want to get at least two eyepieces—one of low power and one of moderate power. A third eyepiece of higher power is useful if you can afford it, but you will find later on that you use it less than the other two.

Set up your telescope in the darkest part of your yard, well away from the glare of streetlights, on a rigid (preferably cement) base. You are now ready to enjoy the magnificent sights of deep space—but in order to find most of them you will need a good star atlas. *Norton's Star Atlas* is excellent; it is a standard guide that has been used by amateurs for years. Another useful atlas—although more expensive—is the larger, more impressive, *Skalnate Pleso Star Atlas*. You may want it later on.

The first objects you will no doubt want to look at are the planets. Their locations are given in astronomy magazines such as *Astronomy* or *Sky and Telescope*. The Messier objects are another group that have challenged amateurs for decades. Details on how to find them are given in most observer handbooks (e.g. *Burnham's Celestial Handbook*, Dover).

As you become increasingly involved in amateur astronomy you may wish to turn to astrophotography. With patience, excellent photographs can be obtained with even a relatively small telescope. A 35 mm camera is perhaps the easiest to use; you can buy an adapter that will allow you to attach the camera body directly into your eyepiece. Although almost any type of film can be used, there is a good selection of fast 400 ASA film now on the market in both color and black and white; it is best to use one of these since they will allow shorter exposure times.

Before you begin, you must equatorially mount your telescope (i.e. align the telescope axis with the earth's axis). Details on how this is done are given in most books on astrophotography (or see *Astronomy,* Aug. 1982). With your telescope equatorially mounted and the clock drive running, the earth's rotation will be compensated for and you will be able to take short exposures; with practice you will be able to get excellent photographs of the moon and planets. For deep space objects (e.g. galaxies), however, you will need longer exposures, and consequently, additional guiding. You will find that because of turbulence in the atmosphere, objects do not remain centered for long, even with the clock drive running. It is essential, therefore, that you be able to speed up or slow down the telescope; this is accomplished with a guiding system (again, see the advertising sections of one of the astronomy magazines). Your first deep sky photos may be disappointing, but they will improve steadily as you master the techniques, until finally you are able to produce excellent photos.

UFOs

The UFO phenomenon is not particularly new; Ezekial's wheel in space (in biblical times), for example, is assumed by many to have been a UFO. Indeed, strange objects in the sky have been reported throughout recorded history.

Note that there is a distinct difference between the terms "UFO" and "flying saucer." UFO refers to any unidentified flying object regardless of shape. The popular term "flying saucer" was coined in the year 1947 just after Kenneth Arnold sighted 9 disk-like objects in the vicinity of Mt. Rainier, Washington. He reported that they were approximately 50 feet in length and travelling at speeds of about 1700 miles per hour (well in excess of speeds that flying crafts were capable of in 1947).

Since 1947 there have been thousands of sightings of UFOs. The sightings seem to come in waves (crests occurred in 1950, 1954, 1965, 1973). A number of scientists have suggested that reports of this type are self-generating. In other words, the first reports cause people to look for UFOs, and of course when they look, they are more likely to see them.

But with so many sightings, surely at least one of these objects has crashed, or perhaps exploded above the earth. And indeed there have been several such reports. One incident of this type occurred in September of 1957 when an object resembling a flying saucer was reported to have exploded in the skies over Brazil. Most of the debris fell into the ocean, but a few pieces were apparently salvaged in the shallows by a fisherman. According to Brazilian officials, tests showed that the material was magnesium of extreme purity and incredible strength, but no tests were ever allowed by persons outside of Brazil, and nothing further was ever heard of the matter. Indeed, in all cases like this, actual evidence seems to be unavailable.

Strange creatures have been reported in many of the sightings. In 1973, for example, Charles Hickson and Calvin Parker stated that as they were fishing near Pascagoula, Mississippi, a spacecraft approached them. Three greyish creatures emerged, according to their report, and floated the two men aboard the spacecraft where (they report) they were examined by a huge electronic eye.

How do astronomers view such reports? As you can likely guess—with extreme skepticism. There is, however, a small minority that take the UFO phenomena seriously. Dr. Allen Hynek has stated that, contrary to public opinion, most reports do not come from so-called "UFO addicts"; most come from people who had little interest in UFOs prior to their sighting. A few, in fact, come from scientifically trained persons who are not likely to mistake Venus for a UFO. Dr. Hynek, unlike most astronomers, believes a serious effort should be put into the study of UFOs.

Carl Sagan, one of the foremost exobiologists in the world, believes there is not yet sufficient evidence for an extraterrestrial explanation. He points out that most of the exotic reports are not sufficiently reliable. On the other hand, the more reliable reports are not, in general, very exotic. He goes on to report another problem. Consider the following: UFO reports flow into the various agencies at the rate of several a day in the United States. Many of these reports (possibly all in a given period of time) are easily explained. However, for even a small fraction of them to be valid (i.e. extraterrestrial), we would have to be an incredibly interesting planet—considering the distances involved. Sagan has shown that the probability that we would be visited within a lifetime is extremely small, even if there were a large number of advanced civilizations within our galaxy. As we saw earlier, the distances between any civilizations within our galaxy (assuming others exist) are likely to be incredibly large, and even at speeds near that of light, exceedingly long times would be needed for interstellar trips.

But perhaps long distances and times are not a barrier. Perhaps extraterrestrials have overcome the speed of light, and, as in "Star Wars," have developed the use of hyperdrive (i.e. a jump to velocities greater than c). They would still need a considerable amount of fuel, however. Could they carry it with them? In 20 meter flying saucers? Not likely. Perhaps they are using the ramscoop technique we discussed earlier, or some variation of it. Or maybe they have somehow overcome the problems associated with black holes and are using them as subways. Interesting as they are, suggestions of this type are sheer speculation at the present time.

It might be argued that just as science in the year 3000 will no doubt have made amazing progress over the science of today, so advanced civilizations may have mastered many of the difficulties we now see in interstellar travel. However, scientists evaluate such speculation according to a principle known as *Occam's Razor*. It states that if we have a number of competing theories or hypotheses, all of which satisfy the data, we must assume the simplest one to be correct. In the case of almost all UFO reports the simpler explanations are well-known phenomena: The object was the planet Venus, a strangely lit cloud, or perhaps a balloon or aircraft.

Because of the tremendous public interest in UFOs, a committee was set up in 1966 to study them. It was headed by Edward U. Condon, a well-known and respected physicist with significant contributions in the area of atomic physics and spectroscopy. The group selected a number of the better-documented cases, and over a period of three years they carefully and exhaustively studied them. Their conclusions were in agreement with those of most astronomers: that the sightings could all be explained in terms of known phenomena.

Astrology

There are two reasons for including an essay on astrology in a book on astronomy. The first is to clear up the apparent confusion that sometimes exists among students (and also the general public) regarding the terms *astronomy* and *astrology*. The second is to point out the fallacies of astrology.

Astrology is a pseudoscience based on present and past positions of the planets, sun, and moon relative to the twelve constellations known as the zodiac. Of particular importance is the position of these celestial objects at your birth; also significant is the "house" (zodiacal constellation) that is then just rising on the eastern horizon. Astrologers use this information to cast a "horoscope" that tells you what your personality is like and what your future holds. Astrology assumes, in effect, what is called predestination: that people's lives are controlled and determined in advance by an external force (some astrologers maintain that the stars create only tendencies, not certainties).

According to astrological doctrine, mysterious forces from celestial objects somehow act on the nuclei of a person's brain cells to implant a type of personality. What are these forces? Astrologers do not know (they are only vaguely defined in astrology), but scientists are familiar with several forces that act over long distances in the universe. Gravitational force is one of the better candidates. But the magnitudes of gravitational forces are easy to calculate, and if the gravitational forces from, say, Mars and Venus were to influence a baby at birth, so would the tables and chairs in the room, and even the doctor and nurses. They exert a greater gravitational effect on the baby than Mars and Venus.

Other forces, such as the electromagnetic force, could be considered. But these fields (from the planets, sun, etc.) are much smaller, and to some degree the electromagnetic force is shielded by buildings. What other possibility does this leave? Nothing that scientists know about.

Astrology developed at a time before the basic forces of nature were understood. It was then, perhaps, reasonable to believe in astrology, since there was no way of proving that it was incorrect. The Babylonians and Egyptians, who were the first astrologers, had no idea how far away the sun, moon, or planets were. The Greeks inherited these early ideas of astrology, and accepted them without question. Much was later added by the Greeks, who we now think of as the "fathers" of astrology. Ptolemy's two books *Tetrabiblos* and *Almagest* are still used by astrologers, and *Tetrabiblos* is considered by many to be the bible of astrology, even though very little was known of astronomy at the time these books were written.

Astrologers pride themselves that such great men of science as Ptolemy, Tycho and Kepler were involved in astrology. True, both Tycho and Kepler cast horoscopes, but Tycho stopped doing so early in life, and later referred to them as foolish. Kepler cast horoscopes primarily because he was forced to do so—the pittance he was paid as an astronomer was not sufficient to provide him even a moderate living.

Many statistical studies of astrology and its predictions have been made over the years, and in each case (assuming legitimate studies only), it has been shown that there is no correlation between date of birth and other factors of later life. For example, Bart J. Bok did a study of scientists to see if they were born in particular months. After allowances were made for usual annual deviations, a uniform spread was found.

Many astrologers have pointed to the numerous correct predictions that they have made. But again, legitimate studies have shown that, in general, the number of correct predictions is no better than a statistical average. If you make enough predictions some of them are bound to come true, and it's easy to forget those that do not.

Why do so many people believe in astrology if there is so much evidence against it? It's human nature, no doubt; after all, who doesn't want to know what the future holds for them? And, it's much easier to "look to the stars" in making important decisions, than it is to sit down and do the necessary homework. It was said during World War II that Churchill was running the war on the basis of astrology—as many had seen a well-known astrologer visit his residence. It was revealed later, however, that Churchill knew that Hitler was an ardent believer in astrology who made few decisions without consulting his astrologer. Churchill was just trying to find out what Hitler was likely to be told.

In 1975 a strongly worded statement against astrology was signed by 192 leading scientists, including 19 Nobel prize winners. The first paragraph from this statement is as follows:

Scientists in a variety of fields have become concerned about the increased acceptance of astrology in many parts of the world. We, the undersigned—astronomers, astrophysicists, and scientists in other fields—wish to caution the public against the unquestioning acceptance of the predictions and advice given privately and publicly by astrologers. Those who wish to believe in astrology should realize that there is no scientific foundation for its tenets.

Appendix 1

Astronomical Quantities and Physical Constants

Angstrom unit (Å) = 10^{-8} cm
Astronomical unit (A.U.) = 1.496×10^{8} km = 9.3×10^{7} mi
Constant of gravitation (G) = 6.668×10^{-8} dyne. cm^2/gm^2
Earth mass ($M_{\oplus}$) = 5.977×10^{27} gm
Earth radius (equatorial) ($R_{\oplus}$) = 6378.16 km = 3963.20 mi
Earth velocity of escape = 11.2 km/sec = 6.94 mi/sec
Electron mass (m_e) = 9.11×10^{-28} gm
Hydrogen atom mass (m_H) = 1.673×10^{-24} gm
Light year (ly) = 9.461×10^{12} km = 5.9×10^{12} mi
Parsec (pc) = 206,265 A.U. = 3.260 ly
Pi (π) = 3.14159 = 22/7
Proton mass (m_P) = 1.67×10^{-24} gm
Sun mass ($M_{\odot}$) = 1.991×10^{33} gm
Sun radius ($R_{\odot}$) = 6.960×10^{5} km = 4.32×10^{5} mi
Velocity of light (c) = 2.99795×10^{5} km/sec = 1.86x 10^{5} mi/sec

Appendix 2

Temperature Scales

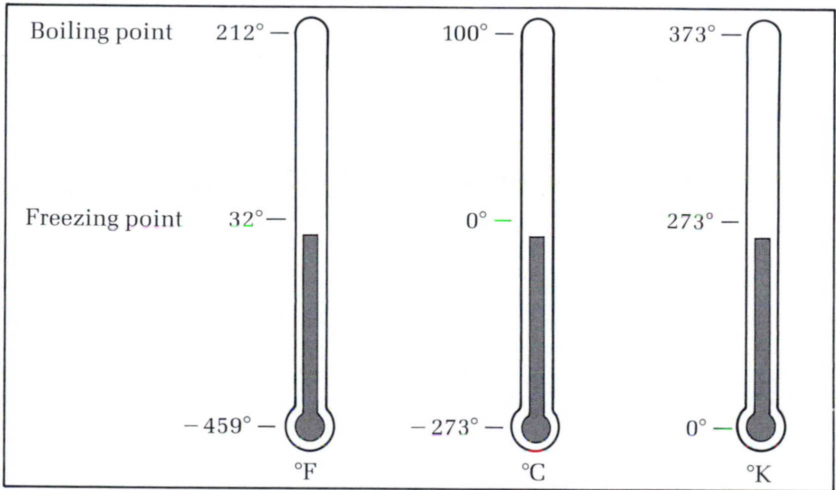

The three major temperature scales: Fahrenheit (°F, at left), Centigrade (°C, at center) and Kelvin (°K, at right). The freezing and boiling points of water are shown on each of them (a division in the Fahrenheit scale is only $\frac{5}{9}$ a division in the other two scales. Zero degrees on the Kelvin scale is the lowest temperature that can, in theory, occur in nature (it is actually unattainable).

Conversion formulas:

$°C = \frac{5}{9}(°F - 32)$

$°F = \frac{9}{5}(°C) + 32$

$°K = \frac{5}{9}(°F + 459)$

$°C = °K - 273$

$°F = \frac{9}{5}(°K) - 459$

$°K = °C + 273$

Appendix 3

The Twenty-Five Brightest Stars

Star	Name	Apparent Visual Magnitude	Spectral Type	Absolute Magnitude	Distance (ly)
1. α CMa A	Sirius	−1.47	A1	+1.45	8.7
2. α Car	Canopus	−0.72	F0	−3.1	98
3. α Boo	Arcturus	−0.06	K2	−0.3	36
4. α Cen A	Rigil Kentaurus	0.01	G2	+4.39	4.3
5. α Lyr	Vega	0.04	A0	+0.5	26.5
6. α Aur	Capella	0.05	G8	−0.6	45
7. β Ori A	Rigel	0.14	B8	−7.1	900
8. α CMi A	Procyon	0.37	F5	+2.7	11.3
9. α Ori	Betelgeuse	0.41	M2	−5.6	520
10. α Eri	Achernar	0.51	B3	−2.3	118
11. β Cen AB	Hadar	0.63	B1	−5.2	490
12. α Aql	Altair	0.77	A7	+2.2	16.5
13. α Tau A	Aldebaran	0.86	K5	−0.7	68
14. α Vir	Spica	0.91	B1	−3.3	220
15. α Sco A	Antares	0.92	M1.5	−5.1	520
16. α PsA	Fomalhaut	1.15	A3	+2.0	22.6
17. β Gem	Pollux	1.16	K0	+1.0	35
18. α Cyg	Deneb	1.26	A2	−7.1	1600
19. β Cru	Beta Crucis	1.28	B0.5	−4.6	490
20. α Leo A	Regulus	1.36	B7	−0.7	87
21. α Cru A	Acrux	1.39	B0.5	−3.9	370
22. α Cen B	Rigel Kentaurus	1.40	K4	+5.8	4.3
23. ϵ CMa A	Adhara	1.48	B2	−5.1	680
24. λ Sco	Shaula	1.60	B1	−3.3	310
25. γ Ori	Bellatrix	1.64	B2	−4.2	470

Appendix 4

Stars Within 15 Light Years

Name	R.A. (1980)	Dec. (1980)	Parallax	Distance (ly)	Spectral Type	Proper Motion (arcsec/y)	Apparent Visual Magnitude	Absolute Visual Magnitude
	h m	° ′						
Sun					G2		−26.72	+4.85
α Cen A	14 38	−60 46	0.760	4.3	G2	3.68	−0.01	4.39
B					K4		1.33	5.73
C	14 28	−62 36			M5e	3.85	11.05	15.45
Barnard's*	17 56	+04 36	.552	5.9	M5	10.61	9.54	13.25
Wolf 359	10 56	+07 10	.431	7.6	M8e	4.71	13.53	16.70
BD+36°2147*	11 03	+36 07	.402	8.1	M2e	4.78	7.50	10.52
Sirius A	6 44	−16 42	.337	8.6	A1	1.33	−1.46	1.42
B					*wd*A		8.7	11.6
Luy 726–8A	1 37	−18 04	.365	8.9	M5e	3.36	12.5	15.3
B					M5e		(13.0)	(15.8)
Ross 154	18 49	−23 50	.345	9.4	M5e	0.72	10.6	13.3
Ross 248	23 40	+44 04	.317	10.3	M6e	1.58	12.29	14.80
ε Eri	3 32	−09 32	.305	10.7	K2e	0.98	3.73	6.15
Luy 789–6	22 38	−15 28	.302	10.8	M7e	3.26	12.18	14.58
Ross 128	11 47	+00 58	.301	10.8	M5	1.37	11.10	13.49
61 Cyg A	21 06	+38 38	.292	11.2	K5e	5.22	5.22	7.55
B*					K7e		6.03	8.36
ε Ind	22 03	−56 52	.291	11.2	K8e	4.69	4.68	7.00
Procyon A	7 39	+05 17	.287	11.4	F5	1.25	0.37	2.66
B					*wd*F		10.7	12.99
Σ 2398 A	18 42	+59 36	.284	11.5	M4	2.28	8.90	11.17
B					M5		9.69	11.96
BD+43°44A	0 18	+43 54	.282	11.6	M1e	2.89	8.07	10.32
B					M6e		11.04	13.29
CD−36°15693	23 05	−35 59	.279	11.7	M2e	6.90	7.36	9.59
τ Ceti	1 43	−16 03	.273	11.9	G8p	1.92	3.50	5.68
G51–15	8 29	+26 51	.273	12.0		0.42	14.81	16.99
BD+5°1668*	7 27	+05 27	.266	12.2	M5	3.73	9.82	11.94
Luy 725–32	1 11	−17 06	.262	12.5	M5e	1.31	11.6	13.7
CD−39°14192	21 16	−38 58	.260	12.6	M0e	3.46	6.67	8.75
Kapteyn's	5 11	−44 59	.256	12.7	M0	8.89	8.81	10.85
Krüger 60 A	22 27	+57 36	.254	12.8	M3	0.86	9.85	11.87
B					M4.5e		(11.3)	(13.3)
Ross 614A	6 28	−02 48	.249	13.1	M7e	0.99	11.07	13.05
B							14.8	16.8
BD−12°4523	16 30	−12 36	.249	13.1	M5	1.18	10.12	12.10
van Maanen's	0 48	+05 19	.234	13.9	*wd*G	2.95	12.37	14.22
Wolf 424A	12 33	+09 09	.229	14.2	M6e	1.75	13.16	14.96
B					M6e		13.4	15.2
G158–27	0 06	−07 38	.226	14.4		2.06	13.73	15.50
CD−37°15492	0 04	−37 27	.225	14.5	M4	6.08	8.63	10.39
BD+50°1725	10 10	+49 33	.217	15.0	K7e	1.45	6.59	8.27

*Star may have an unseen component.

Glossary

aberration (of starlight): The apparent shift in a star's position due to the earth's orbital motion.

absolute magnitude: The magnitude a star would have if it were at a distance of 10 parsecs.

absorption line: A dark spectral line, or line of missing color in the continuous spectrum.

acceleration: The rate of change of velocity.

accretion: The colliding and sticking together of small particles to make larger ones.

aether: A medium that was invented as a propagating medium for light. Does not exist.

albedo: A measure of the reflectivity of a planet or other celestial body.

amino acid: A complex organic molecule. The building block of protein.

amplitude: The amount by which a quantity varies.

angström (Å): A unit of length used in the measurement of light. One centimeter equals 10^8 Å.

annular eclipse (solar): An eclipse in which the moon does not obscure the entire disk of the sun. A ring of light shows around the moon.

anorthosite: A type of lunar rock. Common in the lunar highlands.

aphelion: The point in an orbit around the sun that is farthest from the sun.

apogee: The point in an orbit around the earth that is farthest from the earth.

apparent magnitude: The brightness of a star or other celestial object as it appears as seen from Earth.

asteroid: Small objects that orbit primarily between Mars and Jupiter. A minor planet or planetoid.

astrology: A pseudoscience that relates the positions of heavenly bodies to human affairs. Future events are predicted.

astronomical unit (A.U.): A unit of measure equal to the average distance between the earth and the sun.

asthenosphere: A subsurface layer of considerable plasticity in a planetary body.

atom: The smallest unit of a chemical unit. It consists of a nucleus, which contains protons and neutrons, and electrons, which orbit the nucleus.

aurora: Light emitted in the upper atmosphere as a result of particles from the sun and Van Allen belts colliding with air molecules in the region. Also called Northern (Southern) Lights.

autumnal equinox: A point of intersection of the celestial equator and ecliptic. Autumn begins when the sun passes through this point (moving north to south).

Baily's beads: "Beads" of sunlight that appear at the rim of the moon just before and after totality in a solar eclipse.

Balmer lines: A series of emission or absorption lines that result from transitions between the second energy level and higher levels in hydrogen.

barred spiral: A spiral galaxy with a bright bar running through its nucleus. Trailing arms extend from each end of the bar.

barycenter: The center of mass of two celestial objects that are orbiting one another.

basalt: A type of igneous rock common in lava flows. Found on both the moon and terrestrial planets.

Big Bang Cosmology: A cosmology that assumes the universe began as a gigantic explosion.

binary star: A double star system. Two stars orbiting one another as a result of their mutual gravitational attraction.

black hole: An object with an escape velocity equal to the velocity of light. Once an object passes through its surface, called the event horizon, escape is impossible.

black body: A hypothetical body that absorbs and emits all radiation that falls on it.

black-body radiation: Radiation emitted by a black body.

Bode's law: A numerical relationship giving the approximate distances of the planets from the sun.

bolometric magnitude: The magnitude of a star summing over radiations of all wavelengths.

carbon cycle: A "burning" cycle, or series of nuclear reactions in stars, that converts hydrogen into helium. Carbon acts as a catalyst.

Cassegrain focus: A reflecting telescope in which a small hole is cut through the mirror allowing light reflected from another small mirror to pass through to an eyepiece.

celestial equator: The projection of the earth's equator onto the celestial sphere.

celestial sphere: An immense imaginary sphere that surrounds the earth in the sky. Latitude and longitude projected onto this sphere become declination and right ascension.

center of mass (gravity): The balancing point between two masses (or within a single mass). Mass times distance on one side equals mass times distance on the other side.

cepheid variable: A type of variable star. A supergiant that pulses. It has a period between approximately one and fifty days.

Chandrasekhar limit: The limiting mass for white dwarfs; approximately 1.4 $M_{\odot}$.

chondrite: A stony meteorite containing chondrites, or small round silicate granules.

chromatic aberration: A lens fault. The failure of the lens to bring all wavelengths of light to a common focus. Caused by dispersion.

chromosphere: The layer in the solar atmosphere just above the photosphere.

circumpolar stars: The stars around the celestial poles that never rise or set (always stay above the horizon) as seen by a particular observer.

cluster of galaxies: A number of galaxies that are gravitationally bound and move as a group.

color index: In the simplest case the difference between the photographic and photovisual magnitudes of a star. It gives a measure of the temperature of a star. Various other color indices are defined (e.g. the difference in magnitudes as seen through blue and ultraviolet filters).

coma: The diffuse gas around the head of a comet, given off as it approaches the sun, due to evaporation of materials in the nucleus.

comet: A small member of the solar family consisting of a nucleus, coma and tail. The ices in the nucleus evaporate as it approaches the sun, forming the coma and tail.

conduction: The transfer of heat energy from a hotter to a cooler region via atomic collision.

conjunction: A closely grouped configuration (or time) of two or more celestial bodies in the sky.

constellation: A group of stars, frequently resembling animals or mythological figures. A given constellation is now designated as a definite area in the sky.

continuous spectrum: A spectrum that contains light of all colors in a continuous blend. There are no absorption or emission lines present.

convection: The transfer of heat as a result of the motion of groups of particles (i.e. bulk transfer of heated particles).

corona: The hot, tenuous outer layer of the sun's atmosphere.

cosmic rays: High speed particles from space that strike our atmosphere, producing secondary showers of other kinds of particles.

cosmogony: A theory of the origin and evolution of the universe (or solar system).

cosmology: A theory of the structure of the universe. Theories are usually broadened to include evolutionary developments.

crust: The outermost solid layer of a planet.

dark nebula: An interstellar dust cloud that blocks light of distant stars.

declination: Lines on the celestial sphere analogous to latitude lines on Earth. A star can be located by its declination and right ascension.

deferent: The large circle that the center of an epicycle follows as it moves through space. Important in the Ptolemaic system.

degenerate gas: An extremely dense gas consisting of electrons and nuclei in which the electrons are compressed as close together as possible.

density: Mass per unit volume.

differential rotation: Rotational motion in which material or objects (i.e. stars in a galaxy) at different distances from the center rotate at different rates.

diffraction: A slight bending of light rays that occurs as they pass very close to opaque bodies or through small openings.

DNA: Deoxyribonucleic acid. The basic molecule of life.

Doppler effect: A change in wavelength that occurs when the body emitting the waves is either approaching or receding from the observer.

dwarf: Can pertain to the small red stars at the bottom of the main sequence, or to white dwarfs that lie to the left of the main sequence.

eccentricity: A measure of the elongation of an ellipse. A circle has an eccentricity of zero.

eclipse: The obscuration of light of one celestial body as a result of another passing in front of it (or its shadow passing over it).

eclipsing binary: A binary system in which the orbital plane is seen nearly edge-on so that the stars eclipse one another during each revolution.

ecliptic: The projection of the apparent path of the sun (among the stars) on the celestial sphere.

ecosphere: The life zone around a star; the region

where temperatures are between the freezing and boiling points of water most of the time.

Einstein-Rosen bridge: The wormhole or tunnel in space-time associated with a black hole.

electromagnetic radiation: Radiations associated with the electromagnetic field; includes rays, X-rays, UV, visible, IR and radio waves.

electron: A negatively charged particle that orbits the nucleus in the atom.

element: A basic substance of nature. Cannot be reduced further by chemical means.

ellipse: One of the conic curves. Egg-shaped or elongated and closed.

elliptical galaxy: A galaxy containing old stars that has no arms. Its image in the sky is usually elliptical.

emission spectrum: Bright-line spectrum caused by downward transitions of electrons in the atom.

energy: The capacity for doing work.

energy level: A position or level in the atom at which the electrons orbit; or an overall state of an atom.

epicycle: A small circle whose center traces out a large circle (deferent), usually around the earth. Important in the Ptolemaic system.

equatorial mount: A telescope mount in which the main axis is oriented in the same direction as the earth's axis of spin.

equinox: The points where the ecliptic and the celestial equator cross. Days and nights are of equal length.

escape velocity: The initial velocity required to completely escape the gravitational pull of an object.

evening star: Usually applied incorrectly to Venus when it appears in the evening sky.

event horizon: The surface of a black hole. Escape from the black hole is impossible once this surface is crossed.

evolutionary track: The path along which the point that represents a star in the H-R diagram moves as the star ages

eyepiece: The magnifying lens used to look at the image in a telescope.

filter: An optical device that allows only a narrow range of wavelengths to pass.

fission: The breaking apart or splitting of nuclei (or other body) into two or more smaller parts. Energy is released in nuclear fission.

flash spectrum: The emission spectrum of the chromosphere that appears briefly just before totality.

focal length: The distance from a lens or mirror to its focus.

force: A push or a pull on a body. Produces an acceleration if motion occurs.

frequency: The number of waves passing a given point per second.

fusion: The combining or fusing together of nuclei (or other bodies) to form heavier nuclei (or bodies).

galactic (open) cluster: A small group of gravitationally bound stars. Range is from approximately ten to a few hundred.

galaxy: A large system of stars, usually containing millions to hundreds of billions of stars. Gas and dust are also frequently present.

gamma ray (γ): The electromagnetic radiations that have the highest energy and highest frequency.

gas tail: One of the two tails of a comet. The other is the dust tail.

giant star: A star of large dimension and high luminosity.

globular cluster: A gravitationally bound group of from 10,000 to over a million stars. They form a halo around our galaxy.

globule: A small dark nebula. Will eventually become a protostar.

granulation: Convective cells in the photosphere of the sun that give it a mottled appearance.

gravitation: A basic force of nature. A property of matter such that it attracts other matter.

gravitational wave: A gravitational disturbance that propagates at the speed of light. Generated by oscillating masses.

greenhouse effect: The heating of a planetary atmosphere as a result of radiation trapped and reflected back and forth between the surface and a layer of clouds.

Gregorian calendar: A calendar introduced by Pope Gregory XIII in 1582.

ground state: The lowest energy state of the atom. The electrons are all in their lowest possible energy levels.

H I region: An interstellar region or cloud of neutral hydrogen.

H II region: An interstellar region or cloud of ionized hydrogen.

half-life: The time required for one half of the atoms in a radioactive sample to disintegrate.

halo (of galaxy): Gas, faint stars and globular clusters surrounding the nucleus of a galaxy.

heliocentric: Centered on the sun.

helium flash: The explosion that takes place near the core of a red giant star when the triple alpha nuclear cycle (helium burning) is first triggered.

Hertzprung-Russell (H-R) diagram: A plot of absolute magnitude versus temperature or spectral type for a large number of stars.

Hubble law: The relationship between the velocity of galaxies and their distance. Mathematically: $v = Hd$ where $v =$ velocity, $d =$ distance and $H =$ Hubble's constant.

hydrogen: The lightest element, an atom consisting of a proton as a nucleus with a single electron orbiting it.

hyperbola: One of the open conic curves.

inertia: Resistance to a change in motion.

inferior conjunction: Occurs when an inferior planet such as Venus is directly in line with the sun but between it and the earth.

infrared: The region of electromagnetic radiation that lies between visible and radio wavelengths.

interference: Occurs when two rays of light merge. If the loops and nodes of the waves line up there is constructive interference; if loops are opposite one another there is destructive interference.

interstellar dust: Dust particles lying between the stars.

interstellar gas: Atoms of gas, ions and molecules that lie between the stars.

intrinsic variable: A star that changes its light intensity as a result of some internal phenomena (i.e. pulsation).

ionization: The removal of electrons in the creation of an ionized atom.

ionized atom: An atom that has either more or less than the number of electrons that would make it neutral; i.e., that does not have the same number of electrons as protons.

ionosphere: An atmospheric layer above the earth that contains many ions.

irregular galaxy: A galaxy that does not have a regular shape (not elliptical or spiral).

irregular variable: A variable star that changes magnitude in a random way.

isotope: Atoms that have the same atomic number but different atomic masses.

Jovian planets: Planets like Jupiter, namely Jupiter, Saturn, Uranus and Neptune.

Kelvin scale: The absolute temperature scale. The lowest possible temperature in the universe is zero on this scale.

Kepler's laws: Three laws discovered by Kepler relating to the motions of the planets.

kilo: one thousand.

Kirchhoff's laws: Three laws relating to the way materials produce emission, absorption and continuous spectra.

KREEP: A lunar material consisting of potassium (K), rare earths (REE), and phosphorus (P).

Lagrangian points: The points between celestial bodies where a small body would remain fixed (i.e. the gravitational pull from the two larger bodies are equal).

latitude: The angular distance of a point on Earth from the equator, measured along the meridian.

librations: A small oscillating motion in the moon's motion that allows us to see more than fifty percent of its face.

light: The electromagnetic radiations that are visible to the naked eye.

light curve: A plot of the magnitude of a star or celestial object as a function of time.

light-gathering power: A measure of the ability of a telescope to gather light; depends on the size of the objective.

light year: The distance light travels in one year.

limb: The apparent edge of the disk of a celestial body.

limb darkening: The apparent darkening of the sun near its limb.

line broadening: The broadening of a spectral line due to such things as rotation of the celestial body emitting the line.

Local Group: The group of approximately 30 galaxies surrounding and including the Milky Way Galaxy.

longitude: The angular distance measured on the earth from the Greenwich meridian. Lines of longitude are perpendicular to lines of latitude.

luminosity: The total electromagnetic energy radiated by a celestial object per second.

lunar: Pertaining to the moon.

naked singularity: A singularity that is not surrounded by an event horizon.

magnetosphere: The region around a planet that is under the influence of its magnetic field.

magnifying power: The ratio between the angular diameter of an object seen through a telescope and its angular diameter as seen by the unaided eye.

magnitude: A number giving a measure of the brightness of the star.

main sequence: The diagonal strip in the H-R diagram along which most stars lie.

mantle: The layer in the earth between the crust and core.

maria: Flat, dark plains on the moon. Lunar seas.

mascon: Regions of highly concentrated matter beneath the seas of the moon.

maser: A device that amplifies the emission of microwave radiation.

mass: A measure of the quantity of matter in a body.

mass-luminosity relation: An empirical relationship between the mass and luminosities of main sequence stars.

Maunder minimum: An interval of time (1645 to 1715) during which there were very few sunspots.

mega: one million.

meridian: A great circle that passes through the zenith and the points on the horizon that are directly north and south.

mesosphere: A layer in the earth's atmosphere directly above the stratosphere.

Messier catalogue: A table of fuzzily-seen, celestial objects, compiled by Charles Messier in 1787.

meteor: A falling star. The bright streak seen in the sky when a meteroid strikes our atmosphere.

meteorite: A meteor that has struck our atmosphere and survived the flight to the surface.

meteoroid: A stony or metallic particle that strikes

our atmosphere and glows as it heats. Gives rise to the meteor phenomenon.

microwave: Electromagnetic radiation lying between radio waves and infrared on the spectrum.

Milky Way: The faint diffuse band of light that stretches across the night sky. It consists of billions of stars.

Milky Way Galaxy: The galaxy of which the earth is a part.

minor planet: An asteroid.

model: A theoretical representation of a system, usually based on mathematical equations.

molecule: A combination of atoms, bound together by electromagnetic forces. Smallest unit of a compound.

momentum: A measure of the amount of motion of a body. Mass times velocity.

morning star: A planet, usually Venus, seen just above the eastern horizon in the early morning sky.

multiple system: A bound system of several stars.

mutation: A small change in the hereditary material (DNA).

nebula: A cloud of interstellar gas and dust.

neutrino: An elementary particle produced in certain nuclear reactions. It has no charge and until recently was assumed to have no mass.

neutron: The neutral particle of the atomic nucleus.

neutron star: A star made up mostly of neutrons.

Newtonian focus: A telescope arrangement in which the image from the objective mirror is deflected by a small mirror to an eyepiece set near the upper end of the tube.

Newton's law of gravity: States that the gravitational force between any two objects is proportional to the product of their masses, and inversely proportional to their separation.

NGC: New General Catalog. A catalog of star clusters, nebulae and galaxies. Objects in it are designated as NGC 1, NGC 2 etc.

nodes: The two points where the orbit of the moon crosses the plane of the ecliptic.

nova: A star that suddently increases its brightness by several magnitudes, then returns slowly back to its original magnitude. Believed to occur only in binary systems in which one of the two stars is a white dwarf.

nuclear force: The force that holds the particles of the nucleus together.

objective: The largest lens in a refracting telescope; the mirror in a reflecting telescope.

oblateness: A measure of the equatorial bulge of a rotating sphere.

Oort cloud: The cloud of comets that surrounds the sun from a distance of approximately one light year.

opacity: A measure of the degree to which an object obsorbs light.

open cluster: see galactic cluster.

opposition: The configuration in which a superior planet is opposite the direction of the sun in the sky.

orbit: The path that a celestial object traces out when it is under the gravitational influence of another object.

Oscillating universe: A cosmology in which the universe is presumed to alternately expand and contract. Each new expansion begins explosively from an extremely small nucleus.

parabola: One of the open conic curves.

parallax: The apparent shifting of a nearby star relative to distant background stars as the earth moves around in its orbit.

parsec: An object at one parsec from earth has a heliocentric parallax of one second of arc. Equal to 3.26 light years.

partial eclipse: An eclipse in which only part of the object (the sun in the case of a partial solar eclipse) is obscured.

penumbra: In a sunspot, the less bright region around the umbra. In an eclipse, the region of partial illumination.

perigee: Point in the orbit of the moon that is closest to Earth.

perihelion: Point in the orbit of Earth that is closest to the sun.

period-luminosity relation: A relationship between the period and luminosity of variable stars (usually cepheids).

perturbation: A small deviation from the predicted orbit of a planet or other celestial body caused by the gravitational attraction of another body.

phases: The changing shape of the illuminated portion of the moon as seen from the earth.

photosphere: The atmospheric layer of the sun that is seen from the earth. The layer that emits most of the sun's visible radiation.

photon: A "particle" of electromagnetic radiation.

plage: A bright region of the chromosphere visible in a spectrohelioscope. Associated with sunspots.

Planck's constant: A basic constant of the microcosm. Designated by h, it is the proportionality constant between energy (E) and frequency (f). ($E = hf$).

planetary nebula: A bright gas cloud, spherical in shape, ejected by some stars late in their life cycle.

planetesimals: Small, asteroid-sized objects that form out of the solar nebula. Combine to form protoplanets.

plate tectonics: Motions of the large plates that constitute the crust of the earth.

Population I: Young, generally hot stars, with heavy element abundance similar to our sun, that lie predominantly in the arms of the Galaxy.

Population II: Old stars with low heavy element abundance that lie mostly in the core of the Galaxy, and in globular clusters.

precession: A slow motion around a particular direction in which the axis of rotation traces out a cone.

principle of equivalence: The statement of the equivalence of inertial and gravitational mass.

prograde motion: Rotation or revolution in the same direction as most of the planets in the solar system.

prominence: A luminous gas cloud extending beyond the limb of the sun.

proper motion: The annual rate at which the angular position of a star changes as a result of its motion through space.

proton: An elementary particle found in the nucleus of the atom.

proton-proton cycle: A series of thermonuclear reactions that occurs in the sun (changes hydrogen to helium).

protostar: A contracting mass of gas on its way to becoming a star. A pre-stellar stage.

Ptolemaic system: A geocentric model of the solar system devised by early Greeks and brought to a high degree of perfection by Ptolemy.

pulsar: A radio source that emits radiation in brief pulses. Believed to be a rapidly rotating neutron star that is emitting radiation in two narrow beams.

pulsating variable: A star that changes in magnitude as a result of a pulsation.

quasar: A radio source that looks like a star but is believed to be extragalactic. According to their red shifts most are more distant than the outermost galaxies.

quarks: Elementary particles that make up the baryons and mesons.

radial velocity: Velocity directly away from or directly toward the earth.

radiant: The point in the sky from which meteors appear to radiate.

radiation: Usually refers to the electromagnetic radiations, but can also refer to a mode of transfer of heat.

radioactivity: The disintegration of atomic nuclei with the release of highly energetic radiation or particles.

radio astronomy: The branch of astronomy that is concerned with radio waves from celestial objects.

radio galaxy: A galaxy that emits mostly radio waves.

radius vector: An imaginary line from the sun to a planet that moves with the planet.

red giant: A large, cool, red star. Representative point lies in the upper right region of the H-R diagram.

red shift: A shifting of spectral lines toward the red end of the spectrum that indicates the object emitting the radiation is receding from Earth.

reflecting telescope: A telescope that uses a mirror for its objective.

refracting telescope: A telescope that uses a lens as its objective.

regolith: A fine dust, caused by meteorite bombardment, that covers the moon and some planets.

relativistic: Moving at a speed close to that of light and therfore described adequately only by the theory of relativity.

resolving power: Ability of a telescope to clearly distinguish two closely associated objects (i.e. two stars).

retrograde motion: Motion of a planet in the sky opposite to its usual motion (or the motion of most of the other planets).

revolutionary period: The time to trace out one revolution of the orbit.

right ascension: The lines on the celestial sphere analogous to lines of longitude on Earth. Angular distance from the vernal equinox.

rille: Long cracks or crevices on the moon's surface.

Roche limit: The distance from a planet within which a moon would disintegrate due to tidal forces.

rotationary period: Time taken to rotate or spin once on an axis.

RR Lyra: A type of variable star. Has a period less than one day.

satellite: A moon or small object orbiting another larger one.

Schmidt telescope: A telescope that has a correcting lens in front of its mirror. Wide angle.

Schwarzchild radius: The distance from the singularity to the event horizon.

seeing: Refers to the stability of the Earth's atmosphere.

seismic waves: Vibrations caused by earthquakes that travel through the earth's interior.

Seyfert galaxy: A galaxy with a particularly dense nucleus that exhibits emission lines and frequently is a strong source of radio and infrared radiation.

shower (meteor): Numerous meteors all appearing to radiate from the same point in the sky (the radiant). The shower is usually named after the constellation the radiant is in.

sidereal period: The time of revolution or rotation with respect to the background stars.

singularity: The remnant of the star at the center of a black hole. It has no dimensions and is infinitely dense.

solar wind: The stream of charged particles emitted from the surface of the sun. It consists of protons and electrons.

solstice: The points where the sun is at its maximum distance above and below the celestial equator.

spectral type: A classification based on spectral

characteristics; designated by the letters O B A F G K M. Each of the letters is subdivided from 0 to 9.

spectrograph: The instrument that is used to obtain a spectrogram or photograph of a star's spectrum.

spectroscope: An instrument through which you can observe a spectrum.

spectroscopic binary: A binary star system where only one of the stars is seen, but the presence of the second can be determined as a result of a Doppler shift of the spectral lines.

spectrum: The array of all electromagnetic radiations spread out in order of wavelength (shortest at one end and longest at the other).

spherical aberration: A spherical lens or mirror defect. The light rays striking near the edge of the lens or mirror have a shorter wavelength than those striking near the center.

spicule: A bright, narrow, shortlived jet of gas that extends out of the chromosphere of the sun.

spiral galaxy: A galaxy with spiral arms of young stars and gas surrounding a nucleus of generally old stars.

Steady State Cosmology: A cosmology based on the principle that the universe is the same in all directions to all observers for all time.

Stefan's law: A law stating that the energy emitted per second by a unit area of a black body is proportional to the fourth power of the area's temperature.

stellar evolution: The way a star changes as it ages.

stratosphere: The layer of the earth's atmosphere that lies directly above the troposphere.

sunspot: A dark region on the photosphere of the sun. Cooler than the surrounding regions.

sunspot cycle: The eleven-year cycle in which the number of sunspots varies from a maximum number to a minimum number and back.

supergiant: A massive, highly luminous star.

superior conjunction: Occurs when an inferior planet such as Venus is directly in line with the sun but on the other side of it.

supernova: The explosion of a massive star in which most of its matter is blown off into space. Only a dense core—a neutron star—is left.

synchronous rotation: Orbital motion in which the period of rotation is equal to the period of revolution.

synodic period: The time that it takes to move from an initial sun-earth-planet configuration back to the same configuration.

tangential (transverse) velocity: Velocity of a star perpendicular to the line of sight.

terminator line: The line between darkness and light on the moon.

terrestrial planet: An Earth-like planet. A planet having a solid surface.

thermonuclear reaction: Nuclear reactions that result because of high temperatures.

tidal forces: Deformation forces that result from differences in gravitational pull as a result of different distances across the surface between two gravitating bodies.

Trojans: Asteroids that orbit in Jupiter's orbit. One group is ahead of Jupiter and one group behind.

troposphere: The lowest level of the earth's atmosphere, where most of the weather occurs.

T Tauri stars: Irregular variables, usually surrounded with nebulosity. Pre-main sequence.

UV radiation: The region of electromagnetic radiations between X-rays and visible rays.

umbra: In a sunspot, the darker inner region. In an eclipse, the region where light is completely obscured.

variable star: A star that changes its magnitude either regularly or irregularly (e.g. cepheids, RR Lyraes).

vector: A quantity that has both magnitude and direction. Represented by an arrow.

velocity: Speed in a particular direction.

velocity of escape: The minimum speed that an object needs to completely escape the gravitational pull of another object.

Vernal Equinox: The point where the ecliptic and the celestial equator cross for the sun travelling south to north (occurs approximately Mar. 21).

waning (moon): The sequence of phases from full moon to new moon.

weight: A measure of the gravitational force on a body.

white drawf: A small, extremely dense star near the end of its life cycle. Its representative point lies in the lower left hand region of the H-R diagram.

Wien's law: A formula giving the wavelength at which the maximum amount of radiation is emitted from a hot object.

winter solstice: The point on the ecliptic at which the sun is at its greatest distance south of the celestial equator.

X-rays: Energetic electromagnetic radiations between those of gamma rays and UV.

X-ray star: A star that emits mostly X-rays.

Zeeman effect: The splitting of spectral lines due to the presence of a magnetic field.

zenith: The point on the celestial sphere that is directly overhead.

zodiac: The twelve constellations around the ecliptic.

Illustration Credits

Part 1—Opener, Lick Observatory, University of California

Chapter 1—Opener, top, © B. Kliewe/Jeroboam; center, NASA; bottom, Palomar Observatory Photo

Fig. 1.1, A, © B. Kliewe/Jeroboam; B, © George W. Gardner; C, Landsat Photo; D, NASA; J and L, Palomar Observatory Photos

Figs. 1.2, 1.6 and 1.8, Palomar Observatory Photos

Fig. 1.3, Lund Observatory, Sweden

Figs. 1.5, 1.7 and 1.10, Lick Observatory Photos, University of California

Chapter 2

Fig. 2.2, Yerkes Observatory Photograph, University of Chicago

Figs. 2.4 and 2.11, Adapted from *Realm of the Universe,* 2nd edition by George O. Abell, Copyright © 1980 by George O. Abell, Reprinted by permission of Holt, Rinehart and Winston, CBS College Publishing

Fig. 2.5, Adapted from *Contemporary Astronomy* by Jay M. Pasachoff, Copyright © 1977 by W. B. Saunders Company. Reprinted by permission of Holt, Rinehart and Winston, CBS College Publishing

Boxes, pp. 27, 29, 32 and 36, Yerkes Observatory Photographs, University of Chicago

Fig. 2.7, Adapted from *Drama of the Universe* by George O. Abell. Copyright © 1978 by Georgy O. Abell. Reprinted by permission of Holt, Rinehart and Winston, CBS College Publishing

Fig. 2.12, Yerkes Observatory Photograph, University of Chicago. Box, p. 37, Woolsthorpe National Trust Photo

Chapter 3—Opener, Lick Observatory, University of California

Figs. 3.4, 3.10 and 3.11, from *Astronomy, Fundamentals and Frontiers,* 3rd edition by Robert Jastrow and Malcolm H. Thompson, Copyright © 1977 by Robert Jastrow. Reprinted by permission of John Wiley & Sons, Inc.

Fig. 3.12, left and top right, NASA; lower right, High Altitude Observatory, Boulder, Colorado

Figs. 3.13 and 3.14, Adapted from *Realm of the Universe,* 2nd edition by George O. Abell, Copyright © 1980 by George O. Abell. Reprinted by permission of Holt, Rinehart and Winston, CBS College Publishing

Figs. 3.17 and 3.18, Lick Observatory, University of California

Fig. 3.21, reproduced by permission of AstroMedia Corporation

Figs. 3.25–3.34, Adapted from *Griffith Observer,* Griffith Observatory, Los Angeles, Ca.

Part 2—Opener, Max Planck Institut für Radioastronomie

Chapter 5—Opener, Cornell University

Fig. 5.10, Yerkes Observatory Photograph, University of Chicago

Fig. 5.11, reproduced by permission of AstroMedia Corporation

Fig. 5.13, Courtesy of the Archives, California Institute of Technology

Fig. 5.14, Palomar Observatory Sky Survey

Fig. 5.15, Palomar Observatory Photo

Figs. 5.16 and 5.17, Kitt Peak National Observatory

Fig. 5.18, Bell Laboratories, Inc.

Fig. 5.19, The Max Planck Institut für Radioastronomie

Fig. 5.20, Cornell University

Fig. 5.23, NASA

Chapter 6—Opener, Kitt Peak National Observatory

Figs. 6.3, 6.5 and 6.6, American Institute of Physics, Bohr Library, W. F. Meggars Collection

Fig. 6.9, Palomar Observatory Photo

Fig. 6.13, Kitt Peak National Observatory

Part 3—Opener, NASA

Chapter 7—Opener and Figs 7.1–7.4, Yerkes Observatory Photographs, University of Chicago

Fig. 7.5, from "The Origin and Evolution of the Solar System" by A. G. W. Cameron, Copyright © 1975 by Scientific American, Inc. All rights reserved.

Chapter 8—Opener, H. Zirin, Big Bear Solar Observatory, California Institute of Technology

Fig. 8.1, Kitt Peak National Observatory

Fig. 8.4, Palomar Observatory Photo

Figs. 8.5 and 8.14, from *Astronomy, Fundamentals, and Frontiers,* 3rd edition by Robert Jastrow and Malcolm H. Thompson, Copyright © 1977 by Robert Jastrow. Reprinted by permission of John Wiley & Sons, Inc.

Fig. 8.6, Sacramento Peak Observatory, Association of Universities for Research in Astronomy, Inc.

Fig. 8.7, Kwasan and Hida Observatory, University of Kyoto

Figs. 8.9 and 8.20, H. Zirin, Big Bear Solar Observatory, California Institute of Technology

Fig. 8.10, Martin Schwarzschild, Project Stratoscope of Princeton University

Fig. 8.11, After M. Waldmeier

Fig. 8.12, from *PSSC Physics,* 2nd edition, 1965, D. C. Heath and Company and Educational Development Center, Newton, Mass.

Fig. 8.15, © 1976 by The American Association for the Advancement of Science

Fig. 8.16, NASA

Fig. 8.19, High Altitude Observatory/National Center for Atmospheric Research/National Science Foundation

Chapter 9—Opener, NASA

Figs. 9.2, 9.4, 9.5 and 9.6, Lick Observatory, University of California

Fig. 9.7, NASA

Fig. 9.8, top, Lick Observatory, University of California; bottom, NASA

Fig. 9.12, from "The Moon" by John A. Wood, Copyright © 1975 by Scientific American, Inc. All rights reserved.

Chapter 10—Opener, NASA

Figs. 10.3, 10.4, 10.5 and 10.8, Adapted with the permission of Macmillan Publishing Company from *Astronomy: Structure of the Universe* by William J. Kaufmann III

Fig. 10.7, Adaptation of Figure 2, page 13, from *Astronomy: The Evolving Universe,* 3rd edition by Michael Zeilik. Copyright © 1982 by Michael Zeilik.

Figs. 10.9, 10.11 and 10.14, Adapted with the permission of Macmillan Publishing Company from *Exploring the Solar System* by William J. Kaufmann III, Copyright © William

J. Kaufmann III and Kaufmann Industries, Inc.
Fig. 10.10, from *Astronomy, Fundamentals and Frontiers,* 3rd edition by Robert Jastrow and Malcolm H. Thompson. Copyright © 1977 by Robert Jastrow. Reprinted by permission of John Wiley & Sons, Inc.
Fig. 10.15, Gustav Lamprecht, Physics Institute, University of Alaska

Chapter 11—Opener, 11.1 and 11.3, NASA
Fig. 11.4, from "Mercury" by Bruce Murray. Copyright © 1975 by Scientific American, Inc. All rights observed.
Fig. 11.6, Lowell Observatory
Figs. 11.11–11.14, NASA
Fig. 11.15, Adapted with permission of Macmillan Publishing Company from *Astronomy: Structure of the Universe* by William J. Kaufmann III
Figs. 11.16 and 11.17, Lowell University
Figs. 11.18–11.23, 11.25 and 11.26, NASA

Chapter 12—Opener and Figs. 12.1, 12.2, 12.8 and 12.9, NASA
Figs. 12.3 and 12.4, from "Jupiter" by John A. Wolfe, Copyright © 1975 by Scientific American, Inc. All rights reserved.
Fig. 12.6, Lowell Observatory
Figs. 12.10, 12.12 and 12.13, Lunar and Planetary Observatory, University of Arizona
Fig. 12.14, Yerkes Observatory Photograph, University of Chicago

Chapter 13—Opener, and Fig. 13.8, Yerkes Observatory Photograph, University of Chicago
Figs. 13.1 and 13.4, Palomar Observatory Photos
Fig. 13.7, Adapted from *Astronomy: The Cosmic Journey,* 2nd edition by William K. Hartmann © 1982 by Wadsworth, Inc. Reprinted by permission of the publisher, Wadsworth Publishing Company, Belmont, Ca.
Fig. 13.11, The American Museum—Hayden Planetarium
Fig. 13.12, Meteor Crater, Northern Arizona

Part 4—Opener, Palomar Observatory Photo

Chapter 14—Opener, and Fig. 14.6, Kitt Peak National Observatory
Fig. 14.4, Palomar Observatory Photo
Fig. 14.5, reproduced by permission of AstroMedia Corporation

Chapter 15
Fig. 15.1, American Institute of Physics

Chapter 16—Opener, Figs. 16.5 and 16.7, Palomar Observatory Photos

Chapter 17—Opener, and Fig. 17.10, Lick Observatory, University of California
Fig. 17.2, Photography by Photolabs, Royal Observatorv. Edinburgh. Original negative by U. K. Schmidt Telescope Unit. Copyright © Royal Observatory Edinburgh, 1978
Fig. 17.7, reproduced by permission of AstroMedia Corporation
Fig. 17.14, Palomar Observatory Photo

Part 5—Opener, reproduced by permission of AstroMedia Corporation

Chapter 18—Opener, Figs. 18.2 and 18.6, Palomar Observatory Photos
Fig. 18.3, Museum of Northern Arizona
Fig. 18.9, Kitt Peak National Observatory
Fig. 18.11, reproduced by permission of AstroMedia Corporation
Box, p. 332, Bruce Margon, University of Washington

Chapter 19—Opener, Courtesy of the Archives, California Institute of Technology
Box, p. 340, American Institute of Physics, reproduced with the permission of the Estate of Albert Einstein and The American Friends of The Hebrew University

Chapter 20—Opener and Fig. 20.10, Jerome Kristian, Mount Wilson and Las Companas Observatories, Carnegie Institution of Washington (Palomar Observatory Photograph)

Part 6—Opener, Palomar Observatory Photo

Chapter 21—Opener, Figs. 21.1, 21.4 and 21.13, Palomar Observatory Photos
Fig. 21.2 and box p. 377, Harvard College Observatory
Fig. 21.7, Gart Westerhout, United States Naval Observatory
Fig. 21.11, D. Downes, IRAM, Grenoble

Chapter 22—Opener, Figs. 22.1–22.4, 22.6, 22.7, 22.9, 22.13 and 22.16, Palomar Observatory Photos
Fig. 22.5, The Cerro Tololo Inter-American Observatory
Fig. 22.14, Alar and Juri Toomre
Fig. 22.15, Adapted from *Realm of the Universe,* 2nd edition, by George O. Abell. Copyright © 1980 by George O. Abell, Reprinted by permission of Holt, Rinehart and Winston, CBS College Publishing
Figs. 22.17 and 22.19, Mount Wilson and Las Companas Observatories, Carnegie Institution of Washington
Figs. 22.18 and 22.20, Palomar Observatory Photos
Fig. 22.21, Negative photographs of the Seyfert galaxy NGC 4151 (a) and (b) are from a Mount Wilson Observatory Plate. Print (c) from the National Geographic Society–Palomar Observatory Sky Survey. Yerkes Observatory Photograph, University of Chicago

Chapter 23—Opener, Figs. 23.2 and 23.3, M. Schmidt, Mount Wilson and Las Companas Observatories, Carnegie Institution of Washington
Figs. 23.1 and 23.6, Palomar Observatory Photos
Fig. 23.4, from *Astronomy, Fundamentals and Frontiers,* 3rd edition by Robert Jastrow and Malcolm H. Thompson. Copyright © 1977 by Robert Jastrow. Reprinted by permission of John Wiley & Sons, Inc.
Box, p. 418, Lick Observatory, University of California

Part 7

Chapter 24
Fig. 24.4, Palomar Observatory Photo

Chapter 25
Fig. 25.4, Bell Laboratories
Fig. 25.6, Phys. Rev. Letts
Fig. 25.7, p. 71—bottom of "The Cosmic Background Radiation and the New Aether Drift" by Richard A. Muller, May 1978, Copyright © 1978 by Scientific American, Inc. All rights reserved.
Fig. 25.11, Robert V. Wagoner, Astrophysical Journal 179, p. 349, 1973

Chapter 26
Fig. 26.3, Courtesy Halton Arp

Part 8—Opener, NASA

Chapter 27—Opener, and Fig. 27.4, NASA
Fig. 27.6, Harvard College Observatory

Epilogue—Opener, p. 488 and 490, NASA
p. 491, Paramount Pictures Corporation

Endpapers—from *Griffith Observer,* Griffith Observatory, Los Angeles, CA.

Color Section
pp. C–1 and C–16 © Jerry Schad
Fig. 1.9, © 1959, California Institute of Technology
Fig. 5.18, R. D. Gehrz, University ot Wyoming
Figs. 8.2, 8.17, 10.1, 11.5, 11.24, 12.5 and 12.8, NASA
Fig. 16.6, © 1965, California Institute of Technology
Fig. 17.11, © 1961, California Institute of Technology
Fig. 18.2, © 1959, California Institute of Technology
Fig. 20.12 "Griffith Observatory" by Lois Cohen
Fig. 21.10, © 1961, California Institute of Technology
Fig. 21.12, © 1959, California Institute of Technology

Index

A

B

F

G

H

I

J

K

L

M

N

O

P

Q

R

S

T

U

V

W

X

Y

Z

A 3
B 4
C 5
D 6
E 7
F 8
G 9
H 0
I 1
J 2

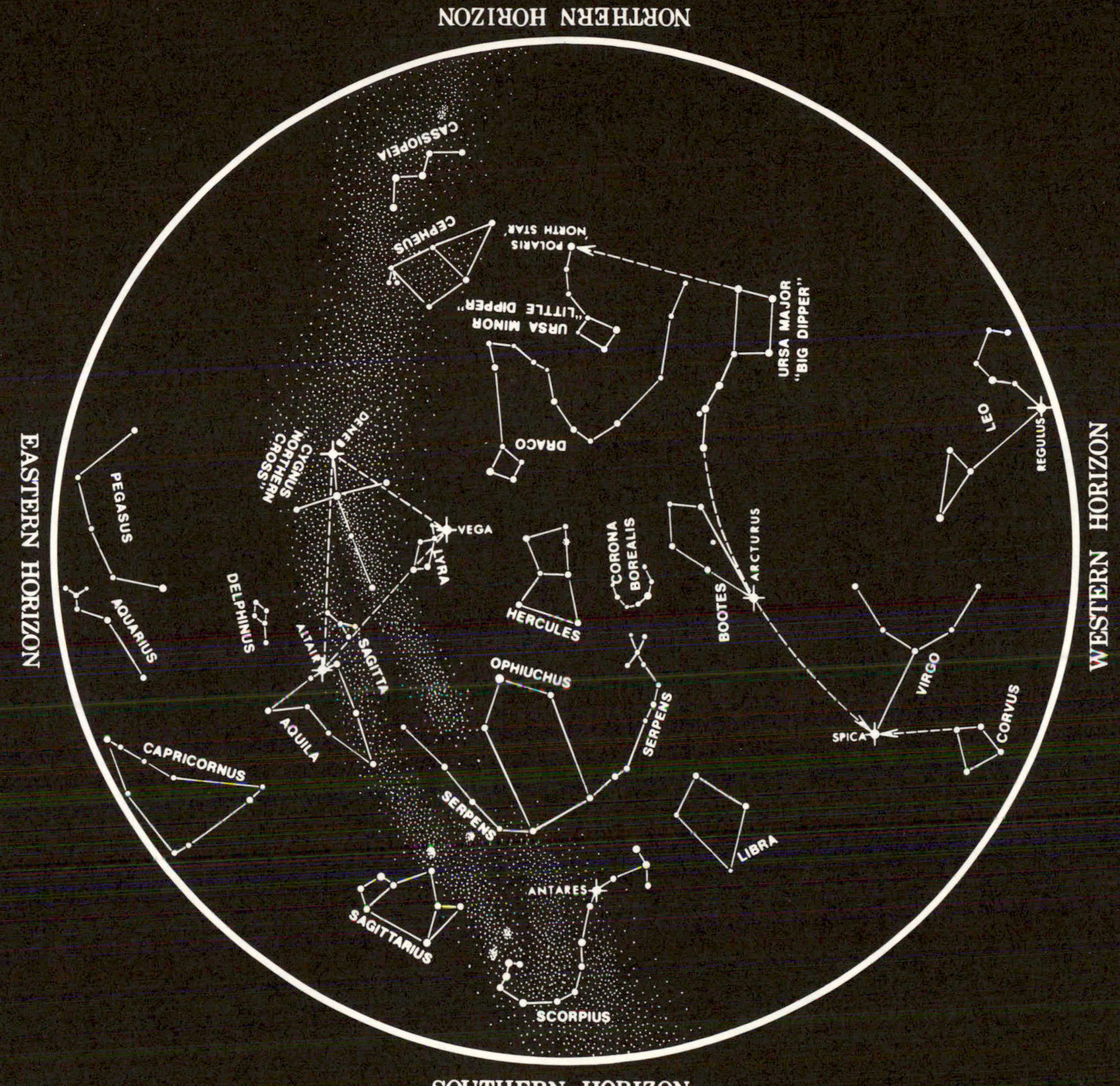

THE NIGHT SKY IN JULY

Chart Time: 9:00 (local standard time)
Middle of month

Chart is also appropriate for: June 11:00
August 7:00

To use chart, hold it so that the direction you are facing shows at the bottom of the chart.

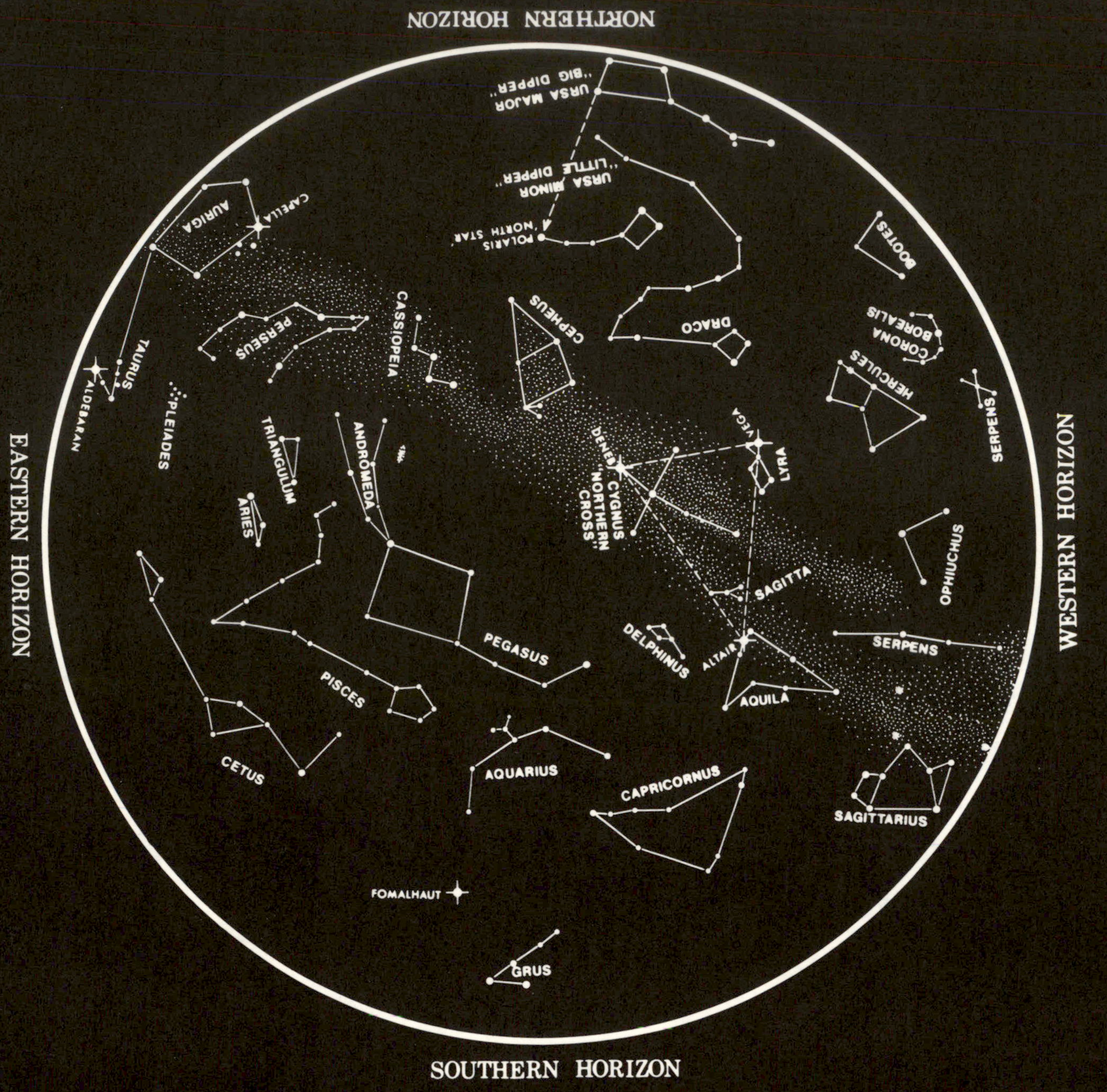

THE NIGHT SKY IN OCTOBER

Chart Time: 9:00 (local standard time)
Middle of month

Chart is also appropriate for: September 11:00
November 7:00

To use chart, hold it so that the direction you are facing shows at the bottom of the chart.

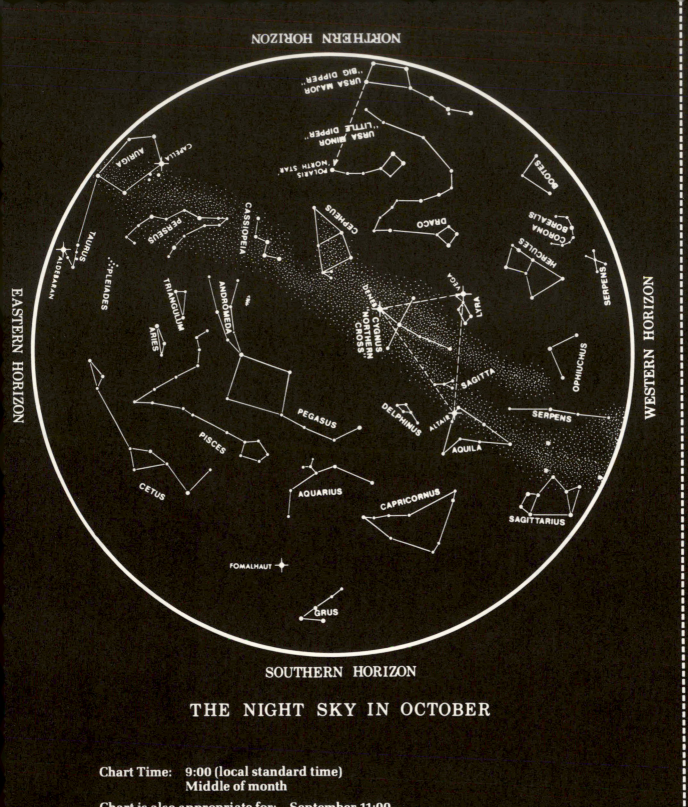